T0137407

Studies in Mechanobiology, Tissue Engineering and Biomaterials

Volume 23

Series Editor

Amit Gefen, Department of Biomedical Engineering, Tel Aviv University, Ramat Aviv, Israel

More information about this series at http://www.springer.com/series/8415

Yanhang Zhang

Editor

Multi-scale Extracellular Matrix Mechanics and Mechanobiology

Springer

Editor
Yanhang Zhang
Department of Mechanical Engineering,
Department of Biomedical Engineering and
Division of Materials Science & Engineering
Boston University
Boston, MA, USA

ISSN 1868-2006   ISSN 1868-2014 (electronic)
Studies in Mechanobiology, Tissue Engineering and Biomaterials
ISBN 978-3-030-20184-5   ISBN 978-3-030-20182-1 (eBook)
https://doi.org/10.1007/978-3-030-20182-1

This Springer imprint is published by the registered company Springer Nature Switzerland AG
The registered company address is: Gewerbestrasse 11, 6330 Cham, Switzerland

Preface

Extracellular matrix (ECM) forms the primary load-bearing component in many connective tissues. In addition to providing structural support, ECM also plays an important role in modulating cell function. Reciprocally, cells can modulate the structure and composition of ECM. Many pathological conditions are associated with loss of organization and function of ECM; however, the mechanisms by which ECM mechanics influence cell and tissue function remain to be fully elucidated. Such understandings require multi-scale approaches since the changes associated with pathological developments span from the tissue to molecular level. Furthermore, ECM has extremely complex hierarchical three-dimensional structures and there exists a tremendous interdependence of ECM compositional, structural, and mechanical properties. This book describes the current state of knowledge in the field of multi-scale ECM mechanics and mechanobiology with a focus on experimental and modeling studies in biomechanical characterization, advanced optical microscopy and imaging, as well as computational modeling. This book also discusses the scale dependency of ECM mechanics, translation of mechanical forces from tissue to cellular level, and advances and challenges in improving our understanding of cellular mechanotransduction in the context of living tissues and organisms.

Boston, MA, USA   Yanhang Zhang

Contents

Abbreviations/Nomenclature

2D	Two-dimensional
3D	Three dimensions
AAA	Abdominal aortic aneurysm
AC	Against curvature
ADAMTS2	A Disinegrin and Metalloproteinase with Thrombospondin Motifs 2
aff	Affine prediction
AFG	Arbitrary function generator
AFM	Atomic force microscopy
AGE	Advanced glycation end product
AI	Alignment index
AngII	Angiotensin II
AOD	Acousto-optic deflectors
AT	Ascending thoracic part
ATA	Ascending thoracic aorta
ATAA	Ascending thoracic aortic aneurysm
AV	Aortic valve
AVIC	Aortic valve interstitial cell
BAV	Bicuspid aortic valve
BGN	Human gene encoding biglycan
BMP	Bone morphogenic protein
BPV	Blood pressure variability
CAVD	Calcific aortic valve disease
cECM	Cardiac extracellular matrix
CF	Cardiac fibroblast
CFPG	Collagen fibril proteoglycan
circ	Circumferential direction
CM	Cardiomyocyte
CMP	Collagen mimetic peptide
COL3A1	Human gene encoding the alpha 1 chain of collagen III
CPC	Cardiac progenitor cell

CRGDS	Cysteine–arginine–glycine–aspartic acid–serine (adhesive peptide)
CRM	Confocal reflection microscopy
CS	Chondroitin sulfate
CT	Curvelet transform
CT-A	Computed tomography-angiography
DAQ	Data acquisition card
DDR	Discoidin domain receptor
dECM	Decellularized extracellular matrix
DHT	Dehydrothermal
DI	Dispersion index
diag	Diagonal direction
DNA	Deoxyribonucleic acid
DOE	Diffractive optical element
DS	Dermatan sulfate
DT	Descending thoracic part
EC	Endothelial cell
ECM	Extracellular matrix
ECs	Endothelial cells
EDC	1-ethyl-3-(3-dimethylaminopropyl)-carbodiimide hydrochloride
EFE	Endocardial fibroelastosis
ELN	Human gene encoding elastin
EMILIN	Elastin microfibril interface-located proteins
EMT	Epithelial-to-mesenchymal transition
ESC-CM	Embryonic-stem-cell-derived cardiomyocyte
ET-1	Endothelin-1
exp	Experimentally observed
FA	Focal adhesion
F-actins	Actin filaments
FAK	Focal adhesion kinase
FB	Fibroblast
FBLN4	Human gene encoding fibulin-4
FBN1	Human gene encoding fibrillin-1
FDA	Food and Drug Administration
FDM	Fluctuation-driven mechanotransduction
FE	Finite element
FEM	Finite element method
FFT	Fast Fourier transform
FGF	Fibroblast growth factor
FIRE	Fiber extraction algorithm
FN	Fibronectin
FRET	Fluorescence resonance energy transfer
FSM	Force spectrum microscopy
G&R	Growth and remodeling
GA	Glutaraldehyde
GAG	Glycosaminoglycan

GelMA	Gelatin methacrylate
HA	Hyaluronic acid
HHL	Histidinyl hydroxylysino-norleucine
HS	Heparan sulfate
HSPG	Heparan sulfate proteoglycan
IAA	Infrarenal abdominal aorta
IACUC	Institutional Animal Care and Use Committee
IGF	Insulin growth factor
ILT	Intraluminal thrombus
IMR	Ischemic mitral valve regurgitation
iPSC-CM	Induced pluripotent stem-cell-derived cardiomyocyte
KCl	Potassium chloride
KO	Transgenic model
LAIR-1	Leukocyte-associated immunoglobulin-like receptor 1
LC-MS/MS	Liquid chromatography tandem mass spectrometry
long	Longitudinal direction
LOR-1	Lysyl oxidase-related protein-1
LOX	Lysyl oxidase
LV	Left ventricular
MA	Micropipette aspiration
MAPK	Mitogen-activated protein kinase
MCP-1	Monocyte chemoattractant protein-1
MEMS	Microelectromechanical systems
MI	Myocardial infarction
MIP	Maximum intensity projection image
MLCK	Myosin light chain kinase
MLU	Medial lamellar unit
MMP	Matrix metalloproteinase
mMP	Mitochondrial membrane potential
MS	Monotonous stretch
MSC	Mesenchymal stem cell
MV	Mitral valve
MVIC	Mitral valve interstitial cell
NA	Non-available
NADH	Nicotinamide adenine dinucleotide
ND	Neutral density
NHS	N-hydroxysuccinimide
NISM	Nonlinear inference stress microscopy
NO	Nitric oxide
NRCM	Neonatal rat cardiomyocyte
OCT	Optical coherence tomography
ODF	Orientation density function
OI	Osteogenesis imperfecta
OS-SIM	Optical sectioning structure illumination microscopy
PAA	Polyacrylamide

PBS	Phosphate-buffered saline
PDGF	Platelet-derived growth factor
PDMS	Polydimethylsiloxane
PEG	Poly (ethylene glycol)
PG	Proteoglycan
phMLC	Phosphorylated myosin light chain
PKC	Protein kinase c
PLM	Polarized light microscopy
PSD	Power spectral density
PSF	Point spread function
PV	Pulmonary valve
PVIC	Pulmonary valve interstitial cell
QPD	Quadrant photodiode
Rag1$^{-/-}$	Mouse knockout of the gene encoding recombination-activating gene 1
RAGE	Receptor for advanced glycation end products
RBCs	Red blood cells
ref	Related to the reference configuration
RESA	Ring-infected erythrocyte surface antigen
RGD	Arginine–glycine–aspartic acid
rhBMP-2	Recombinant human bone morphogenetic protein
rhCOL	Recombinant human collagen
RNA	Ribonucleic acid
ROIs	Regions of interest
ROS	Reactive oxygen species
RVE	Representative volume element
SEM	Scanning electron microscopy
SF	Stress fiber
SHG	Second-harmonic generation
SIS	Small intestine submucosa
SMC	Smooth muscle cell
SNARE	Soluble N-ethylmaleimide-sensitive factor attachment protein receptor
TAA	Thoracic aortic aneurysm
TACS	Tumor-associated collagen signatures
TAV	Tricuspid aortic valve
TEM	Transmission electron microscopy
TFM	Traction force microscopy
TGF-β	Transforming growth factor beta
TGF-β1	Transforming growth factor beta one
TIRF	Total internal reflection fluorescence
TNC	Tenascin c
TPEF	Two-photon-excited fluorescence
TPF	Two-photon fluorescence
tr	Transverse direction

TV	Tricuspid valve
TVV	Tidal volume variability
URS	Universal rotary stage
US	Unstretched
UV	Ultraviolet
VEGF	Vascular endothelial growth factor
VIC	Valve interstitial cell
VIFs	Vimentin intermediate filaments
VS	Variable stretch
VSMC	Vascular smooth muscle cell
WASP	Wiskott–Aldrich syndrome protein
WC	With curvature
WSS	Wall shear stress
WT	Wild-type
α–SMA	Alpha smooth muscle actin
αSMA	α smooth muscle actin
λ	Stretch

Biomechanics and Mechanobiology of Extracellular Matrix Remodeling

Jay D. Humphrey and Marcos Latorre

Abstract Biomechanics is the development, extension, and application of the principles of mechanics for the purposes of understanding better both biology and medicine. Mechanobiology is the study of biological responses of cells to mechanical stimuli. These two fields must be considered together when studying the extracellular matrix of load-bearing tissues and organs, particularly, how the matrix is established, maintained, remodeled, and repaired. In this chapter, we will illustrate a few of the myriad aspects of matrix biology and mechanics by focusing on the arterial wall. All three primary cell types of the arterial wall are exquisitely sensitive to changes in their mechanical environment and, together, they work to establish and promote mechanical homeostasis, loss of which results in diverse pathologies, some of which have life threatening consequences. There is, therefore, a pressing need to understand better the intricate inter-relations between the biomechanics and the mechanobiology of arteries and so too for many other tissues and organs.

1 Introduction

The extracellular matrix (ECM) consists of myriad proteins, glycoproteins, and glycosaminoglycans (GAGs) that collectively endow healthy tissues and organs with both appropriate structural properties and critical instructional information. In particular, it is the ECM that provides much of the mechanical stiffness (a measure of how stress changes with changes in strain) and mechanical strength (a measure of the maximum stress that can be tolerated prior to failure) that are needed for many tissues and organs to function properly under the action of applied loads. It is also the ECM that sequesters and presents to cells diverse growth factors (to promote cell proliferation and matrix production), cytokines (secreted by inflammatory cells), proteases (which degrade proteins), and other biomolecules that are needed to inform or enable cell-mediated functions. The specific composition and architecture of the ECM

J. D. Humphrey (✉) · M. Latorre
Department of Biomedical Engineering, Yale University, New Haven, CT 06520, USA
e-mail: jay.humphrey@yale.edu

© Springer Nature Switzerland AG 2020
Y. Zhang (ed.), *Multi-scale Extracellular Matrix Mechanics and Mechanobiology*,
Studies in Mechanobiology, Tissue Engineering and Biomaterials 23,
https://doi.org/10.1007/978-3-030-20182-1_1

dictates its different roles in different tissues, thus it should not be surprising that cells within the ECM establish, maintain, remodel, and repair the matrix in response to local biochemomechanical signals [21]. That is, the structure and properties of a particular tissue or organ are designed by the resident cells to achieve specific functions under normal conditions. A lost ability by cells to accomplish these tasks often leads to disease progression or an inability to repair after injury.

Because of the important mechanical roles of many soft tissues and organs, we will review some of the basic mechanics and the associated mechanobiology. By mechanics, we mean the study of motions and the applied loads that cause them; by mechanobiology, we mean the biological responses by cells to mechanical stimuli. Clearly these fields should be studied together for the mechanics influences the cellular responses, which in turn dictate the geometry and mechanical properties of the tissues and thus how they respond to applied loads structurally. For illustrative purposes, we will also use a series of computational simulations to emphasize a number of key factors that determine tissue properties, particularly during ECM growth and remodeling (G&R). By growth we mean a change in mass; by remodeling we mean a change in microstructure. Although most of our discussion is meant to be general, we focus on mechanobiological or biomechanical simulations for central arteries, such as the mammalian aorta, which will illustrate salient considerations. These vessels serve as deformable conduits for blood flow, which because of the pulsatile system and associated pressure waves must have appropriate resilience as well as stiffness and strength [14]. Hence, we begin our discussion with a brief review of central arterial structure.

2   Arterial Composition and Microstructure

Central arteries comprise three basic layers: the intima (innermost), media (middle), and adventitia (outermost). The intima consists primarily of a monolayer of endothelial cells attached to a basement membrane that consists primarily of laminin and the network forming collagen IV, and in some conditions the glycoprotein fibronectin. The media consists primarily of smooth muscle cells that are embedded within a lamellar structure consisting primarily of elastic fibers, fibrillar collagen (mainly type III), and aggregated glycosaminoglycans (GAGs). In particular, the elastic fibers consist of about 90% elastin in addition to multiple elastin-associated glycoproteins, including the fibulins and fibrillins. The fibulins appear to play important roles in elastogenesis whereas the fibrillins play important roles in long-term biological stability, among others. These elastic fibers form nearly concentric laminae within which the smooth muscle cells reside along with the fibrillar collagen III and the proteoglycan versican (note: a proteoglycan consists of a protein core and attached negatively charged GAGs such as chondroitin sulfate). The adventitia consists primarily of fibrillar collagen I with embedded fibroblasts, though admixed elastin and other constituents as well. In particular, the proteoglycans biglycan and decorin play important roles in collagen assembly within the adventitia and so too the matricellular

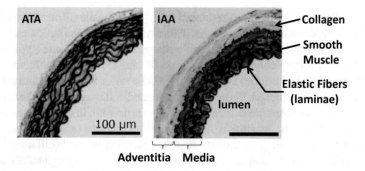

Fig. 1 Geometry and composition of the adult murine ascending thoracic aorta (ATA) and infrarenal abdominal aorta (IAA), showing three of the primary structural constituents [elastic fibers -> laminae (black), collagen fibers (brownish-grey), and smooth muscle (red)] and two of the three primary layers (media and adventitia, noting that the third layer, the intima, is too thin to visualize on these sections). Note the greater number of elastic laminae in the ATA, which is closer to the heart, and its thinner adventitia. Note, too, that vessel sections were stained with Verhoeff-Van Gieson (VVG) after fixation of the vessel in the unloaded state, which provides a reliable state for comparison across regions and mouse models wherein segments can necessarily experience different axial loads and blood pressures

protein thrombospondin-2. See Fig. 1 for the basic structure of two central arteries, the ascending thoracic aorta (ATA—the most proximal segment of the aorta) and the infrarenal abdominal aorta (IAA—the most distal segment) from a normal mouse. Consistent with a strong structure-function relationship, the former contains more elastic fibers (~7 laminae in the mouse) and the latter fewer elastic fibers (~4 laminae) but more collagen fibers (thicker adventitia). That is, elastic arteries are found closer to the source of the pulsatile blood flow, the heart.

The elastin-rich elastic fibers endow the arterial wall with its resilience, that is, its ability to store energy elastically upon deformation. Elastin is a unique protein within the vasculature for it has an extremely long half-life (50+ years) under normal conditions. That is, elastic fibers are typically deposited, organized, and cross-linked during the perinatal period; they are then deformed significantly during somatic growth, which results in a pre-stretch that stores elastic energy—hence the terminology "elastic fibers". It appears that the natural tendency of an elastic fiber to recoil in its homeostatic configuration is offset, in part, by neighboring structural constituents, including fibrillar collagens and GAGs. Indeed, the elastic pre-stretch may help to build in a natural undulation in the collagen fibers [10], which increases overall compliance of the arterial wall under normal pressures. Note that one of the primary mechanical functions of central arteries is to use energy stored elastically during systole to work on the blood and adjacent soft tissues during diastole to augment blood flow. Interestingly, species with higher heart rates tend to have higher ratios of elastin to collagen [20]. The mouse, with a heart rate ~600 bpm, has an arterial elastin:collagen ratio >1 whereas rats, rabbits, dogs, pigs, and humans have progressively smaller ratios < 1.

The fibrillar collagens endow the arterial wall with its tensile stiffness and strength, both of which are needed to maintain mechanical functionality and structural integrity under the incessant loading due to pulsatile changes in blood pressure (typically ~11 to 16 kPa under resting conditions). It appears that the overall circumferential material stiffness of central arteries is ~500 to 1000 kPa under normal conditions across many species [3, 34]. Indeed, the normally undulated collagen fibers at normal blood pressures allow considerable deformation of the aortic wall; in contrast, these fibers become straightened in response to abrupt increases in blood pressure, which limits deformation of the wall. It is thought that straightening of the fibrillar collagen I, mainly in the adventitia, thereby protects the underlying more fragile smooth muscle cells, elastic fibers, and fibrillar collagen III of the media during marked, acute increases in blood pressure, as, for example, in weight lifting wherein pressures can transiently exceed 200 mmHg [2]. Importantly, the normal aortic wall can withstand quasi-static increases in pressure in vitro up to ~1500 mmHg, evidence of the remarkable strength of engaged collagen fibers in health. In contrast, aggregating GAGs endow the arterial wall with its compressive stiffness. That is, the negatively charged GAGs attract Na^+ ions to ensure electroneutrality, which via osmosis sequesters water, thus hydrating the wall and providing a modest Gibbs-Donnan-swelling pressure. It has been suggested that, although GAGs constitute only ~3 to 5% of the arterial wall by wet weight, the associated normal swelling pressure may help maintain preferred separation distances between the elastic laminae despite the overall compressive radial stress on the wall; such separation between laminae may facilitate certain aspects of smooth muscle cell mechano-sensing of its mechanical environment [32]. More detail on arterial wall composition can be found in [36]. Discussion of pathologic consequences of excessive accumulations of intramural GAGs can be found in [32].

Importantly, each of the three primary cell types of the arterial wall—endothelial, smooth muscle, and fibroblast—are highly sensitive to changes in their local mechanical environment [16]. For example, increased blood pressure-induced wall stresses can increase the production of the peptide angiotensin II (AngII) by smooth muscle cells, which in turn can stimulate a host of downstream consequences: synthesis of collagens or GAGs, in part through the up-regulation of transforming growth factor beta (TGF-β), as well as degradation of such matrix constituents via both up-regulation of matrix metalloproteinases (MMPs) and up-regulation of monocyte chemoattractant protein-1 (MCP-1), which recruits monocytes/macrophages that also secrete various proteinases. That is, in many cases there is a delicate balance between ECM production and removal in G&R processes, not just in normalcy. Indeed, even the activity of the MMPs is counteracted by tissue inhibitors of MMPs, known as TIMPs, noting that increased mechanical stress can both stimulate the production of MMPs and TIMPs as well as sterically reduce the binding sites available on a protein to enable MMP-mediated degradation. As another example, increased blood-flow induced wall shear stresses can increase the production of nitric oxide (NO) by endothelial cells, with NO both causing relaxation of smooth muscle cells (and thus vasodilatation) and attenuating their production of ECM. In contrast, decreased blood-flow induced wall shear stresses can increase the production of endothelin-1 (ET-1) by endothelial cells, with ET-1 causing contraction of smooth muscle cells (and thus vasoconstriction) and heightening their production of ECM. Clearly, therefore, understanding better the mechanobiological responses of multiple cell types, and their interactions, is critical to understanding tissue structure and function.

Towards this end, it is important to note that cells tend to interact mechanically with the ECM in which they reside via specialized transmembrane receptors, especially integrins. These heterodimeric receptors are typically denoted as $\alpha_x \beta_y$, as, for example, $\alpha_1 \beta_1$ and $\alpha_2 \beta_1$ which bind collagen, $\alpha_5 \beta_1$ and $\alpha_v \beta_3$ which bind fibronectin, and $\alpha_7 \beta_1$ which binds laminin. Although these integrins have preference for particular components of the ECM, they are not specific to single constituents. For example, $\alpha_5 \beta_1$ and $\alpha_v \beta_3$ also bind fibrillin-1. Importantly, intracellular domains of these transmembrane receptors couple to the cytoskeleton, particularly actin and myosin. In this way, cells can directly probe or assess their local mechanical environment by reaching out and pulling on the ECM. Many studies have sought to quantify precisely what the cells "feel" and thus respond to mechanobiologically. Although stress and strain, and metrics derived from them, have emerged as convenient metrics for correlating cellular responses to changes in their mechanical environment, these quantities are conceptual and not physical. Hence, actual mechanosensing is probably via forces and associated displacements that result in conformational changes in proteins and other biomolecules. Nevertheless, because of the continual evidence that the continuum assumption is valid (that length scales associated with the inherent microstructure are well less than the physical length scale of interest) and that continuum biomechanics is useful, basing biomechanical and mechanobiological studies on metrics such as stress, stiffness, and strength is appropriate in most cases [15].

Advances in genetics over the past few decades have revealed critical new insights into the roles of many of the constituents that form the ECM. For example, mutations to the gene that encodes elastin (*ELN*) can lead to supravalvular aortic stenosis (i.e., narrowing of the aorta, particularly at the aortic root) as in Williams syndrome; mutations to the genes that encode fibrillin-1 and fibulin-4 (*FBN1* and *FBLN4*) lead to aneurysms (i.e., local dilatations) of the aortic root and/or ascending aorta, as in Marfan syndrome; mutations to the genes that encode collagen III (*COL3A1*) and biglycan (*BGN*) lead to fragile arteries, highly susceptible to dissection or rupture as in vascular Ehlers-Danlos syndrome. These and many other syndromic and non-syndromic diseases arising from genetic mutations emphasize the need to link the genetics with both the clinical and biomechanical phenotypes, noting that defects in either the primary structural proteins (e.g., elastin or collagen) or their accessory proteins and glycoproteins (e.g., fibrillin-1 and biglycan) can have dire consequences and thus deserve our very best attention. Again, the interested reader is referred to [36] for more on genetic drivers of changes in ECM composition and function.

3 Quantification of Mechanical Properties

A constitutive relation describes the response of a material to applied loads under conditions of interest, noting that the term constitutive is used since these material responses depend on the internal constitution, or make-up, of the particular material. Soft tissues, in general [11], and arteries, in particular [14], exhibit complex mechanical responses to applied loads that result in large part from the specific ECM that was fashioned and remodeled by the resident cells. In particular, these tissues tend to exhibit nonlinearly, inelastic, and anisotropic behaviors. The nonlinearity appears to result from the gradual recruitment of previously undulated fibers, mainly collagens; the inelasticity appears to result from the GAG-sequestered water within the tissue, often referred to as a ground substance matrix; and the anisotropy tends to arise from non-uniform spatial distributions of the primary structural constituents, often elastic fibers and collagen fibers, but also differences in prestretches at which the constituents are deposited into and incorporated within extant tissue. In addition, soft tissues are often heterogeneous in both composition and behavior, as, for example, due to the medial and adventitial layers of the aorta (Fig. 1).

Under many physiological conditions of interest, however, these tissues often exhibit a nearly elastic response, hence they are frequently described using a concept of pseudoelasticity [11] or hyperelasticity. In hyperelasticity, one introduces a stored energy function W that depends of the state of strain, that is, a Helmholtz potential under isothermal conditions. The classical model of soft tissue pseudo-elasticity (wherein best-fit material parameters are given separately for loading and unloading curves) is the Fung-exponential,

$$W = c(e^Q - 1)$$

where Q is quadratic in the Green strain, $E = (F^T F - I)/2$ with F the deformation gradient tensor (note by the polar decomposition theorem, $F = RU = VR$ where R is a rotation tensor and U and V are stretch tensors, thus $F^T F$ and E are properly insensitive to rigid body motions). For an orthotropic behavior (with respect to the reference configuration, and referring to physical components of the tensor E), Q can be written

$$Q = c_1 E_{11}^2 + c_2 E_{22}^2 + c_3 E_{33}^2 + 2c_4 E_{11} E_{22} + 2c_5 E_{22} E_{33} + 2c_6 E_{33} E_{11}$$
$$+ c_7 (E_{12}^2 + E_{21}^2) + c_8 (E_{23}^2 + E_{32}^2) + c_9 (E_{13}^2 + E_{31}^2)$$

with c_i the material parameters that need to be determined from data, namely via nonlinear regressions. An advantage of this form of Q, with 9 material parameters for orthotropy, is that it can also be used to describe either transversely isotropy, by collecting together select terms and reducing the number of parameters to 5, or isotropy, by collecting together more terms and reducing the number of parameters to 2. Additionally, a 2-D theory can be constructed easily from this 3D form [13].

Regardless, given a specific form for W, one can then compute the 2nd Piola-Kirchhoff stress directly, or using the Piola transformation, the Cauchy stress t, which in the case of incompressibility (often a good assumption because of the GAG-sequestered water) can be written:

$$t = -p\mathbf{I} + 2\mathbf{F}\frac{\partial W}{\partial \mathbf{C}}\mathbf{F}^T$$

where p is a Lagrange multiplier (not hydrostatic pressure, in general) that enforces the kinematic constraint of incompressibility, $det\,\mathbf{F} = 1$. Note that the second Piola-Kirchhoff stress is $\mathbf{S} = 2\partial W/\partial \mathbf{C}$. If the response is visco-hyperelastic, one can add to the general expression for Cauchy stress an additional (linear) term $2\mu\mathbf{D}$, where μ is a viscosity and the stretching tensor $\mathbf{D} = \left(\dot{\mathbf{F}}\mathbf{F}^{-1} + \mathbf{F}^{-T}\dot{\mathbf{F}}^T\right)/2$ with the over-dot denoting a rate of change. Of course, given an expression for Cauchy stress, one must then satisfy linear momentum balance,

$$div\,t + \rho\mathbf{b} = \rho\mathbf{a}$$

where ρ is the mass density, $\mathbf{b}$ the body force, and $\mathbf{a}$ the acceleration. In the case of equilibrium (i.e., static or quasi-static processes), the acceleration is zero and the equation reduces in complexity. Because of the generally low mass density and low accelerations, most of arterial mechanics can be studied within a quasi-static framework [17] whereby one need only enforce $div\,t = 0$, except in cases such as deceleration injuries in vehicular accidents.

An advantage of the Fung-exponential relation is that it has proven successful time and time again in describing well the available experimental data from multiaxial tests on many different soft tissues, including arteries. A disadvantage, however, is that this relation is purely phenomenological. For this reason, many have sought more microstructurally motivated relations. Among the first to pursue this was Lanir [25], whose work continues to inspire many studies to this day. Of note here, Holzapfel et al. [12] presented a two-fiber family model for the arterial wall that was later extended to a four-fiber family model by Baek et al. [1]. Both have proven useful in many studies, noting that the latter tends to fit multiaxial data better (cf. [9, 33]). Nevertheless, these fiber-based models can be written

$$W = c(I_C - 3) + \sum_{k=1}^{n}\frac{c_1^k}{4c_2^k}(\exp[c_2^k(IV_C^k - 1)^2] - 1)$$

where $I_C = tr(\mathbf{C}) = tr(\mathbf{F}^T\mathbf{F})$ and $IV_C^k = \mathbf{M}\cdot\mathbf{CM}$ are coordinate invariant measures of the deformation, with $\mathbf{M}$ a unit vector denoting the direction of locally parallel fiber family in a reference configuration; finally, c, c_1^k, and c_2^k are material parameters. Because the values of these material parameters need to be determined via nonlinear regressions of data, this relation is also phenomenological, though structurally motivated: the first term is motivated by a contribution due to an amorphous (isotropic) matrix while the second set of summed terms is motivated by multiple directional contributions due to n locally parallel families of fibers, with $n = 2$ or 4 usually. If two families of fibers are directed circumferentially and axially in a cylindrical artery, one obtains an orthotropic model with 5 parameters (in contrast to the 10 parameters in the Fung model, which describes shearing behaviors better). It appears, however, that two symmetric diagonally oriented families of fibers reflects better the imaging data on collagen fiber distributions in some arteries [12], consistent with concepts from composite materials on optimal fiber orientations [7]. The four-fiber family model, combining circumferential, axial, and symmetric diagonal families of fibers yet appears to describe multiaxial data better [33] because it captures the different prominent fiber orientations seen in some arteries [9, 37] and it can also account phenomenologically for the effects of unmeasured cross-links or physical entanglements within a complex ECM. Importantly, these and other structurally motivated (not structurally based since no model yet accounts for all microstructural complexities of soft tissues, including the ECM and its interactions with cells) models have also helped to advance our understanding of tissue adaptations and disease progression.

4   Quantification of Growth and Remodeling

Recalling that an individual soft tissue is defined largely by its particular ECM and embedded cells that produce, maintain, remodel, and repair it, we emphasize that the myriad structural constituents found within a given tissue can exhibit individual mechanical properties (stiffness and strength), individual stress-free (i.e., natural) configurations, and individual rates of turnover (production and removal). Hence, although multiple theories exist for describing growth (change in mass) and remodeling (change in microstructure), mixture-based models can be particularly useful because one can account for the different mechanics and biology of the individual constituents. Indeed, mixture approaches can even be used to model the G&R of tissue engineered constructs that consist, at different times, of changing percentages of synthetic (polymeric) and natural (cells and ECM) constituents [31]. That said, using a full mixture theory for mass and linear momentum balance, even in isothermal situations, poses significant challenges. Notably, it is difficult to identify appropriate constitutive relations for momentum exchanges between multiple constituents, particularly as they evolve, and it is difficult to prescribe traction boundary conditions, which requires rules for how tractions partition on boundaries. For these and other reasons, we proposed and advocated for a constrained mixture theory [18] to describe soft tissue G&R, that is, adaptations to perturbations in loading as well as disease progression.

Briefly, one satisfies a full mixture relation for mass balance but a classical relation for linear momentum balance, with stored energy written in terms of a simple mass averaged rule-of-mixtures (i.e., the total energy is the sum of the energies of the constituent parts, each multiplied by their respective mass fractions). Mass balance can be written in spatial form as

$$\frac{\partial \rho^\alpha}{\partial s} + div(\rho^\alpha v^\alpha) = \overline{m}^\alpha$$

where ρ^α is the apparent mass density for constituent $\alpha (= 1, 2, \ldots, N)$, with v^α its velocity and $\overline{m}^\alpha$ its mass exchange (i.e., net rate of mass density production/removal), where s denotes the current G&R time. Three key constitutive assumptions that have proven useful are: each constituent can possess an individual natural configuration ($X^\alpha \neq X$, which denotes original positions), but is otherwise constrained to move with the mixture as a whole (namely $x^\alpha = x$, which denotes current positions, and thus requires $v^\alpha \equiv v$); the G&R process is sufficiently slow so that the tissue can be assumed to exist in a sequence of quasi-static equilibria ($v = 0$); and the net rate of mass density production/removal $\overline{m}^\alpha$ can be written as a multiplicative decomposition in terms of the true rate of production $m^\alpha > 0$ and a survival function $q^\alpha(s, \tau) \in [0, 1]$, which accounts for that fraction of constituent α produced at G&R time $\tau \in [0, s]$ that survives to the current time s. The resulting mass balance for this open system is integrable, namely one can find [26, 35]

$$\rho_R^\alpha(s) = \rho^\alpha(0) Q^\alpha(s) + \int_0^s m_R^\alpha(\tau) q^\alpha(s, \tau) d\tau$$

where the subscript R refers to quantities defined per unit reference volume (e.g., $\rho_R^\alpha = J\rho^\alpha$, with J the Jacobian of the mixture), hence $\rho_R^\alpha(0) \equiv \rho^\alpha(0)$), and $Q^\alpha(s) = q^\alpha(s, 0)$ is a useful special case. Because most removal (i.e., degradation of ECM or death of cells) follows first order-type kinetics, we can often let

$$q^\alpha(s, \tau) = \exp\left(-\int_\tau^s k^\alpha(t) dt\right)$$

where k^α is a rate-type parameter that may depend on biochemomechanical stimuli. Nonetheless, a useful form for the stored energy function of the mixture (tissue) per unit reference volume can be written,

$$\rho W_R^\alpha(s) = \rho^\alpha(0) Q^\alpha(s) \widehat{W}^\alpha\left(F_{n(0)}^\alpha(s)\right) + \int_0^s m_R^\alpha(\tau) q^\alpha(s, \tau) \widehat{W}^\alpha\left(F_{n(\tau)}^\alpha(s)\right) d\tau$$

where $\rho = \sum \rho^\alpha$ is the mass density of the mixture (tissue) which typically remains constant, and $\widehat{W}^\alpha$ are stored energy functions for individual structurally significant constituents that depend on constituent-specific deformations $F_{n(\tau)}^\alpha(s)$ that are measured relative to individual (potentially evolving) natural configurations $\kappa_{n(\tau)}^\alpha \equiv n(\tau)$.

(a)

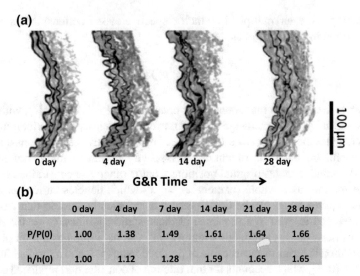

(b)

	0 day	4 day	7 day	14 day	21 day	28 day
P/P(0)	1.00	1.38	1.49	1.61	1.64	1.66
h/h(0)	1.00	1.12	1.28	1.59	1.65	1.65

Fig. 2 Evolution of the composition and geometry of the infrarenal abdominal aorta from an adult male *Apoe⁻/⁻* mouse over 28 days of hypertension induced with angiotensin II infused at 1000 ng/kg/min. **a** The histological stain is similar to that used in Fig. 1, hence the elastin stains black, though in this case the smooth muscle stains purple and the collagen pink. **b** Note that overall wall thickness h increases in response to the fold-increase in pressure ($P/P(0)$) so as to nearly mechano-adapt by 14 days (theoretically requiring $h \to \varepsilon^{1/3}\gamma h(0)$, where ε is the fold-change in blood flow, which in this case was unity, and γ is the fold-change in blood pressure). This adaptation appeared to be driven by mechano-mediated processes alone. Data from [5]

Note that $W_R^\alpha(0) \equiv W^\alpha(0) = \phi^\alpha(0)\widehat{W}^\alpha(0)$ where $\phi^\alpha = \rho^\alpha/\rho$ is a mass fraction, hence recovering a standard rule-of-mixtures prior to G&R as desired. Note, too, that in the case of tissue turnover within an unchanging configuration, $n(0) \equiv n(\tau)$, the full mixture relation again recovers the simple rule-of-mixtures. This model has proven useful in modeling diverse conditions, including arterial adaptations to altered mechanical loading [35], the enlargement of aneurysms [38], and inflammation-mediated aortic fibrosis in hypertension [28] (Figs. 2 and 3), among others. Of course, the total stored energy is simply assumed to be $W = \sum W^\alpha$ at any G&R time s, which can then be used in a standard way to compute Cauchy stress and satisfy linear momentum balance given appropriate constitutive equations.

Importantly, this theory reveals the need for three constituent-specific constitutive relations: true mass production $m^\alpha(\tau)$, mass removal $q^\alpha(s,\tau)$, and stored energy $\widehat{W}^\alpha\big(F_{n(\tau)}^\alpha(s)\big)$. As noted above, the survival function is often assumed to follow first-order type kinetics, hence necessitating the prescription of the rate parameter/function in terms of biochemomechanical stimuli, often time varying. The true rate of mass production similarly needs to be prescribed in terms of biochemomechanical stimuli whereas the stored energy function for individual constituents can often be prescribed similar to select terms in the structurally motivated fiber-based models. In other words, the greatest challenge tends to be identification of constitutive relations

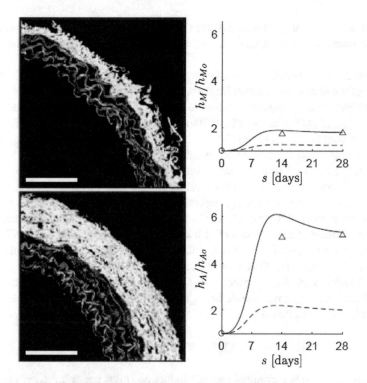

Fig. 3 Illustrative polarized light microscopic images of picro-sirius red stained sections from the descending thoracic aorta (DTA) after 14 days of hypertension induced with angiotensin II infused at 490 ng/kg/min, with marked immuno-mediated adventitial fibrosis (see bottom left image relative to the top left image). Shown, too, are predictions from a constrained mixture model of the primary hypertensive response, namely, medial (top right) and adventitial (bottom right) thickening. The solid line is the prediction when accounting for both mechano- and immuno-mediated G&R; the dashed line shows the prediction when accounting for mechano-mediated G&R alone; the open triangles show experimental data at 14 and 28 days, which were taken from [4]

for tissue turnover, namely, ECM synthesis and degradation and cell division and apoptosis.

Notwithstanding the challenges of solving initial-boundary value problems on complex domains, as for an asymmetric aortic aneurysm, identification of appropriate constitutive relations is generally the most difficult challenge of biomechanics, particularly in biosolid mechanics. Whereas standard methods now exist to identify best-fit material parameters for common stored energy functions for tissues, much remains to be learned regarding the best functional forms of these relations and the related parameters for individual constituents, initially elastic fibers versus collagen fibers but eventually elastin versus fibrillin-1, collagen III versus collagen I, and so on. Currently, most relations are restricted to "constituent-dominated" phenomenological behaviors, as, for example, describing elastin-dominated amorphous behaviors with a neo-Hookean form of $\widehat{W}^{\alpha}$ or describing collagen fiber-dominated directional behaviors with a Fung-exponential form of $\widehat{W}^{\alpha}$. Although associated simulations capture many salient features of tissue behavior, much more work is needed in this regard.

Even more challenging, however, is identification of the functional forms and values of the related parameters for mass production and removal of different structurally significant constituents. It is here that the importance of mechanobiology is most evident (cf. [16]). Early (mid-1970s) and continuing studies reveal that endothelial cell production of NO and ET-1 exhibits a sigmoidal relationship (increasing for NO and decreasing for ET-1) with increasing flow-induced wall shear stress [14]. Considering relations near the homeostatic value of shear stress allows one to use a linear approximation, however. Recalling from above that increases in NO and ET-1 tend to attenuate and hasten, respectively, the production of collagen and GAGs by smooth muscle cells and fibroblasts reveals an important paracrine factor that must be included in most vascular G&R modeling. Similarly, early (mid-1970s) and continuing studies reveal that smooth muscle cells and fibroblasts increase their production of collagen and GAGs in response to increasing pressure-induced normal stresses, again sigmoidal in relation but approximately linear about a homeostatic state. Consequently, a reasonable first approximation for a stress-based relation for mass density production is

$$m_R^{\alpha}(\tau) = m_N^{\alpha}(\tau)\left(1 + K_{\sigma}^{\alpha}\Delta\sigma^{\alpha}(\tau) - K_{\tau_w}^{\alpha}\Delta\tau_w^{\alpha}(\tau)\right)$$

where $m_N^{\alpha}(\tau) > 0$ is a nominal rate of production (mass per volume per time) that may change over G&R time, with $m_N^{\alpha}(0) \equiv m_o^{\alpha}$ the original basal rate; K_{σ}^{α} and $K_{\tau_w}^{\alpha}$ are positive gain-type parameters, and $\Delta\sigma^{\alpha}$ and $\Delta\tau_w^{\alpha}$ are normalized differences in current values of normal or shear stress relative to homeostatic targets. In most cases, increases in intramural stresses above homeostatic values increase production rates whereas increases in wall shear stress above homeostatic decrease production rates. That such targets are typically on the order of 150 kPa versus 1.5 Pa for intramural normal and luminal wall shear stress, respectively, is a strong reminder that although one component of stress may be negligible with respect to another in terms of the mechanics, all components can be critical mechanobiologically depending on the cell types (endothelial versus smooth muscle) and homeostatic targets. Importantly, however, we have mentioned but a few responses by endothelial and intramural cells to changes in their local mechanical environment, as indicated by changes in stress. There are likely changes in the expression of literally hundreds of genes with sustained changes in mechanical stimuli, affecting additional vasoactive molecules, structural proteins, growth factors, chemokines, cytokines, and proteases, among others. Much more remains to be discovered with regard to the associated "mechanobiological dose response curves" which will be needed to build improved models, particularly as we move to the age of precision medicine.

5 Mechanobiological Equilibrium and Stability

Concepts of equilibrium and stability are well established in mechanics, and are equally fundamental in biomechanics. For example, one can consider an intracranial saccular aneurysm to exist in quasi-equilibrated states at different times during a cardiac cycle and examine whether it is mechanically stable at any such state, as, for example, if a limit point (static) instability exists—which does not [24]. Indeed, one could assess such mechanical stability within any quasi-equilibrated state at any time of G&R, treating the lesion at each stage of enlargement as a mechanical structure subjected to mechanical loading. One can similarly examine the dynamic stability of an aneurysm that is subjected to a time varying load, determining for example when an attractive limit cycle exists—which does exist [8]. Identifying whether a tissue is mechanically unstable (i.e., allows marked changes in state in response to a perturbation) or stable (insensitive to the action of a perturbation) is thus important.

In parallel, one can define and should consider concepts of mechanobiological equilibrium and stability. Recalling the aforementioned form for the survival function $q^\alpha(s, \tau)$, if we assume that the original homeostatic value of the rate parameter is given by k_o^α and the original homeostatic production rate is m_o^α, then it can be shown that, in this case, $\rho^\alpha(0) = m_o^\alpha / k_o^\alpha$ for all $\alpha = 1, 2, \ldots, N$ structurally significant constituents [35]. In other words, appropriately balanced rates of production and removal exist in normal tissues in mechanobiological equilibrium, as promoted by the process of mechanical homeostasis, regardless of the specific functional forms of the rates of change in cases of adaptation, disease, or injury. Other terms have been used to describe this mechanobiological equilibrium, which in the most basic case exists in "tissue maintenance" wherein balanced turnover in an unchanging state preserves the geometry and properties of the tissue despite continual mass removal and mass production. Indeed, fully balanced rates of production and removal must exist in mechanobiologically equilibrated states that evolve in response to sustained stimuli [26], for which, equivalent to the original homeostatic case, it can be shown that $\rho_h^\alpha = m_h^\alpha / k_h^\alpha$, with subscript h used for evolved homeostatic quantities. Interestingly, one can even generalize this concept to find that a nearly balanced turnover $\rho_R^\alpha(s) \approx m_R^\alpha(s)/k^\alpha(s)$ also holds in evolution form if the characteristic time of the external loading stimulus is much longer than the characteristic time of the G&R process itself, that is, for mechanobiologically quasi-equilibrated states [27] that may progress, for instance, during slow hypertensive or aneurysmal G&R processes. Assessing mechanobiological stability is more difficult and much remains to be accomplished in this regard. The interested reader is referred to, for example, the work by Cyron and Humphrey [6] or Latorre and Humphrey [29] for an introduction to this important topic.

6 Illustrative Results

Here, we provide a few illustrative examples for arterial G&R in response to sustained increases in pressure, noting that elevated blood pressure (hypertension) is one of the primary risk factors for many cardiovascular diseases. To do so, and because we are often most interested qualitatively in fully resolved G&R responses, we do not consider transient effects, but rather compute mechanobiologically equilibrated solutions directly for each pressure increment. Hence, the following results hold equally for fully equilibrated states that are reached after a sustained application of each pressure increment (i.e., on a point-by-point basis [26]) or quasi-equilibrated states computed for slow increases in pressure (i.e., in evolution form [27]). The reader is referred to the original papers for detailed explanations of these simplified, yet useful, G&R frameworks. Table 1 lists representative values of key geometric, elastic, and G&R parameters for a single-layered, thin-walled model of the murine descending thoracic aorta (DTA).

In particular, we address arterial adaptations to increases in pressure for different properties of the collagen fiber families within the ECM, namely cross-linking between different families (Fig. 4), orientation of diagonal fibers (Fig. 5), and overall fiber undulations that manifest from different deposition stretches (Fig. 6). We let the strain energy function for collagen-dominated behaviors be described by a Fung-type relation, which, under mechanobiological equilibrium, specializes to

$$\widehat{W}^c\left(G^c\right) = \frac{c_1^c}{4c_2^c}\left(e^{c_2^c\left(G^{c2}-1\right)^2} - 1\right)$$

Table 1 Representative baseline material parameters for a mouse descending thoracic aorta (adapted from the bilayered model in [27])

Arterial mass density	ρ	1050 kg/m
Inner radius, thickness	a, h	647μm, 40 μm
Constituent mass fractions	$\phi_o^e, \phi_o^m, \phi_o^c$	0.3, 0.35, 0.35
Collagen relative fractions	$\beta^\theta, \beta^z, \beta^d$	0.07, 0.35, 0.58
Diagonal collagen orientation	α_0	45°
Elastic material parameters	$c^e, c_1^m, c_2^m, c_1^c, c_2^c$	115 kPa, 400 kPa, 1, 400 kPa, 5
Deposition stretches	$G_\theta^e, G_z^e, G^m = G^c$	1.9, 1.6, 1.15
Mass production gains	$K_\sigma^m = K_\sigma^c, K_\tau^m = K_\tau^c$	1, 1
Mass removal rates	$k_o^m = k_o^c$	1/15 day^{-1}

Superscripts e, m, and c denote elastin, smooth muscle, and collagen. Superscripts/subscripts θ, z, and d denote circumferential, axial, and symmetric diagonal directions. Subscript o denotes original homeostatic values whereas subscripts σ and τ denote intramural and wall shear stresses, respectively

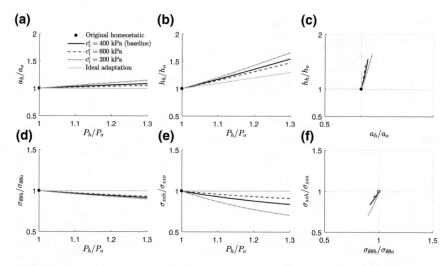

Fig. 4 Mechanobiologically stable (static) equilibrium responses illustrated for the DTA for different degrees of collagen cross-linking, represented by different strain energy constants for collagen c_1^c. Note that increasing values of c_1^c tend to stabilize the equilibrated (fully grown and remodeled) response, which approaches the ideal mechano-adaptive limit (dotted line). Note that the baseline results described the experimental data well [28]

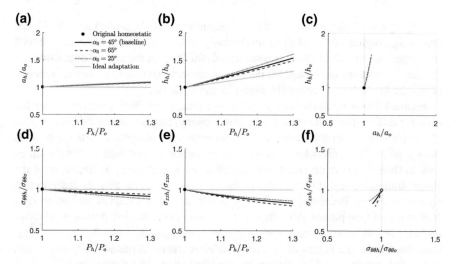

Fig. 5 Similar to Fig. 4, but for different orientations α_0 of the symmetric diagonal families of collagen fibers, with the angle calculated with respect to the axial direction in the reference configuration. Note that the baseline results described the experimental data well [28]

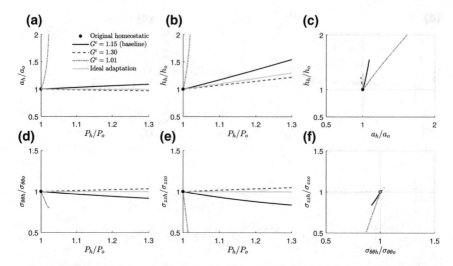

Fig. 6 Similar to Fig. 4, but with clearly differentiated mechanobiologically stable ($G^c =$ 1.15, 1.30) and unstable ($G^c = 1.01$) static equilibrium responses computed for different collagen undulations, represented by different deposition stretches of collagen G^c. Note that the baseline results described the experimental data well [28]

with c_1^c, c_2^c and G^c the associated material parameters and deposition stretch. We can thus assign higher values of c_1^c to increasingly cross-linked fiber networks, as well as interpret lower values of G^c as more undulated fibers at the time of deposition. Of course, the orientation of diagonal fibers with respect to the axial direction is ideally described in our four-fiber family model by the angle α_0.

Figure 4 shows mechanobiologically stable (static) equilibrium responses for the DTA for different degrees of collagen cross-linking, represented by different values of the constant c_1^c in the stored energy function for collagen. Panels (a, b, d, e) show equilibrium values for (bounded) inner radius a_h and wall thickness h_h, as well as (bounded) circumferential $\sigma_{\theta\theta h}$ and axial σ_{zzh} stresses, as functions of the stimulation-driver for different elevations of blood pressure, namely the pressure ratio $\gamma_h = P_h/P_o$. Panels (c) and (f) show the associated evolution of the homeostatic state in phase-type planes. Also shown for comparison is an ideal mechano-adaptive response for which $\sigma_{\theta\theta h}/\sigma_{\theta\theta o} = \sigma_{zzh}/\sigma_{zzo} = 1$, $a_h/a_o = \varepsilon_h^{1/3}$ and $h_h/h_o = \varepsilon_h^{1/3}\gamma_h$. Note that increasing values of c_1^c tend to stabilize the equilibrated (fully grown and remodeled) response, which approaches the ideal mechano-adaptive limit.

Similar to the prior figure, Fig. 5 shows effects of different orientations α_0 (measured with respect to the axial direction) for the symmetric-diagonal families of collagen fibers in the four-fiber family constitutive model. Note that wall thickness and circumferential stress (and, slightly, inner radius) approach their ideal mechano-adaptative responses for increasing values of α_0 (i.e., with diagonal fibers oriented more towards the circumferential direction), although at the expense of a more pronounced drop in axial stress. Interesting is the finding in panel (c) that the equilibrium relation $h_h/h_o(a_h/a_o)$ does not depend on changes in α_0 (with other orientations tested but not shown).

Finally, Fig. 6 reveals clearly differentiated mechanobiologically stable ($G^c = 1.15, 1.30$) and unstable ($G^c = 1.01$) static equilibrium responses resulting for different degrees of collagen fiber undulation, represented by different values of deposition stretch G^c, with respective bounded ($G^c = 1.15, 1.30$) and unbounded ($G^c = 1.01$) inner radius a_h and wall thickness h_h, as well as bounded (all cases) circumferential $\sigma_{\theta\theta h}$ and axial σ_{zzh} stresses. Thus, increasing values of G^c tend to stabilize the equilibrated response, with the (perhaps, excessively high) value $G^c = 1.30$ providing even lower (a_h and h_h) and higher ($\sigma_{\theta\theta h}$ and σ_{zzh}) values than the respective ideal ones and the (perhaps, excessively low) value $G^c = 1.01$ yielding an asymptotic growth response for $\gamma_h \to 1.025$, for which $\sigma_{zzh} > 0$ (not shown).

In summary, these simple examples show that ECM parameters such as collagen cross-linking, orientation, and undulation play key roles in potential G&R responses to sustained changes in mechanical loading, here illustrated for elevated pressure loading of a central artery. Perhaps as one would expect, collagen undulation (reflected by the value of the deposition stretch parameter) emerged as a particularly sensitive parameter, hence suggesting that greater attention should be directed to understanding how synthetic cells regulate this parameter in health and especially in disease. There is also a need to parameterize factors such as collagen type (I vs. III) and similarly contributions by accessory constituents such as biglycan and thrombospondin-2, which affect collagen fibrillogenesis and thus mechanics. Much remains to be discovered and then modeled.

7 Discussion

Biological soft tissues manifest wonderfully diverse characteristics, most importantly an ability to grow, remodel, adapt, and repair in response to myriad changes in conditions. This homeostatic process depends on changing gene expression, sometimes reflecting a phenotypic modulation of the resident or infiltrating cells. Much has been learned, but much remains unknown. Continuing mechanobiological experiments promise to provide the information that is needed both to increase our understanding and to inform our increasingly sophisticated models, both murine and computational. Whereas considerable success has been gained over the past few decades in describing the complex (nonlinear, anisotropic, nonhomogenous) mechanical behaviors of soft tissues, our understanding of mechanobiologically controlled growth and remodeling processes remains limited. There is, therefore, a pressing need for new data-driven theoretical frameworks and constitutive frameworks that can describe better the complex ability of cells to mechano-sense and mechano-regulate the ECM [22], thereby controlling tissue and organ size, shape, properties, and multiple functions. We briefly reviewed a constrained mixture approach that has proven successful in capturing many salient features of vascular G&R and suggest that continued advancements in the associated constituent-specific constitutive relations promises even greater insight into tissue and organ physiology and pathophysiology. Although we did not discuss the associated hemodynamics beyond wall shear stress control of endothelial gene expression leading to changes in NO or ET-1 production, there is a pressing need for combining advances in modeling fluid-solid interactions and wall growth and remodeling, resulting in coupled fluid-solid-growth (FSG) models [19]. Indeed, understanding the interactions between wall G&R and blood pressure pulse waves demands increased attention to the coupling between local mechanobiological and global physiological responses, particularly in aging and hypertension [23].

Notwithstanding the importance of the biomechanics and mechanobiology, recent studies have revealed an often critical role of immunobiology as well. Blood borne immune cells (i.e., white blood cells, including monocytes/macrophages of the innate immune system and T-cells and B-cells of the adaptive immune system) can also play critical roles in the G&R of diverse tissues, including arteries. For example, Wu et al. [39] showed that aortic fibrosis (i.e., excessive deposition of ECM, mainly collagens in the adventitia) resulted primarily from T-cell activity in a wild-type mouse model of hypertension induced using continuous angiotensin II infusion. Specifically, they showed that fibrosis did not develop in $Rag1^{-/-}$ mice, which lack mature T-cells and B-cells, but that the fibrosis re-emerged when T-cells were adoptively transferred into these mice. This fibrosis resulted in a mechano-maladaptation, meaning that mean circumferential wall stress ($\sigma_{\theta\theta} = Pa/h$, where P is the distending pressure, a the luminal radius, and h the wall thickness) was not maintained at its original homeostatic value. Bersi et al. [4] quantified associated changes in the compositional and mechanical properties in the fibrotic wild-type mice, and Latorre and Humphrey [28] showed that this maladaptation could be modeled well with the aforementioned constrained mixture G&R model if one adds inflammatory burden to the mass density production function and lets the mechanical properties of smooth muscle and collagen evolve accordingly. Indeed, the maladaptation could not be captured with the purely mechano-driven production function, regardless of the values of the associated material parameters. Importantly, Louis et al. [30] had also shown a loss of fibrosis in mice without the α_1 integrin subunit, which binds fibrillar collagens, hence affirming the importance of mechanical loads in stimulating the hypertensive phenotype. Thus, experiment and simulation agree—one should account for both mechano- and immuno-mediated turnover of ECM in general. This general concept is also reinforced by G&R simulations of the in vivo development of a tissue engineered vascular graft that begins as a polymeric scaffold. This graft elicits a strong foreign body response, and thus immuno-driven ECM production, and the associated neovessel development could only be described and predicted when both immuno-driven and mechano-mediated mechanisms were included in the model [31].

In summary, the production and removal of ECM in potentially evolving states enables tissue maintenance but also drives tissue growth, remodeling, repair, and even many diseases. In many cases, such production and removal is driven by the associated cells sensing and regulating this matrix, with signals arising from biological, chemical, and mechanical processes. Although significant understanding continues to come from clever and careful experiments, computational models are expected to play increasingly greater roles in advancing our understanding.

Acknowledgements This work was supported, in part, by current grants from the US National Institutes of Health: R01 HL086418 on abdominal aortic aneurysms, R01 HL105297 on arterial stiffening in hypertension and aging, U01 HL116323 on aortic dissection, R01 HL128602 and R01 HL139796 on tissue engineered vascular grafts, and P01 HL134605 on thoracic aortic aneurysms. We acknowledge Dr. M. R. Bersi for his outstanding experimental work, upon which many of the simulations herein were based. Finally, we acknowledge the many, many authors who have contributed so much to our general understanding of tissue mechanics over the years. In order to reduce the length of this work, we have cited primarily our own papers, in which copious references can be found of others' work, particularly in the many review articles and books that are cited.

References

1. Baek, S., Gleason, R.L., Rajagopal, K.R., Humphrey, J.D.: Theory of small on large: potential utility in computations of fluid–solid interactions in arteries. Comput. Methods Appl. Mech. Eng. **196**(31–32), 3070–3078 (2007)
2. Bellini, C., Ferruzzi, J., Roccabianca, S., Di Martino, E.S., Humphrey, J.D.: A microstructurally motivated model of arterial wall mechanics with mechanobiological implications. Ann. Biomed. Eng. **42**(3), 488–502 (2014)
3. Bersi, M.R., Ferruzzi, J., Eberth, J.F., Gleason, R.L., Humphrey, J.D.: Consistent biomechanical phenotyping of common carotid arteries from seven genetic, pharmacological, and surgical mouse models. Ann. Biomed. Eng. **42**(6), 1207–1223 (2014)
4. Bersi, M.R., Bellini, C., Wu, J., Montaniel, K.R., Harrison, D.G., Humphrey, J.D.: Excessive adventitial remodeling leads to early aortic maladaptation in angiotensin-induced hypertension. Hypertension **67**(5), 890–896 (2016)
5. Bersi, M.R., Khosravi, R., Wujciak, A.J., Harrison, D.G., Humphrey, J.D.: Differential cell-matrix mechanoadaptations and inflammation drive regional propensities to aortic fibrosis, aneurysm or dissection in hypertension. J. R. Soc. Interface **14**(136), 20170327 (2017)
6. Cyron, C.J., Humphrey, J.D.: Vascular homeostasis and the concept of mechanobiological stability. Int. J. Eng. Sci. **85**, 203–223 (2014)
7. Cyron, C.J., Humphrey, J.D.: Preferred fiber orientations in healthy arteries and veins understood from netting analysis. Math. Mech. Solids **20**(6), 680–696 (2015)
8. David, G., Humphrey, J.D.: Further evidence for the dynamic stability of intracranial saccular aneurysms. J. Biomech. **36**(8), 1143–1150 (2003)
9. Ferruzzi, J., Vorp, D.A., Humphrey, J.D.: On constitutive descriptors of the biaxial mechanical behaviour of human abdominal aorta and aneurysms. J. R. Soc. Interface **8**(56), 435–450 (2011)
10. Ferruzzi, J., Collins, M.J., Yeh, A.T., Humphrey, J.D.: Mechanical assessment of elastin integrity in fibrillin-1-deficient carotid arteries: implications for Marfan syndrome. Cardiovasc. Res. **92**(2), 287–295 (2011)
11. Fung, Y.C.: Biomechanics: Mechanical Properties of Living Tissues. Springer, NY (1981)
12. Holzapfel, G.A., Gasser, T.C., Ogden, R.W.: A new constitutive framework for arterial wall mechanics and a comparative study of material models. J. Elast. **61**(1–3), 1–48 (2000)
13. Humphrey, J.D.: Computer methods in membrane biomechanics. Comput. Methods Biomech. Biomed. Engin. **1**(3), 171–210 (1998)
14. Humphrey, J.D.: Cardiovascular Solid Mechanics: Cells, Tissues, and Organs. Springer, NY (2002)
15. Humphrey, J.D.: Continuum biomechanics of soft biological tissues. Proc. R. Soc. A **459**, 3–46 (2003)
16. Humphrey, J.D.: Vascular adaptation and mechanical homeostasis at tissue, cellular, and subcellular levels. Cell Biochem. Biophys. **50**(2), 53–78 (2008)
17. Humphrey, J.D., Na, S.: Elastodynamics and arterial wall stress. Ann. Biomed. Eng. **30**(4), 509–523 (2002)

18. Humphrey, J.D., Rajagopal, K.R.: A constrained mixture model for growth and remodeling of soft tissues. Math. Models Methods Appl. Sci. **12**(03), 407–430 (2002)
19. Humphrey, J.D., Taylor, C.A.: Intracranial and abdominal aortic aneurysms: similarities, differences, and need for a new class of computational models. Annu. Rev. Biomed. Eng. **10**, 221–246 (2008)
20. Humphrey, J.D., Eberth, J.F., Dye, W.W., Gleason, R.L.: Fundamental role of axial stress in compensatory adaptations by arteries. J. Biomech. **42**(1), 1–8 (2009)
21. Humphrey, J.D., Dufresne, E.R., Schwartz, M.A.: Mechanotransduction and extracellular matrix homeostasis. Nat. Rev. Mol. Cell Biol. **15**(12), 802–812 (2014)
22. Humphrey, J.D., Milewicz, D.M., Tellides, G., Schwartz, M.A.: Dysfunctional mechanosensing in aneurysms. Science **344**(6183), 477–479 (2014)
23. Humphrey, J.D., Harrison, D.G., Figueroa, C.A., Lacolley, P., Laurent, S.: Central artery stiffness in hypertension and aging: a problem with cause and consequence. Circ. Res. **118**(3), 379–381 (2016)
24. Kyriacou, S.K., Humphrey, J.D.: Influence of size, shape and properties on the mechanics of axisymmetric saccular aneurysms. J. Biomech. **29**(8), 1015–1022 (1996). Erratum 30, 761
25. Lanir, Y.T.: Constitutive equations for fibrous connective tissues. J. Biomech. **16**(1), 1–12 (1983)
26. Latorre, M., Humphrey, J.D.: A mechanobiologically equilibrated constrained mixture model for growth and remodeling of soft tissues. Z. Angew. Math. Mech. **98**(12), 2048–2071 (2018)
27. Latorre, M., Humphrey, J.D.: Critical roles of time-scales in soft tissue growth and remodeling. APL Bioeng. **2**(2), 026108 (2018)
28. Latorre, M., Humphrey, J.D.: Modeling mechano-driven and immuno-mediated aortic maladaptation in hypertension. Biomech. Model. Mechanobiol. **17**(5), 1497–1511 (2018)
29. Latorre, M., Humphrey, J.D.: Mechanobiological stability of biological soft tissues. J. Mech. Phys. Solids **125**, 298–325 (2019)
30. Louis, H., Kakou, A., Regnault, V., Labat, C., Bressenot, A., Gao-Li, J., Gardner, H., Thornton, S.N., Challande, P., Li, Z., Lacolley, P.: Role of α1β1-integrin in arterial stiffness and angiotensin-induced arterial wall hypertrophy in mice. Am. J. Physiol. Heart Circ. Physiol. **293**(4), H2597–H2604 (2007)
31. Miller, K.S., Khosravi, R., Breuer, C.K., Humphrey, J.D.: A hypothesis-driven parametric study of effects of polymeric scaffold properties on tissue engineered neovessel formation. Acta Biomater. **11**, 283–294 (2015)
32. Roccabianca, S., Bellini, C., Humphrey, J.D.: Computational modelling suggests good, bad and ugly roles of glycosaminoglycans in arterial wall mechanics and mechanobiology. J. R. Soc. Interface **11**(97), 20140397 (2014)
33. Schroeder, F., Polzer, S., Slažanský, M., Man, V., Skácel, P.: Predictive capabilities of various constitutive models for arterial tissue. J. Mech. Behav. Biomed. Mater. **78**, 369–380 (2018)
34. Shadwick, R.E.: Mechanical design in arteries. J. Exp. Biol. **202**, 3305–3313 (1999)
35. Valentin, A., Humphrey, J.D.: Evaluation of fundamental hypotheses underlying constrained mixture models of arterial growth and remodelling. Phil. Trans. R. Soc. Lond. A **367**(1902), 3585–3606 (2009)
36. Wagenseil, J.E., Mecham, R.P.: Vascular extracellular matrix and arterial mechanics. Physiol. Rev. **89**(3), 957–989 (2009)
37. Wicker, B.K., Hutchens, H.P., Wu, Q., Yeh, A.T., Humphrey, J.D.: Normal basilar artery structure and biaxial mechanical behaviour. Comput. Methods Biomech. Biomed. Engin. **11**(5), 539–551 (2008)
38. Wilson, J.S., Baek, S., Humphrey, J.D.: Parametric study of effects of collagen turnover on the natural history of abdominal aortic aneurysms. Proc. R. Soc. A **469**(2150), 20120556 (2013)
39. Wu, J., Thabet, S.R., Kirabo, A., Trott, D.W., Saleh, M.A., Xiao, L., Madhur, M.S., Chen, W., Harrison, D.G.: Inflammation and mechanical stretch promote aortic stiffening in hypertension through activation of p38 MAP Kinase. Circ. Res. **114**, 616–625 (2013)

Multi-scale Modeling of the Heart Valve Interstitial Cell

Alex Khang, Daniel P. Howsmon, Emma Lejeune and Michael S. Sacks

Abstract Heart valve interstitial cells (VIC) are fibroblast–like cells that reside within the interstitium of heart valve leaflets. The biosynthetic activity of VICs is highly dependent upon the the mechanical demands of the extracellular environment. Thus, regular deformation of the leaflets throughout the cardiac cycle provides the mechanical stimulation that is necessary for VICs to maintain homeostasis of the valve and manage normal turnover of extracellular matrix constituents. When the deformation pattern of the VICs is altered during periods of growth or disease, VICs can undergo cellular activation and remodel the ECM of the valve to re-establish homeostasis. In order to better engineer treatments for heart valve diseases, it is of great importance to delineate the underlying mechanisms governing this crucial remodeling process. In this chapter, we present current experimental and computational modeling approaches used to study the complex multi-scale mechanical relationship between the valve leaflets and the underlying VICs. In addition, we discuss future directions toward modeling VIC signaling pathways and developing improved 3D multi-scale models of VICs.

A. Khang · D. P. Howsmon · E. Lejeune · M. S. Sacks (✉)
James T. Willerson Center for Cardiovascular Modeling and Simulation,
Institute for Computational Engineering and Sciences and the Department of Biomedical
Engineering, The University of Texas at Austin, Austin, TX, USA
e-mail: msacks@ices.utexas.edu

A. Khang
e-mail: ak36582@utexas.edu

D. P. Howsmon
e-mail: daniel.howsmon@utexas.edu

E. Lejeune
e-mail: elejeune@utexas.edu

© Springer Nature Switzerland AG 2020
Y. Zhang (ed.), *Multi-scale Extracellular Matrix Mechanics and Mechanobiology*,
Studies in Mechanobiology, Tissue Engineering and Biomaterials 23,
https://doi.org/10.1007/978-3-030-20182-1_2

1 Introduction

Heart valves are dynamic living tissues whose physiological function is quite simple: to prevent the retrograde flow of blood within the heart. The mammalian heart has four valves: the aortic valve (AV), mitral valve (MV), pulmonary valve (PV), and tricuspid valve (TV). All valve leaflets are multi-layered structures composed of a heterogenous mixture of extracellular matrix (ECM) components with a specific structure that allows proper opening and closure of the valves. For instance, in the AV, the fibrosa layer contains mainly collagen and provides the majority of mechanical support in the valve while the ventricularis layer contains radially oriented elastin fiber networks that act as a low-energy recoiling mechanism during valve closure. In between these two layers, the spongiosa layer contains mainly proteoglycans and glycosaminoglycans that aid in compression of the valve leaflets during coaptation.

All valves undergo a complex time varying pattern of stretch and flexure that in turn deform the underlying valve interstitial cells (VIC). VICs are fibroblast-like mechanocytes that comprise $\sim 10\%$ of the total valve leaflet volume and respond to tissue-level stresses and deformations [5, 20]. In order to maintain the intricate and precise structure of the valve, VICs sense their mechanical environment and remodel the ECM constituents accordingly. In the quiescent state, VICs maintain the normal upkeep of ECM components to ensure valve homeostasis. During periods of both growth and disease, VICs are known to transition to an activated, myofibroblast-like phenotype, where they display increased contractility, prominent alpha smooth muscle actin (α-SMA) stress fibers, and increased ECM remodeling activity [4, 13]. VIC-ECM coupling is made possible through the VICs binding via integrins containing α and β subunits to ECM components such as fibronectin, laminin, and collagen [23, 43].

Remodeling of heart valve leaflet ECM can occur due to intrinsic and extrinsic factors. For instance, during pregnancy, the heart undergoes a drastic pressure over-load, which drastically increases ventricular size and the mechanical loading of the MV [38]. Initially, this causes a reduction in collagen crimp which mechanically compresses the mitral VICs (MVICs). This in turn induces growth and remodeling mechanisms in an attempt to reestablish a normal homeostatic cell shape. It is thus hypothesized that in late pregnancy, homeostasis is reestablished through remodeling processes that decrease collagen fiber alignment and restore the collagen crimp, relieving the compression of MVICs.

Another example, for a diseased state, is ischemic MV regurgitation (IMR) which arises from cardiomyopathy and/or coronary heart disease that induces MV regurgitation as a secondary complication [7, 8]. This is driven by the displacement of papillary muscles, which assist in tethering the MV via chordae tendineae [21], during left ventricular (LV) remodeling. This process can cause the MV to dilate and prevent proper closing. To compensate, the MV enlarges and becomes stiffer through remodeling processes. Despite this, the MVICs fail to reestablish a homeostatic loading state within the MV because the altered loading patterns persist. As a result, MV repair and replacement procedures are commonly performed in these scenarios [1].

Pregnancy and IMR are two compelling examples of how tissue-scale mechanics can drive the cellular-scale deformation of MVICs and in turn induce drastic remodeling mechanisms. In the AV, calcific aortic valve disease (CAVD) is a major driver of aortic stenosis in elderly individuals and those with bicuspid aortic valves. During the onset of CAVD, AVICs form calcium nodules which drastically stiffen the valve leaflets, leading to stenosis. AV replacements with heterograft tissues or bioprosthetic valves are commonly performed to combat these symptoms.

These phenomena highlight the importance of multi-scale and holistic approaches to studying the mechanobiology of VICs and how the mechanobiology ultimately determines the tissue-level response of the valves. Despite our current understanding of the relationship between VIC mechanobiology and valve mechanics, little is known about the underlying cellular processes that govern it. In addition, little is known about layer-specific properties of VICs and how this connects to the formation of different valve layers with vastly different mechanical function [37]. In this chapter, we discuss current experimental and computational modeling techniques to study the mechanics of isolated VICs. We then go on to discuss the in-situ properties of VICs and preliminary investigations into layer-specific mechanical responses. We end by summarizing the use of synthetic hydrogel environments to study VIC mechanobiology and provide future directions towards developing cell signaling models that can be used to explain cellular mechanisms that allow for VIC regulation of valve ECM and refining existing models to more accurately elucidate VIC and ECM mechanical properties.

2 Isolated Cell Studies

2.1 Overview

Previous isolated cell studies have shown that VICs from the AV and MV are intrinsically stiffer than VICs from the PV and TV owing to the larger mechanical demands of the left side of the heart in which the AV and MV reside [33, 34]. In this more mechanically demanding environment, the VICs of the AV and MV undergo larger cytoplasmic and nuclear deformations which drives an up-regulation of α-SMA expression and subsequently increases VIC stiffness. In addition, VICs from the AV and MV display an up-regulation of ECM biosynthesis, resulting in thicker and stiffer valve leaflets overall.

Micropipette aspiration (MA) [34] and atomic force microscopy (AFM) [33] are valuable techniques that have been used to study the mechanics of isolated VICs. However, AFM measurements of VIC stiffness are 10–100 fold larger in magnitude than MA measurements. This is thought to be caused by the difference in internal stress fiber composition and architecture between free-floating, suspended VICs in MA experiments, and VICs seeded on a planar substrate in the AFM experiments. In suspension, VICs are thought to be in an inactivated state and do not express prominent α-SMA fibers. This contrasts with VICs seeded on a 2D substrate that are in an activated state and therefore generate stronger contractile forces. In addition to the differences in VIC subcellular structures between the two tests, differences in effective stiffness values may arise because the organelles and cell structures are being measured at different length scales with each test. MA experiments measure a broader region on the cell while AFM is used to assess more local properties at the periphery or near the nucleus. A continuum mixture finite element model of an isolated VIC was developed to gain a better understanding of the role that organelles and especially stress fibers play in VIC effective stiffness.

2.2 Model Formulation

A first mechanical model of the VIC was developed through integrating the experimental data from both MA and AFM measurements on VICs from the AV (AVIC) and PV (PVIC) [39]. In the formulation of the model, consideration was given to the organelles most likely to contribute during a specific test. For example, during analysis of the MA results, the passive properties of both the basal cytoskeleton and cytoplasm (all organelles except for the stress fibers and the nucleus) and the stress fibers were thought to contribute the most and therefore their shear moduli were estimated from the MA data. Then, data from AFM measurements at the periphery of VICs and directly above the nucleus was used to calibrate the contractile strength of the α-SMA stress fibers and the shear modulus of the nucleus, respectively.

The finite element models were developed in the open source software FEBio [30]. In the MA simulations, the initial shape of the VIC was modeled as a sphere and the boundary between the cell and the micropipette was subjected to a contact–traction to induce friction-less sliding between the two surfaces [39]. In the AFM experiments, a representative VIC was modeled on a flat surface with two separate subdomains for the nucleus and the cytoplasm [40]. A no-slip boundary condition was applied on the bottom surface of the VIC and the contact between the VIC and the rigid indenter was modeled with a no-penetration, no-slip boundary condition.

2.2.1 Calibration of the Cytoplasm, Stress Fibers, and Nuclear Mechanics

The initial step in estimating the mechanical properties of the VIC uses MA data to extract the shear modulus of the cytoplasm (μ^{cyto}) and the stress fibers (μ^{sf}) (Fig. 1a). The total Cauchy stress $\mathbf{T}$ is the sum of both components

$$\mathbf{T} = \mathbf{T}_{cyto} + \mathbf{T}_{sf}. \tag{1}$$

The cytoplasm is modeled as a nearly incompressible, isotropic, hyperelastic material with a neoHookean material model with a transversely isotropic contribution from

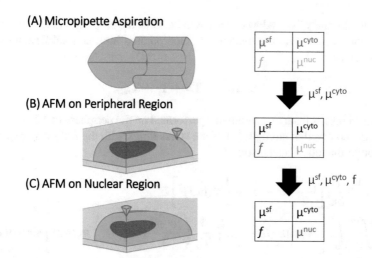

Fig. 1 **a** First, micropipette aspiration data is used to calibrate the shear modulus of the stress fibers (μ^{sf}) and the cytoplasm (μ^{cyto}). **b** Next, AFM measurements made at the periphery of VICs is used to estimate the stress fiber contraction strength (f). **C** Finally, AFM measurements made above the nuclear region are used to back out the shear modulus of the nucleus (μ^{nuc}). Figure adapted from [39]

the passive stress fibers. The full explicit Cauchy stress used to analyze the MA data is as follows

$$
\mathbf{T} = 2\frac{1}{J}\mathbf{F}\frac{\partial}{\partial \mathbf{C}}\left[\frac{\mu^{cyto}}{2}(\bar{I}_1 - 3) + \frac{1}{2}K(\ln J)^2\right]\mathbf{F}^T
$$
$$
+ \frac{1}{4\pi}\int_0^{2\pi}\int_0^{\pi}\left[H(I_4 - 1)2\frac{I_4}{J}\frac{\partial}{\partial I_4}\left(\frac{\mu^{sf}\bar{\phi}^{sf}}{2}(I_4 - 1)^2\right)\mathbf{m}\otimes\mathbf{m}\right]\sin\theta\, d\theta d\phi.
$$

$$(2)$$

where $\bar{I}_1 = J^{\frac{-2}{3}}I_1$, I_1 is the first invariant of the deformation gradient tensor $\mathbf{F}$ ($I_1 = \text{tr}(\mathbf{F})$), J is the jacobian ($J = \det(\mathbf{F})$), $\mathbf{C}$ is the right Cauchy-Green deformation tensor ($\mathbf{C} = \mathbf{F}^T\mathbf{F}$), K is the bulk modulus, H is a heavy side step-function introduced to enforce the passive stress that only arises from the fiber stretch, I_4 is the stress fiber stretch along the initial direction $\mathbf{m}_0$, $\mathbf{m}$ is the current state of the fibers initially oriented along $\mathbf{m}_0$ ($\mathbf{m} = \lambda^{-1}\mathbf{F}\cdot\mathbf{m}_0$, where $\lambda = \sqrt{I_4}$), and $\bar{\phi}^{sf}$ is the normalized α-SMA expression level. Direction $\mathbf{m}_0$ is characterized using two angles: $\theta \in [0, \pi]$ and $\phi \in [0, 2\pi]$ and $\mathbf{m}_0(\theta, \phi) = \sin\theta\cos\phi\mathbf{e}_1 + \sin\theta\sin\phi\mathbf{e}_2 + \cos\theta\mathbf{e}_3$ where $\mathbf{e}_1$, $\mathbf{e}_2$, and $\mathbf{e}_3$ are the unit normal basis vector for the Cartesian coordinates. The parameters μ^{cyto} and μ^{sf} are determined through minimization of the least square error between the simulated and experimental aspiration lengths with respect to the applied aspiration pressure of the VICs in the MA experiments.

During the analysis of AFM experimental results, the properties of the stress fiber contraction strength and the nucleus are incorporated into the total Cauchy stress (Fig. 1b, c) as:

$$\mathbf{T} = \mathbf{T}_{cyto} + \mathbf{T}_{sf} + \mathbf{T}_f + \mathbf{T}_{nuc}. \tag{3}$$

First, AFM results from measurements performed on the periphery of VICs are used to find the contraction strength (f) of the stress fibers with the following expanded equation of the total Cauchy stress:

$$\mathbf{T} = 2\frac{1}{J}\mathbf{F}\frac{\partial}{\partial \mathbf{C}}\left[\frac{\mu^{cyto}}{2}(\bar{I}_1 - 3) + \frac{1}{2}K(\ln J)^2\right]\mathbf{F}^T$$
$$+ \int_0^{2\pi}\int_0^{\pi}\left[\Gamma_t(\mathbf{m})H(I_4 - 1)2\frac{I_4}{J}\frac{\partial}{\partial I_4}\left(\frac{\mu^{sf}\bar{\phi}^{sf}}{2}(I_4 - 1)^2\right)\mathbf{m}\otimes\mathbf{m}\right]\sin\theta\,d\theta d\phi.$$
$$+ \int_0^{2\pi}\int_0^{\pi}\left[\Gamma_t(\mathbf{m})f\cdot\bar{\phi}^{sf}\frac{I_4}{J}\mathbf{m}\otimes\mathbf{m}\right]\sin\theta\,d\theta d\phi. \tag{4}$$

where $\Gamma_t(\mathbf{m})$ is the constrained von-Mises distribution [17] which represents the stress fiber orientation distribution with respect to $\mathbf{m}$, the current fiber direction. Then, the shear modulus of the nucleus μ^{nuc} was determined by using AFM data from measurements made above the nucleus and through the following equation for Cauchy's stress:

$$\mathbf{T} = 2\frac{1}{J}\mathbf{F}\frac{\partial}{\partial \mathbf{C}}\left[\frac{\mu^{nuc}}{2}(\bar{I}_1 - 3) + \frac{1}{2}K(\ln J)^2\right]\mathbf{F}^T. \tag{5}$$

The parameters f and μ^{nuc} are calibrated through minimization of the least square error between the experimental and simulated indentation depth with respect to applied force at the cell periphery and above the nucleus, respectively.

From the constitutive model (Eqs. 4 and 5), the shear modulus of the cytoplasm, stress fibers, the nucleus and the stress fiber contraction strength (f) are determined (Table 1). The model predicts that the intrinsic stiffness of the cellular sub-components is conserved between VICs from the AV and PV. However, a drastic difference in the contraction strength between AVICs and PVICs is predicted. AVIC contraction strength is reported to be $\sim$10 times stronger than PVIC contraction, reinforcing that the stiffness of VICs is positively correlated with α-SMA expression levels [34]. This is consistent with the previous discussion of why AFM measurements are 10–100 fold larger than MA measurements.

Table 1 Parameter values determined from the constitutive model for both AVICs and PVICs. Table adapted from [39]

	μ^{cyto} (Pa)	μ^{sf} (Pa)	f (kPa)	μ^{nuc} (kPa)
AVIC	5	390	35.27 ± 3.42	15.64 ± 2.46
PVIC	5	390	3.86 ± 0.61	16.06 ± 2.19

2.2.2 Incorporation of Stress Fiber Force-Length Relations, Expression Levels, and Strain Rate

In an extended mechanical model of the VIC [40], stress fiber force-length relations, expression levels of F-actin and α-SMA stess fibers, and the strain rate sensitivity of VICs in response to AFM indentation are incorporated. Briefly, VICs were seeded onto 2D collagen coated cover slips and were subjected to five separate treatments before AFM measurements were made:

1. **CytoD**: Cytochalasin D was used to disassemble the actin fiber network of the VICs. This group served as a negative control.
2. **C5 and C90**: The C5 VIC group was subjected to testing within 5 mM potassium chloride (KCl) to emulate normal physiological conditions. The C90 group was tested within 90 mM KCl (induces cell activate contraction) and is representative of a hypertensive cell state. These groups served as a negative control for the TGF-β1 group to follow.
3. **T5 and T90**: TGF-β1 is a potent cytokine that effectively induces the activation of VICs and has been shown to lead to complications in vivo such as calcification and valve stenosis [12, 36]. TGF-β1 has been shown to induce VICs to transition to a myofibroblast-like phenotype which increases α-SMA expression and contractility. The T5 group is treated with both TGF-β1 and 5 mM KCl to allow for measurements on VICs in the myofibroblast state and under normal conditions. The T90 group is treated with TGF β1 and 90 mM KCl to study myofibroblast-like VICs in a hypertensive state.

For model formulation and subsequent analysis of the AFM results, the F-actin and α-SMA stress fiber expression levels of the VICs under each chemical treatment group were quantified from analyzing fluorescence images (Fig. 2a, b). Only α-SMA stress fibers that co-localized with F-actin stress fibers are considered to avoid accounting for α-SMA subunits that exist within the cytoplasm and do not contribute to active contraction. Consistent with the first pass model, the total Cauchy stress is taken to be the sum of the contributions from the cytoplasm and the stress fibers (Eq. 1). The cytoplasm is modeled as a nearly incompressible neo-Hookean material as follows:

$$\mathbf{T}^{cyto} = 2\frac{1}{J}\mathbf{F}\frac{\partial}{\partial \mathbf{C}}\left[\frac{\mu^{cyto}}{2}(\bar{I}_1 - 3) + \frac{1}{2}K(\ln J)^2\right]\mathbf{F}^T. \tag{6}$$

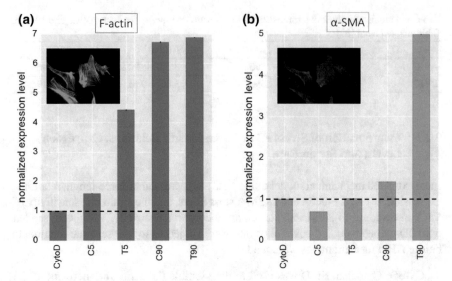

Fig. 2 a F-actin expression levels of VICs with respect to the chemical treatment conditions. **b** Expression level of α-SMA stress fibers that co-localize with F-actin fibers under the different chemical treatments. Figure adapted from [40]

The stress fibers were modeled as

$$\mathbf{T}^{sf} = \frac{1}{\mathbf{J}}\mathbf{F}\left[\int_{-\frac{\pi}{2}}^{\frac{\pi}{2}} \Gamma(\mathbf{m}_0)[H(I_4 - 1)T^p(\mathbf{m}_0) + T^a(\mathbf{m}_0) + T^v(\mathbf{m}_0)] \times (\mathbf{m}_0 \otimes \mathbf{m}_0)d\theta\right]\mathbf{F}^T \tag{7}$$

where $\mathbf{m}_0$ defines the inital direction of the stress fibers and T^p, T^a, and T^v are the one dimensional passive, active, and viscous contributions to stress, respectively. The passive component of the stress fibers are captured by

$$T^p = 2\mu_{sf}\bar{\phi}_{F-actin}(I_4 - 1) \tag{8}$$

where μ_{sf} is the shear modulus of the stress fibers and $\bar{\phi}_{F-actin}$ is the expression level of F-actin (Fig. 2a). The active portion of the stress fibers is modeled as

$$T^a\left(\bar{\phi}_{\alpha-SMA}, \bar{\phi}_{F-actin}, \lambda\right) = f_0\left(\bar{\phi}_{\alpha-SMA}, \bar{\phi}_{F-actin}\right)f_1(\lambda) \tag{9}$$

where f_0 is the maximum amount of contraction possible and is a function of the α-SMA and F-actin expression levels (Fig. 2a and b), and f_1 is the length-tension relationship which is a function of the stretch (λ). The maximum contraction is thus modeled as:

$$f_0\left(\bar{\phi}_{\alpha-SMA}, \bar{\phi}_{F-actin}\right) = f_{\alpha-SMA}\bar{\phi}_{\alpha-SMA} + f_{F-actin}\bar{\phi}_{F-actin}, \tag{10}$$

where $f_{\alpha-SMA}$ and $f_{F-actin}$ are the contraction strength of each stress fiber per unit expression level and are kept constant regardless of activation level. The length-tension relation is adapted from [44] and is implemented as:

$$f_1(\varepsilon, \varepsilon_0, \varepsilon^*) = \exp -((\varepsilon - \varepsilon^*)/\varepsilon_0)^2$$
$$\varepsilon = (I_4 - 1)/2 \qquad (11)$$

where ε is the fiber strain, ε^* is the strain level where maximum contraction occurs, and ε_0 is the rate of decay of the contractile strength with respect to ε^*. From simulation, it is noted that the value of ε^* had no substantial effect on the results when $|\varepsilon^*| < \varepsilon_0$. Therefore ε^* and ε_0 are set to 0 and 0.1, respectively, producing a similar length-tension relationship to one used previously to model muscle fiber contraction [15, 44]. From the indentation experiments, it is noted that the stiffness of the VICs is highly dependent upon levels of α-SMA. Thus, it is assumed that the viscous response of the stress fibers should be related to the viscosity of α-SMA ($\eta_{\alpha-SMA}$), α-SMA expression level, and strain rate ($\dot{\varepsilon}$) as follows:

$$T^v = \eta_{\alpha-SMA} \bar{\phi}_{\alpha-SMA} \dot{\varepsilon}. \qquad (12)$$

In its entirety, the updated constitutive model of the VIC is as follows:

$$\mathbf{T} = 2\frac{1}{J}\mathbf{F}\frac{\partial}{\partial \mathbf{C}}\left[\frac{\mu^{cyto}}{2}(\bar{I}_1 - 3) + \frac{1}{2}K(\ln J)^2\right]\mathbf{F}^T$$
$$+ \frac{1}{J}\mathbf{F}[\int_{\theta_p-\frac{\pi}{2}}^{\theta_p+\frac{\pi}{2}} \Gamma(\mathbf{m}_0)[H(I_4 - 1)2\mu_{sf}\bar{\phi}_{F-actin}(I_4 - 1) \qquad (13)$$
$$+ f_0\left(\bar{\phi}_{\alpha-SMA}, \bar{\phi}_{F-actin}\right)f_1(\lambda) + \eta_{\alpha-SMA}\bar{\phi}_{\alpha-SMA}\dot{\varepsilon}]$$
$$\times (\mathbf{m}_0 \otimes \mathbf{m}_0)d\theta]\mathbf{F}^T.$$

where the first term represents the contributions of the cytoplasm, and the second through fourth terms are the passive, active contraction, and viscous contributions of the stress fibers, respectively.

2.3 Model Results

The system of partial differential equations presented previously was incorporated into FEBio to simulate the MA and AFM experiments (Fig. 3a and b). All parameters of interest were computed using non-linear least squares with the Levenberg-Marquardt algorithm to match the experimental and simulated indentation depth versus force data obtained from AFM measurements. From the constitutive model, the shear modulus (μ_{sf}), contraction strength (f_0), and viscosity ($\eta_{\alpha-SMA}$) of the stress fibers are determined. From this, the total stress fiber shear modulus

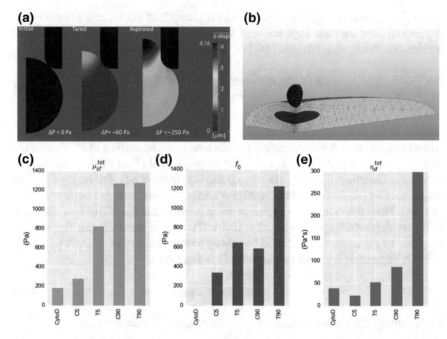

Fig. 3 Finite element models of the **a** MA and **b** AFM experiments, respectively. The **c** total shear modulus (μ_{sf}^{tot}), **d** contraction strength (f_0), and **e** total viscosity (η_{sf}^{tot}) of the stress fibers for each treatment condition. Higher expression of α-SMA in the T90 group largely increases the contraction strength and viscosity of the stress fibers. Figures taken from [39, 40]

($\mu_{sf}^{tot} = \mu_{sf}\bar{\phi}_{F-actin}$) and total viscosity of the stress fibers ($\eta_{sf}^{tot} = \eta_{\alpha-SMA}\bar{\phi}_{\alpha-SMA}$) are easily computed (Fig. 3c–e).

Results of the simulation clearly show that as VICs become more activated, the contraction strength (f_0) of the stress fibers increases (Fig. 3d). In addition, the viscosity of the stress fibers increases drastically (Fig. 3e, T90 group). This finding indicates that the difference in mechanical response when VICs in the T90 group were subjected to fast and slow strain rates (Fig. 4) is due to an increase in α-SMA stress fiber viscosity. No strain-rate sensitivity was observed in any of the other treatment groups.

The computed contraction strength (f_0) is found to be non-zero for the C5, T5, C90, and T90 groups (Fig. 3d). From these values, the contractile force arising separately from α-SMA and F-actin stress fibers is calculated using Eq. 10, resulting in four linear equations and two unknowns. From this, $f_{\alpha-SMA}$ and $f_{F-actin}$ are estimated to be 96.8 and 76.7 Pa, respectively. This result indicates that the incorporation of α-SMA in the stress fibers increases the contraction of VICs, consistent with previously reported experimental findings [10, 19, 31, 46]. The separate contributions of F-actin and α-SMA stress fibers to the overall contraction strength the estimated total contraction strength (f_0), and the expression levels of α-SMA ($\bar{\phi}_{\alpha-SMA}$) and F-actin ($\bar{\phi}_{F-actin}$) are computed by using Eq. 10 (Fig. 5). It is observed that within

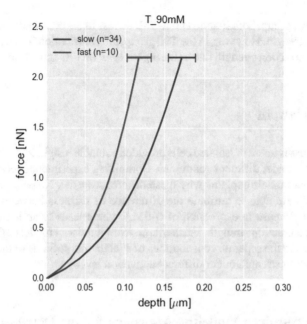

Fig. 4 Indentation depth versus force plots from AFM measurements performed on VICs within the T90 treatment group. Strain rate sensitivity is only observed within this group. The model predicts that an increase in the total viscosity of the α-SMA fibers (η_{sf}^{tot}) is likely the source of the observed strain-rate sensitivity. Figure adapted from [40]

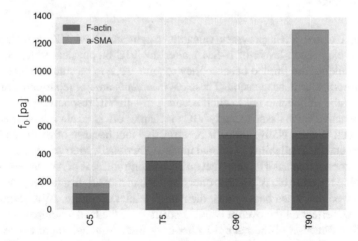

Fig. 5 The contributions of F-actin and α-SMA stress fibers to the overall estimated contraction strength. Note how every group contains contributions from both stress fibers but the T90 group is largely dominated by the effects of α-SMA. Figure taken from [40]

each treatment group, both F-actin and α-SMA stress fibers contributed to the overall contraction strength. However, in the T90 group, the contribution from α-SMA dominates the contraction strength due to the higher level of VIC activation in this group.

2.4 Conclusions

Mechanical assessment of isolated cells provided valuable insight into the differences between VICs under different conditions. Combining experimental techniques with computational models explains why these differences exist. Namely, the computational models are able to attribute the difference in stiffness between AVICs and PVICs to an increase in expression of α-SMA which leads to an increase in total stress fiber contraction strength. Furthermore, computational models of the VIC are able to delineate the separate contributions of F-actin and α-SMA to the total stress fiber contraction strength under different activation levels.

3 A Macro-micro Modeling Approach for the Down-Scale Estimation of AVIC Mechanics

3.1 Overview

Studying isolated VICs provides valuable insight into how VIC activation and increased expression levels of α-SMA alter the VIC biophysical state. However, these techniques are limited because they require VICs to be studied either on 2D glass coverslips or in suspension. These environments are not representative of the native cellular milieu and fail to elicit accurate in-situ VIC response.

Using native valve tissue to study VIC mechanics offers an alternative method to isolated cell studies. Native tissue is an attractive tool because of its physiological relevance and the availability of animal models. Previously, Merryman et al. showed that VIC contraction has a direct effect on the bending stiffness of AV leaflets and that the bending response of AV tissue is direction dependent [32]. In addition, Merryman et al. showed that when bent against the natural leaflet curvature, AVIC contraction had a larger effect on the overall bending stiffness of the leaflet compared to being bent in the direction of the natural leaflet curvature. Although these experimental results are insightful, it is difficult to delineate the underlying causes. Thus, the development of computational models is needed to further our understanding of the role that the underlying AVICs play in AV micromechanics. However, the AV leaflet tissue does not satisfy conventional multi-scale requirements due to the similarity in scale between the average leaflet thickness ($500\,\mu$m) and the average size of an AVIC ($15\,\mu$m). This similarity in scale (10:1) is not appropriate for multi-scale modeling methods and thus an alternative approach must be utilized. To circumvent these issues, a macro-micro modeling approach is established to directly map the boundary conditions on the macro level to the micro level.

3.2 Summary of Experimental Methods

Fresh native porcine AV leaflets were excised and subjected to three-point bending tests (Fig. 6a, b) [32]. Each leaflet was tested under normal, hypertensive, and inactive conditions. This was done by incubating the test specimens within 5 mM KCl, 90 mM of KCl, and thapsigargin (a known potent inhibitor of calcium ATPase), respectively. From the tests, moment versus curvature plots (M/I vs. $\Delta\kappa$) are generated. Then, the effective stiffness (E_{eff}) of the leaflets under each contractile level and with respect to testing with the natural leaflet curvature (WC - ventricularis under tension and fibrosa under compression) and against the natural leaflet curvature (AC - fibrosa under tension and ventricularis under compression) are calculated using the Euler-Bernoulli relation as follows:

$$\frac{M}{I} = E_{eff}\,\Delta\kappa \qquad (14a)$$

$$E_{eff} = \frac{M}{I\,\Delta\kappa}. \qquad (14b)$$

Values of E_{eff} calculated from Eqs. 14a and 14b for the AV leaflet test specimens are reported in Fig. 6c.

From the experimental results, it was noted that only in the AC direction showed a substantial increase in E_{eff} after treatment with 90 mM KCl (Fig. 6b). In addition, a decrease in E_{eff} was noted in all groups treated with thapsigargin, which completely inhibits the contraction of the underlying AVICs. Although these experimental results provide insight into how AVIC contraction modulates the flexural stiffness of the AV leaflet, they fail to meaningfully delineate the underlying mechanisms that cause a drastic change in E_{eff} in the AC curvature direction. In addition, they fail to elucidate the lack of significant changes observed in the WC direction despite treatment with 90 mM KCl. To further investigate the underlying phenomenon responsible for this behavior, a finite-element model of the bending experiment is developed and the displacements of the macro level representative volume element (RVE) are mapped to a refined micro level RVE to determine the underlying parameters that govern AVIC in-situ contractile effects on AV leaflet flexural stiffness [2].

3.3 Macro Model Formulation

Previously, it has been shown that the spongiosa layer behaves mechanically as a contiguous extension of the fibrosa and ventriculars layers [3]. Therefore, the

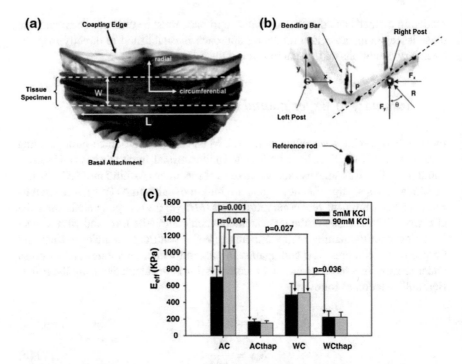

Fig. 6 **a** An excised AV leaflet. The test specimens were consistently excised just below the nodulus of Arantus. **b** Photograph of an AV test specimen undergoing three-point bending. **c** The effective stiffness (E_{eff}) of the AV leaflets before (5 mM KCl) and after (90 mM KCl) exposure to the active contraction treatment under normal (no thapsigargin) and inactive (thapsigargin) conditions. E_{eff} of the leaflets not treated with thapsigargin increased drastically in the AC direction after treatment with 90 mM KCl. The introduction of thapsigargin decreased E_{eff} in all groups. Figure adapted from [32]

AV leaflet is modeled as a bilayer, bonded beam with the finite-element software ABAQUS (Dassault Systemes, Johnston, RI, USA) using brick elements. Both layers were assigned a bimodular neo-Hookean isotropic nearly incompressible material model as follows:

$$W^d = \frac{_{state}\mu^{l,d}_{Macro}}{2}(I_1 - 1) - p(I_3 - 1) \tag{15}$$

where $_{state}\mu^{l,d}_{Macro}$ is the state dependent shear modulus of the layer (l = fibrosa (F) or ventricularis (V)) with respect to the testing direction (d = under tension (+) or compression (−)), $I_3 = det(\mathbf{C})$ is the third invariant of the right Cauchy-Green deformation tensor, and p is the Lagrange multiplier that enforces incompressibility. The macro-level simulation boundary conditions are a pin constraint at the left edge of the specimen with small and controlled displacements in the x-direction applied at the right edge of the specimen. The dimensions of the simulated geometry are

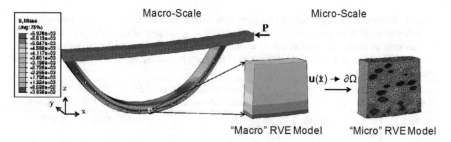

Fig. 7 Macro-level FE simulation of the bending experiments. The displacements of the "Macro" RVE model are mapped as boundary conditions to the "Micro" RVE model. Figure taken from [2]

kept consistent with that of the experimental test specimens (15 mm × 4 mm × 0.462 mm). The bilayer is divided in the thickness direction as 60 % fibrosa and 40 % ventricularis based on measurements from native AV tissue [5] (Fig. 7).

3.4 Macro Model Parameter Estimation

The tissue parameters are estimated using the experimental bending data. From a previous study [3], the moduli ratios for each layer under tension and compression were determined and used to estimate the inactive tissue parameters by fitting the FE simulation in the WC and AC bending directions simultaneously. The simulated moment-curvature plots are generated using the nodal locations along the edge of the geometry and the resulting reaction force on the left side of the geometry due to end displacement.

Next, the tissue parameter estimation for both the hypertensive and normal cases is performed. It is assumed that under compression, AVICs do not contribute to the stiffening of the valve tissues due to lack of tension from the surrounding ECM, an integral component needed for actin filament assembly and thus cell contraction. Therefore, the inactive compressive moduli are held constant and the tensile moduli are determined for both layers under normal and hypertensive conditions by fitting the simulated results to the experimental data for the AC and WC cases. Twelve total parameters are estimated from the macro RVE model (2 (layers) × 2 (tensile and compressible moduli) × 3 (each contractile state - "inactive", "normal", and "hyper") = 12 total parameters) and are reported in Table 2.

3.5 Micro Model Formulation

The micro model contains discrete ellipsoidal AVIC inclusions with distribution, size, and density determined from histological slices (Fig. 8) [5, 20]. The AVICs are modeled using a prolate spheroid, and are 14.16 μm long and 7.49 μm wide. The

Table 2 Estimated tissue parameters for each layer in tension and compression and under inactive, normal, and hypertensive contractile states. In addition, goodness of fit to the experimental data (R^2) as well as thickness normalized (0 = start of ventricularis, 1 = end of fibrosa) neutral axis locations (where axial stretch = 1) are reported for each activation state and test direction. Table taken from [2]

Activation state	Layer	μ_{Macro}^+	μ_{Macro}^-	Tension/compression Layer Moduli Ratios	R^2	Neutral axis (WC)	Neutral axis (AC)
Inactive	F	206.42	43.76	~5:1	0.990	0.36	0.75
	V	96.19	24.77	~4:1			
Normal	F	1496.67	43.76	~34:1	0.998	0.28	0.87
	V	311.93	24.77	~13:1			
Hyper	F	2250.78	43.76	~51:1	0.989	0.28	0.88
	V	311.93	24.77	~13:1			

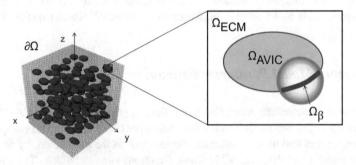

Fig. 8 Mico RVE model with ellipsoidal AVIC inclusions. Ω_{ECM}, Ω_{AVIC}, and Ω_β represent the sub-domains of the ECM, AVIC, and inter-facial boundary, respectively. $\delta\Omega$ represents the boundary of the domain facing outwards. Figure taken from [2]

density of the AVICs is kept consistent with native tissues at 270 million AVICs per ml, and the AVICs are assigned a preferred direction along the circumferential axis and are oriented at normally distributed angles of 9.7° and 5.3°.

3.5.1 Modeling of the ECM Subdomain

The micro RVE model is divided into three subdomains: the ECM (Ω_{ECM}), the AVICs (Ω_{AVIC}), and the boundary between the two (Ω_β) (Fig. 8). A neo-Hookean material model is used to model the ECM subdomain (Ω_{ECM}) with the Cauchy stress defined by

$$\mathbf{T} = 2\frac{1}{J}\mathbf{F}\frac{\partial}{\partial \mathbf{C}}\left[\frac{\mu_{micro}^{l,d}}{2}(\bar{I}_1 - 3) + \frac{1}{2}K(\ln J)^2\right]\mathbf{F}^T. \tag{16}$$

3.5.2 Modeling of the AVIC Subdomain

For the AVIC subdomain (Ω_{AVIC}), a modified version of the VIC mechanical model formulated from the isolated cell studies discussed in the earlier section is used with a combined Cauchy stress of

$$\mathbf{T} = \mathbf{T}_{cyto} + \mathbf{T}_{sf} + \mathbf{T}_{active} \tag{17}$$

where $\mathbf{T}_{cyto}$, $\mathbf{T}_{sf}$, and $\mathbf{T}_{active}$ are contributions from the cytoplasm, passive stress fibers, and active contractile response of the stress fibers, respectively. The constitutive model used for the isolated VIC (Eq. 13) was adapted to remove any contributions from the viscous effects of the stress fibers since strain rate studies were not performed under bending. The final AVIC constitutive model is as follows:

$$\mathbf{T} = 2\frac{1}{J}\mathbf{F}\frac{\partial}{\partial \mathbf{C}}\left[\frac{\mu_{cyto}}{2}(\bar{I}_1 - 3) + \frac{1}{2}K(\ln J)^2\right]\mathbf{F}^T$$
$$+ \frac{1}{J}\mathbf{F}\left[\int_0^{2\pi}\int_0^{\pi}\Gamma(\mathbf{m}_0(\psi,\theta))H(I_4 - 1)2\mu_{sf}\bar{\phi}_{F-actin}(I_4 - 1)\right.$$
$$\left.\mathbf{m}_0(\psi,\theta) \otimes \mathbf{m}_0(\psi,\theta)\sin\psi\,d\psi\,d\theta\right]\mathbf{F}^T$$
$$+ \frac{1}{J}\mathbf{F}\left[\int_0^{2\pi}\int_0^{\pi}\Gamma(\mathbf{m}_0(\psi,\theta))f_0 f_1(\lambda)\mathbf{m}_0(\psi,\theta) \otimes \mathbf{m}_0(\psi,\theta)\sin\psi\,d\psi\,d\theta\right]\mathbf{F}^T. \tag{18}$$

A π periodic von Mises distribution [29] (the 3D counterpart to the constrained von-Mises distribution used for the isolated VIC models [39, 40]) is used to model the 3D stress fiber orientation distribution. The function takes the form:

$$\Gamma(b, \mathbf{m}) = \frac{1}{\pi}\sqrt{\frac{b}{2\pi}}\frac{\exp(2bm_1^2)}{\text{erfi}(\sqrt{2b})} \tag{19}$$

where $\mathbf{m} = (m1, m2, m3)$ is the orientation vector of the stress fibers and $b > 0$ is a shape concentration parameter. The function erfi is defined as

$$\text{erfi}(b) = -i\text{erf}(ib) = \frac{2}{\sqrt{\pi}}\int_0^b \exp(t^2)dt. \tag{20}$$

Only the case in which b is real is considered. This produces a transversely isotropic fiber orientation because Eq. 19 only contains the first component of $\mathbf{m}$ (i.e. m_1). In Eq. 20, the parameter b is tunable such that the lower limit produces an isotropic orientation ($b = 0$) and the upper limit produces a completely one dimensional orientation ($b = \infty$).

3.5.3 Modeling of the Boundary Interface

The complete microstructure of each layer is not explicitly modeled due to lack of feasibility and computational cost. Instead, a thin interface boundary (Ω_β) around every AVIC inclusion is incorporated to allow for the control of the AVIC-ECM connectivity level. This boundary is 0.650 μm in thickness and is modeled using explicit boundary elements that replaced a small volume of the ECM around each AVIC. The interface boundary is modeled with the following constitutive equation

$$\mathbf{T} = 2\frac{1}{J}\mathbf{F}\frac{\partial}{\partial \mathbf{C}}\left[\frac{(\beta^l_{AVIC})(\mu^{l,d}_{micro})}{2}(\bar{I}_1 - 3) + \frac{1}{2}K(\ln J)^2\right]\mathbf{F}^T. \tag{21}$$

When $\beta^l_{AVIC} = 1$, the AVICs are completely connected to the ECM and the interface boundary has the same mechanical properties as the ECM. As β^l_{AVIC} approaches 0, the softer interfacial layer causes the AVICs to become less connected to the ECM.

3.6 Macro and Micro Model Homogenization Scheme

After the tissue scale parameters are determined from the macro RVE model, the displacements of the RVEs from each layer are mapped to the micro RVE model as boundary conditions (Fig. 7). This essentially imposes a boundary value problem on the micro model. The parameters within the micro model are then determined by minimizing the difference in the von Mises stress index between the macro and micro RVE models to maintain continuity between them using

$$\overline{T}^{VM}_{MacroRVE} = \sqrt{\frac{3}{2}T'_{Macro} : T'_{Macro}} \tag{22a}$$

$$\overline{T}^{VM}_{MicroRVE} = \sqrt{\frac{3}{2}T'_{Micro} : T'_{Micro}} \tag{22b}$$

where T'_{Macro} and T'_{Micro} are the deviatoric stresses of the volume averaged Cauchy stresses in the macro and micro RVE models, respectively, and the : operator symbolizes the double dot product of the two second order tensors. The difference between $\overline{T}^{VM}_{MacroRVE}$ and $\overline{T}^{VM}_{MicroRVE}$ are calculated and minimized using

$$\Delta = \overline{T}^{VM}_{MacroRVE} - \overline{T}^{VM}_{MicroRVE} \tag{23a}$$

$$\| \Delta \| = \sqrt{\Delta : \Delta} \mid_{min}. \tag{23b}$$

The computed von Mises stress for the macro RVE are reported in Table 3. The residual von Mises stress between the macro and micro RVE models was minimized to enforce consistency between both models and to ensure proper mapping to the down–scale model.

Table 3 The von Mises stress index ($\overline{T}^{VM}_{MacroRVE}$) used for macro-mico model coupling. Values are reported for both layers (Fibrosa–F, Ventricularis–V) in tension (+) and compression (−) and under inactive, normal and hypertensive states. Table adapted from [2]

Activation State	F+ (kPa)	F− (kPa)	V+ (kPa)	V− (kPa)
Inactive	4.58	3.08	3.46	2.76
Normal	6.92	3.29	4.90	2.89
Hyper	8.18	3.29	4.90	2.99

3.7 Micro Model Parameter Estimation

The AVIC cytoplasm stiffness (μ^{cyto}) and ECM stiffness ($\mu^{l,d}_{ECM}$) are defined as fixed values within the micro RVE model. Cytoplasm stiffness (μ^{cyto}) is set to 18 kPa based on AFM measurements performed on AVICs [33]. ECM stiffness ($\mu^{l,d}_{ECM}$) is computed using the tissue parameters determined from the macro RVE model for the inactive treatment group to systematically exclude contributions from either AVIC basal tonus or contraction. Using the rule of mixtures, the relation between ECM stiffness and AVIC cytoplasm stiffness is formulated as:

$$_{inactive}\mu^{l,d}_{Macro} = \mu^{l,d}_{ECM} \phi_{ECM} + \mu^{cyto} \phi_{AVIC} \qquad (24)$$

where $_{inactive}\mu^{l,d}_{Macro}$ are the inactive layer tissue parameters from the macro RVE model and ϕ_{ECM} and ϕ_{AVIC} are the volume fractions of the ECM and AVICs set at 0.90 and 0.10, respectively. Using Eq. 24, the stiffness of the ECM for each layer under tension and compression ($\mu^{F+}_{ECM}, \mu^{F-}_{ECM}, \mu^{V+}_{ECM}$ and μ^{V-}_{ECM}) are determined and input into the micro RVE model.

After μ^{cyto} and $\mu^{l,d}_{ECM}$ are determined, the following parameters are estimated from the micro RVE model through the macro-micro coupling approach:

1. $_{state}\mu^{l,d}_{SF}$: Stress fiber stiffness
2. $_{state}f^{l,d}_{0}$: Stress fiber contraction strength
3. β^{l}_{AVIC} : AVIC connectivity to the ECM.

The remainder of this section describes the influence of each parameter.

3.7.1 The Effect of Stress Fiber Stiffness on the Micro RVE Response

The inactive state volume average von Mises stress in the fibrosa was used to evaluate the effect of the stress fiber shear modulus on tissue mechanical response. Stress fiber stiffness (μ_{sf}) is varied from 0–50 kPa and it is observed that this had little effect on the micro RVE model volume averaged von Mises stress (Fig. 9a). Therefore, μ_{sf} was determined to be 1 kPa from previous 2D microindentation studies and was fixed at this value for the remainder of the simulations.

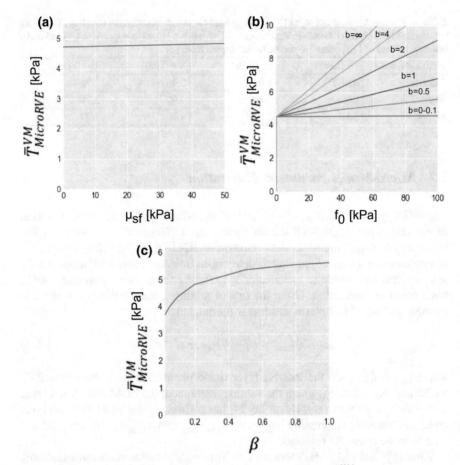

Fig. 9 **a** The stress fiber shear modulus (μ_{sf}) has little to no effect on $\overline{T}^{VM}_{MicroRVE}$. **b** The effect of the stress fiber orientation parameter b on $\overline{T}^{VM}_{MicroRVE}$. **c** The effect of the AVIC-ECM connectivity parameter β on the volume averaged von Mises stress. As connectivity increases, $\overline{T}^{VM}_{MicroRVE}$ increases as well. Figure taken from [2]

3.7.2 Determination of AVIC In-Situ Contraction Strength

To estimate the contraction strength, an appropriate stress fiber orientation parameter b is first determined. A parametric study of the parameter b is performed where the value for b is set to $0, 0.5, 1, 2, 4$, and ∞ (Fig. 9b) for the fibrosa layer in the inactivated state. It was observed that when $b = 0$, the AVIC contraction did not contribute towards the von Mises stress of the micro RVE and it remained at 4.58 kPa as determined from the macro RVE (Table 3). As b increased, a linear trend between the contraction strength f_0 and $\overline{T}^{VM}_{MicroRVE}$ emerged. It is noted that as b increases, less contraction

Table 4 Parametric study of the parameter b. The contraction force required to match the RVE_{macro} von Mises stress decreased as b increased for both the normal and hypertensive cases

b	0	0.1	0.5	1	2	4	∞
f_0–normal [kPa]	N/A	N/A	223	105.5	52.1	35.1	27.4
f_0–hypter [kPa]	N/A	N/A	306	156.3	79.9	53.8	42.4

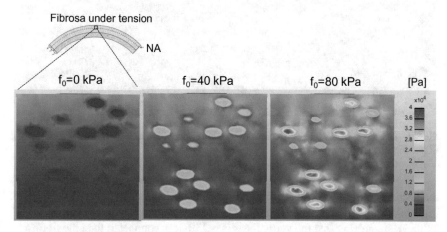

Fig. 10 The effect of AVIC contraction strength (f_0) on the surrounding von Mises stress of the ECM. Figure taken from [2]

force is needed to reach the same $\overline{T}^{VM}_{MicroRVE}$ (Table 4). This is because increasing b essentially increases the alignment of the 3D stress fiber orientation, allowing the AVIC inclusions to contribute in concert and as a result decrease the total level of f_0 needed per AVIC to reach the same outcomes.

From comparison of the π-periodic von Mises distribution with different b values to the constrained 2D von Mises distribution quantified previously [39, 40], a b value of 2.0 was chosen as the estimate of the 3D stress fiber orientation distribution. Shown in Table 4, $f_0 = 52.1$ and 79.9 kPa are used here on out for the normal and hypertensive simulations, respectively. With $b = 2.0$, simulations are implemented to assess how the magnitude of the contraction strength alters the surrounding ECM (Fig. 10). When the contraction strength is set to 0 kPa, the von Mises stress of the ECM is greater than that of the AVIC inclusions due to the lack of contraction. As contraction is increased to 40 and 80 kPa, the von Mises stress of the AVIC inclusions surpassed that of the ECM.

3.7.3 Layer-Specific AVIC-ECM Connectivity

With the b and f_0 parameters determined, the AVIC-ECM connectivity (β) within the ventricularis is estimated using Eq. 21. The β term is varied and the effect on

Ventricularis under tension

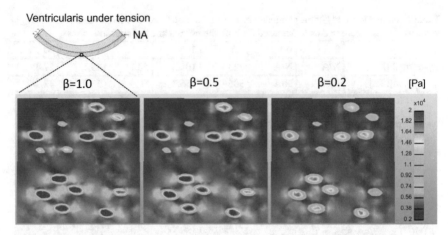

Fig. 11 The effect of AVIC-ECM connectivity (β) on the von Mises stress of the surrounding ECM. Figure taken from [2]

the resulting von Mises stress is plotted for the ventricularis under tension (Fig. 9c). Figure 9c, shows that a β value of 0.24 best matches the von Mises stress (4.90 kPa, Table 3), suggesting that the vetricularis is $\sim$75% less connected than the fibrosa.

It is noted that as the value of β increases, the total von Mises stress increases as well (Fig. 11). This is because the mechanical connection between the AVICs and the ECM governs the net effect of AVIC contraction on the von Mises stress of the surrounding ECM. Not only does the von Mises stress within the AVIC inclusions increase due to an increase in connectivity, but the von Mises stress of the immediately surrounding ECM also increases.

3.8 Differences in Cell Total Traction Forces in 2D Versus 3D

To better understand the difference in AVIC biophysical state in 2D vs 3D, the total force generated by each individual AVIC is computed through the expression

$$F_{tot} = \int_{\Gamma} |T(x)| dA \qquad (25)$$

where Γ represents the boundary of the AVIC inclusions and $T(x)$ is the traction along the boundary. It is found that in-situ, AVICs generate 0.2 and 0.35 μN of force in the normal and hypertensive states, respectively. This is more than twice the force that is generated by AVICs seeded on 2D substrates as determined from microindentation studies (normal $= 0.07\,\mu$N and hypertensive $= 0.16\,\mu$N) [40]. The average traction force over the boundary is computed as

$$T_{avg} = \frac{1}{SA} \int_{\Gamma} |T(x)| dA. \tag{26}$$

In-situ, the average traction over the AVIC boundary is 429 Pa and 752 Pa for the normal and hypertensive states, respectively. In the 2D micro-indentation studies, the average traction over the boundary is computed as 300 and 450 pa for the normal and hypertensive state, respectively, which is substantially lower than what was found for AVICs in-situ.

3.9   Conclusions

The insights gained from the isolated cell studies were successfully incorporated into a more physiologically relevant model of the AVIC in-situ. From the macro-micro modelling approach, the total force and average traction of the AVICs in-situ is determined. In 3D, AVICs generate greater total and average forces per unit area. This is hypothesized to be driven by the increased attachment sites available in 3D compared to 2D cultures.

The layer-specific connectivity of the AVICs is also estimated. Most notably, AVICs within the ventricularis are estimated to be $\sim$75% less connected to the ECM than AVICs in the fibrosa. This offers a possible explanation as to why no significant change in E_{eff} was observed from the AV bending experiments when the AVICs were in a hypertensive state and the AV test specimens were bent in the WC direction (ventricularis under tension) (Fig. 6b). In contrast, the AVICs in the fibrosa layer are predicted to be more connected to the ECM which potentially explains the significant increase in E_{eff} observed when the AV leaflets are tested in the AC direction under hypertensive treatments.

The macro-micro model of the AV leaflet described here should be considered a first pass towards modeling in-situ AVIC mechanical behavior. Recent advances in micromechanical continuum mechanics offers a potential strategy to solve the micro RVE model analytically which will reduce computational costs [35]. In the future, the model should also be refined to include realistic cell geometries, and incorporate the separate contributions of the stress fiber networks (namely α-SMA and F-actin). In addition, future efforts should also focus on modeling the structural differences between the fibrosa and ventricularis layers.

4   An Approach to Studying VIC Mechanobiology Using Tunable Hydrogels

4.1   Overview

The majority of VIC mechanical and mechanobiological studies have been performed either on 2D substrates or in suspension. Although insightful, these techniques vastly underestimate the complexity of the native heart valves and do not recapitulate the in-vivo cell environment. Using native tissue for experimentation offers an alternative to 2D cultures but is also limited in that they cannot be tuned to answer specific mechanobiological questions nor can the subcellular components of the AVICs be directly and conveniently visualized. In response to these limitations, synthetic hydrogels have recently grown in popularity as an alternative method to study cells within highly tunable and transparent 3D environments. Synthetic hydrogels also open up the possibility of conducting high-throughput studies using healthy and diseased VICs isolated from human tissue samples. Studies have been conducted to assess AVIC contractile responses within peptide-modified, poly (ethylene glycol) (PEG) hydrogels [22] and are summarized in this section.

4.2 Experimental Methods

4.2.1 AVIC-Hydrogel Fabrication

Porcine AVICs were suspended in a hydrogel pre-cursor solution consisting of norbornene-functionalized PEG, matrix metalloproteinase (MMP)-degradable crosslinking peptides, CRGDS adhesive peptides, lithium phenyl- 2,4,6 -trimethyl-benzolphosphinate photoinitiator, and PBS at a concentration of 10 millions cells/ml. The cell seeded hydrogel polymer solution was then pipetted into 10×5 mm molds and AVIC-hydrogels were cured under UV light for 3 min (Fig. 12). The amount of MMP-degradable crosslinks and CRGDS peptides were tuned to study the effect of hydrogel stiffness and adhesive ligand density, respectively, on AVIC contraction.

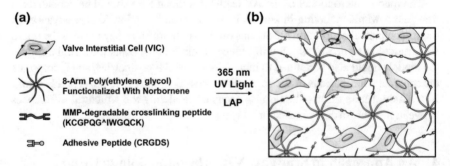

(a)

Valve Interstitial Cell (VIC)

8-Arm Poly(ethylene glycol) Functionalized With Norbornene

MMP-degradable crosslinking peptide (KCGPQG^IWGQCK)

Adhesive Peptide (CRGDS)

(b)

365 nm UV Light

LAP

Fig. 12 a The components of the hydrogel pre-cursor solution. **b** Schematic of the theorized AVIC-hydrogel internal structure. Figure taken from [22]

4.2.2 End-Loading, Flexural Deformation Testing

Similar to the bending tests performed on native AV leaflets [32], the AVIC hydrogels were subjected to repeated flexural deformation tests after treatment with 5 mM KCl (normal), 90 mM KCl (hypertensive), and 70% methanol (inactive) for a total of three flexural tests per specimen (Fig. 13).

The Euler-Bernoulli relation was used for analysis. The change in curvature of the specimen ($\Delta\kappa$) was computed using the position of the fiducial markers. The moment (M) at the center of the specimen was computed as:

$$M = P * y \tag{27}$$

where P is the end-loading force required to bend the specimen and y is the deflection of the center marker in the y-direction. From these calculations, moment versus curvature plots were generated for the AVIC-hydrogel specimens (Fig. 14).

4.3 Finite Element Modeling

From the experimental data, it was observed that the moment-curvature plots for the AVIC hydrogels were consistently non-linear (Fig. 14). This causes E_{eff} to be non-constant and a function of the change in curvature. This posed difficulty towards selecting an appropriate E_{eff} to compare between hydrogel groups with both different compositions and different contractile states of the underlying AVICs. To mitigate this, a finite element model of the bending experiment was developed in FEniCS

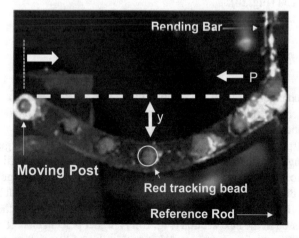

Fig. 13 Flexural deformation testing of an AVIC-hydrogel. Red fiducial markers were placed along the length of the hydrogel to track the deformation of the test specimen. Figure taken from [22]

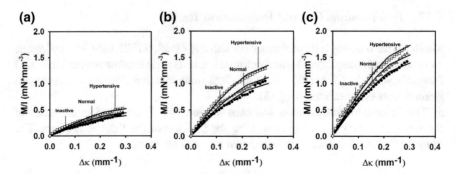

Fig. 14 Experimental (dotted line) and simulated (solid line) moment versus curvature plots for **a** 2.5, **b** 5, and **c** 10 kPa hydrogels under normal, hypertensive, and inactive conditions. Figure taken from [22]

[27, 28], and the AVIC-hydrogels were modeled with a neo-Hookean material model with the following strain energy density function:

$$W = \frac{\mu}{2}(I_1 - 3) + \frac{1}{2}K(J - 1)^2 \tag{28}$$

where μ serves as a convenient material parameter to compare between hydrogel groups and contractile levels and is the large deformation equivalent of the shear modulus and K is the bulk modulus of the material. To enforce incompressibility, $K = \mu * 10,000$ and the change in volume ratio was subsequently confirmed to be insignificant. Shear modulus (μ) was determined using a gradient descent algorithm to minimize the least-square errors between the simulated and experimental moment vs curvature data (Fig. 14).

4.4 Modeling Results

The percent change in μ from the inactive to the normal state and from the inactive to the hypertensive state is reported for both the stiffness modulated and CRGDS adhesive peptide modulated AVIC-hydrogels in Fig. 15.

In the stiffness modulated group, AVICs seeded within 2.5 kPa gels displayed a larger change in stiffness from the inactive to the normal condition and from the inactive to the hypertensive condition than in the 5 and 10 kPa gels (Fig. 15a). This may be due to an increase in AVIC activation within softer hydrogels and the development of prominent α-SMA fiber networks that are seen in the fluorescent images discussed in the next section (Fig. 16). A higher percent change in μ from the inactive condition to the normal condition within the 2.5 kPa gels suggests that the basal tonus of the AVICs contributes more to the overall construct stiffness than in

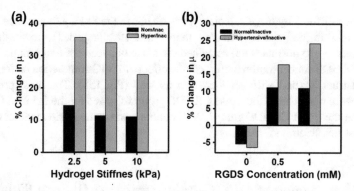

Fig. 15 **a** The percent change in μ for the stiffness modulated AVIC-hydrogels. **b** The percent change in μ for AVIC-hydrogels with varying levels of CRGDS adhesive peptide concentration. Figure adapted from [22]

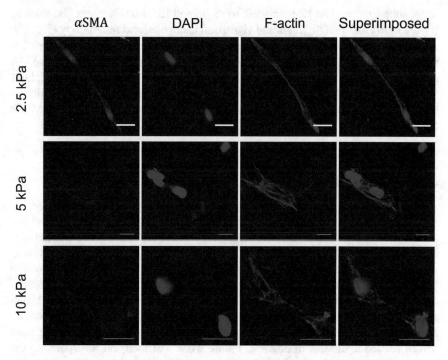

Fig. 16 Subcellular AVIC structures imaged directly within the hydrogel using confocal microscopy (scale bar = 20 μm). Figure taken from [22]

the higher stiffness hydrogels. Notably, the AVICs in the 5 kPa gels produce a larger change in μ in the hypertensive state than in the 10 kPa gels. However, the basal tonus (inactive to normal state) remained the same between the two groups.

As CRGDS concentration increased, the effect of AVIC contraction on the hydrogel construct flexural stiffness increased as well (Fig. 15b). The most prominent effects were observed in the 1 mM CRGDS group followed by the 0.5 mM CRGDS group. In the 0 mM CRGDS group, AVIC contraction had no positive effect towards the construct flexural stiffness.

4.5   Visualization of AVIC Cellular Structures Directly Within the Hydrogel Environment

A separate group of AVIC-hydrogels were stained for α-SMA fibers, the nucleus (DAPI stain), and F-actin fibers and visualized using confocal microscopy (Fig. 16). AVICs seeded within softer hydrogels seemed to elongate more than in higher stiffness hydrogels. In addition, AVICs within 2.5 kPa gels displayed detectable α-SMA expression whereas AVICs within 5 and 10 kPa gels did not. This potentially explains the larger influence of AVIC contraction on hydrogel construct flexural stiffness within the 2.5 kPa group. Complex F-actin fiber networks were present within AVICs among all hydrogel stiffness groups.

4.6   Conclusions

AVICs establish significant mechanical coupling to the synthetic hydrogel environment and can effect the overall flexural stiffness of these constructs. The gel system is an extremely attractive tool for future mechanobiological studies involving mechanical conditioning of the AVIC-hydrogels within uniaxial and biaxial stretch bioreactors. Current work is underway to use the gel system to study the effects of cyclic uniaxial loading on human AVICs from healthy patients and those with bicuspid AVs to investigate the effects of cell deformation in a 3D microenvironment on AVIC biosynthesis. This is a first pass attempt at modeling the macro level respose of cell seeded hydrogels in bending. Future work will focus on incorporating the explicit contraction effects of AVICs within the hydrogel through adapting the current AV leaflet macro-micro modeling approach to the hydrogel system and utilizing the transparency of the gels to inform the model with realistic cell and stress fiber geometries.

5 Future Directions

5.1 Modeling of VIC Cell Signaling Pathways

The macro-micro modeling approach described in Sect. 3 uses the parameter β_{AVIC}^l to describe the degree of connectivity between the AVIC and the ECM and this approach has been able to describe some key aspects of how VICs sense their surrounding ECM. However, a mechanistic view that couples the binding of AVIC receptors to the ECM and how these events propagate to altered transcription is lacking. Therefore, future research will be directed toward elucidating the intracellular mechanotransduction pathways that allow VICs to sense and respond to altered ECM concentrations and mechanical stimuli via mathematical modeling.

As the major receptors that allow cells to adhere to substrates such as ECM, integrins are key players in VIC mechanotransduction processes. In addition to binding the ECM, integrins also bind a variety of adapter proteins that provide a structural link to the actin cytoskeleton, connecting external and internal cellular forces. This structural linkage has been described as a "mechanical clutch" and was initially modeled by Chan and Odde [6] and has been recently extended by others (e.g., [11, 16, 47]). These models explicitly incorporate the force on the ECM-integrin-adapter-actin structural linkage through force-dependent bond lifetimes and the unveiling of binding sites for structural reinforcement in the adapter. These models are able to relate substrate rigidity to the maturity of focal adhesion structures developed.

In addition to their structural properties, integrins also signal through proteins such as focal adhesion kinase (FAK). FAK interacts with many downstream pathways, including (a) Rho/Rac signaling to alter the actin cytoskeleton [24], (b) Ras/ERK pathways to dismantle focal adhesions [41], and (c) PI3K/Akt pathways to facilitate the myofibroblast transition and αSMA production [9, 45]. These "pathways" are not linear cascades; rather, they are influenced by many cross-talk mechanisms, instead creating a complex signaling network.

The complexity of this integrin-mediated signaling network obfuscates the role of individual proteins in overall VIC phenotypic transitions. However, reconstructing these pathways through mathematical modeling can provide information on the most influential states or parameters governing these transitions, directing future experiments. Furthermore, such a model can be used to influence cellular biomechanics models within a multiscale framework.

5.2 A Flexible and Extensible 3D Multi-scale Computational Modeling Framework

Future computational research will be geared towards robust mechanistic modeling of the VICs seeded in the 3D hydrogel systems described in Sect. 4. The tunable hydrogel system and the multi-scale, multi-resolution modeling framework will work in concert to elucidate experimentally observed cell and system behavior and ultimately aid in comparing between the synthetic and native VIC environments. Direct enhancements to the VIC-ECM model described in Sect. 3 include the implementation of a more realistic cellular geometry, implementation of additional sub-cellular features such as cell nuclei, a more complex strategy for capturing the cell-ECM attachment zone, and the consideration of aleatoric uncertainty. With regard to the ECM fiber network, sub-cellular scale mechanism based material models for ECM and cell cytoplasm remodeling will be explored [14]. Finally, integrating aspects of the cell signaling models described in Sect. 5.1 into the finite element framework will offer a controllable platform to explore potential non-linearities and feedback loops that rely on spatial information.

With this significant added model complexity, model interpretability will suffer. And, more computationally expensive models will pose feasibility challenges for robust simulations at larger spatial scales. To mitigate this, novel strategies for model interpretability are required [25]. One key strategy will be defining experimentally relevant quantities of interest, and running global sensitivity analysis to understand the relative contributions of different model input parameters [18, 26, 42]. Another key strategy will be developing and incorporating appropriate methods for multi-scale and multi-resolution model integration. Looking forward, the computational model and associated techniques developed to meet these challenges should be extensible to modeling other experimental systems.

References

1. Acker, M.A., Parides, M.K., Perrault, L.P., Moskowitz, A.J., Gelijns, A.C., Voisine, P., Smith, P.K., Hung, J.W., Blackstone, E.H., Puskas, J.D.: Mitral-valve repair versus replacement for severe ischemic mitral regurgitation. N. Engl. J. Med. **370**(1), 23–32 (2014)
2. Buchanan, R.M.: A An integrated computational-experimental approach for the in situ estimation of valve interstitial cell biomechanical state. Doctoral Thesis. The University of Texas at Austin (2016). https://repositories.lib.utexas.edu/handle/2152/39463
3. Buchanan, R.M., Sacks, M.S.: Interlayer micromechanics of the aortic heart valve leaflet. Biomech Model Mechanobiol (2013). https://doi.org/10.1007/s10237-013-0536-6. http://www.ncbi.nlm.nih.gov/pubmed/24292631
4. Butcher, J.T., Simmons, C.A., Warnock, J.N.: Mechanobiology of the aortic heart valve. J. Heart Valve. Dis. **17**(1), 62–73 (2008). http://www.ncbi.nlm.nih.gov/entrez/query.fcgi?cmd=Retrieve&db=PubMed&dopt=Citation&list_uids=18365571
5. Carruthers, C.A., Alfieri, C.M., Joyce, E.M., Watkins, S.C., Yutzey, K.E., Sacks, M.S.: Gene Expression and Collagen Fiber Micromechanical Interactions of the Semilunar Heart Valve Interstitial Cell. Cell. Mol. Bioeng. **5**(3), 254–265 (2012). https://doi.org/10.1007/s12195-012-0230-2
6. Chan, C.E., Odde, D.J.: Traction dynamics of filopodia on compliant substrates. Science **322**(5908), 1687–1691 (2008). https://doi.org/10.1126/science.1163595
7. Chaput, M., Handschumacher, M.D., Guerrero, J.L., Holmvang, G., Dal-Bianco, J.P., Sullivan, S., Vlahakes, G.J., Hung, J., Levine, R.A.: Mitral leaflet adaptation to ventricular remodeling: prospective changes in a model of ischemic mitral regurgitation. Circulation **120**(11 Suppl), S99–103 (2009). https://doi.org/10.1161/CIRCULATIONAHA.109.844019. http://www.ncbi.nlm.nih.gov/pubmed/19752393

8. Chaput, M., Handschumacher, M.D., Tournoux, F., Hua, L., Guerrero, J.L., Vlahakes, G.J., Levine, R.A.: Mitral leaflet adaptation to ventricular remodeling: occurrence and adequacy in patients with functional mitral regurgitation. Circulation **118**(8), 845–52 (2008). https://doi.org/10.1161/circulationaha.107.749440. http://circ.ahajournals.org/content/118/8/845.full.pdf

9. Chen, H.C., Appeddu, P.A., Isoda, H., Guan, J.L.: Phosphorylation of tyrosine 397 in focal adhesion kinase is required for binding phosphatidylinositol 3-kinase. J. Biol. Chem. **271**(42), 26329–26334 (1996). https://doi.org/10.1074/jbc.271.42.26329

10. Chen, J., Li, H., SundarRaj, N., Wang, J.H.: Alpha-smooth muscle actin expression enhances cell traction force. Cell Motil Cytoskelet. **64**(4), 248–57 (2007). https://doi.org/10.1002/cm.20178. http://www.ncbi.nlm.nih.gov/pubmed/17183543

11. Cheng, B., Lin, M., Li, Y., Huang, G., Yang, H., Genin, G.M., Deshpande, V.S., Lu, T.J., Xu, F.: An integrated stochastic model of matrix-stiffness-dependent filopodial dynamics. Biophys. J. **111**(9), 2051–2061 (2016). https://doi.org/10.1016/j.bpj.2016.09.026

12. Clark-Greuel, J.N., Connolly, J.M., Sorichillo, E., Narula, N.R., Rapoport, H.S., Mohler 3rd, E.R., Gorman 3rd, J.H., Gorman, R.C., Levy, R.J.: Transforming growth factor-beta1 mechanisms in aortic valve calcification: increased alkaline phosphatase and related events. Ann. Thorac. Surg. **83**(3), 946–53 (2007). https://doi.org/10.1016/j.athoracsur.2006.10.026. http://www.ncbi.nlm.nih.gov/pubmed/17307438

13. Cushing, M.C., Liao, J.T., Anseth, K.S.: Activation of valvular interstitial cells is mediated by transforming growth factor-beta1 interactions with matrix molecules. Matrix Biol. **24**(6), 428–37 (2005). http://www.ncbi.nlm.nih.gov/entrez/query.fcgi?cmd=Retrieve&db=PubMed&dopt=Citation&list_uids=16055320

14. Deshpande, V.S., Mrksich, M., McMeeking, R.M., Evans, A.G.: A bio-mechanical model for coupling cell contractility with focal adhesion formation. J. Mech. Phys. Solids **56**(4), 1484–1510 (2008)

15. Edman, K.A.: The relation between sarcomere length and active tension in isolated semi-tendinosus fibres of the frog. J. Physiol. **183**(2), 407–17 (1966). http://www.ncbi.nlm.nih.gov/pubmed/5942818

16. Elosegui-Artola, A., Oria, R., Chen, Y., Kosmalska, A., Pérez-González, C., Castro, N., Zhu, C., Trepat, X., Roca-Cusachs, P.: Mechanical regulation of a molecular clutch defines force transmission and transduction in response to matrix rigidity. Nat. Cell Biol. **18**(5), 540–548 (2016). https://doi.org/10.1038/ncb3336

17. Gouget, C.L., Girard, M.J., Ethier, C.R.: A constrained von Mises distribution to describe fiber organization in thin soft tissues. Biomech Model Mechanobiol **11**(3–4), 475–82 (2012). https://doi.org/10.1007/s10237-011-0326-y. http://www.ncbi.nlm.nih.gov/pubmed/21739088

18. Hart, J., Alexanderian, A., Gremaud, P.: Efficient computation of sobol' indices for stochastic models. SIAM J. Sci. Comput. **39**(4), 1514–1539 (2017)

19. Hinz, B., Phan, S.H., Thannickal, V.J., Prunotto, M., Desmouliere, A., Varga, J., De Wever, O., Mareel, M., Gabbiani, G.: Recent developments in myofibroblast biology: paradigms for connective tissue remodeling. Am J Pathol **180**(4), 1340–55 (2012). https://doi.org/10.1016/j.ajpath.2012.02.004. http://www.ncbi.nlm.nih.gov/pubmed/22387320

20. Huang, H.Y., Liao, J., Sacks, M.S.: In-situ deformation of the aortic valve interstitial cell nucleus under diastolic loading. J Biomech Eng **129**(6), 880–89 (2007). https://doi.org/10.1115/1.2801670. http://www.ncbi.nlm.nih.gov/entrez/query.fcgi?cmd=Retrieve&db=PubMed&dopt=Citation&list_uids=18067392

21. Khalighi, A.H., Drach, A., Bloodworth, C.H., Pierce, E.L., Yoganathan, A.P., Gorman, R.C., Gorman, J.H., Sacks, M.S.: Mitral valve chordae tendineae: Topological and geometrical characterization. Ann. Biomed. Eng. **45**, 378–393 (2017). https://doi.org/10.1007/s10439-016-1775-3

22. Khang, A., Gonzalex, A.G., Schroeder, M.E., Sansom, J., Anseth, K.S., Sacks, M.S.: An approach to quantify valve interstitial cell biophysical state using highly tunable poly(ethylene) glycol hydrogels. Manuscript in preparation (2018)

23. Latif, N., Sarathchandra, P., Taylor, P., Antoniw, J., Yacoub, M.: Molecules mediating cell-ECM and cell-cell communication in human heart valves. Cell Biochemistry and Biophysics **43**(2), 275–287 (2005). https://doi.org/10.1385/CBB:43:2:275

24. Lawson, C.D., Burridge, K.: The on-off relationship of Rho and Rac during integrin-mediated adhesion and cell migration. Small GTPases **5**(1), e27958 (2014). https://doi.org/10.4161/sgtp. 27958
25. Lejeune, E., Linder, C.: Quantifying the relationship between cell division angle and morphogenesis through computational modeling. J. Theor. Biol. **418**, 1–7 (2017)
26. Lejeune, E., Linder, C.: Understanding the relationship between cell death and tissue shrinkage via a stochastic agent-based model. J. Biomech. **73**, 9–17 (2018)
27. Logg, A., Mardal, K.A., Wells, G.: Automated Solution of Differential Equations by the Finite Element Method the FEniCS Book, 2012. edn. Lecture Notes in Computational Science and Engineering,. Springer Berlin Heidelberg, Berlin, Heidelberg (2012). http://UTXA.eblib.com/ patron/FullRecord.aspx?p=885214
28. Logg, A., Wells, G.N.: DOLFIN: Automated Finite Element Computing. Acm Transactions on Mathematical Software **37**(2) (2010). https://doi.org/10.1145/1731022.1731030. http://WOS:000277057400008
29. Maas, S., Rawlins, D., Weiss, J., Weiss, J.: FEBio User's Manual Version 2.4 (2015). http:// febio.org/download/febio-2-4-users-manual/
30. Maas, S.A., Ellis, B.J., Ateshian, G.A., Weiss, J.A.: FEBio: finite elements for biomechanics. J Biomech Eng **134**(1), 011005 (2012). https://doi.org/10.1115/1.4005694. http://www.ncbi. nlm.nih.gov/pubmed/22482660
31. Merryman, W.D.: Mechanobiology of the aortic valve interstitial cell. Doctoral Dissertation, University of Pittsburgh, Doctoral Dissertation (2007)
32. Merryman, W.D., Huang, H.Y.S., Schoen, F.J., Sacks, M.S.: The effects of cellular contraction on aortic valve leaflet flexural stiffness. J Biomech **39**(1), 88–96 (2006). http://www.ncbi.nlm. nih.gov/entrez/query.fcgi?cmd=Retrieve&db=PubMed&dopt=Citation&list_uids=16271591
33. Merryman, W.D., Liao, J., Parekh, A., Candiello, J.E., Lin, H., Sacks, M.S.: Differences in tissue-remodeling potential of aortic and pulmonary heart valve interstitial cells. Tissue Eng. **13**(9), 2281–9 (2007). http://www.ncbi.nlm.nih.gov/entrez/query.fcgi?cmd=Retrieve& db=PubMed&dopt=Citation&list_uids=17596117
34. Merryman, W.D., Youn, I., Lukoff, H.D., Krueger, P.M., Guilak, F., Hopkins, R.A., Sacks, M.S.: Correlation between heart valve interstitial cell stiffness and transvalvular pressure: implications for collagen biosynthesis. Am. J. Physiol. Heart Circ. Physiol. **290**(1), H224–31 (2006). http://www.ncbi.nlm.nih.gov/entrez/query.fcgi?cmd=Retrieve& db=PubMed&dopt=Citation&list_uids=16126816
35. Morin, C., Avril, S., Hellmich, C.: Non-affine fiber kinematics in arterial mechanics: a continuummicromechanical investigation. J. Appl. Math. Mech. (2018)
36. Osman, L., Yacoub, M.H., Latif, N., Amrani, M., Chester, A.H.: Role of human valve interstitial cells in valve calcification and their response to atorvastatin. Circulation **114**(1 Suppl), 1547–1552 (2006). http://www.ncbi.nlm.nih.gov/entrez/query.fcgi?cmd=Retrieve& db=PubMed&dopt=Citation&list_uids=16820635
37. Rego, B.V., Sacks, M.S.: A functionally graded material model for the transmural stress distribution of the aortic valve leaflet. J. Biomech. **54**, 88–95 (2017). https://doi.org/10.1016/j. jbiomech.2017.01.039
38. Rego, B.V., Wells, S.M., Lee, C.H., Sacks, M.S.: Mitral valve leaflet remodelling during pregnancy: insights into cell-mediated recovery of tissue homeostasis. J. R. Soc., Interface **13**(125) (2016). https://doi.org/10.1098/rsif.2016.0709. http://rsif.royalsocietypublishing.org/content/ 13/125/20160709
39. Sakamoto, Y., Buchanan, R.M., Sacks, M.S.: On intrinsic stress fiber contractile forces in semilunar heart valve interstitial cells using a continuum mixture model. J Mech Behav Biomed Mater **54**, 244–58 (2016). https://doi.org/10.1016/j.jmbbm.2015.09.027. http://www.ncbi.nlm. nih.gov/pubmed/26476967
40. Sakamoto, Y., Buchanan, R.M., Sanchez-Adams, J., Guilak, F., Sacks, M.S.: On the functional role of valve interstitial cell stress fibers: a continuum modeling approach. J. Biomech. Eng. **139**, (2017). https://doi.org/10.1115/1.4035557

41. Schlaepfer, D.D., Hanks, S.K., Hunter, T., van der Geer, P.: Integrin-mediated signal transduction linked to Ras pathway by GRB2 binding to focal adhesion kinase. Nature **372**, 786–791 (1994). https://doi.org/10.1038/372786a0

42. Sobol, I.: Global sensitivity indices for nonlinear mathematical models and their monte carlo estimates. Math. Comput. Simul. **55**(1), 271–280 (2001)

43. Stephens, E., Durst, C., Swanson, J., Grande-Allen, K., Ingels, N., Miller, D.: Functional Coupling of Valvular Interstitial Cells and Collagen Via a2b1 Integrins in the Mitral Leaflet. Cell Mol. Bioeng. pp. 1–10 (2010). https://doi.org/10.1007/s12195-010-0139-6

44. Vernerey, F.J., Farsad, M.: A constrained mixture approach to mechano-sensing and force generation in contractile cells. J. Mech. Behav. Biomed. Mater. **4**(8), 1683–1699 (2011). https://doi.org/10.1016/j.jmbbm.2011.05.022. http://WOS:000298764700012

45. Wang, H., Tibbitt, M.W., Langer, S.J., Leinwand, L.A., Anseth, K.S.: Hydrogels preserve native phenotypes of valvular fibroblasts through an elasticity-regulated PI3K/AKT pathway. Proc. Natl. Acad. Sci. **110**(48), 19336–19341 (2013). https://doi.org/10.1073/pnas.1306369110

46. Wang, J., Zohar, R., McCulloch, C.A.: Multiple roles of alpha-smooth muscle actin in mechanotransduction. Exp. Cell Res. **312**(3), 205–14 (2006). http://www.ncbi.nlm.nih.gov/entrez/query.fcgi?cmd=Retrieve&db=PubMed&dopt=Citation&list_uids=16325810

47. Welf, E.S., Johnson, H.E., Haugh, J.M.: Bidirectional coupling between integrin-mediated signaling and actomyosin mechanics explains matrix-dependent intermittency of leading-edge motility. Mol. Biol. Cell **24**(24), 3945–3955 (2013). https://doi.org/10.1091/mbc.e13-06-0311

Modeling the Structural and Mechanical Properties of the Normal and Aneurysmatic Aortic Wall

T. Christian Gasser

Abstract The structural properties of the Extracellular Matrix change in response to many factors, such as mechanical stress, age, disease and lifestyle. The ECM ensures not only the vessel wall's structural integrity, but it also defines the micro-mechanical environment within which vascular cells are embedded and to which they respond. Its mechanical properties are governed by the delicate interaction of elastin, collagen, ProteoGlycans, fibronectin, fibrilin and other, constituents which are synthesized by vascular cells and degraded, mainly by Matrix MetalloProteinases. The present chapter discusses the structural organization of the vessel wall towards the multi-scale mechanical characterization of the aneurysmatic aorta. It is assumed that aneurysmatic vessel wall properties are mainly governed by collagen fibrils, with their undulation and orientation being the most influential micro-histological parameters. Purely passive constitutive descriptions are further complemented by collagen turnover kinetics, and all models are set-up such that they may be used for organ-level vascular biomechanics simulations.

1 Introduction

Vascular diseases are among the leading causes of death in industrialized countries and associated risk factors, such as obesity, diabetes, and life expectancy are increasing. Current clinical options are somewhat limited, and therefore there is clearly an urgent need for multi-disciplinary approaches to improve our current understanding of vascular diseases. *Biomechanical conditions* play a key role in the genesis and development of vascular diseases [1] and the identification of the specific causative links between biomechanics and biochemistry may help advance our current view of physiology and pathology.

T. C. Gasser (✉)
KTH Solid Mechanics, KTH Royal Institute of Technology,
Teknikringen 8D, 100 44 Stockholm, Sweden
e-mail: gasser@kth.se

© Springer Nature Switzerland AG 2020
Y. Zhang (ed.), *Multi-scale Extracellular Matrix Mechanics and Mechanobiology*,
Studies in Mechanobiology, Tissue Engineering and Biomaterials 23,
https://doi.org/10.1007/978-3-030-20182-1_3

Continued advances in computer technology and computational methods allow us to nowadays model patient-specific problems, where the nonlinear Finite Element Method (FEM) effectively solves the governing, and often strongly coupled mechanical problems. The FEM allows for the incorporation of the inherent non-linearity of related problems and also combines synergetically with medical imaging. Consequently, *computer simulations* have become highly significant in the exploration of vascular biomechanics phenomena. Although, to some extent, traditional mechanics concepts are directly applicable to the solution of such problems, they remain a modeling challenge due to complex spatial domains, constitutive nonlinearities, and coupling among structural, fluid, chemical and electrical fields. Specifically, the inherent property of biological tissue to *adapt* to mechanical and chemical environments remains a challenging modeling task.

The present chapter discusses the mechanical characterization of the normal and the aneurysmatic vessel wall. During the first part, the histology and the physiological function of vessel wall layers are reviewed, with the focus on the Extracellular Matrix (ECM) components *elastin* and *collagen*. Specifically, their structure organization within the vessel wall, their synthesis by vascular cells, and their degradation by Matrix MetalloProteinases (MMPs) are of particular interest. Subsequently, some macroscopic properties of the normal and the aneurysmatic aorta are reported, always in relation to the vessel wall's histology and morphology. The last part of this chapter is dedicated to the *mechanical modeling* of the aneurysmatic aorta, and follows LANIR'S general framework of fibrous biological tissues [2]. Here, the macroscopic mechanical properties are governed by collagen fibers, with their *undulation* and *orientation* being the most influential micro-histological parameters. Together with the fibers' mechanical properties, they determine the macroscopic tissue properties, which are derived through two nested integrations: one over the undulation and another one over the fibers' spatial orientation. Under certain assumptions, time-consuming numerical integrations can be avoided, which in turn allows such models to be used for organ-level vascular simulations. The initially purely passive description of the aneurysmatic aortic wall is enriched further through the modeling of *collagen turnover*, and the expansion of an *Abdominal Aortic Aneurysm (AAA)* over time is simulated.

2 Histology and Morphology of the Vessel Wall

ECM components, such as elastin, collagen, ProteoGlycans (PGs), fibronectin and fibrilin ensure the vessel wall's *structural integrity*, whilst vascular cells, such as Endothelial Cells (EC), Smooth Muscle Cells (SMC), fibroblasts and myofibroblasts maintain its *metabolism*, and thus generate building blocks that in turn form the ECM. In parallel, ECM components in the vessel wall are continuously degraded, mostly by MMPs [3]. Both processes together determine the continuous turnover of tissue constituents, which in turn establishes the growth and remodeling of the vessel wall. The vessel's geometrical, histological and mechanical properties change along the vascular tree, likely in an effort to maintain conditions for optimal mechanical operation [4]. However, vessel wall properties also alter in response to a plethora of other factors, such as age, disease and lifestyle.

2.1 Vessel Wall Layers

The vessel wall is built-up by *intimal, medial* and *adventitial* layers that are separated by the *elastica lamina interna* and the *elastica lamina externa*, see Fig. 1.

Situated inner to the elastica lamina interna, the *intima* is the innermost layer, see Fig. 1. It is formed by the *endothelium* resting on a thin *basal membrane* and a subendothelial layer made of ECM. The basal membrane provides physical support for ECs and mechanically *decouples* their shear deformation from the deformation of the vessel wall. This allows ECs to sense and in turn respond to Wall Shear Stress (WSS). In addition, ECs provide an anti-thrombogenic and low-resistance lining between blood and vessel tissue. In non-diseased large vessels, the structural impact of the intima is often negligible.

Situated between the elastica lamina interna and the elastica lamina externa, the *media* is the middle layer of the vessel wall, see Fig. 1. It consists of a complex 3D network of *SMCs, elastin, collagen* fibers and fibrils, and other connective tissue. These structural components are preferentially aligned along the circumferential vessel direction [5, 6] and organized in repeating Medial Lamellar Units (MLUs) of 13–15 (μm) in thickness [6–8]. The thickness of MLUs is independent of the radial location in the wall and the number of MLUs increases with increasing vessel diameter. The tension carried by a single MLU in the normal vessel wall remains constant at approximately 1.6–2.4 (N m^{-1}) [7]. The media's layered structure is gradually lost towards the periphery, and a discrete laminated architecture is hardly present in muscular arteries or smaller veins. The high SMC content in the media equips it with excellent vasoactivity, a property that is especially important in smaller vessels in order to regulate blood streams. Additionally, differentiated SMCs in the media synthesize ECM constituents.

Situated outside of the elastica lamina externa, the *adventitia* is the outermost vessel wall layer, see Fig. 1. It consists mainly of FibroBlasts (FBs) embedded in an ECM of mainly thick bundles of *collagen* fibrils and some other connective tissue. The adventitia *shields* the vital medial layer from overstretching and *anchors* the vessel to surrounding tissue. FBs in the adventitia continuously synthesis ECM. The adventitia also hosts *nerves* running to the SMCs in the medial layer, as well as the *vasa vasorum*, as a means to perfuse not only the adventitia itself, but also the outer media. The intima and the inner media are perfused by the radial convection of fluid that arises due to the pressure gradient between the circulation and the interstitial pressure in the adventitia.

The thickness of the media and the adventitia depends strongly on the physiological function of the blood vessel and its topographical site. The thickness of vessel wall layers, especially the one of the intima is influenced by factors, such as atherosclerosis and age.

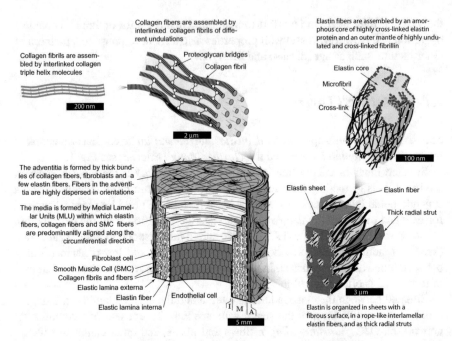

Fig. 1 Histological idealization of the normal aorta. It is composed of three layers: intima (I), media (M), adventitia (A). The intima is the innermost layer consisting of a single layer of endothelial cells, a thin basal membrane and a subendothelial layer. Smooth Muscle Cells (SMCs), elastin and collagen are key mechanical constituents in the media, and in the human aorta these constituents are arranged in up to 60 Medial Lamellar Units (MLUs) of the thickness of 13–15 (nm). In the image, only three MLUs are shown. In the adventitia the primary constituents are collagen fibers and fibroblasts. Collagen fibers with a thickness in the range of micrometers are assembled by collagen fibrils (50–300 (μm) thick) of different undulations. Load transition between collagen fibrils is maintained by Proteoglycan (PG) bridges. Elastin fibers with a thickness of hundreds of nanometers are formed by an amorphous core of highly cross-linked elastin protein that is encapsulated by 5 (μm) thick microfibrils. Elastin fibers are organized in thin concentric elastic sheets, in a rope-like interlamellar elastin fibers, and as thick radial struts

2.2 Cells in the Vessel Wall

Vascular cells, such as ECs, SMCs, and FBs, *sense* and *respond* to mechanical loads and thus allow the vessel to undergo changes during normal development, ageing, and in response to disease or to implanted devices.

ECs have a half-life time of one to three years. They are constantly exposed to WSS, in response to which they secrete vasoactive agents that control the tonus of adjacent contractile SMCs. ECs present as a non-thrombotic surface to prevent blood from clotting.

SMCs are layered within each MLU in the media. They are aligned with the circumferential direction and at a radial tilt of approximately 20° [6, 9]. SMCs are involved in many vascular diseases, such as artherosclerosis, restenosis, hypertension and aneurysm disease. In their contractile phenotype, vascular SMCs contract in response to electrical, chemical or mechanical stimuli, and actively influence aortic diameter [10]. Whilst in their synthetic phenotype, SMCs respond to injury, diseases or remodeling through migration, proliferation, and different kinds of proinflamatory and secretary responses. SMCs switch between phenotypes in response to *environmental cues*, where pulsatility and blood pressure appear to determine SMC phenotype in the aorta. Contractile SMCs are quiescent, and thus do not proliferate. In contrast, at their synthetic phenotype, SMCs have a high migration and proliferation rate, which enables them to effectively respond to injury and the like.

FBs and SMCs can produce ECM proteins, mainly collagen and elastin, and secrete MMPs to facilitate remodeling. Additionally, ECs and SMCs are able to produce a variety of immune and inflamatory mediators, and thus stimulate the migration of immune cells and inflamatory cells from the blood into the vessel wall tissue.

3 The ECM's Role in Vessel Wall Mechanics

The ECM provides an essential supporting scaffold for the *structural* and *functional* properties of vessel walls. The three-dimensional organization of ECM constituents, such as elastin and collagen is vital to the accomplishment of proper physiological functionality. Therefore, the ECM, rather than being merely a system of scaffolding for the surrounding cells, is an actively evolving mechanical structure that controls the *micro-mechanical* and *macro-mechanical* environments to which vascular tissue is exposed. Consequently, the ECM's mechanical properties control the amount of stress and strain that is transmitted from the macroscopic to the cellular levels of vascular tissue.

Collagen is one of the most dominant structural proteins in the ECM and has a large impact upon its mechanical properties. Sixty years ago ROACH AND BURTON [11] reported that *collagen* is the main determinant of the mechanical properties of arterial tissue at *high* strain levels. In contrast to collagen, *elastin* determines the mechanical properties of arterial tissue at *low* strain levels and plays an important role in the *recoiling* of the vessel wall during the diastolic phase of the cardiac cycle.

3.1 Collagen Structure and Function

Collagen of the types I, III, and IV are found in the vessel wall, where veins tend to have higher collagen contents than arteries. The fibrilar types I and III constitute most of the collagen in the wall, of which type I accounts for 50–70%. Type IV is mainly seen in the basal membrane and surrounding SMCs. Collagen is in a continuous state of deposition and degradation, with a normal half-life of 60–70 days [12]. Physiological maintenance of the collagen structure relies upon a delicate (coupled) balance between degradation (mainly through MMPs, collectively known as *collagenases*) and synthesis by cells, such as SMCs, FBs and myofibroblasts [13]. Collagen is synthesized throughout the lifespan, and in the normal vessel, it leads to collagen of stable quality.

Collagen gives *stiffness, strength* and *toughness* to the vascular wall. Earlier observations indicated that the collagen-rich abdominal aorta is stiffer than the collagen-poor thoracic aorta [14, 15], and later, the regional variations of aortic properties has been specifically documented [16]. Numerous further references can be found in the seminal works of FUNG [17] and HUMPHREY [18]. Other than the amount of collagen in the wall, its spatial orientation [19], as well as the variation in orientations [20], strongly influences the macroscopic mechanical properties of vascular tissue. At Mean Arterial Pressure (MAP), only about 6–7% of collagen fibers are mechanically engaged [21, 22].

Collagen fibrils range from fifty to a few hundreds of nanometers in diameters. They are the basic building block of many fibrous collagenous tissues [19], and their organization into *suprafibrilar structures* highly influences the vessel wall's macroscopic mechanical properties. Within the MLU, collagen fibrils or bundles of fibrils (10–40 fibrils per bundle) run in parallel, closely enveloping the SMCs [6]. The collagen fibers are *not* woven together but aligned in parallel, very much like in a tendon or ligament, a factor that most likely enables the vessel wall to better cope with mechanical load [6].

To ensure the collagen fibers' structural integrity, the fibrils within it have to be mechanically interlinked. Currently it is unclear how fibrils are interlinked, but it is thought that *PG-bridges* [23, 24] could potentially support interfibrillar load transition. Specifically, small PGs, such as decorin, bind noncovalently but specifically to collagen fibrils and cross-link adjacent collagen fibrils at about 60 (nm) intervals [23]. Reversible deformability of the PG bridges is crucial to them serving as shape-maintaining modules [23] and, *fast* and *slow* deformation mechanisms have been identified. The fast (elastic) deformation is supported by the sudden extension of about 10% of the L-iduronate (an elastic sugar) at a critical load of about 200 (pN) [25]. The slow (viscous) deformation is based on a sliding filament mechanism of the twofold helix of the glycan [23], and may explain the large portion of macroscopic visco-elasticity observed from collagen. PG-based cross-linking is supported by many experimental studies which show that PGs play a direct role in inter-fibril load sharing [23, 26–28], and this has been further verified through theoretical investigations [29–31]. However, the biomechanical role of PGs is still somewhat uncertain, and some data indicates minimal PG contribution to the tensile properties of the tissue [29, 32, 33]. Whilst all these information relates to tendon and ligaments, no such information of vascular tissue has been reported.

3.2 Elastin Structure and Function

Elastin functions in partnership with collagen, and mainly determines the mechanical properties of the vessel wall at *low* strain levels [11]. In arteries, elastin is important

(a)   **(b)**

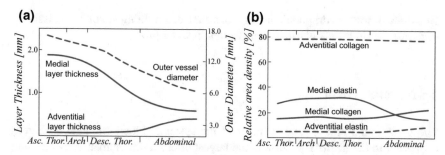

Fig. 2 Variation of geometrical properties and histological composition of the porcine aorta reported elsewhere [16]. **a** Change of outer aortic diameter together with the thicknesses of the medial and the adventitial layers. The intimal layer thickness covers less than 1% of the total wall thickness. **b** Relative area density of elastin and collagen in the medial and the adventitial layers. The relative area density represents the area that is covered by a constituent in histological stains. (Asc.Thor. - Ascending thoracic aorta; Arch. - Aortic arch; Desc.Thor. - Descending thoracic aorta; Abdominal - Abdominal aorta)

to *recoil* the vessel during each pulse cycle. Elastin is predominantly seen in the *media*, see Fig. 2b. It is organized as 1.0–2.0 (μm) thick concentric sheets (71%), 100–500 (nm) thick rope-like interlamellar elastin fibers (27%) and about 1.5 (μm) thick radial struts (2%) [6, 8, 34]. The elastin sheets encapsulate the MLUs, but they are perforated and gusseted by elastin fibers.

Microscopy studies also indicate that elastin is made-up of repeating self-similar structures at many length-scales [35]. Elastin fibers are composed of two significant components: 90% of which is an amorphous core of highly cross-linked elastin protein and the remaining 10% is a fibrillar mantle of about 5 (nm) thick microfibrils [36, 37]. Many elastin molecules are cross-linked and connected to each other and other molecules, including microfibrils, fibulins, and collagen.

Elastin is synthesized and secreted by vascular SMCs and FBs, a process that normally stops soon after puberty once the body reaches maturity. Although the dense lysyl cross linking makes elastin fibrils extremely insoluble and stable (half-life times of *tens of years* [38]), elastin may be degraded by selective MMPs, collectively known as *elastases*. Elastases cause disruption of elastin fiber integrity and subsequently diminishes tissue mechanical properties. Elastin is a critical autocrine factor that maintains vascular homeostasis through a combination of biomechanical support and biologic signaling, see [1] and references therein. Whilst elastin degradation is related to several diseases, such as atherosclerosis, Marfan syndrome, Cutis laxa, it is also important for many physiological processes. Amongst them are growth, wound healing, pregnancy and tissue remodeling [39], and thus the proteolytic degradation of elastin may have important consequences for normal elastogenesis and repair processes [40]. Repair of protease-damaged elastin can occur, but does not appear to produce elastin of the same quality as when originally laid down during primary vascular growth [41].

Elastin and rubber share some mechanical similarities. For example, both materials are highly deformable, entropic-elastic, and go through a glassy transition.

However, elastin's hydrophobic interactions are a determining factor in its elasticity, such that elastin is *only* elastic when swollen in water [40].

4 Normal Aorta

The *aorta* is the first arterial segment of the systemic blood circulation, directly connected to the heart. The aorta is the largest artery in the human body and may be divided into the ascending and descending thoracic as well as abdominal aortic segments, respectively. At its origin the aorta has a diameter of about 3 (cm), which reduces to about 1.8–2 (cm) in the abdomen. Due to its prominent role, the aorta is one of the most investigated vessels.

The aorta's pressure-diameter property is of critical importance to the entire cardiovascular system and determines the non-linearity of its pressure-flow relationship [42]. The aorta contributes almost the entire capacity of the cardiovascular system, and thus defines its Windkessel properties, see [43] and references therein. The aortic volume compliance $C = \Delta V / \Delta p$ is constant over a wide range of pressures p, and the thoracic aorta alone contributes 85% to the overall aortic compliance [44].

The properties of aortic wall tissue vary depending upon its location within the vasculature. Figure 2 illustrates the variation of some properties from the ascending to the abdominal aortic segments. The aorta's circumferential stiffness is highest at the level of the diaphragm [44, 45] and might also be higher in males than in females [46]. The thick and collagen-rich adventitial layer of the abdominal aorta leads to the observed heightened stiffness of the abdominal aorta in comparison to the thoracic aorta. The aorta's elastic properties as well as its strength are discussed in relation to aneurysmal disease in Sect. 5. Analyzing in-vivo pressure-diameter properties of the abdominal aorta revealed that the load-bearing fraction of collagen between diastole and systole oscillates between 10 and 30%, respectively [47]. In humans, the aorta is axially pre-stretched at about 5% [47], a property that, however, changes with age [48] and between the different aortic segments [44].

The aorta is highly vulnerable to *aging*, such that age-related increase in diameter and stiffness [13, 47, 49, 50] are much more pronounced in the aorta than in other vessels. Specifically, the amount of elastin in the wall decreases, while collagen and MMP-2 increases. In addition, the aged aortic wall is thinner, exhibits splitting and the fragmentation of the MLUs, and increased levels of glycation of elastin and collagen cross-linking.

5 Aneurysmatic Aorta

An *aneurysm* is a local dilatation of an artery by at least 1.5 times its normal diameter. Although every artery can become aneurysmatic, the infrarenal abdominal aorta seems to be most vulnerable to this disease, and frequently develops an *AAA*. The formation of AAAs is promoted by factors such as age, male gender, family susceptibility, high mean arterial blood pressure and smoking. If left untreated, an AAA progresses in time until the wall stress eventually exceeds the failure strength of the degenerated aortic tissue, and the AAA *ruptures*.

AAAs are the end-result of irreversible pathological remodeling of the ECM [51, 52]. Specifically, the walls of larger AAAs show [1, 51, 53–55] (i) degradation of the elastin, (ii) compensatory increased collagen synthesis and content, (iii) excessive inflammatory infiltration, and (iv) apoptosis of vascular SMC. It is widely accepted that loss of elastin (and possibly SMC) triggers initial dilatation, while collagen turnover promotes enlargement and local wall weakening that eventually leads to wall rupture [51].

Due to its high clinical relevance, the aneurysmatic infrarenal aorta has been extensively studied. Along with the development of aneurysm disease, the well-defined organization of the normal vessel wall shown in Fig. 1, is lost. Often in larger AAA it is not even possible to distinguish between individual vessel wall layers, such as intima, media and adventitia. The entire wall seems to resemble a fibrous collagenous tissue similar to the adventitial layer in the normal aorta [5]. In nearly all AAAs of a clinically relevant size, an *Intra-Luminal Thrombus (ILT)* forms [56]. The ILT is a pseudo-tissue that develops from coagulated blood and has solid-like properties [57, 58]. It is composed of a fibrin mesh, traversed by a continuous network of interconnected canaliculi and contains blood cells, such as erythrocytes and neutrophils as well as aggregated platelets, blood proteins, and cellular debris [59, 60]. It creates an environment for *increased proteolytic activity* [61, 62] that may be linked to the observed weakening [63] and thinning [53] of the vessel wall.

As already indicated by the AAA wall's inhomogeneous patho-histology [64], its strength also shows significant inter-patient and intra-patient variabilities [65]. Whilst some of the factors influencing aortic wall strength are known [66–70], much remains unclear and the experimentally measured strength is highly scattered. Table 1 illustrates this by summarizing some of the reported AAA wall strength and thickness data. On average, the wall withstands a stress of 865 (SD 390) (kPa) and it is 1.6 (SD 0.6) (mm) thick. The aneurysmatic aorta is weaker than the normal infrarenal aorta, which is reported to have a strength of 1210 (SD 330) (kPa) [71] or even 1710 (SD 140) (kPa) [72]. The normal thoracic aorta is slightly stronger, with strengthes of 1950 (SD 600) (kPa) [73] and 1470 (SD 910) (kPa) [74] reported for the descending thoracic and midthoracic aortic segments, respectively. However, very limited experimental data is available for the Thoracic Aortic Aneurysm (TAA) wall, some of which is listed in Table 2. The aforementioned wall strength information is, with very few exceptions [75], commonly derived from uniaxial tensile testing, which unfortunately does not reflect the biaxial in-vivo loading of the AAA wall.

AAA wall strength and thickness are strongly negatively correlated [65, 67], which is also seen from the data in Table 1. Consequently, the local variations of wall thickness and strength may at least partly compensate for each other, which could explain the reasoning behind why wall stress multiplied with wall thickness has been suggested to be a more robust rupture risk predictor [86]. The inverse

Table 1 Abdominal Aortic Aneurysm (AAA) wall thickness and wall strength measured from in-vitro tensile testing. The number of test samples is denoted by N and some further cohort specification is listed.

Reference	Sample specification	Thickness (mm)	Strength (MPa)
[76]	N = 31; fibrous	1.2	1.2
	N = 38; partly calcified	1.5	0.87
[77]	N = 28	1.18	–
[78]	N = 83	–	0.81
[79]	N = 26	1.32	–
[80]	N = 25; AAA diam. <55 (mm)	1.53	0.77
	N = 65; AAA diam. >55 (mm)	1.58	1.03
[81]	N = 76	–	Female: 0.68
			Male: 0.88
[65]	N = 163	1.57	1.42
[82]	N = 374/48	1.48	1.26
[83]	N = 14	1.5–1.9	Long.: 0.93
			Circ.: 1.15
[66]	N = 16	2.06	0.57
[84]	Anterior: N = 29	2.73	Long.: 0.38
			Circ.: 0.52
	Lateral: N = 9	2.52	Long.: 0.51
			Circ.: 0.73
	Posterior: N = 9	2.09	Long.: 0.47
			Circ.: 0.45
[85]	Intact AAA: N = 26	2.5	0.82
	Ruptured AAA: N = 13	3.6	0.54
[86]	Intact: N = 278/56	1.5	0.98
	Ruptured: N = 141/21	1.7	0.95
[87]	Long.: N = 45	–	0.86
	Circ.: N = 19	–	1.02
[63]	ILT layer thick. >4 (mm): N = 7	–	1.38
	ILT layer thick. <4 (mm): N = 7	–	2.16

Circ. - circumferential; Long. - longitudinal; ILT - intra-luminal thrombus

correlation between wall strength and thickness may also justify the commonly-used uniform wall thickness employed in AAA biomechanics simulations.

Besides its strength, the elastic properties of the AAA wall also differ considerably from the normal aorta. Figure 3 shows equi-biaxial tensile properties extracted from data reported elsewhere [79]. It illustrate that the AAA wall is considerably stiffer than the aged but still normal abdominal aorta.

Table 2 Thoracic aortic aneurysm (TAA) wall thickness and wall strength measured from in-vitro tensile testing. The number of test samples is denoted by N and data for Tricuspid aortic valve (TAV) and Bicuspid aortic valve (BAV) anatomy is reported separately.

Reference	Sample specification	Aortic valve anatomy	Strength (MPa)
[88]	N = 163	TAV	Long.: 0.54
			Circ.: 0.961
		BAV	Long.: 0.698
			Circ.: 1.656
[67]	N = 27	TAV	0.878
		BAV	1.310

Circ. - circumferential; Long. - longitudinal

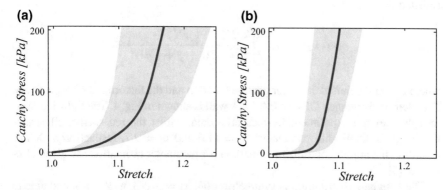

Fig. 3 Elastic properties of the normal aorta wall (**a**) and the Abdominal Aortic Aneurysm (AAA) wall (**b**). Data reflects equi-biaxial tension and has been extracted from in-vitro experimental tissue characterization reported in the literature [79]. The thick solid line and the shaded area represents the median and the domain within which approximately 75% of the data falls, respectively

5.1 Collagen Fiber Orientation in the AAA Wall

In large, and thus clinically relevant AAAs, elastin is degraded and fragmented, and *collagen* is the only remaining protein able to carry the mechanical load. Consequently, the structural organization of collagen in the AAA wall is of primary importance to AAA wall mechanics. Collagen is intrinsically birefringent and Polarized Light Microscopy (PLM) provides an ideal method for its detection and analysis [89–91]. Enhanced by picrosirius red staining, PLM provides a clear qualitative image of the organization of collagen. Combined with a Universal Rotary Stage (URS), it even allows a *quantitative* representation of the 3D collagen fiber organization [5, 92, 93].

The orientation of a collagen fiber, or a coherent bundle of fibers, in the 3D space is uniquely defined by its *azimuthal* angle θ and its *elevation* angle ϕ, which are measured by (in-plane) rotating and (out-of-plane) tilting the URS, respectively. In order to acquire robust experimental data, it is required to measure the collagen orientation of a large number of data points covering a representative section of the vessel wall. This makes 3D PLM measurements very labor intensive. However, in combination with image processing, the acquisition of the azimuthal collagen fiber distribution can be highly automatized [94].

The measurements from PLM may be represented by frequency plots, and then fitted to an Orientation Density Function (ODF), such as the *Bingham ODF* [5, 95, 96]

$$g(\theta, \phi) = c^{-1} \exp[\kappa_1 (\cos \theta \cos \phi)^2 + \kappa_2 (\cos \phi \sin \theta)^2]. \qquad (1)$$

Here, κ_1 and κ_2 are parameters that determine the shape of the ODF, whilst c is used to normalize it, such that $\int_{\phi=-\pi/2}^{\pi/2} \int_{\theta=-\pi/2}^{\pi/2} g \cos \phi \, d\phi d\theta = 1$ holds. This condition is satisfied for

$$c(\kappa_1, \kappa_2) = \sqrt{\pi} \sum_{i,j=0}^{\infty} \frac{\Gamma(i + \frac{1}{2})\Gamma(j + \frac{1}{2})\kappa_1^i \kappa_2^j}{\Gamma(i + j + \frac{3}{2})i!j!} \qquad (2)$$

where Γ and $i!$ denoted the Euler gamma function and the factorial of i, respectively. The identified collagen ODF of the AAA wall is shown in Fig. 4. The light-blue surface denotes the experimentally-measured data, whilst the red surface illustrates the Bingham ODF with parameters $\kappa_1 = 11.6$ and $\kappa_2 = 9.7$, respectively. Details regarding the applied optimization method for parameter estimation are given elsewhere [5].

The Bingham distribution is symmetric $g(\phi, \theta) = g(\phi + \pi, \theta + \pi)$, and able to capture a large spectrum of distributions. Specifically, it accounts for different fiber dispersions in the tangential and cross-sectional planes, and it has more flexibility than a transversely isotropic distribution, suggested earlier to model the collagen fiber organization in the arterial wall [20].

Fig. 4 Bingham Orientation Density Function (ODF) with $\kappa_1 = 11.6$ and $\kappa_2 = 9.7$ (red) fitted to the experimentally-measured fiber orientation distribution (light-blue) in the Abdominal Aortic Aneurysm (AAA) wall

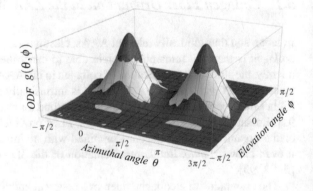

6 Constitutive Modeling of AAA Tissue

Constitutive modeling of vascular tissue is an active field of research and numerous descriptions have been reported. The phenomenological approaches [97–101] have been successfully used to fit experimental data, but *cannot* allocate stress or strain to the different histological constituents in the vascular wall. Structural constitutive descriptions [2, 20, 102–107] overcome this limitation and integrate histological and mechanical information of the arterial wall.

The AAA wall is essentially a collagenous tissue, and LANIR'S constitutive framework [2] seems ideally suited to describe the mechanics of such a tissue. This powerful approach allows the integration of the *undulation, orientation* and *stiffness* of collagen fibers. It assume that the tissue's Cauchy stress is the superposition of the individual collagen fiber contributions, according to

$$\sigma = \frac{2}{\pi} \int_{\phi=0}^{\pi/2} \int_{\theta=0}^{\pi/2} g(\phi, \theta) \sigma(\lambda) \mathrm{dev}(\mathbf{m} \otimes \mathbf{m}) \cos \phi \mathrm{d}\phi \mathrm{d}\theta + p\mathbf{I}. \tag{3}$$

Here, $\mathbf{m} = \mathbf{FM}/|\mathbf{FM}|$ denotes the spatial orientation vector of the collagen fiber, and $p\mathbf{I}$ being the hydrostatic stress contribution with the hydrostatic pressure p serving as a Lagrange parameter to enforce incompressibility. In equation (3) the constitution of the collagen fiber is incorporated through its Cauchy stress $\sigma(\lambda)$, and the integration is taken over the unit sphere, see Fig. 5.

Equation (3) can be numerically integrated by *spherical t-designs*, and thus

$$\int_{\phi=0}^{\pi/2} \int_{\theta=0}^{\pi/2} f(\theta, \phi) \mathrm{d}\phi \mathrm{d}\theta \approx \frac{4\pi}{l_{\mathrm{int}}} \sum_{i=1}^{l_{\mathrm{int}}} f(\theta_i, \phi_i)$$

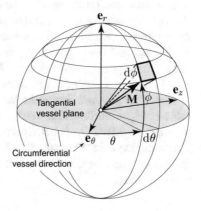

Fig. 5 Integration over the unit sphere. The Cartesian base vectors $\mathbf{e}_\theta, \mathbf{e}_z, \mathbf{e}_r$ denote the circumferential, axial and radial vessel wall directions, respectively. The tangential plane is formed by the vessel's circumferential $\mathbf{e}_\theta$ and axial $\mathbf{e}_z$ directions. Elevation ϕ and azimuthal θ angles determine the fiber direction $\mathbf{M}$ in the vessel wall

with l_{int} denoting the total number of integration points distributed over the unit-sphere. A spherical t-design integrates a polynomial expression $f(\theta, \phi)$ of degree $\leq t$ exactly [108], with further details regarding the numerical integration given elsewhere [109].

Finally, collagen fibers are embedded in an isotropic and soft *matrix* captured by a neoHookean strain energy $\psi = \mu(I_1 - 3)/2$ per unit reference tissue volume. Here, $I_1 = \text{tr}\mathbf{C}$ denotes the first invariant of the right Cauchy-Green strain $\mathbf{C}$, and μ is a stress-like material parameter. In larger AAA, the matrix contribution is almost negligible, but helped to stabilize the numerical computations.

6.1 A Passive Histomechanical AAA Wall Model

Each collagen fiber within the AAA wall is assembled by a bundle of collagen fibrils mutually interconnected by PG bridges [23, 24] that provide interfibrillar load transition, see Sect. 3.1. This structural view defines a basic load carrying unit, termed the Collagen Fibril ProteoGlycan-complex (CFPG-complex). Stretching a collagen fiber involves the continuous recruitment of collagen fibrils, such that they gradually start carrying a load. A straightening stretch λ_{st} defines the stretch, beyond which the collagen fibril is stretched elastically and elastic energy is stored in the CFPG-complex.

6.1.1 Finite Strain Kinematics

Finite deformation kinematics was considered, where the unit direction vector $\mathbf{M}$ denotes the local collagen fiber direction in the reference configuration Ω_0, see Fig. 6. The deformation gradient $\mathbf{F}_{st\,i}$ relates to the *straightened* i-th collagen fibril: it mapped its crimped referential configuration into a straight, but still unstressed intermediate configuration $\Omega_{st\,i}$. In contrast the deformation gradient $\mathbf{F}_{c\,i}$ records *deformation* relative to $\Omega_{st\,i}$ and maps the fibril to its spatial configuration Ω. Here, the intermediate configuration serves as a local reference configuration, with fibril stretch $\lambda_{c\,i}$ and fibril tension $T_{c\,i} = 0$, relative to which the fibril deforms elastically. Consequently, *multiplicative kinematics* relates the continuum deformation $\mathbf{F} = \mathbf{F}_{c\,i}\mathbf{F}_{st\,i}$ to the sub-deformations $\mathbf{F}_{c\,i}$ and $\mathbf{F}_{st\,i}$, respectively.

For simplicity and due to the lack of micro-structural data it was assumed that those collagen fibrils that formed a collagen fiber straightened according to a *symmetric triangular PDF* [110]. Specifically, the first and last fibrils within a collagen fiber straightened at fiber stretches λ_{min} and λ_{max}, respectively. Finally, affine deformation between the continuum and the collagen fiber, such that $\lambda = |\mathbf{FM}| = |\mathbf{m}|$, was considered.

Fig. 6 Multiplicative
kinematics of the collagen
fiber reinforced tissue.
Configurational map, where
the intermediate
configuration $\Omega_{\text{st}\,i}$ separates
the straightening and the
stretching of the i-th
collagen fibril. A collagen
fiber is thought to be
assembled by a number of
undulated collagen fibrils

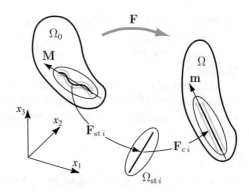

6.1.2 Constitutive Description of the CFPG-Complex

The first Piola-Kirchhoff stress in the intermediate configuration is the most convenient metric to formulate a collagen fibril constitutive model [112]. Collagen fibrils have an approximately *linear stress-stretch property* [113, 114], which is captured by the first Piola-Kirchhoff stress $T_{c\,i} = k\log\lambda_{c\,i}$ in the logarithmic stretch domain. Here, $\lambda_{c\,i} = \lambda/\lambda_s$ denotes the stretch of the i-th fibril with respect to its intermediate configuration $\Omega_{\text{st}\,i}$, see Fig. 6. It was assumed that the same constitutive relation also describes the i-th CFPG-complex. Considering the continuous recruitment of collagen fibrils, the first Piola-Kirchhoff stress may be expressed by

$$T(\lambda) = \int\limits_0^\lambda \frac{k}{x}\text{CDF}(x)\mathrm{d}x, \tag{4}$$

where $\text{CDF}(x)$ denotes the Cumulative Density Function of the triangular PDF. Integrating this expression yields piece-wise analytic expressions for the first Piola-Kirchhoff stress $T(\lambda)$ of a collagen fiber, and so far we have not imposed incompressibility of the collagen fiber.

Limiting the fiber model to incompressibility, the expression (4) defines the Cauchy stress $\sigma(\lambda) = \int_0^\lambda k\text{CDF}(\xi)\mathrm{d}\xi$, through which integration led to the very simple analytical expressions

$$\sigma(\lambda) = \begin{cases} 0, & 0 < \lambda \le 1, \\ k\frac{2}{3a^2}(\lambda - 1)^3, & 1 < \lambda \le b, \\ k[\lambda - \frac{2}{3a^2}(\lambda - \lambda_{\max})^3 - b], & b < \lambda \le \lambda_{\max}, \\ k(\lambda - b), & \lambda_{\max} < \lambda \le \infty. \end{cases} \tag{5}$$

Here, the abbreviations $a = \lambda_{\max} - 1$ and $b = (\lambda_{\max} + 1)/2$ have been used, and the expression 5 exhibits the typically non-linear property of vascular wall tissues. Whilst the derivation of 5 followed the framework proposed elsewhere [112], an earlier derivation [104] used different assumptions and found the identical analytical expressions, but instead of the Cauchy stress, they defined the first Piola-Kirchhoff stress of the collagen fiber. In any case, with the application to AAA tissue, which is known to deforms in the range of 5%, the distinction between $T(\lambda)$ and $\sigma(\lambda)$ is negligible.

The outlined model has been extended to supra-physiological loading and applied to tendon failure [112]. To this end, it considered the rupture of collagen fibrils once a predefined failure stretch λ_f is reached, which is thought to be another material property.

6.1.3 Parameter Identification and Model Validation

The proposed model for vascular tissue used the parameters κ_1, κ_2 to describe the orientation, and the parameters $k, \lambda_{min}, \lambda_{max}$ to describe the stretch-stress properties of a collagen fiber. These parameters have been identified from macroscopic planar biaxial testing of AAA wall tissue [79], where λ_c and λ_a denotes the circumferential and axial tissue stretches, respectively. Introducing the deformation gradient $\mathbf{F} = \mathrm{diag}[\lambda_c, \lambda_a, (\lambda_c\lambda_a)^{-1}]$ of biaxial deformation of an incompressible solid, and expressing the orientation of the collagen fiber $\mathbf{M} = [\cos\phi\cos\theta \quad \sin\theta \quad \sin\phi\cos\theta]^T$ through the azimuthal and elevation angles, the circumferential σ_c and axial σ_a Cauchy stresses read

$$\sigma_i = \frac{2}{\pi} \int\limits_{\phi=0}^{\pi/2} \int\limits_{\theta=0}^{\pi/2} g(\phi,\theta)\sigma(\lambda)(a_{ii} - a_{rr})\cos\phi\, d\phi d\theta; \quad i = c, a. \tag{6}$$

Here, a_{cc}, a_{aa} and a_{rr} were the diagonal coefficients of the spatial tensor $\mathbf{a} = \mathbf{m} \otimes \mathbf{m} = (\mathbf{FM}) \otimes (\mathbf{FM})$. The data from biaxial AAA tissue characterization [79] has been complemented with the Bingham ODF $g(\phi,\theta)$ using the parameters $\kappa_1 = 11.6$ and $\kappa_2 = 9.7$, see Fig. 4. Consequently, the expressions (6) permitted the estimation of $k, \lambda_{min}, \lambda_{max}$ from the reported experimental data [79] through least-square optimization. All details are given elsewhere [5].

In order to achieve a reasonable agreement with the experimental AAA wall data [79], it was essential that collagen fibers *changed* their properties with respect to their *azimuthal* alignment, and the upper undulation limit λ_{max} of a circumferentially-aligned collagen fiber was assumed to be about twice the one of an axially-aligned fiber [5]. This seems reasonable, since the undulation of collagen fibrils is likely to be determined by the *pulsatility* of the vessel wall, to which collagen is continually deposited. Consequently, the higher undulation limit of collagen fibrils aligned with the circumferential direction may be the direct consequence of the higher pulsating strains in the circumferential direction, compared to the axial vessel direction.

In another vascular biomechanical study, the collagen fiber ODF $g(\phi,\theta)$ has been identified together with the sample's biaxial tensile properties [115]. Consequently, complete sets of data for all individual aorta samples have been acquired, which allowed a more rigorous validation of the proposed constitutive model. This study concluded that the model showed very good *extrapolative* capability, and the tissue's biaxal mechanical properties could be accurately predicted from uniaxial tensile tests [115].

6.2 An Adaptation Model for the AAA Wall

Vascular tissue responds to mechanical stimuli, a mechanism necessary for the *optimization* of cardio-vascular function. It is likely that the vessel's geometrical, histological and mechanical properties change along the vascular tree, in an effort to maintain conditions for optimal mechanical operation [4]. However, malfunction of such vascular adaption could potentially lead to *pathologies*, such as aneurysm formation. Besides the size of an AAA, its growth rate is used to assess its risk for rupture. An increase of the AAA's largest diameter by 10 (mm) per year is a common indication for progressive clinical treatment [116]. However, current understanding of AAA growth is somewhat limited, and AAAs are seen to grow *inhomogenously* in both time [117] and space [118], which is thought to be influenced by *many* factors [119]. The reliable prediction of AAA growth could potentially improve their clinical management. This explains as to why several AAA growth models have been reported, base on several different conceptual frameworks [120], see among others [121–126]. However, these models require significant further development if they are to successfully augment clinical decisions.

This section includes a description of collagen turnover to the purely passive AAA wall model introduced in Sect. 6.1. Fibrobast-based synthesis and MMP-based degradation of collagen is considered, such that the ODF $g(\phi, \theta)$ and the undulation limits λ_{min} and λ_{max} of collagen may develop towards potentially homeostatic targets.

6.2.1 Collagen Turnover Model

The present model assumes that fibroblast senses the state of strain and pre-stretches collagen fibrils prior to their deposition. This is formulated through three distinct sub-models, denoted as *the sensing model, the collagen turnover model* and *the structural update model*, respectively. The sensing model defines the physical quantity that stimulates collagen synthesis, the collagen turnover model quantifies the relation between the sensed stimulus and the change of collagen mass, and the structural update model details the manner in which collagen is integrated/disintegrated into/from the existing collagen structure.

The *sensing model* defines the physical quantity ξ that stimulates collagen synthesis, and thus the production of new collagen. For $\xi > 1$, the existing collagen is stretched too much, such that in total more collagen is required to reach homeostasis. In this case, the rate of collagen turnover needs to be elevated in order to increase the total collagen density in the tissue. Equivalently, for $\xi < 1$, the rate of collagen turnover needs to decrease in order to that homeostasis is reached.

Specifically, the model assumes that collagen stretch tends towards its homeostatic value of λ_{ph} by satisfying the optimality condition

$$\xi(\mathbf{M}) = \frac{\lambda(\mathbf{M})}{\lambda_{ph}} \frac{\overline{\rho}}{\rho} \to 1 . \tag{7}$$

Here, ρ denotes the specific collagen density that aims at approaching the target density of $\overline{\rho}$. The use of $\overline{\rho}$ in expression 7 replaces the maximum collagen turnover rate, which was introduced previously [127]. Whilst in the normal artery wall only about 6–7% of collagen is engaged at MAP, we consider the homeostatic value of λ_{ph} that corresponds to 10% of engaged collagen fibrils in the AAA wall [127]. Finally, it is emphasized that ξ depends on the orientation, through the unit direction vector $\mathbf{M}$ that defines the orientation of the particular collagen fiber, see Fig. 5.

The *mass turnover model* quantifies the relation between the sensed stimulus $\xi(\mathbf{M})$ and the change in the specific collagen density, i.e. the relation between degraded $\dot{\rho}^-$ and synthesized $\dot{\rho}^+$ collagen density rates, respectively. Despite experimental data suggesting a stretch-based degradation of collagen (see [128] and references therein), in the present model collagen is considered to degrade independently relative to the orientation $\mathbf{M}$. The rate equation $\dot{\rho}^- = -\eta\rho$ is used, where η defines the time-scale of the degradation process. In contrast, collagen synthesis is linked to the stimulus $\xi(\mathbf{M})$ and follows $\dot{\rho}(\mathbf{M})^+ = \eta\rho\xi(\mathbf{M})$.

Finally, the *structural update model* specifies how collagen fibrils are integrated and removed over time. The model assumes that collagen fibrils are removed without changing their undulation characteristics. In contrast, synthesized collagen fibrils are integrated at a certain *distribution of pre-stretches*, and follow a triangular PDF with pre-defined $\overline{\lambda}_{min}$ and $\overline{\lambda}_{max}$; details are reported elsewhere [127]. Note that the triangular PDF, which describes the integration of collagen also explicitly defines the aforementioned homeostatic stretch λ_{ph}. Specifically, for a 10% recruitment of collagen fibrils, this stretch reads $\lambda_{ph} = \overline{\lambda}_{min} + 0.224(\overline{\lambda}_{max} - \overline{\lambda}_{min})$ [127].

Although the outlined structural update model assumes a pre-stretched deposition of collagen fibrils, one could also consider collagen fibrils being stretched by fibroblasts subsequent to their deposition. It is widely accepted that fibroblasts impose tensile forces upon the collagen network to which they are attached [38]. However, Fig. 5 in [129] (taken from [130]) nicely shows that collagen fibrils may already be in a state of tension when synthesized.

6.2.2 Results

The model has been implemented in the finite element environment FEAP (University of California at Berkeley, US) [131] and a patient-specific AAA was reconstructed from standard Computed Tomography-Angiography (CT-A) images using A4clinics Research Edition (VASCOPS GmbH, Austria) [132, 133]. Although the intra-luminal thrombus is known to be an important solid structure [57] that increases the predictability of biomechanical AAA models [134], it was not considered in the present study.

The individual AAA model was exposed to a constant blood pressure of 100 (mmHg), and the parameters that determined collagen turnover, have been set to match the measured growth of the AAA's maximum diameter reported elsewhere [118]. Figure 7 shows the computed AAA expansion over a time period of four years based on the parameters summarized in Table 3. In order to ensure the sufficiently accurate integration of governing equations, a 'look-ahead' technique was used. Specifically, the largest collagen density increment $\max[\dot{\rho}^-, \dot{\rho}^+]\Delta t$ over all Gauss points of all finite elements was used to control the time step through FEAP's *AUTO MATErial* command [131]. Overall, the results showed promising correlations with experimental data. Simulations are able to replicate the coupled circumferential and longitudinal growth of the AAA wall, and the AAA developed into reasonable geometries, see Fig. 7. In addition, the adaptation model avoided high stress gradients across the vessel wall [135], which are thought to be non-physiological. Other than

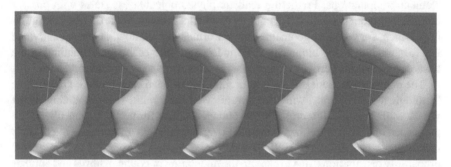

Fig. 7 Evolution of the shape of an Abdominal Aortic Aneurysm (AAA), as predicted by the adaptive AAA wall descriptions described in Sect. 6.2. The simulation covered a period of four years of AAA growth and used the parameters listed in Table 3

Table 3 Parameters used to simulate Abdominal aortic aneurysm (AAA) growth according to constitutive model described in Sect. 6.2

Parameter	Value	Description
μ	21.0 (kPa)	neoHookean parameter of the matrix
k	60.0 (MPa)	CFPG-complex stiffness
λ_{min}	1.045	Initial collagen minimum straightening stretch
λ_{max}	1.117	Initial collagen maximum straightening stretch
$\bar{\rho}$	0.4	Specific collagen target mass density
$\bar{\lambda}_{min}$	0.94	Minimum straightening stretch of deposited collagen
$\bar{\lambda}_{max}$	1.117	Maximum straightening stretch of deposited collagen
η	0.167 (years)	Collagen turn over rate

the discussed macroscopic consequences of collagen turnover, the initially isotropic collagen orientation density shifted into a locally orthotropic distribution, which correlates well with histological studies of the AAA wall, see Sect. 5.1.

6.3 Discussion and Conclusion

Vascular biomechanics is critical in order to define new diagnostic and therapeutic methods that could have a significant influence on our medical understanding and even on the lifestyle of human beings. Vascular biomechanical simulations critically depend on an accurate constitutive description of vascular tissue. As with other biological tissues, the vascular wall responds to its mechanical environment and predictions based on passive constitutive models, i.e. suppressing tissue remodeling and growth, can only cover a limited time period. Vascular tissue develops at a loaded in-vivo configuration, which implicitly defines *residual strains* in its (hypothetical) load-free configuration, the setting that typically serves as a reference for finite element computations. Predicting realistic physiological stress states with *passive* constitutive models commonly requires residual strains in the load-free configuration, which, for complex geometries are unfortunately unknown. Consequently, the key towards improving biomechanical models is to further the understanding of the tissue's inherent properties to *adapt* to its mechanical and other environments.

The constitutive models discussed in this chapter have been based on reasonable histological assumptions, which in turn allowed for a *multi-scale* modeling of the vessel wall. Specifically, cross-linked collagen fibrils formed collagen fibers, the properties of which were then integrated over the unit-sphere to define the tissue's macroscopic properties [2, 109]. The proposed models introduced two statistical probability distributions, one to capture the undulation of collagen fibrils and another one to describe the orientation of the collagen fibers. Vascular tissue is heterogenous and the parameters defining such probability distributions will likely change over the vessel wall, very much like constitutive parameters of any other description. Whilst direct, especially in-vivo measurements of the parameters defining such probability distributions seems unfeasible, their physical interpretation may support their *indirect* estimation. The probability of collagen fibers aligned along directions that do not experience tension should disappear, and the collagen straightening stretches are tightly linked to the strains experienced during the cardiac cycle.

It is well understood that *inter-fiber* and *inter-fibril* sliding plays a significant role in tendon deformation [136]. Whilst the present collagen fiber model could account for inter-fibril sliding, any inter-fiber sliding was suppressed through the prescribed affine kinematics between collagen fiber and tissue levels. A sound experimental investigation of this point requires a strict definition of a mechanical fiber, to be distinguished from a fibril, and tracking the the fiber under conditions that reflect the in-vivo mechanical environment of the vessel. In addition, experimental data should be provided from a large number of tissue samples in order to allow the definition of a Representative Volume Element (RVE). Each of which is challenging and current microscopy techniques are not able to image collagen fibers dynamically during a cardiac cycle. Whilst recent data rejects the affine transformation of collagen fibers in the adventitia [111], it is clearly not representative for in-vivo loading conditions of vascular tissue. In the case inter-fiber sliding would also constitute an important element in vascular tissue biomechanics, the applied affine transformation should be replaced. An interface model, explicitly accounting for the *non-rigid interface* between the fibers and the matrix, could be introduced instead. Unlike the affine fiber-continuum kinematics employed in the present work, fiber-reinforced tissues have been modeled by interlinked network structures [137]. Such an approach is clearly justified for hydrated collagen networks, where the absence of inter-fibrillar (solid) material allows a largely unconstraint motion of collagen fibrils. In contrast, the inter-fibrillar material in the AAA wall is expected to define a rather affine transformation between the collagen fibers and the continuum. Likewise, in the media of the normal aorta, bundles of collagen are not woven together [6], i.e. cross-linking among fibers (but not between fibers and matrix) is missing, which further reinforces the affine kinematics approach. Whilst there is still no experimental evidence available concerning whether or not affine kinematics holds for the AAA wall tissue, a recent study questions affine kinematics in the adventitial layer of carotid arteries [111, 138].

The suggested adaptation model was able to predict the growth of an ILT-free aneurysm, where the undulation limits $\bar{\lambda}_{min}$ and $\bar{\lambda}_{max}$ of the newly-formed collagen seemed to be the most influential parameters. However, their direct experimental identification is not straightforward.

The collagen target density $\bar{\rho}$, another model parameter likely plays an important role in the final phase of aneurysm disease. Through equation (7), it gradually limits the synthesis of collagen at wall segments that already have a high density of collagen $\rho > \bar{\rho}$. Consequently, the collagen synthesis can no longer keep-up with the increase of wall stress, which in turn accelerates AAA growth until it ruptures.

The present AAA wall adaptation model did not explicitly prescribe how an increase of collagen mass, and thus collagen volume, updates the vessel wall's geometry. For simplicity, isochoric growth kinematics have been used, such that only very moderate AAA expansion may be simulated. Explicit assumptions regarding the underlying growth kinematics would be required to generalize the model towards 3D growth [121].

It is likely that synthesis, deposition and degradation of collagen fibers are also influenced by factors, such as *systolic-diastolic pulsatility* of the vessel wall. The present version of the model was not able to consider such factors, but assumed that MAP-based mechanical loading is the dominating factor in collagen turnover. Much more experimental evidence is needed in order to better understand such mechanisms in the vessel wall, and in doing so, form the basis of more sophisticated biomechanical modeling.

The present chapter focused on the structural aspects of vascular tissue, whilst it is known that aneurysm disease is also greatly influenced by the blood flow through the aorta [139]. The widened lumen lowers the WSS in AAAs [140] and vortex dynamics differ remarkably from the normal aorta [141]. These hemodynamic alterations dictate the distribution of chemical species, such as thrombin [142], and their associated biological consequences are as yet unknown. Finally, although the constitutive models presented were able to successfully capture some features of an AAA wall, a rigorous validation against experimental data is crucial toward the evaluation of its descriptive and predictive capabilities. In this respect, a mixed experimental numerical approach that accounts for tissue growth and remodeling seems to be the most appropriate.

Acknowledgements The author would like to express his very great appreciation to Christopher Miller for his valuable and constructive feed-back in writing this text. This work has been financially supported by the Project Grant No. 2015-04476 provided by the Swedish Research Council, and the Hälsa, Medicin och Teknik Project Grant No. 20150916 provided by Stockholm county.

References

1. Bäck, M., Gasser, T., Michel, J.B., Caligiuri, G.: Review. biomechanical factors in the biology of aortic wall and aortic valve diseases. Cardiovasc. Res. **99**, 232–241 (2013)
2. Lanir, Y.: Constitutive equations for fibrous connective tissues. J. Biomech. **16**, 1–12 (1983)
3. McDonald, D.A.: Blood Flow in Arteries, 6th edn. Edward Arnold, London (2011)
4. Rachev, A., Greenwald, S.E., Shazly, T.: Are geometrical and structural variations along the length of the aorta governed by a principle of Optimal Mechanical Operation? J. Biomech. Eng. **135** (2013). https://doi.org/10.1115/SBC2013-14427
5. Gasser, T.C., Gallinetti, S., Xing, X., Forsell, C., Swedenborg, J., Roy, J.: Spatial orientation of collagen fibers in the Abdominal Aortic Aneurysm wall and its relation to wall mechanics. Acta Biomater. **8**, 3091–3103 (2012)
6. O'Connell, M., Murthy, S., Phan, S., Xu, C., Buchanan, J., Spilker, R., Dalman, R., Zarins, C., Denk, W., Taylor, C.: The three-dimensional micro- and nanostructure of the aortic medial lamellar unit measured using 3d confocal and electron microscopy imaging. Matrix Biol. **27**, 171–181 (2008)
7. Clark, J.M., Glagov, S.: Transmural organization of the arterial media: the lamellar unit revisited. Arteriosclerosis **5**, 19–34 (1985)
8. Dingemans, K.P., Teeling, P., Lagendijk, J.H., Becker, A.E.: Extracellular matrix of the human aortic media: an ultrastructural histochemical and immunohistochemical study of the adult aortic media. Anat. Rec. **258**, 1–14 (2000)
9. Fujiwara, T., Uehara, Y.: The cytoarchitecture of the medial layer in rat thoracic aorta: a scanning electron-microscopic study. Cell Tissue Res. **270**, 165–172 (1992)
10. Milewicz, D.M., Kwartler, C.S., Papke, C.L., Regalado, E.S., Cao, J., Reid, A.J.: Genetic variants promoting smooth muscle cell proliferation can result in diffuse and diverse vascular diseases: Evidence for a hyperplastic vasculomyopathy. Genet. Med. **12**, 196–203 (2010)
11. Roach, M.R., Burton, A.C.: The reason for the shape of the distensibility curve of arteries. Canad. J. Biochem. Physiol. **35**, 681–690 (1957)
12. Nissen, R., Cardinale, G., Udenfriend, S.: Increased turnover of arterial collagen in hypertensive rats. Proc. Natl. Acad. Sci. USA **75**, 451–453 (1978)
13. Nichols, W.W., O'Rourke, M.F., Vlachopoulos, C.: McDonald's Blood Flow in Arteries, Sixth Edition. Theoretical, experimental and clinical principles, 6th edn. Arnold, London (2011)
14. Bergel, D.H.: The static elastic properties of the arterial wall. J. Physiol. **156**, 445–457 (1961)
15. Langewouters, G.J., Wesseling, K.H., Goedhard, W.J.A.: The static elastic properties of 45 human thoracic and 20 abdominal aortas in vitro and the parameters of a new model. J. Biomech. **17**, 425–435 (1984)

16. Sokolis, D.P.: Passive mechanical properties and structure of the aorta: segmental analysis. Acta Physiol. **190**, 277–289 (2007)
17. Fung, Y.C.: Biomechanics. Mechanical Properties of Living Tissues, 2nd edn. Springer-Verlag, New York (1993)
18. Humphrey, J.D.: Cardiovascular Solid Mechanics. Cells, Tissues, and Organs. Springer-Verlag, New York (2002)
19. Fratzl, P. (ed.): Collagen - Structure and Mechanics. Springer-Verlag, New York (2008)
20. Gasser, T.C., Ogden, R.W., Holzapfel, G.A.: Hyperelastic modelling of arterial layers with distributed collagen fibre orientations. J. R. Soc. Interface **3**, 15–35 (2006)
21. Armentano, R.L., Levenson, J., Barra, J.G., Fischer, E.I., Breitbart, G.J., Pichel, R.H., Simon, A.: Assessment of elastin and collagen contribution to aortic elasticity in conscious dogs. Am. J. Physiol. **260**, H1870–H1877 (1991)
22. Greenwald, S.E., Moore Jr., J.E., Rachev, A., Kane, T.P.C., Meister, J.J.: Experimental investigation of the distribution of residual strains in the artery wall. J. Biomech. Eng. **119**, 438–444 (1997)
23. Scott, J.E.: Elasticity in extracellular matrix 'shape modules' of tendon, cartilage, etc. A sliding proteoglycan-filament model. J. Physiol. **553**(2), 335–343 (2003)
24. Scott, J.E.: Cartilage is held together by elastic glycan strings. Physiological and pathological implications. Biorheology **45**, 209–217 (2008)
25. Haverkamp, R., Williams, M.W., Scott, J.E.: Stretching single molecules of connective tissue glycans to characterize their shape-maintaining elasticity. Biomacromols **6**, 1816–1818 (2005)
26. Liao, J., Vesely, I.: Skewness angle of interfibrillar proteoglycans increases with applied load on mitral valve chordae tendineae. J. Biomech. **40**, 390–398 (2007)
27. Robinson, P.S., Huang, T.F., Kazam, E., Iozzo, R., Birk, D.E., Soslowsky, L.J.: Influence of decorin and biglycan on mechanical properties of multiple tendons in knockout mice. J. Biomech. Eng. **127**, 181–185 (2005)
28. Sasaki, N., Odajima, S.: Elongation mechanism of collagen fibrils and force-strain relations of tendon at each level of the structural hierarchy. J. Biomech. **29**, 1131–1136 (1996)
29. Fessel, G., Snedeker, J.G.: Equivalent stiffness after glycosaminoglycan depletion in tendon-an ultra-structural finite element model and corresponding experiments. J. Theor. Biol. **268**, 77–83 (2011)
30. Redaelli, A., Vesentini, S., Soncini, M., Vena, P., Mantero, S., Montevecchi, F.M.: Possible role of decorin glycosaminoglycans in fibril to fibril force transfer in relative mature tendons-a computational study from molecular to microstructural level. J. Biomech. **36**, 1555–1569 (2003)
31. Vesentini, S., Redaelli, A., Montevecchi, F.M.: Estimation of the binding force of the collagen molecule-decorin core protein complex in collagen fibril. J. Biomech. **38**, 433–443 (2005)
32. Rigozzi, S., Mueller, R., Snedeker, J.: Local strain measurement reveals a varied regional dependence of tensile tendon mechanics on glycosaminoglycan content. J. Biomech. **42**, 1547–1552 (2009)
33. Rigozzi, S., Mueller, R., Snedeker, J.: Collagen fibril morphology and mechanical properties of the achilles tendon in two inbred mouse strains. J. Anat. **216**, 724–731 (2010)
34. Berry, C.L., Greenwald, S.E.: Effect of hypertension on the static mechanical properties and chemical composition of the rat aorta. Cardiovasc. Res. **10**, 437–451 (1976)
35. Tamburro, A.M., DeStradis, A., D'Alessio, L.: Fractal aspects of elastin supramolecular structure. J. Biomol. Struct. Dyn. **12**, 1161–1172 (1995)
36. Cleary, E.G.: The microfibrillar component of the elastic fibers. morphology and biochemistry. In: Uitto, J., Perejda A.J. (eds.) Connective Tissue Disease. Molecular Pathology of the Extracellular Matrix, vol. 12, pp. 55–81. Dekker, New York (1987)
37. R. Ross, R., Bornstein, P.: The elastic fiber: I. the separation and partial characterization of its macromolecular components. J. Cell Biol. **40**, 366–381 (1969)
38. Alberts, B., Bray, D., Lewis, J., Raff, M., Roberts, K., Watson, J.: Molecular Biology of the Cell. Garland Publishing, New York (1994)

39. Werb, Z., Banda, M.J., McKerrow, J.H., Sandhaus, R.A.: Elastases and elastin degradation. J. Invest. Dermatol. **79**, 154–159 (1982)
40. Vrhovski, B., Weiss, A.: Biochemistry of tropoelastin. Eur. J. Biochem. **258**, 1–18 (1998)
41. Soskel, N.T., B.Sandberg, L.: Pulmonary emphysema. from animal models to human diseases. In: Uitto, J., Perejda, A.J. (eds.) Connective tissue disease. Molecular pathology of the extracellular matrix, vol. 12, pp. 423 – 453. Dekker, New York (1987)
42. Fung, Y.C.: Biomechanics. Motion, Flow Stress, and Growth. Springer-Verlag, New York (1990)
43. Westerhof, N., Lankhaar, J.W., Westerhof, B.E.: The arterial windkessel. Med. & Biol. Eng. & Comput. **47**, 131–141 (2009)
44. Guo, X., Kassab, G.: Variation of mechanical properties along the length of the aorta. Am. J. Physiol. Heart Circ. Physiol. **285**, H2614–H2622 (2003)
45. Tanaka, T.T., Fung, Y.C.: Elastic and inelastic properties of the canine aorta and their variation along the aortic tree. J. Biomech. **7**, 357–370 (1974)
46. Sonesson, B., Hansen, F., Stale, H., Länne, T.: Compliance and diameter in the huma abdomial aorta - The influence of age and sex. Eur. J. Vasc. Surg. **7**, 690–697 (1993)
47. Astrand, H., Stålhand, J., Karlsson, J., Karlsson, M., Länne, B.S.B.T.: In vivo estimation of the contribution of elastin and collagen to the mechanical properties in the human abdominal aorta: effect of age and sex. J. Appl. Physiol. **110**, 176–187 (2011)
48. Horny, L., Adamek, T., Zitny, R.: Age-related changes in longitudinal prestress in human abdominal aorta. Archive Appl. Mech. **83**, 875–888 (2013)
49. Bailey, A., Paul, R., Knott, L.: Mechanism of maturation and ageing of collagen. Mech. Ageing Dev. **106**, 1–56 (1998)
50. Länne, T., Sonesson, B., Bergqvist, D., Bengtsson, H., Gustafsson, D.: Diameter and compliance in the male human abdominal aorta: influence of age and aortic aneurysm. Eur. J. Vasc. Surg. **6**, 178–184 (1992)
51. Choke, E., Cockerill, G., Wilson, W.R., Sayed, S., Dawson, J., Loftus, I., Thompson, M.M.: A review of biological factors implicated in Abdominal Aortic Aneurysm rupture. Eur. J. Vasc. Endovasc. Surg. **30**, 227–244 (2005)
52. Davis, M.J.: Aortic aneurysm formation: lessons from human studies and experimental models. Circulation **98**, 193–195 (1998)
53. Kazi, M., Thyberg, J., Religa, P., Roy, J., Eriksson, P., Hedin, U., Swedenborg, J.: Influence of intraluminal thrombus on structural and cellular composition of abdominal aortic aneurysm wall. J. Vasc. Surg. **38**, 1283–1292 (2003)
54. Michel, J.B., Martin-Ventura, J.L., Egido, J., Sakalihasan, N., Treska, V., Lindholt, J., Allaire, E., Thorsteinsdottir, U., Cockerill, G., Swedenborg, J.: Novel aspects of the pathogenesis of aneurysms of the abdominal aorta in humans. Cardiovasc. Res. **90**, 18–27 (2011)
55. Rizzo, R., McCarthy, W., Dixit, S., Lilly, M., Shively, V., Flinn, W., Yao, J.: Collagen types and matrix protein content in human abdominal aortic aneurysms. J. Vasc. Surg. **10**, 365–373 (1989)
56. Hans, S.S., Jareunpoon, O., Balasubramaniam, M., Zelenock, G.B.: Size and location of thrombus in intact and ruptured abdominal aortic aneurysms. J. Vasc. Surg. **41**, 584–588 (2005)
57. Gasser, T.C., Görgülü, G., Folkesson, M., Swedenborg, J.: Failure properties of intra-luminal thrombus in Abdominal Aortic Aneurysm under static and pulsating mechanical loads. J. Vasc. Surg. **48**, 179–188 (2008)
58. Vande Geest, J.P., Sacks, M.S., Vorp, D.A.: A planar biaxial constitutive relation for the luminal layer of intra-luminal thrombus in abdominal aortic aneurysms. J. Biomech. **39**, 2347–2354 (2006)
59. Adolph, R., Vorp, D.A., Steed, D.L., Webster, M.W., Kameneva, M.V., Watkins, S.C.: Cellular content and permeability of intraluminal thrombus in Abdominal Aortic Aneurysm. J. Vasc. Surg. **25**, 916–926 (1997)
60. Gasser, T., Martufi, G., Auer, M., Folkesson, M., Swedenborg, J.: Micro-mechanical characterization of intra-luminal thrombus tissue from abdominal aortic aneurysms. Ann. Biomed. Eng. **38**, 371–379 (2010)

61. Folkesson, M., Silveira, A., Eriksson, P., Swedenborg, J.: Protease activity in the multi-layered intra-luminal thrombus of abdominal aortic aneurysms. Atherosclerosis **218**, 294–299 (2011)

62. Swedenborg, J., Eriksson, P.: The intraluminal thrombus as a source of proteolytic activity. Ann. N.Y. Acad. Sci. **1085**, 133–138 (2006)

63. Vorp, D.A., Lee, P.C., Wang, D.H., Makaroun, M.S., Nemoto, E.M., Ogawa, S., Webster, M.W.: Association of intraluminal thrombus in abdominal aortic aneurysm with local hypoxia and wall weakening. J. Vasc. Surg. **34**, 291–299 (2001)

64. Erhart, P., Grond-Ginsbach, C., Hakimi, M., Lasitschka, F., Dihlmann, S., Böckler, D., Hyhlik-Dürr, A.: Finite element analysis of abdominal aortic aneurysms: predicted rupture risk correlates with aortic wall histology in individual patients. J. Endovas. Ther. **21**, 556–564 (2014)

65. Reeps, C., Maier, A., Pelisek, J., Hartl, F., Grabher-Maier, V., Wall, W., Essler, M., Eckstein, H.H., Gee, M.: Measuring and modeling patient-specific distributions of material properties in abdominal aortic wall. Biomech. Model. Mechanobio. **12**, 717–733 (2013)

66. Forsell, C., Gasser, T.C., Swedenborg, J., Roy, J.: The quasi-static failure properties of the abdominal aortic aneurysm wall estimated by a mixed experimental-numerical approach. Ann. Biomed. Eng. **11**, (2012). https://doi.org/10.1007/s10,439-012-0712-3

67. Forsell, C., Björck, H.M., Eriksson, P., Franco-Cereceda, A., Gasser, T.C.: Biomechanical properties of the thoracic aneurysmal wall; differences between bicuspid aortic valve (BAV) and tricuspid aortic valve (TAV) patients. Ann. Thorac Surg. **98**, 65–71 (2014)

68. Maier, A., Essler, M., Gee, M.W., Eckstein, H.H., Wall, W.A., Reeps, C.: Correlation of biomechanics to tissue reaction in aortic aneurysms assessed by finite elements and [18f]-fluorodeoxyglucose-pet/ct. Int. J. Numer. Meth. Biomed. Eng. **28**, 456–471 (2012)

69. Marque, V., Kieffer, P., Gayraud, B., Lartaud-Idjouadiene, I., Ramirez, F., Atkinson, J.: Aortic wall mechanics and composition in a transgenic mouse model of Marfan syndrome. Arterioscl. Thromb. and Vasc. Biol. **21**, 1184–1189 (2001)

70. Shim, C., Cho, I., Yang, W.I., Kang, M.K., Park, S., Ha, J.W., Jang, Y., Chung, N.: Central aortic stiffness and its association with ascending aorta dilation in subjects with a bicuspid aortic valve. J. Am. Soc. Echoradiogr. **24**, 847–852 (2011)

71. Vorp, D.A., Raghavan, M.L., Muluk, S.C., Makaroun, M.S., Steed, D.L., Shapiro, R., Webster, M.W.: Wall strength and stiffness of aneurysmal and nonaneurysmal abdominal aorta. Ann. N.Y. Acad. Sci. **800**, 274–276 (1996)

72. Vorp, D.A., Schiro, B.J., Ehrlich, M.P., Juvonen, T.S., Ergin, M.A., Griffith, B.P.: Effect of aneurysm on the tensile strength and biomechanical behavior of the ascending thoracic aorta. Ann. Thorac Surg. **800**, 1210–1214 (2003)

73. Adham, M., Gournier, J.P., Favre, J.P., De La Roche, E., Ducerf, C., Baulieux, J., Barral, X., Pouyet, M.: Mechanical characteristics of fresh and frozen human descending thoracic aorta. J. Surg. Res. **64**, 32–34 (1996)

74. Mohan, D., Melvin, J.W.: Failure properties of passive human aortic tissue I - uniaxial tension tests. J. Biomech. **15**, 887–902 (1982)

75. Romo, A., Badel, P., Duprey, A., Favre, J.P., Avril, S.: In vitro analysis of localized aneurysm rupture. J. Biomech. **189**, 607–616 (2014)

76. O'Leary, S., Mulvihill, J., Barrett, H., Kavanagh, E., Walsh, M., McGloughlin, T., Doyle, B.: Determining the influence of calcification on the failure properties of abdominal aortic aneurysm (AAA) tissue. J. Mech. Behav. Biomed. Mater. **42**, 154–167 (2015)

77. O'Leary, S., Healy, D., Kavanagh, E., Walsh, M., McGloughlin, T., Doyle, B.: The biaxial biomechanical behavior of abdominal aortic aneurysm tissue. Ann. Biomed. Eng. **42**, 2440–2450 (2014)

78. Vande Geest, J.P., Wang, D.H.J., Wisniewski, S.R., Makaroun, M.S., Vorp, D.A.: Towards a noninvasive method for determination of patient-specific wall strength distribution in abdominal aortic aneurysms. Ann. Biomed. Eng. **34**, 1098–1106 (2006)

79. Vande Geest, J.P., Sacks, M.S., Vorp, D.A.: The effects of aneurysm on the biaxial mechanical behavior of human abdominal aorta. J. Biomech. **39**, 1324–1334 (2006)

80. Monteiro, J.T., da Silva, E., Raghavan, M., Puech-Leão, P., de Lourdes Higuchi, M., Otoch, J.: Histologic, histochemical, and biomechanical properties of fragments isolated from the anterior wall of abdominal aortic aneurysms. J. Vasc. Surg. **59**, 1393–1401 (2013)

81. Vande Geest, J.P., Dillavou, E.D., DiMartino, E.S., Oberdier, M., Bohra, A., Makaroun, M.S., Vorp, D.A.: Gender-related differences in the tensile strength of Abdominal Aortic Aneurysm. Ann. N.Y. Acad. Sci. **1085**, 400–402 (2006)
82. Raghavan, M.L., Kratzberg, J., de Tolosa, E.C., Hanaoka, M., Walker, P., da Silva, E.: Regional distribution of wall thickness and failure properties of human abdominal aortic aneurysm. J. Biomech. **39**, 3010–3016 (2006)
83. Xiong, J., Wang, S., Zhou, W., Wu, J.: Measurement and analysis of ultimate mechanical properties, stress-strain curve fit, and elastic modulus formula of human abdominal aortic aneurysm and nonaneurysmal abdominal aorta. J. Vasc. Surg. **48**, 189–195 (2008)
84. Thubrikar, M., Labrosse, M., Robicsek, F., Al-Soudi, J., Fowler, B.: Mechanical properties of abdominal aortic aneurysm wall. J. Med. Eng. Technol. **25**, 133–142 (2001)
85. DiMartino, E.S., Bohra, A., Geest, J.P.V., Gupta, N., Makaroun, M.S., Vorp, D.A.: Biomechanical properties of ruptured versus electively repaired abdominal aortic aneurysm wall tissue. J. Vasc. Surg. **43**, 570–576 (2006)
86. Raghavan, M., Hanaoka, M., Kratzberg, J., de Lourdes Higuchi, M., da Silva, E.: Biomechanical failure properties and microstructural content of ruptured and unruptured abdominal aortic aneurysms. J. Biomech. **44**, 2501–2507 (2011)
87. Raghavan, M.L., Webster, M.W., Vorp, D.A.: Ex vivo biomechanical behavior of abdominal aortic aneurysm: assesment using a new mathematical model. Ann. Biomed. Eng. **24**, 573–582 (1996)
88. Pichamuthu, J., Phillippi, J., Cleary, D., Chew, D., Hempel, J., Vorp, D., Gleason, T.: Differential tensile strength and collagen composition in ascending aortic aneurysms by aortic valve phenotype. Ann. Thorac Surg. **96**, 2147–2154 (2013)
89. Lindeman, J.H.N., Ashcroft, B.A., Beenakker, J.W.M., van Es, M., Koekkoek, N.B.R., Prins, F.A., Tielemans, J.F., Abdul-Hussien, H., Bank, R.A., Oosterkamp., T.H.: Distinct defects in collagen microarchitecture underlie vessel-wall failure in advanced abdominal aneurysms and aneurysms in Marfan syndrome. Proc. Natl. Acad. Sci. USA **107**, 862–865 (2009)
90. Vidal, B.C., Mello, M.L.S., Pimentel, E.R.: Polarization microscopy and microspectrophotometry of sirius red, picrosirius and chlorantine fast red aggregates and of their complexes with collagen. Histochem. J. **14**, 857–878 (1982)
91. Weber, K.T., Pick, R., Silver, M.A., Moe, G.W., Janicki, J.S., Zucker, I.H., Armstrong, P.W.: Fibrillar collagen and remodeling of dilated canine left ventricle. Circulation **82**, 1387–1401 (1990)
92. Canham, P.B., Finlay, H.M.: Morphometry of medial gaps of human brain artery branches. Stroke **35**, 1153–1157 (2004)
93. Canham, P.B., Finlay, H.M., Dixon, J.G., Boughner, D.R., Chen, A.: Measurements from light and polarised light microscopy of human coronary arteries fixed at distending pressure. Cardiovasc. Res. **23**, 973–982 (1989)
94. Polzer, S., Gasser, T., Forsell, C., Druckmullerova, H., Tichy, M., Vlachovsky, R., Bursa, J.: Automatic identification and validation of planar collagen organization in the aorta wall with application to abdominal aortic aneurysm. Micros. Microanal. **19**, 1395–1404 (2013)
95. Alastrué, V., Saez, P., Martínez, M.A., Doblaré, M.: On the use of the bingham statistical distribution in microsphere-based constitutive models for arterial tissue. Mech. Res. Commun. **37**, 700–706 (2010)
96. Bingham, C.: An antipodally symmetric distribution on the sphere. Ann. Statist. **2**, 1201–1225 (1974)
97. Chuong, C.J., Fung, Y.C.: Three-dimensional stress distribution in arteries. J. Biomed. Eng. **105**, 268–274 (1983)
98. Fung, Y.C., Fronek, K., Patitucci, P.: Pseudoelasticity of arteries and the choice of its mathematical expression. Am. J. Physiol. **237**, H620–H631 (1979)
99. Humphrey, J.D., Strumpf, R.K., Yin, F.C.P.: Determination of constitutive relation for passive myocardium - Part I and II. J. Biomech. Eng. **112**, 333–346 (1990)
100. Takamizawa, K., Hayashi, K.: Strain energy density function and uniform strain hypothesis for arterial mechanics. J. Biomech. **20**, 7–17 (1987)

101. Vaishnav, R.N., Young, J.T., Janicki, J.S., Patel, D.J.: Nonlinear anisotropic elastic properties of the canine aorta. Biophys. J. **12**, 1008–1027 (1972)
102. Gasser, T.C.: An irreversible constitutive model for fibrous soft biological tissue: a 3d microfiber approach with demonstrative application to Abdominal Aortic Aneurysms. Acta Biomaterialia **7**, 2457–2466 (2011)
103. Holzapfel, G.A., Gasser, T.C., Ogden, R.W.: A new constitutive framework for arterial wall mechanics and a comparative study of material models. J. Elasticity **61**, 1–48 (2000)
104. Martufi, G., Gasser, T.C.: A constitutive model for vascular tissue that integrates fibril, fiber and continuum levels. J. Biomech. **44**, 2544–2550 (2011)
105. Pena, J., Martínez, M.A., Pena, E.: A formulation to model the nonlinear viscoelastic properties of the vascular tissue. Acta Mech. **217**, 63–74 (2011)
106. Wuyts, F.L., Vanhuyse, V.J., Langewouters, G.J., Decraemer, W.F., Raman, E.R., Buyle, S.: Elastic properties of human aortas in relation to age and atherosclerosis: a structural model. Phys. Med. Biol. **40**, 1577–1597 (1995)
107. Zulliger, M.A., Fridez, P., Hayashi, K., Stergiopulos, N.: A strain energy function for arteries accounting for wall composition and structure. J. Biomech. **37**, 989–1000 (2004)
108. Hardin, R.H., Sloane, N.J.A.: McLaren's improved snub cube and other new spherical designs in three dimensions. Discrete Comput. Geom. **15**, 429–441 (1996)
109. Federico, S., Gasser, T.C.: Non-linear elasticity of biological tissues with statistical fibre orientation. J. R. Soc. Interface **7**, 955–966 (2010)
110. Kotz, S., vanDorp, J.: Beyond beta: other continuous families of distributions with bounded support. World Scientific (2004)
111. Krasny, W., Magoariec, H., Morin, C., Avril, S.: Kinematics of collagen fibers in carotid arteries under tension-inflation loading. J. Mech. Behav. Biomed. Mater. **77**, 718–726 (2018)
112. Hamedzadeh, A., Gasser, T., Federico, S.: On the constitutive modelling of recruitment and damage of collagen fibres in soft biological tissues. Eur. J. Mech. A/Solids **72**, 483–496 (2018)
113. Miyazaki, H., Hayashi, K.: Tensile tests of collagen fibers obtained from the rabbit patellar tendon. Biomed. Microdevices **2**, 151–157 (1999)
114. Shen, Z.L., Dodge, M.R., Kahn, H., Ballarini, R., Eppell, S.J.: Stress-strain experiments on individual collagen fibrils. Biophys. J. **95**, 3956–3963 (2008)
115. Polzer, S., Gasser, T., Novak, K., Man, V., Tichy, M., Skacel, P., Bursa, J.: Structure-based constitutive model can accurately predict planar biaxial properties of aortic wall tissue. Acta Biomater. **14**, 133–145 (2015)
116. Moll, F.L., Powell, J.T., Fraedrich, G., Verzini, F., Haulon, S., Waltham, M., van Herwaarden, J.A., Holt, P.J.E., van Keulen, J.W., Rantner, B., Schloesser, F.J.V., Setacci, F., Rica, J.B.: Management of abdominal aortic aneurysms clinical practice guidelines of the European society for vascular surgery. Eur. J. Vasc. Endovasc Surg. **41**, S1–S58 (2011)
117. Brady, A., Thompson, S., Fowkes, F., Greenhalgh, R., Powell, J.: Uk small aneurysm trial participants abdominal aortic aneurysm expansion: risk factors and time intervals for surveillance. Circulation pp. 16–21 (2004)
118. Martufi, G., Auer, M., Roy, J., Swedenborg, J., Sakalihasan, N., Panuccio, G., Gasser, T.C.: Multidimensional growth measurements of abdominal aortic aneurysms. J. Vasc. Surg. **58**, 748–755 (2013)
119. Martufi, G., Liljeqvist, M.L., Sakalihasan, N., Panuccio, G., Hultgren, R., Roy, J., Gasser, T.: Local diameter, wall stress and thrombus thickness influence the local growth of abdominal aortic aneurysms. J. Endovas. Ther. **23**, 957–966 (2016)
120. Gasser, T., Grytsan, A.: Biomechanical modeling the adaptation of soft biological tissue. Cur. Opinion Biomed. Engrg. **1**, 71–77 (2017)
121. Grytsan, A., Eriksson, T., Watton, P., Gasser, T.: Growth description for vessel wall adaptation: a thick-walled mixture model of abdominal aortic aneurysm evolution. Materials **10**, 994 (2017)
122. Kroon, M., Holzapfel, G.: A theoretical model for fibroblast-controlled growth of saccular cerebral aneurysms. J. Theor. Biol. **257**, 73–83 (2009)

123. Volokh, K., Vorp, D.A.: A model of growth and rupture of abdominal aortic aneurysm. J. Biomech. **41**, 1015–1021 (2008)
124. Watton, P.N., Hill, N.A.: Evolving mechanical properties of a model of abdominal aortic aneurysm. Biomech. Model. Mechanobio. **8**, 25–42 (2009)
125. Wilson, J.S., Baek, S., Humphrey, J.D.: Importance of initial aortic properties on the evolving regional anisotropy, stiffness and wall thickness of human abdominal aortic aneurysms. J. R. Soc. Interface **9**, 2047 (2012)
126. Zeinali-Davarani, S., Baek, S.: Medical image-based simulation of abdominal aortic aneurysm growth. Mech. Res. Commun. **42**, 107–117 (2012)
127. Martufi, G., Gasser, T.C.: Turnover of fibrillar collagen in soft biological tissue with application to the expansion of abdominal aortic aneurysms. J. R. Soc. Interface **9**, 3366–3377 (2012)
128. Loerakker, S., Obbink-Huizer, C., Baaijens, F.: A physically motivated constitutive model for cell-mediated compaction and collagen remodeling in soft tissues. Biomech. Model. Mechanobio. **13**, 985–1001 (2014)
129. Silver, F.H., Freeman, J.W., Seehra, G.P.: Collagen self-assembly and the development of tendon mechanical properties. J. Biomech. **36**, 1529–1553 (2003)
130. McBridge, D.: Hind limb extensor tendon development in the chick: a light and transmission electron microscopic study. Master's thesis, Rutgers University, New Jersey (1984)
131. Taylor, R.L.: FEAP – A Finite Element Analysis Program, Version 8.2 User Manual. University of California at Berkeley, Berkeley, California (2007)
132. Auer, M., Gasser, T.C.: Reconstruction and finite element mesh generation of abdominal aortic aneurysms from computerized tomography angiography data with minimal user interaction. IEEE T. Med. Imaging **29**, 1022–1028 (2010)
133. Gasser, T.: Patient-specific computational modeling. Lecture Notes in Computational Vision and Biomechanics, chap. Bringing Vascular Biomechanics into Clinical Practice. Simulation-Based Decisions for Elective Abdominal Aortic Aneurysms Repair, pp. 1–37. Springer (2012)
134. Gasser, T.C., Auer, M., Labruto, F., Swedenborg, J., Roy, J.: Biomechanical rupture risk assessment of Abdominal Aortic Aneurysms. Model complexity versus predictability of finite element simulations. Eur. J. Vasc. Endovasc. Surg. **40**, 176–185 (2010)
135. Martufi, G., Gasser, T.C.: Review: The role of biomechanical modeling in the rupture risk assessment for abdominal aortic aneurysms. J. Biomed. Eng. **135**, 021,010 (2013)
136. Gupta, H.S., Seto, J., Krauss, S., Boesecke, P., Screen, H.R.C.: In situ multi-level analysis of viscoelastic deformation mechanisms in tendon collagen. J. Struct. Biol. **169**, 1183–1191 (2010)
137. Chandran, P., Barocas, V.: Affine versus non-affine fibril kinematics in collagen networks: theoretical studies of network behavior. J. Biomed. Eng. **128**, 259–270 (2006)
138. Krasny, W., Morin, C., Magoariec, H., Avril, S.: A comprehensive study of layer-specific morphological changes in the microstructure of carotid arteries under uniaxial load. Acta Biomater. **57**, 342–351 (2017)
139. Taylor, C., Humphrey, J.: Open problems in computational vascular biomechanics: hemodynamics and arterial wall mechanics. Comput. Meth. Appl. Mech. Eng. **198**, 3514–3523 (2009)
140. Biasetti, J., Gasser, T.C., Auer, M., Hedin, U., Labruto, F.: Hemodynamics conditions of the normal aorta compared to fusiform and saccular abdominal aortic aneurysms with emphasize on thrombus formation. Ann. Biomed. Eng. **38**, 380–390 (2009)
141. Biasetti, J., Hussain, F., Gasser, T.C.: Blood flow and coherent vortices in the normal and aneurysmatic aortas. A fluid dynamical approach to Intra-Luminal Thrombus formation. J. R. Soc. Interface **8**, 1449–1461 (2011)
142. Biasetti, J., Spazzini, P., Gasser, T.C.: An integrated fluido-chemical model towards modeling the formation of intra-luminal thrombus in abdominal aortic aneurysms. Front. Physiol. **3**, article 266 (2011)

Cellular and Extracellular Homeostasis in Fluctuating Mechanical Environments

Béla Suki, Harikrishnan Parameswaran, Calebe Alves, Ascânio D. Araújo and Erzsébet Bartolák-Suki

Abstract Homeostasis is considered to be a cellular feedback mechanism that maintains a target such as mean blood pressure at a well-defined level, which is called upward causation. In this chapter, we consider a downward causation in which higher level properties are not only sensed by cells, but they also cause changes in cellular level behavior. Specifically, we examine how fluctuations at the level of the target, such as beat-to-beat blood pressure variability, present themselves as boundary conditions at the level of the cell and how this affects cellular behavior such cytoskeletal structure or homeostasis of bioenergetics. The changes in low level mechanisms may feed back to the regulation of the target until some homeostasis is achieved in which the system is under far from equilibrium conditions but without ever reaching a steady state. Consequently, the target is allowed to fluctuate within homeostatic limits. We will first summarize current concepts in conventional mechanotransduction and then discuss the dynamic aspects of mechanotransduction in the presence of noisy mechanical inputs, called fluctuation-driven mechanotransduction (FDM). Next, we will demonstrate how FDM alters cytoskeletal organization using a computational model of the actin-myosin network under time-varying strain boundary conditions in which peak strains vary from cycle to cycle. The simulation results imply that FDM should be considered as an emergent multiscale network phenomenon. In vascular smooth muscle cells, cytoskeletal structural alterations also lead to mitochondrial

B. Suki (✉) · E. Bartolák-Suki
Department of Biomedical Engineering, Boston University, Boston, MA, USA
e-mail: bsuki@bu.edu

E. Bartolák-Suki
e-mail: ebartola@bu.edu

H. Parameswaran
Department of Biomedical Engineering, Northeastern University, Boston, MA, USA
e-mail: h.parameswaran@northeastern.edu

C. Alves · A. D. Araújo
Departamento de Física, Universidade Federal do Ceará, Fortaleza, CE 60451-970, Brazil
e-mail: calebe.alves@fisica.ufc.br

A. D. Araújo
e-mail: ascanio@fisica.ufc.br

© Springer Nature Switzerland AG 2020
Y. Zhang (ed.), *Multi-scale Extracellular Matrix Mechanics and Mechanobiology*,
Studies in Mechanobiology, Tissue Engineering and Biomaterials 23,
https://doi.org/10.1007/978-3-030-20182-1_4

remodeling with subsequent changes in ATP production. The latter affects cell contractility as well as bioenergetics which in turn feed back to collagen maintenance, vascular wall stiffness and ultimately blood pressure regulation. We argue that FDM is a general phenomenon that other cell types also exhibit such as enhanced surfactant production by lung epithelial cells due to tidal volume variability. Following some speculation on the possible roles of fluctuations in diseases and aging, we will offer a general picture of how the breakdown of FDM disturbs homeostasis which can lead to the pathogenesis of diseases.

1 Introduction

The idea of homeostasis, loosely stated as the ability of organisms to exhibit long-term stability in an ever-changing environment, has long been recognized as a driving force of life. Perhaps the oldest known expression of homeostasis was due to Hippocrates who said that "Natural forces within us are the true healers of disease". The term homeostasis, however, was only coined in 1926 by Cannon [1] who defined it as follows: "The highly developed living being is an open system having many relations to its surroundings ... Changes in the surroundings excite reactions in this system, or affect it directly, so that internal disturbances of the system are produced. Such disturbances are normally kept within narrow limits, because automatic adjustments within the system are brought into action and thereby wide oscillations are prevented and the internal conditions are held fairly constant." Classic examples can be found in the vascular literature, which has mostly focused on blood pressure, salt and electrolyte homeostasis [2–4].

More recently, the possibility of a mechanical homeostasis began to emerge based on the discovery that endothelial cells feel and respond to blood flow-induced shear stresses in a way to control vessel caliber by inducing vascular smooth muscle cell (VSMC) relaxation or contraction [5]. Using a previously developed optimal relationship between vessel diameter and flow, now called Murray's law [6], Zamir proposed [7] that during laminar flow a linear relation between blood flow and the cube of vessel diameter can result in a constant shear stress. This constancy of the shear stress may in turn be the physiological target of vascular mechanical homeostasis. The mechanisms driving such mechanical homeostasis may involve the full repertoire of mechanotransduction, the conversion of mechanical stimulus to biochemical signals [8, 9], including mechanosensing by surface receptors, intracellular signaling and changes in gene expression [10]. These findings led to the notion of a multiscale version of mechanical homeostasis that attempts to maintain a preferred mechanical state spanning scales from subcellular structures through cells to tissues and organs [11]. Interestingly, in isolated endothelial cells, tensional homeostasis is maintained only when cells organize themselves into multicellular objects [12]; this breakdown of mechanical homeostasis, however, appears to be limited to endothelial cells [13].

More generally, cells in the body are continuously exposed to various mechanical factors including osmotic stress, external pressure, shear stress and tensile stress. For homeostasis to function, deviation from the target value must be tracked and hence, cells must also serve as mechanical sensors. Indeed, cilia in kidney [14], bone [15], cartilage [16] and eye [17] cells are sensitive to pressure. During movement, muscles contract and exert stresses on themselves [18] as well as on nerves [19]. As the heart ejects blood into the circulation, the propagating blood pressure (BP) waves generate shear stresses on endothelial cells [20] and circumferential stresses on VSMCs [21]. During breathing, contraction of the respiratory muscles produce rhythmic stretch of the lung that is transmitted along the extracellular matrix (ECM), including mostly collagen and elastin [22], to all resident cells of the lung [23]. In all these cases, cells respond by altering signaling, and eventually changing the production of some factors to restore homeostasis. For example, a critical response to stretching the lung is the release of surfactant by type II epithelial cells at the air-liquid interface [24] which in turn increases lung compliance to reduce the work of breathing, as part of the mechanical homeostasis in the lung.

Homeostasis, however, should not be conceptualized as a regulatory mechanism that operates similar to a household thermostat whose sole function is to minimize fluctuations and maintain temperature at a well-defined set point. Indeed, Zamir raised the possibility that certain variations in shear stress on the endothelium should be acceptable [7]. As Torday argued, homeostasis is everything but static and if viewed through the lens of multiple generations across large time scales, homeostasis is an essential component of biological evolution [25]. This idea then brings forth the necessity that a target physiological parameter must be controlled within allowable limits over both short and long time scales. As an example, consider the cardiovascular system. While the mean heart rate is ~70/min in healthy adults, normal values can fluctuate between about 50 and 115 corresponding to sleep and daily activity, respectively [26]. What came as a surprise, however, is the fact that cardiovascular health is associated with long term correlated fluctuations in heart rate while too much irregularity, as in fibrillation, or too much regularity, such as mode locking into a highly periodic behavior, signify severe pathology [27, 28]. Just as heart rate, BP must also be under dynamic homeostatic control because blood pressure variability (BPV) is significant even in health [29]. Similarly, tidal volume (TV) even during quiet breathing exhibits substantial variations [30]. The question is how the complex time varying mechanical stresses due for example to BPV or tidal volume variability (TVV) affect the ability of cells to maintain dynamic mechanical homeostasis.

Besides the magnitude of mechanical stresses, both their exposure time and frequency composition contribute to the actual signaling response [31]. The dynamic effects of time varying mechanical stimuli (e.g., BPV or TVV) on mechanotransduction are rarely examined in the laboratory. Instead, mechanotransduction is studied using static or cyclic but monotonous stretch (MS) patterns. In a recent study, we presented evidence that fluctuations in peak cycle-by-cycle strain, called variable stretch (VS), applied to VSMCs fundamentally alter cellular bioenergetics, cytoskeletal organization and signaling [32]. In this chapter, we will describe how fluctuations in mechanical stimuli regulate mechanotransduction in the context of homeostasis. While homeostasis is invariably considered to be a cellular feedback mechanism that maintains a target such as BP, here we also consider a downward causation in which higher level properties are not only sensed by cells, but they also cause changes in cellular level behavior [33]. Specifically, we will examine how fluctuations at the level of the target, namely BPV or TVV, present themselves as boundary conditions at the level of the cell and how this affects cellular behavior such cytoskeletal structure or homeostasis of bioenergetics. The changes in low level mechanisms may feed back to the regulation of the target until some homeostasis is achieved in which the system is under non-equilibrium conditions but without ever reaching a steady state. Consequently, the target is allowed to fluctuate within homeostatic limits, a phenomenon that has been termed homeokinesis [34]. We will first summarize current concepts in conventional mechanotransduction and then discuss the dynamic aspects of mechanotransduction in the presence of noisy mechanical inputs, called fluctuation-driven mechanotransduction (FDM), closely following our previous review [35]. Next, we will demonstrate how FDM alters cytoskeletal organization using a computational model of the actin-myosin network during VS in which peak strains vary from cycle to cycle while maintaining the same mean strain as MS. The simulation results imply that FDM should be considered as an emergent multiscale network phenomenon. Cytoskeletal structural alterations can affect VSMC contractility as well as bioenergetics which feeds back to collagen maintenance, vascular wall stiffness and ultimately BP. Following some speculation on the possible roles of fluctuations in diseases and aging, we will offer a general picture of how the breakdown of FDM disturbs homeostasis which can lead to the pathogenesis of diseases.

2 Conventional Mechanotransduction

Mechanotransduction plays fundamental roles in the genesis and maintenance of normal tissue structure and function and hence tissue homeostasis [8]. Being a complex signaling process, mechanotransduction influences many cell functions such as migration, contraction and production of ECM molecules [31, 36–43]. Cells interact with the ECM through adhesion receptors such as integrins which in turn structurally and functionally regulate cell membrane proteins and their linkages to the cytoskeleton including actin microfilaments, microtubules and intermediate filaments. These events involve multi-molecular conformational changes, biochemical reactions as well as whole organelle responses such as the reorganization of the cytoskeleton, organelle biogenesis or changes in organelle structure and function.

A key element in mechanotransduction is the coupling of the cell to the ECM by multi-molecular complexes, the focal adhesions (FAs) which serve as dynamic biochemical signaling hubs capable of regulating signaling proteins, adhesion receptors and their binding to the ECM as well as receptors on other cells [44–47]. Among these, the bidirectional integrin mediated processes have been studied extensively [48]. Mechanical signals arriving from the ECM to integrins reorganize FAs, which are also regulated by intracellular biochemical events. Nearly all cells express integrins with cell type specificity that is often further regulated by structural and mechanical features of the cell's microenvironment. Integrins form heterodimers which consist of α and β subunits. Ligands such as the RGD (Arg-Gly-Asp) sequence along collagen, fibronectin or laminin bind to integrins depending on subunit combination [36, 49, 50]. Following ligand binding, integrin subunits form domains and clusters [51] and undergo conformational changes from a low to a high affinity state [52]. G-protein coupled receptors and Ras-related small GTPases and their effectors modulate integrin binding affinity [52, 53]. The external [54] and internal [55] integrin domains interact with other proteins which regulate the bidirectional signaling events. For example, talin and α-actinin bind to the intracellular domains connecting thus integrins to vinculin, paxillin and tensin as well as FA kinase (FAK) which, via tyrosine phosphorylation, regulate the assembly of actin filaments [56]. Following integrin-mediated cell adhesion to the ECM, FAK is rapidly phosphorylated to recruit Src and phosphorylate FAK, which then regulate FA assembly/disassembly with additional proteins such as Rho and mitogen-activated protein kinase (MAPK). The MAPK can also be activated by integrins without FAK. After integrin activation, MAPK translocates to the nucleus to regulate transcription factors [57] leading to fast expression of vinculin, actin, α-actinin and β1 integrin [58]. Of note is that MAPK can activate myosin light chain kinase (MLCK) and hence influence cell contractility and motility [59]. Furthermore, MAPK can also physically associate with microtubules and alter their assembly [60]. The other major signaling pathway is through the Rho family of GTPases which regulates actin stress fiber formation and anchorage for integrin signaling [61]. Similar to MAPK, the Rho kinase can also phosphorylate MLC which affects cell contractility [62].

3   The Concept of Fluctuation-Driven Mechanotransduction

3.1   Physiological Variability in Stresses and Strains

The above mechanisms related to conventional mechanotransduction mostly considered the type of proteins and their interactions, not the types of mechanical stimulus. In other words, information on signaling was obtained under static or cyclic but monotonous stimuli. However, mechanical stresses and strains exhibit fluctuations throughout the body in an organ and tissue specific manner. The mechanical conditions at the organ level result in microscopic level stresses and strains acting on cells. The actual micro-strains depend on the structural organization of the ECM and the distribution of stiffness at different length scales. For example, tidal breathing generates regional mechanical stresses in the lung parenchyma which in turn stretch the alveolar septal walls straightening up collagen fibrils [22]. This chain of force transmission eventually determines part of the time-varying boundary conditions which cells are exposed to and respond by mechanotransduction.

In the lung, mechanical fluctuations are due to breath-to-breath variability of tidal volume [30] as well as regional variations in tissue stiffness. Tidal stretches are superimposed on a pre-existing stress corresponding to functional residual capacity. Additionally, gravity generates a gradient in both the pre-stress and regional stretch due to tidal breathing which results in a larger static stretch and a smaller dynamic tidal stretch in the upper regions of the lung. At the macroscale, the breath-to-breath coefficient of variation (CV) of regional tidal volume varies between 0.28 and 0.39 [30]. The microscale linear strains vary considerably in the lung and are estimated to range between 0.1 and 0.4 when the lung is inflated from 40 to 80% of total lung capacity [63]. Epithelial cells covering the surface of the alveoli experience approximately 2-dimensional strains and undergo 10–15% cyclic area strain during breathing with a maximum of ~37% corresponding to total lung capacity [64]. Due to the thin membranous structure of alveoli, fibroblast in the septal walls should receive strains similar to those seen by epithelial cells. In vitro analysis suggests that airway smooth muscle cells experience ~5% circumferential strains during quiet breathing but deep inspirations can lead to ~25% strain in the larger and 40% strain in the smaller airways [65]. These values are generally consistent with in vivo imaging in rats, but the data also suggest substantial heterogeneities both in circumferential and axial strains [66].

In the vasculature, both heart rate and stroke volume show significant beat-to-beat variabilities [67] and so does blood pressure [29]. The range of fluctuations in circumferential strain can be estimated from BPV and the stress-strain curve of the vessel wall. Accordingly, VSMCs in the rat aorta experience a static strain of ~70% with the peak strain amplitudes distributed between 77.5 and 92.5% with frequencies ranging from 2 to 6 Hz [32]. In that in vitro study, the strains and frequencies were chosen on a cycle-by-cycle basis that maintained a constant strain rate. However, the in vivo strain rate is not constant due to natural heart rate variability which follows a long-range correlated time series with a non-Gaussian distribution [68]. Furthermore, heart rate variability is altered by vascular disease [69]. Shear stresses on the endothelium vary largely along the arterial tree [70–72] with peak shear stresses between about 7 and 18 dyn/cm^2 during a 12 h period [73]. While blood pressures are similar in magnitude across mammals, heart rate (f) varies between ~30 and ~600 beats per minute across the spectrum of mammals [74]. Furthermore, hear rate f_H is said to follow an allometric scaling relation with body mass M [75]: $f_H \sim M^{-0.25}$. Consequently, if wall strain depends less on M than f_H, strain rates in a mouse would be much higher than in an elephant. A similar analysis is likely valid for the lung [75]. It is currently not known how such different strain rates affect mechanotransduction in various species.

3.2 Effects of Fluctuating Boundary Conditions

Little is known about the mechanisms of how fluctuations influence mechanotransduction. Our recent study demonstrated in VSMCs that VS alters many cellular processes when compared to MS [32]. In the following discussion, we will consider several possible mechanisms that could lead to different signaling during MS and VS. Since cells adhere to collagen or fibronectin through integrins, stretching of ECM fibers will be sensed by the integrins that are linked intracellularly to talin and, through vinculin, to the cytoskeletal protein network. Forces of different magnitudes may be sensed by multimodular proteins along the chain of mechanotransduction triggered by conformational changes and unfolding motifs [76]. Thus, mechanotransduction may be governed by switch-like mechanisms so that stronger and weaker forces can be detected by unfolding different protein domains. During MS, domain folding might undergo cyclic variations eventually leading to a non-equilibrium steady state. However, the occasionally larger than average forces during VS can unfold different protein domains. If this happens a sufficient number of times, VS will induce mechano-activation by phosphorylation and binding of a different set of signaling molecules than MS, which in turn is equivalent to exciting a different signaling pathway. While VS also maintains non-equilibrium conditions, cells and the stretch-sensitive subcellular structures may never reach a true steady state because the incessant fluctuations always kick the molecular arrangement out of the steady state. Furthermore, because the average time for a motif to be in the unfolded state is expected to be different during VS than MS, biochemical signaling pathways with different time constants can be selectively triggered by FDM. There are additional mechanisms not based on switch-like conformational changes that may be involved in FDM. For example, since bond dissociation depends on the force on a bond [77], the binding and unbinding rates can also be affected by the amount of fluctuations in VS. A possible mechanism is the force-rate sensitivity of bond dissociation. Small amplitude fluctuations can trigger the release of slip bonds, while catch bonds are reinforced until a larger than average stretch arrives causing the catch bonds to cross the catch-slip threshold [78]. Taken together, the reinforcement of FAs by mechanical forces are expected to be different during MS and VS, which results in stretch pattern-dependent cytoskeletal mechanosensing via myosin motors (see below).

We can conceptually demonstrate why FDM can initiate different signaling pathways compared to conventional mechanotransduction. Consider a nonlinear viscoelastic Kelvin body characterized by a nonlinear spring in parallel with a linear spring connected in series with a linear dashpot (Fig. 1). The nonlinear spring has a force (F)-displacement (x) relation given by $F_1 = k_1 x + b x^2$ where k_1 and b are the linear and nonlinear spring constants, respectively. The linear spring is described by $F_2 = k x_k$ and the linear dashpot by $F_2 = r \dot{x}_r$ where k and r are the linear spring constant and the dashpot coefficient, respectively. The x_k and x_r denote the displacements of the linear spring and dashpot, respectively, and the dot over a variable represents differentiation with respect to time. The total displacement of the

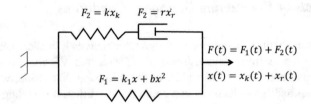

Fig. 1 A schematic diagram of the nonlinear Kelvin body. The system is driven by the input displacement x and the output is the total force F that is the sum of the forces F_1 in the lower branch and F_2 in the upper branch. The constitutive equation for each element is given above the element. The linear spring with constant k experiences a displacement x_k while the dashpot with coefficient r experiences a displacement x_r.

unit is $x = x_r + x_k$ while the total force is $F = F_1 + F_2$. After some algebraic manipulation, we can obtain the following nonlinear differential equation:

$$\dot{F} + \frac{k}{r}F = (k + k_1 + 2bx)\dot{x} + \frac{k}{r}(k_1 + bx)x \tag{1}$$

Notice that if $r \to \infty$ and $b = 0$, we obtain a linear elastic system whose solution is $F = (k_1 + k)x$ for which we can easily calculate the response to VS. We assume that x consists of N sine waves each with a different peak-to-peak amplitude x_i stitched together. We have shown that if $x_i = x_0 + u_i (i = 1, \ldots, N)$ where x_0 is the mean amplitude and u_i are drawn from a zero mean distribution, the total elastic energy E_{VS} delivered to the system after N cycles [35]:

$$E_{VS} = E_{MS} + \frac{1}{2}(k_1 + k)N\sigma^2 \tag{2}$$

where $E_{MS} = \frac{1}{2}N(k_1 + k)x_0^2$ is the total energy following N identical cycles during MS and σ^2 is the variance of the time series of u_i. Since both terms in Eq. 2 contain N, we can see that the average energy stored in the system per cycle (E_{VS}/N) during VS increases linearly with the variance of the fluctuations in the input stimulus. While this insight from linear analysis is useful, it is limited because cells and tissues are nonlinear and viscoelastic.

For the case of MS, or when $r \to \infty$ during VS, Eq. 1 can be solved analytically. However, when the input is a complex, fluctuating displacement signal, numerical solution of the general case is easier. For VS, the input displacement is a sequence of N sine waves of one period each stitched to the end of the previous cycle (Fig. 2a) with the ith segment defined as

$$x_i(t) = -(x_0 + u_i)\cos\{2\pi(f + \Delta f_i)t\} + x_s \tag{3}$$

Here, Δf_i is the random component of the frequency f chosen such that the strain rate is the same for all cycles and x_s is the static component of the displacement added when the effect of pre-stress is investigated. The results in Fig. 2b compare several solutions with different parameters. In the case of a linear elastic system (green, $b = 0$, $r = 10^8$), the force F is also sinusoidal with its amplitude and frequency varying from cycle to cycle exactly following the input displacement. When nonlinearity is added (black, $b = 0.5$), both the magnitude and the fluctuations of F increase. Allowing for linear viscoelasticity (blue, $r = 2$), slightly increases the fluctuations, but notice that F no longer returns to zero at the end of the cycle because the time constant of the system is comparable to the time period of oscillations. Finally, combining nonlinearity and viscoelasticity (red, $b = 0.5$, $r = 2$) further increases the fluctuations. The energy taken up is calculated by integrating the force over the displacement for only the loading parts and averaging over 1000 cycles. The energy due only to fluctuations, $\Delta E_{VS} = E_{VS} - E_{MS}$, corresponding to the 4 cases are compared as a function σ^2 in Fig. 2c. Notice that for the linear elastic case (black), this energy linearly increases with σ^2 in agreement with Eq. 2. For the nonlinear elastic case, the energy still linearly increases with σ^2; however, due to fluctuations, the curve is no longer smooth. Interestingly, adding viscoelasticity introduces a nonlinear relationship between ΔE_{VS} and σ^2 and the combination of nonlinearity and viscoelasticity leads to the highest energy uptake for large fluctuations.

The above results imply that fluctuations generally increase the energy taken up by cells the amount of which is modulated by nonlinearity and viscoelasticity. This energy surplus is then converted to chemical energy. One could argue that if we elevate the extension x during MS in Eq. 1 so that $E_{MS} = E_{VS}$, the difference between the two stretching modes should disappears. However, this is not necessarily the case. Consider for example a signaling protein that unfolds at an extension $x_i > x_0$. If $x_i < x_0 + \sigma$, then VS can occasionally unfold the protein, while MS won't be able to do that. Gradually increasing x_0 in MS so that $E_{MS} = E_{VS}$, with MS now having a different x_0 than VS, MS can be made to unfold the same protein. However, MS will unfold the protein in every cycle while VS only occasionally achieves that despite the total energy delivered by the two stretching modes being the same. As a consequence, during VS, the fraction of time the protein is in the phosphorylated state is different than during MS. If the boundary conditions are force or stress, the analysis is slightly different, but it remains true that MS and VS would take up different amounts of energy from the environment. An interesting implication is that by varying σ^2, different signaling pathways may be selected. Nonlinearity and viscoelasticity can also amplify the difference between MS and VS in terms of the elastic energy uptake by the cell (Fig. 2c). Therefore, due either to the surplus energy or the different phosphorylation state of the cell, VS should result in different signaling mechanisms eventually leading to altered gene and protein expressions compared to MS [79].

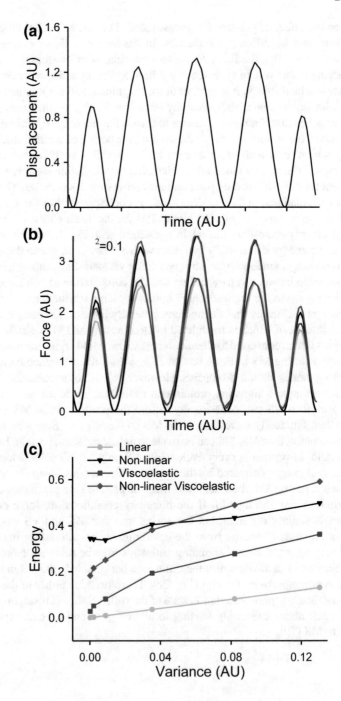

◀Fig. 2 Numerical solutions of the nonlinear Kelvin body in Fig. 1. **a** A short segment of the time varying input displacement composed of single sine waves of different amplitudes and frequencies stitched together to form a time series. **b** Comparison of the solutions for the linear, non-linear, linear viscoelastic and nonlinear viscoelastic cases. The variance σ^2 of the peak amplitude is 0.1. Note that for the elastic systems, the force always returns to zero (green and black) when the displacement is zero, whereas in the presence of viscous losses (blue and red), the force does not reach zero especially after a small-amplitude faster cycle. **c** The extra energy taken up by each model from panel (**b**) due to fluctuations as a function of σ^2

3.3 Contractile Cytoskeletal Responses to Fluctuations in Stretch

In the above analysis, we assumed that cell stiffness remains constant during stretch. However, in response to MS and VS, cell stiffness will likely change. There is data suggesting that during MS, cell stiffness declines [80]. Following VS, the organization of the actin cytoskeleton, the most important contributor to cell stiffness, exhibits a stronger network structure [32] which should result in a higher stiffness. To test this prediction, we used a previously developed active network model of the actin-myosin cytoskeletal network [81]. Briefly, a 2-dimensional network is composed of randomly oriented springs in a square lattice which represents the sub-cortical actin of the cell. Actin filaments are discretized as a set of springs of identical length and spring constant. The end points of each spring along a filament also serve as binding sites for myosin crosslinks which we model as linear elastic springs with a different spring constant. The myosin springs can attach to binding sites on two distinct but neighboring actin filaments. The number of actin filaments is approximately equal to the number of myosin crosslinks. At every point along the square boundary where actin filaments are attached, an additional linear elastic spring is added to represent the combined effective stiffness of FAs and the ECM. These springs connect to an actin filament at one end while the other end is held fixed. The non-muscle myosin II motor proteins use energy from ATP hydrolysis to repetitively bind to actin and undergo conformational changes that result in the molecule walking along the actin filament towards its plus (barbed) end which is implemented by moving each end of the myosin crosslink to the next binding site along the filament. Consistent with experimental observations [82, 83], when the network carries no force, the time interval between two walk steps is the same for all myosin crosslinks. However, once the myosin motors start walking, the network develops tension and the time interval between walk steps increases in proportion to the force that the crosslink carries. If the load on a myosin cross link exceeds the stalling force, the motor stops. After each walk step, the network rearranges itself to balance local forces and the new equilibrium configuration is determined by minimizing the total elastic energy of the network given by the sum of the elastic energy of the actin, myosin and the ECM springs (Fig. 3a). The total traction force is obtained as the sum of the forces carried by the ECM springs whereas the stiffness of the network is computed by first

(a)   **(b)**

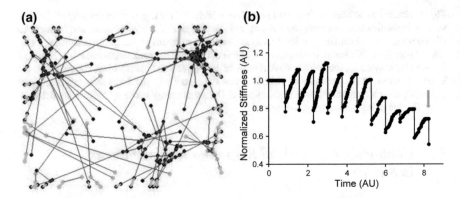

Fig. 3 A computational model of the cytoskeletal actin-myosin network. **a** Network configuration following 10 cycles of monotonous stretch (MS) modeled as a sequence of identical pulse waves. The green segments denote ECM attachments, the blue and red lines are actin and myosin filaments, respectively. Notice the "star"-shaped highly clustered actin configuration. **b** The normalized cell stiffness gradually decreases during MS. Blue arrow shows the time point that corresponds to the network configuration on panel (**a**). During the rising edge of the square wave, stretching breaks myosin bonds and the stiffness drops (red arrow) whereas during the constant part of the square wave, myosin bonds reattach and the stiffness builds up (green arrow)

applying a small incremental equi-biaxial strain on the network and calculating the ratio of the change in force and the incremental strain.

Next, the network is subjected to equi-biaxial cyclic stretch using repetitive pulses. In MS, the pulses are the same while in VS, the height and width of the pulses vary from cycle to cycle similarly to the stitched together sine wave time series in Fig. 2a. During the stretch phase of the pulse, an actin-myosin crosslink can rupture if the load on the myosin exceeds a predefined threshold force. Following a rupture event, both ends of the cross-link detach and the myosin diffuses through the network. The myosin can form a new cross-link at another location if the binding sites on two neighboring actin filaments are sufficiently close. In between the stretch phases, the attached cross-links ratchet up the internal tension and the network gradually recovers its force and stiffness. Since traction force and stiffness are linearly related in the model, as well as in experiments [84], the results shown for stiffness are also valid for contractile force. During MS both the traction force and stiffness start to quickly decline after only 3 stretch cycles (Fig. 3b). Interestingly, stable "star" like junctions develop that persist through further stretch cycles. These star junctions are created by geometrical configurations in the network where the barbed end of different actin filaments are brought together by myosin motors. The formation of star junctions causes a decline in network force and stiffness because force generation depends on the ratcheting motion of myosin cross links along actin filaments and when myosins reattach close to a star junction, the average distance available for a crosslink to ratchet up an actin filament decreases, reducing the available increase in network force. When the network is stretched with VS, the situation is drastically

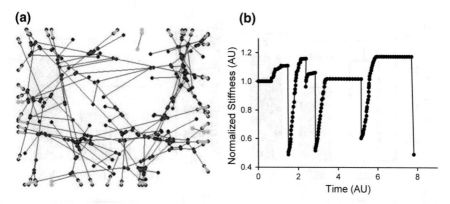

Fig. 4 **a** Network configuration following 5 cycles of variable stretch (VS) modeled as a sequence of pulse waves of different height and width. The green segments denote ECM attachments, the blue and red lines are actin and myosin filaments, respectively. Notice the lack of star-shaped actin clusters. **b** The normalized cell stiffness fluctuates during VS, but does not decline over the same simulation time as during MS in Fig. 3b

different. The occasional larger than average pulses in VS are able to break up the star formation prohibiting the network from settling in a non-equilibrium steady state (Fig. 4a). Since the network is often kicked out of a developing steady state, the myosin motors constantly ratchet up tension but rarely reach the stalling force. Consequently, both the traction force and the network stiffness are higher during VS than MS showing no sign of decline over the same simulation time (Fig. 4b). An interesting consequence of this adaptive response is that if the boundary conditions are nearly independent of the increasing network stiffness, then the energy taken up during VS is maximized by the dynamic adaptation of network structure. The unsteady nature of the actin-myosin network dynamics should thus result in more ATP consumption than the steady state during MS. Consequently, it is reasonable to conclude that this excited state of the cytoskeletal network responds to additional mechanical or chemical stimuli differently than during MS, which is the essence of FDM.

The predictions of the active network model can be contrasted with our previous work [32] in which we reported that the fractal dimension, a measure of complexity and space filling capacity, of actin was higher following 4 h of equi-biaxially delivered VS than MS in VSMCs in culture (Fig. 5a). Besides other cytoskeletal changes including microtubule and vimentin, a key observation was that ATP production rate (Fig. 5b), assayed by the mitochondrial membrane potential, as well as total ATP amount (Fig. 5c) in the cell was also higher during VS than MS. Because both the production and amount of ATP is higher during VS, the consumption must also be higher. These results suggest that FDM stimulates the cytoskeleton to be in a much more active state which allows different signaling that can be seen from the higher overall tyrosine phosphorylation of the cell (Fig. 5d). Although measurement of cell stiffness under VS still needs to be carried out experimentally, the phosphorylated

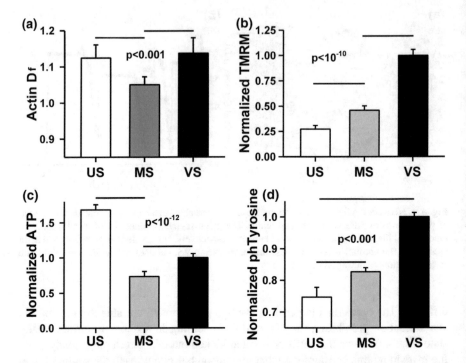

Fig. 5 Comparison of the fractal dimensions D_f of the actin cytoskeleton (**a**), ATP production rate measured by TMRM intensity (**b**), total ATP content (**c**) and tyrosine phosphorylation (**d**) in vascular smooth muscle cells in culture following 4 h in the unstretched (US) condition, and 4 h of monotonous stretch (MS) or variable stretch (VS). Except for D_f, all data are normalized to the level under VS. Data were obtained from Bartolak-Suki et al. [32] with permission

myosin light chain (phMLC), which regulates cell contractility, was significantly higher following VS [32]. Figure 6 demonstrates that these cell culture-based results are also maintained at the aorta tissue level. Here, the mean and amplitude of circumferential cyclic strain for MS was obtained from measured blood pressure recordings and pressure-radius curves in rats whereas the variability in VS was computed from beat-to-beat blood pressure fluctuations. The mean force generated by aorta rings during 4 h of MS or VS (Fig. 6a, b) shows that contractility, assessed at times 0 and 4 h using KCl challenge, declined by ~25% during MS compared to VS which maintained contractility after 4 h (Fig. 6c). Recalling that stiffness and traction force are linearly related, these numbers compare well with the ~30% reduction in stiffness of the network following MS (Fig. 3b) and no change in stiffness following VS (Fig. 4b). To test how ATP availability influences contractility, oligomycin, an inhibitor of the ATP synthase, was applied for an additional hour followed by another KCl challenge demonstrating that contractility decreased significantly under both MS and VS. The phMLC also decreased by 28% after MS compared to VS which was further reduced in both MS and VS following oligomycin (Fig. 6d). Furthermore, immunohistochemical expression of phMLC in muscle cells within the aorta wall

(Fig. 6e) and the corresponding statistics (MS 22% lower than VS; Fig. 6f) closely mirrored both the biochemically determined phMLC (Fig. 6d) and tissue contractility (Fig. 6c). Taken together, these results demonstrate that tissue level phMLC and vascular contraction are regulated by stretch pattern in a mitochondrial ATP-dependent manner. Thus, we see that downward causation from BPV-induced fluctuations in the boundary conditions applied to cells regulates microscale contractility which in turn influences macroscale BP via maintaining arterial wall stiffness [85]. We thus conclude that FDM plays a role in a bi-directional dynamic homeostatic maintenance of BP regulation.

4   Examples of Fluctuation-Driven Mechanotransduction

A prime example of FDM is our recent study [32] part of which is presented in Fig. 6. In this study, we found that ATP production rate in VSMCs in culture stretched equibiaxially was higher by about a factor of 2 following 4 h of VS compared to MS [32]. The ATP production was quantified by the inner mitochondrial membrane potential (mMP) [86] that is a measure of ATP production [87]. The FDM-induced increase in ATP production was accompanied by upregulation of key components of the electron transport chain including the ATP synthase. Using a set of 15 inhibitors, the underlying mechanism can be described as follows [32]. When actin polymerization, microtubule depolymerization, the ATP synthase, FAK or calcium availability were inhibited individually, ATP production was altered, but VS maintained a higher ATP production than MS. Two cytoskeletal structures, microtubules and vimentin, however, play a key role in mitochondrial ATP-related FDM since inhibiting their polymerization eliminated the difference between MS and VS cells. Blocking nonmuscle myosin II, the dynamin-related protein 1, which regulates mitochondrial fission [88], or mitotic kinesin-like protein 2, which reduces mitochondrial movement along microtubules, all reduced ATP production in VS cells to the levels found in MS cells. Since changes in ATP production are associated with alterations in mitochondrial network complexity [89], VSMCs appear to utilize the elastic energy surplus available during VS (Fig. 2c) via remodeling the cytoskeletal network and convert it to chemical energy stored in ATP via reorganization of the mitochondrial network. FDM should thus play crucial roles in many essential cell functions that require ATP-based energy supply. Hence, we conjecture that all cellular processes that participate in homeostatic control of some physiological parameter are dynamically also regulated by organ level mechanical fluctuations.

Surfactant in the lung, secreted by type II epithelial cells, is a key component of alveolar surface film stability. A potent stimulant of surfactant production and secretion is a single large stretch [24]. A period of 30 min of cyclic stretch with 50% area strain applied to epithelial cells in culture showed first a decline in surfactant secretion after an additional incubation time of 30 min, but resulted in increasing levels of secretion after further incubations for 60 and 210 min [90]. Another study found an increase in secretion at 60 min when type II cells were stretched for 15 min, but a

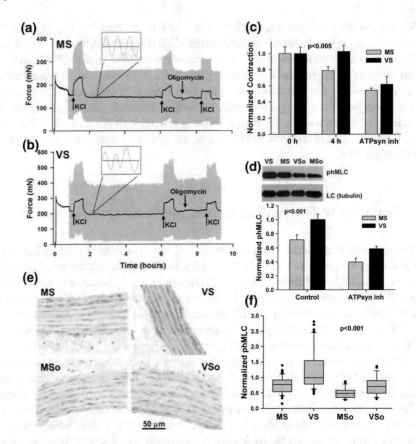

Fig. 6 Contractility of aorta rings is regulated by stretch pattern. Rat aorta rings were stretched first to a circumferential strain of 70% and allowed to relax for 45 min. Tissues were then cyclically stretched with MS (**a**; 30% sinusoidal strain amplitude with a fixed peak strain of 85%, 3 Hz) or VS (**b**; peak strains distributed between 77.5 and 92.5%, mean frequency of 3 Hz). After 10 min of stabilization, KCl was added and 30 min later washed out. The KCl protocol was repeated after 4 h stretching. In a subset of tissues, oligomycin was added to block the ATP synthase and the KCl challenge was repeated after 1 h. The instantaneous force fluctuations (gray lines) were low-pass filtered (black lines). The insets show a short segment of the wave forms for MS and VS. **c** Contractility is defined as the difference in force between the force at 30 min after adding KCl and the force just before adding KCl. The contractility normalized to its value at time 0 drops significantly in MS samples (n = 10) but not in VS samples (n = 9) and this difference disappears following oligomycin treatment (n = 9, two-way ANOVA, $p < 0.005$). **d** Example western blots and loading controls (LC) of phosphorylated myosin light chain (phMLC, top row) in aorta segments after 4 h of MS or VS and following 1 additional hour of exposure to oligomycin (MSo or VSo) to inhibit the ATP synthase. The effect of stretch is significant ($p < 0.001$), the effect of oligomycin is significant ($p < 0.001$) and all groups are different. **e** Immunohistochemical images of phMLC in aorta walls with blue and pink denoting phMLC and nuclei, respectively. **f** Statistical analysis of phMLC levels from immunohystochemical images after MS (n = 54), VS (n = 42), MSo (n = 24) and VSo (n = 24). Data are normalized to the median of the VS group. The effect of stretch is significant ($p < 0.001$), the effect of oligomycin is significant ($p < 0.001$) and all groups are different except MS and VSo. Data were obtained from Bartolak-Suki et al. [32] with permission

decrease when these cells were stretched for 30 or 60 min [91]. The delayed responses are consistent with the notion that stretch increases surfactant transport from the interior to the exterior through a surface layer with a limited number of secretory pores which temporarily hinder secretion due to an interaction among the surfactant containing lamellar bodies, effectively creating a "traffic" jam [92]. When the stretch stops, the jam takes additional time to clear implying that surfactant secretion is a fundamentally nonlinear process with memory representing collective behavior at the level of a single cell. The jam is likely due to the organization of the cytoskeletal network around fusion pores since depolymerizing actin generally enhances surfactant secretion [93]. Additionally, surfactant secretion is tightly regulated by membrane-bound target SNARE (soluble N-ethylmaleimide-sensitive factor attachment protein receptors) proteins called SNAP-23 and syntaxin-2 [94]. Syntaxin in the plasma membrane also self-organize into clusters due to weak inter-molecular attractive interactions with cluster sizes limited by molecular crowding-induced repulsion [95]. The available syntaxins in the membrane may group into a finite-number of docking sites, allowing jamming to occur when secretion is stimulated. How does FDM affect the jamming process? When a small amount of cycle-by-cycle variability (5%) was added to the peak amplitude stretching type II epithelial cells in culture, surfactant secretion was doubled [91] suggesting that VS reduces jamming. While the precise mechanism is not known, several possibilities arise. First, as Fig. 4a demonstrates, VS can have significant impact on subcortical cytoskeletal organization which in turn should alter the cluster distribution of docking proteins and hence the fusion pore opening dynamics. The opening of calcium channels required for secretion [24] is a nonlinear stretch-dependent process [96]. Since fluctuations can enhance signal transduction even in the simplest ion channel [97], the VS-induced cytoskeletal changes can increase intracellular calcium levels too. Finally, if ATP production is upregulated by VS in type II cells similarly to VSMCs [32], the excess ATP can depolymerize actin and act as a strong surfactant stimulant. Since surfactant is essential for maintaining normal lung elasticity and stability [98], it can be concluded that natural TVV at the level of the organ alters cellular processes to produce proper amounts of surfactant which in turn contributes to the homeostasis of macroscopic lung mechanical function.

Endothelial cells lining the vascular walls provide a selective tight-junction barrier to circulating blood cells and macromolecules in the presence of shear stresses due to blood flow. Countless studies have demonstrated that shear stress regulates various genes [10] which in turn control the production of vasoactive materials such as nitric oxide (NO), which relaxes VSMCs, and endothelin-1, which contracts VSMCs. While this knowledge comes from countless experiments using steady shear flow, only one study examined how physiological variability in shear flow based on observed in vivo flow patterns might influence vascular signaling [73]. Both steady and variable shear flows upregulated superoxide dismutase that protects against oxidative stress; however, variable shear flow also upregulated NO synthase activity and NO production as well as down-regulated TNF-α-induced leukocyte adhesion which implies a more quiescent endothelium. Thus, while FDM seems to lower endothelial inflammation, little is known about the underlying mechanism. It is likely that the glycocalix, comprised of proteoglycans and glycoproteins, on the cell surface mediates such mechanotransduction [99]. For short time scales, mechanical noise in shear stress due to turbulent flow may contribute to selective activation of signal transduction pathways [100]. For longer time scales, slowly varying shear stresses may modulate glycocalix synthesis and degradation [101]. The amount and clustering of glycocalix in turn can influence how instantaneous shear stress activates G-protein signaling which regulates the burst-like production and release of NO [102]. Although uncovering the exact mechanisms requires further studies, NO release significantly influences vascular wall stiffness [85] which is a key determinant of overall BP. FDM therefore may regulate general endothelial function which feeds back to organ level BP regulation.

Several studies also attempted to incorporate variability into tissue engineering approaches. Using a novel stretcher, Gelfoam samples were seeded with neonatal rat lung fibroblasts and uniaxially stretched with different levels of variability (0, 25, 50 and 75%) superimposed on 20% mean strain [79]. While the data showed that VS increased the mRNA expression of collagen 1α and lysyl oxidase, the mRNA expression of syndecan-4 was maximized at 25% variability in strain amplitude. Further studies are required to assess any functional consequences of the altered expression profile. In another study using a novel bioreactor, VS with 25% variability was used in cardiac tissue engineering with a small but significant increase in twitch force compared to static stretch [103]. There is only one study that tested the effects of varying the frequency alone [104]. Cardiomyocyte constructs from neonatal rats in fibrin hydrogels were stimulated with constant, Gaussian or uniform distribution of frequencies for 2 weeks with 10 or 20% variations around the mean. The observed differences in protein expressions suggest that variable-frequency stimulation affected cell–cell coupling and growth pathway activation. Although these studies are more descriptive, they present new opportunities to selectively tune the composition of tissue constructs and better understand ECM homeostasis.

5 Homeostatic Maintenance of the Extracellular Matrix

Most cells in the body continually probe the mechanical properties of the ECM through integrin receptors and the active network of actin and myosin [105] and respond by balancing enzymatic degradation with replacement of the digested fragments with newly synthesized molecules [106]. This mechano-regulatory process shapes and maintains the ECM in a homeostatic manner [8]. The strong links among the cytoskeleton, mitochondria and mechanical factors [89] also raise the possibility that beyond conventional mechanotransduction, the FDM-induced bioenergetic stimulation of cells will also affect ATP-dependent cellular processes responsible for the production and secretion of ECM material and their remodeling enzymes. We attempt to address this question by separately examining the cell-ECM interactions at two different time scales. At long time scales, the maintenance of the fibrous structure of tissues may be more influenced by the average stresses than their fluctuations whereas at shorter time scales FDM may be more directly involved in the homeostatic maintenance including response to injury.

5.1 ECM Maintenance at Long Time Scales

In several organs and tissues such as the vasculature, skin, heart and periodontium, a quick turnover of collagen with half-lives between 20 and 250 days have been observed [107, 108]. Interestingly, the mechanical stresses of exercise and the associated metabolic activity were found to induce a rapid and significant increase in collagen turnover within 72 h in tendon [109], a tissue that exhibits a very long collagen turnover [110]. Despite the continual cellular maintenance, collagen structure appears to remain in a stable homeostatic state throughout most of adult life even when demanding mechanical stresses lead to strong remodeling. For example, while the diameter distribution of collagen in mouse tail tendon undergoes major changes in the first 3 months of life due to development, this distribution remains nearly independent of age between 4 and 23 months [111], a range that spans the adult life of the mouse and corresponds approximately to the span between 10 and 60 years in humans [112]. Collagen morphometry is also nearly constant throughout adult life in skin [113]. The functional consequence of the nearly constant diameter distribution is a stable strain energy density that maintains proper mechanical function and resistance to rupture [111]. An important question of mechanobiology is then: "How are cells able to maintain such a homeostatic structure over a period that corresponds to decades of human life"?

In an effort to answer this question at least partially, we developed a computational model of the long-term maintenance of collagen fibers under tensile stress [106]. Briefly, the model attempts to account for the ability of cells to balance enzymatic digestion and repair of a single collagen fiber such that the fiber diameter remains constant over time as well as along the fiber. The cell is able to probe the stiffness of the fiber using its acto-myosin network (not explicitly included in the model) and respond to changes in overall stiffness by releasing regenerative (R) or degradative (D) particles when the overall fiber stiffness decreases below or increases above a threshold, respectively. The R and D particles represent collagen monomers and collagen-cleaving enzymes (e.g., MMPs) responsible for the local fiber repair and digestion, respectively. The fiber consists of a one-dimensional chain of linearly elastic springs in series surrounded by two layers of sites along which the R and D particles can diffuse and bind to the fiber at discrete locations corresponding to the springs (Fig. 7a). Both ends of the chain are subject to a constant force that mimics tension on ECM fibers.

When a D or an R particle unbinds from the fiber, the local spring constant k is reduced ($k \to \gamma k$) or increased ($k \to k/\gamma$), respectively, by a constant factor $\gamma < 1$. If the fiber was a regular array of molecules, cleavage would simply decrease the number of molecules in parallel. This would then result in a linear decrease in the spring constant during subsequent cleavages and it is possible that over a long time scale, the local spring constant reaches zero effectively rupturing the chain. In reality, the molecules overlap and are connected via multiple inter-molecular as well as inter-fibril cross-links in a complex manner (Fig. 7b). Hence, the degradation of the fiber is also a complex process and a single cleaving event does not eliminate a complete monomer from the chain. As Fig. 7b demonstrates, following the cleavage of a monomer, part of this monomer can still carry a force. Thus, subtracting one unit from k following unbinding of a D particle is not appropriate; instead, the cleavage process is modeled by assuming that the decrease in k is proportional to its actual value, which leads to a multiplicative degradation process. The above consideration is valid only when the process of degradation is far away from the rupture threshold and under this assumption, this model provided an accurate description of stress relaxation and fragment release from elastin fibers during digestion [114]. While

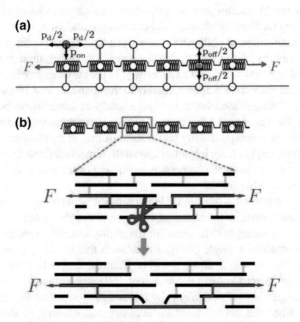

Fig. 7 Schematic diagram of a fiber during cellular maintenance. **a** The fiber is a chain of springs with binding sites represented by small open circles. The two layers of sites surrounding the fiber are represented by the big open circles and the two types of particles are shown as filled circles with the red and blue circles corresponding to a diffusing degradative and a bound regenerative particle, respectively. A particle at the top, can move down with an on-binding probability (p_{on}), left or right with a diffusing probability ($p_d/2$) whereas a particle at the bottom can similarly move up, left, or right. When a particle binds to a spring, it can stay there or move up or down with an off-binding probability ($p_{off}/2$). The ends of the chain are submitted to a constant force F. A periodic boundary condition was applied along the axis of the chain. **b** A zoom into one of the springs representing collagen monomers in parallel (black) reinforced by cross-links (gray). Red shows the force transmission pathways and the blue scissors represent a bound enzyme to a collagen monomer (middle diagram). Once the enzyme unbinds, the monomer is cleaved, but part of the monomer still participates in force transmission (bottom diagram). Images were obtained from Alves et al. [106] with permission

this model was able to achieve a stable total stiffness of the chain (K), the control mechanism failed to produce a constant variance of k along the chain because, for long time scales, the variance increased nearly linearly with time, similar to a random walk, producing a highly heterogeneous fiber not observed in experiments [106].

The only solution to avoid the divergence of the variance of k is to introduce local control that can increase the probability of binding one of the particles to a binding site that requires a change. For example, a random local decrease or increase in k requires that a R or a D particle, respectively, fixes the local departure from the average stiffness. This means that either the cell "figures out" the local inhomogeneity of the fiber and releases the proper particle at the exact location of repair, or the already released particles are biased to bind to the location with an altered k. Since both the enzyme and the collagen monomer are very small compared to the cell (~100 nm vs. ~10 μm), this would require an extreme precision from the cell. It seems more plausible that the cell senses the stiffness at a scale of its own size, since stiffness sensing requires larger scale structural transitions within the actin network [115]. The other possibility is that the binding probability itself is biased such that if a local region is too stiff or too soft, the binding sites in that region attract D or R particles, respectively. Since the force on a fiber can be taken to be constant as long as the diameter to length ratio is small, a soft region would stretch more than a stiff region allowing different binding sites to open up or unfold as a function of local stretch [116, 117] and hence stiffness. Accordingly, we can introduce stiffness-dependent on-binding probabilities for R and D particles as follows:

$$p_{on,R} = 1 - \frac{2}{3}e^{-F/k} \quad \text{and} \quad p_{on,D} = 1 - \frac{2}{3}e^{-k/F} \tag{4}$$

where F is the force per unit length on the fiber and k is the local stiffness. Notice that if $k \to 0$, $p_{on,R} \to 1$ and $p_{on,D} \to 1/3$, whereas if $k \to \infty$, $p_{on,R} = 1/3$ and $p_{on,D} \to 1$. In other words, the on-binding probabilities approaches 1 for R or D particles when the site locally needs to increase or decrease its stiffness, respectively. For homeostasis to be functional, it is sufficient if one of these mechanisms is active. Indeed, implementing $p_{on,R}$ in the model successfully achieves homeostatic control for both K and the spatial homogeneity of k along the fiber [106]. Moreover, the model suggests that there is an optimum force on the fiber that minimizes the variance of k.

Can we experimentally validate these model predictions? Atomic Force Microscopy measurements along type I collagen fibril in indentation mode showed significant axial heterogeneity of stiffness [118]; however, the indentation induces both compression and shear and since collagen is anisotropic with different axial and lateral moduli, such data are not directly applicable to estimate the Young's modulus in axial stretch. A positive correlation between average fiber diameter and the low strain modulus has been reported for artificial assemblies of type I collagen fibers [119]. We can extend this result to obtain a relation between local diameter and local stiffness. First, note that the stiffness at a given location should be related to the num-ber of collagen monomers N_c that pass the cross section A of the fiber. If the cross-link density is constant along the fiber, the local stiffness should be proportional to N_c.

Since A is the sum of the cross sectional areas of individual monomers, A is also proportional to N_c implying that fiber diameter is proportional to the square root of N_c. Thus, the distribution of $\sqrt{k}$ can be compared with the distribution of measured diameters along individual collagen fibrils. Scanning electron microscopic images of the thoracic aorta of normal rats [106] were processed (Fig. 8a) and segments were selected if their borders could be identified properly (Fig. 8b). The distribution of diameters exhibits two peaks (Fig. 8c) similar to the results obtained by measuring cross sectional areas on images perpendicular to the fiber axis [111]. When each diameter value along a fiber is normalized with the median diameter of the fiber, the distribution becomes unimodal which is compared to the distributions of $\sqrt{k}$ of the model for 3 values of F in Fig. 8d. There is a good agreement between the experimental results and the model simulations when $F = 0.6$, which corresponds to the minimum of the variance of k or the maximum homogeneity of the fiber [106]. Despite several limitations such the 2-dimensional system, the lack of convective particle transport via interstitial flows, or volume exclusion, this modeling exercise suggests the possibility that the ECM evolved together with cells in a manner to actively participate in its own homeostatic maintenance. Whether FDM contributes to such long-term processes is an open question.

5.2 ECM Maintenance at Short Time Scales

Could fluctuations in stretch or mechanical force at a shorter time scale play a role in ECM maintenance? With regard to FDM, we argued above that the cytoskeleton exhibits a strong response to VS which is also reflected in gene expressions [79] and ultimately the production and secretion of ECM maintaining R and D particles such as collagens and MMPs. Indeed, our preliminary data suggest that collagen secretion by mouse lung fibroblasts in culture is sensitive to fluctuations in stretch amplitude [120]. Furthermore, we recently also reported that MS and VS differentially regulate the organization of the ECM [121]. Briefly, representative images of elastin, collagen and β-actin within the aorta wall soon after isolation (termed unstretched, US) and following 4 h of MS or VS are shown in Fig. 9a. For elastin, there are strong wavy elastic lamellae within the media layer which appear dark blue and are interspersed with bundles of dark red/pink VSMCs. The thick lamellar fibers and the thin inter-lamellar fibrils are wider and looser and hence darker during MS. The thin network of collagen fibrils appears as bright blue in the media and adventitia. Immunohistochemistry also revealed a significant stretch dependence of the thin network of β-actin in the smooth muscle layer (labeled blue) with visibly more β-actin following VS. The complex space-filling capacity of these structures was characterized by the fractal dimension D_f from ~50 images (Fig. 9b). For elastin, D_f was higher after MS than VS ($p < 0.001$) whereas for collagen, D_f significantly dropped following MS ($p < 0.001$). Compared to US, D_f of β-actin did not change following MS but significantly increased following VS ($p < 0.001$), the latter being consistent with the VSMC culture results in Fig. 5a. The changes in ECM struc-

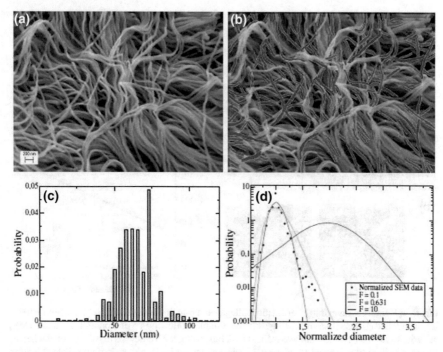

Fig. 8 Panel **a** shows part of the collagen fiber structure of the thoracic aorta of a healthy adult rat obtained using scanning electron microscopy. The highlighted parts in **b** are the segments selected for measurement of diameters along the contour length of the fibers. Panel **c** presents the distribution of diameters exhibiting a bi-modal shape. In **d** the diameters corresponding to each fiber are normalized by the median of the fiber diameter (filled black circles). Note that the distribution now shows a single peak. Also shown are several model simulations of predicted diameter distributions corresponding to 3 values of the force F on the fiber in the model. Note that for $F = 0.6$, the simulated distribution matches the experimental one. Data were obtained from Alves et al. [106] with permission.

ture imply a looser elastic lamella and more compact collagen fibers following MS. However, in the muscle layer, it is likely that following isolation of the aorta, actin is quickly depolymerized in the US samples because of the lack of ~30 min stretch before fixation. Thus, the larger β-actin amount in VS suggests that stretch pattern affects β-actin more than stretch itself. These differences in turn influence cortical actin organization in agreement with Figs. 3 and 4, and hence ECM stiffness sensing. Since D_f and other ECM structural parameters (not shown) following VS were closer to US than MS [121], these results are consistent with the notion that in vitro, VS better maintains the natural organization of the ECM than MS. Lamellar fibers were also segmented on the elastin images and their waviness was determined (Fig. 9c). Compared to US, both MS and VS resulted in a higher waviness which was highest after VS ($p < 0.001$, not shown). Interestingly, many of the structural parameters were changed and the difference between MS and VS disappeared after treatment of the aorta with oligomycin A, an inhibitor of ATP production, and blebbistatin, an

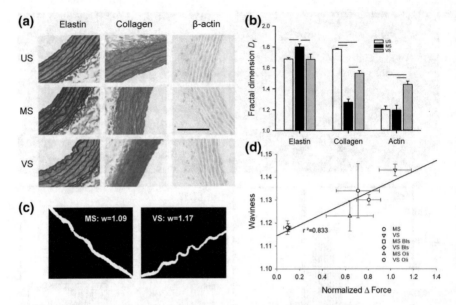

Fig. 9 **a** Representative images of elastin, collagen and β-actin for experimental conditions including unstreched (US) samples following isolation, samples following 4 h of monotonous stretch (MS) or variable stretch (VS). Elastin was stained dark blue using the Verhoeff's elastic staining whereas collagen is light blue stained by Masson's trichrome method. The network of cytoskeletal β-actin within the smooth muscle layer is shown in blue by immunohistochemistry with pink counterstained nuclei. The scale bar on MS β-actin corresponds to 100 μm. **b** Statistics of the fractal dimension D_f of elastin, collagen and β-actin in US, MS and VS samples. One-way ANOVA showed that except for US and VS for elastin and US versus MS for β-actin, all conditions were significantly different from each other ($p < 0.001$). **c** Example binary images of lamellar elastin after MS (left) and VS (right). The corresponding values of waviness, defined as the ratio of contour length and end-to-end distance of a fiber, are given in each image. **d** Regression between the median waviness and normalized contractile force from Fig. 6c including the data corresponding to the ATP synthase inhibitor oligomycin (Oli) and the β-actin polymerization inhibitor blebbistatin (Bls). Images and data were obtained from Imsirovic et al [121] with permission

inhibitor of non-muscle myosin II which hinders cortical actin network formation. However, there was an unexpectedly strong correlation between contractile force and waviness with an $r^2 = 0.833$ (Fig. 9d). Taken together, these results suggest that muscle tone regulated by FDM is necessary for proper homeostatic ECM organization even at a relatively short time scale, a notion that is in agreement with a recent report that mechanical homeostasis appears to restore collagen fibril waviness in tendon via active fibroblast contraction [122].

6 Homeostasis and Disease

In disease, there are characteristic changes in stretch pattern that should influence FDM. Breath-to-breath variability in respiratory variables has been reported in both health and disease [30, 123]. For example, while the CV of TVV is ~26% in normal subjects, it is 43% in obstructive and only 17% in restrictive lung diseases [124]. Although the local mechanical properties of the lung tissue can influence the actual strains delivered to lung cells by a tidal breath, FDM is primarily determined by TVV. If, as part of the downward causation, TVV is altered by a disease, the homeostatic regulation of some lung physiological parameter through FDM should also be affected. The upward causation should modulate TVV via alterations by the disease of tissue mechanical properties or their nervous control. Indeed, when variability was partitioned into random and non-random components, healthy subjects under elastic loading mimicking restrictive diseases decreased the random component of TVV [125]. Furthermore, increasing variability in breathing pattern also induced dyspnea in patients with restricted lung disease, indicating that these patients decreased TVV in order to minimize dyspnea [126]. Whether changes in breathing pattern are inherent to the disease or adopted by the patient to alleviate breathing problems, the increase or decrease in TVV will result in abnormal strains delivered to cells which should lead to pathological FDM and possibly contribute to the progression of the disease in a feedback loop. Although currently little experimental data are available, a significant shift in FDM should affect fibroblasts' ability to maintain a dynamic homeostatic control of the lung ECM. An extreme reduction in TVV occurs during mechanical ventilation of patients which eliminates all variability. The monotonous stretching of the lung during conventional mechanical ventilation has been shown to impair surfactant metabolism and promote cytokine release while adding variability to tidal volume can be used as a potential treatment in animal models of lung injury [127–129]. Furthermore, in prematurely born lambs with surfactant deficiency, 3 h of ventilation with TVV was able to differentially regulate surfactant protein expressions and reduce lung inflammation and alveolar leakage compared to conventional ventilation with no variability [130]. This suggests the intriguing possibility that evolution has built fluctuations into the maturation of the surfactant system of the lung.

In the cardiovascular system, heart rate and BPV arises from the numerous feedback systems interacting in a noisy environment [131]. While such fluctuations are necessarily present in healthy subjects, BPV increases in hypertensive patients and animals [132, 133]. Furthermore, BPV is known to occur at widely different time scales ranging from beat-to-beat, to 24 h, and to periodic doctor visits [134, 135]. While mean BP as well as systolic BP are considered primary diagnostic indexes of cardiovascular disease, various measures of increased BPV on the time scale of periodic doctor visits was identified as an independent predictor of stroke [135]. Similarly, lower 24 h BPV correlated with lesser prevalence of target organ damage in patients with hypertension [134]. An increase in BPV elevates the maximum strain acting on endothelial cells as well as VSMCs, both of which are exquisitely sensitive to fluctuations in mechanical stimuli exhibiting thus FDM [32, 73]. An interesting animal model of high BPV is the sinoaortic denervation which increases BPV and leads to aortic and left ventricular hypertrophy without an increase in mean BP [136]. Alternatively, long-term treatment of spontaneously hypertensive rats with low-dose ketanserine, an anti-hypertensive agent, reduced BPV but not mean BP and less-ened end-organ damage [137]. The biological outcome of decoupling mean BP from BPV is important. Indeed, cells in the vessel wall are likely able to accommodate an increase in static stretch associated with increased mean BP leading to a non-equilibrium steady state. However, cells may react unexpectedly to a pathological increase in BPV which maintains an unsteady non-equilibrium condition. Increased BPV in humans is also associated with arterial wall remodeling and elevated arterial wall stiffness [138], a strong and independent predictor of cardiovascular risk [139]. Thus, the increased BPV in cardiovascular diseases is perhaps not just a symptom of the disease, but may also be a mechanical signal that contributes to the breakdown of the healthy homeostatic regulation of BP. Indeed, in hypertensive patients under-going 3 months of antihypertensive treatment, the reduction in BPV by amlodipine, a calcium channel blocker, was also associated with a reduction in mean blood pres-sure [140]. At the other end of the spectrum, too low BPV can occur under certain conditions such as anesthesia [141] or brain death [142] suggesting a role for the central nervous system in the regulation of normal BPV. Complete elimination of circumferential strain variability in in vitro experiments in otherwise normal VSMCs in culture demonstrated a substantial shift in metabolic activity, reduced contractility and increased reactive oxygen species (ROS) production [32]. While it is possible that similar to high BPV, long-term pathologically low BPV results in loss of home-ostatic control, further experiments are required to determine how FDM changes and whether such changes have detrimental effects on tissue remodeling.

7 Homeostasis and Aging

Before concluding, some remarks on the possible role of FDM in aging are due. Although many molecular theories of cellular aging have been put forth since the middle of the last century [143–145], little is known about how mechanotransduction itself ages or contributes to aging. While organs may age differently, there is a trend that collagen processing becomes slower leading to increased cross-linking and stiffer collagen as we age [146, 147]. This phenomenon represents a gradual shift in the homeostatic maintenance of collagen. Since many cell functions are sensitive to the local stiffness of the ECM to which they attach [148], cells are likely to respond to the gradual changes in the physical properties of the ECM in aging [149]. For example, VSMCs cultured on proteolysis-resistant collagen display early aging with reduced replicative life-span that can be blocked by vitronectin, a ligand for α5β3 integrin. Alternatively, senescence can be triggered in cells grown on normal collagen by inhibiting α5β3 integrin binding [150]. It seems reasonable to assume that mechanotransduction plays a role in aging. Below, we discuss three possible factors related to mechanotransduction that could contribute to the observed shift in mechanical homeostasis in tissue aging.

The first is related to the aging of the mitochondrial-nuclear (mitonuclear) communication. Mitochondria have been at the forefront of aging theories since the introduction of the idea that ROS from mitochondria are a primary cause of aging [151]. Although there is a steady decline in respiratory chain activity during aging [152], this idea has been questioned because reducing the gene expression of the electron transport chain results in longer life in flies [153]. However, aging also induces significant oxidative/nitrosative stresses in mitochondria which is accompanied by a substantially reduced activity of the respiratory chain and ATP production that could be mitigated by administration of melatonin [154]. Since the production of ECM molecules and remodeling enzymes requires ATP, the aging of mitochondria in fibroblasts should play an important role in how the homeostatic maintenance of the ECM ages. For proper ATP generation, the nucleus and the mitochondria must continuously communicate [155]. During aging, there is a specific loss of mitochondrial subunits of the electron transport chain [156]. Interestingly, the underlying cause of the reduced mitonuclear communication is a decline in nuclear NAD^+ and the accumulation of hypoxia-inducible factor HIF-1α under normoxic conditions. Deleting sirtuin 1 accelerates aging, whereas restoring normal NAD^+ levels in old mice elevates mitochondrial function to that of young mice. It is plausible that the decline in mitonuclear communication contributes to reduced mitochondrial ATP production which in turn redistributes the energy budget of fibroblasts such that their cytoskeletal organization and capacity to maintain the ECM via mechanotransduction becomes impaired, a prediction that needs experimental verification.

The second mechanotransduction-related factor involved in aging is the stiffness of the ECM. Aging alters the composition of lung ECM [157] including the pulmonary vessels [158]. The morphology of collagen and elastin, the two major determinants of ECM stiffness, changes in aging with the fibers becoming straighter and thicker [159]. Both the turnover of collagen [160, 161] and the activity of lysyl oxidase, a key enzyme in collagen and elastin processing, decrease with age [162]. The less stable reducible cross-links gradually decline whereas the more stable non-reducible cross-links increase throughout life [147]. The non-enzymatic cross-linking of collagens by sugars through Maillard glycation results in advanced glycation end-products [163] which gradually accumulate during aging [164]. Cross-linking and denaturing of proteins caused by glycation are key factors in early aging of the lung [165, 166], which should increase fiber stiffness although no direct data are available. Cross-linking not only stiffens the tissue, but makes collagen more resistant to proteolysis, a key process of normal homeostatic ECM maintenance by fibroblasts. Furthermore, resistance to proteolysis has been shown to drive the aging process in vascular cells [150]. Thus, fibroblasts' ability to maintain the ECM is reduced during aging due to increases in cross-linking and stiffness of ECM fibrils to which fibroblasts attach. These pathological shifts disturb normal homeostasis by inhibiting normal mechanosensitive processes with the possibility of reinforcing a mild fibrotic phenotype.

Third, healthy FDM itself likely suffers from aging. As discussed above, cells respond to external cues [167] utilizing ATP produced mostly in mitochondria. Mitochondria form a tubular network [168] and the inner membrane potential, oxidative respiration and ATP production are down regulated when this network is fragmented [169]. Mitochondrial network formation is also related to the ability of mitochondria to move along microtubules [170]. Macroscopic physiological processes such BPV [171] and TVV [30] maintain and drive cellular homeostasis utilizing FDM such that cells can harness the fluctuations in external mechanical stimuli by altering their cytoskeletal and mitochondrial network leading to altered membrane potential and hence ATP production [32]. Specifically, VS enhances ATP production by increasing the expression of ATP-synthase's catalytic domain, cytochrome C oxidase and its tyrosine phosphorylation, mitofusins and PGC-1α [32], the master regulator of mitochondrial biogenesis under external stimuli [172]. FDM is also mediated by motor proteins and by the enhancement of microtubule-, actin- and mitochondrial-network complexity measured by the fractal dimension [32]. Since these mechanisms are key to proper ATP production, mechanosensitive and ATP-dependent cellular processes should be sensitive to alterations in the level of fluctuations that occur with advancing age. For example, the coefficient of variation of TVV increases from 33% at a mean age of 29 years to 44% at a mean age of 69 years [123]. On the other hand, the literature on BPV is slightly controversial showing only a mild negative dependence of diastolic BPV on age [173]. Nevertheless, visit-to-visit diastolic BPV is significantly associated with mortality risk in the elderly [174]. In a recent preliminary study, we found that TVV significantly impacts mitochondrial structure and function in lung fibroblasts in an age dependent manner [175]. When young fibroblasts were stretched with VS corresponding to young TVV, mitochondrial structure was more complex and ATP production rate was higher than those in old fibroblasts stretched with VS corresponding to an old age. Interestingly, these age-related structural and functional changes in old cells could be partially restored to the levels seen in young cells when the old cells were exposed to the physiologic TVV of young cells. Although more experiments will be needed to shed light on the relevant molecular details, the reduction in ATP availability due to age-associated changes in TVV and BPV should contribute to the altered fibroblast phenotype responsible for the gradual loss of proper ECM sensing and maintenance characteristic of the young age.

Finally, it is likely that the above 3 mechanisms are not independent of each other. ATP is a common link because it is related to mitonuclear function, ECM remodeling as well as FDM. Downward causation may come from aging of the central nervous regulation of TVV and BPV leading to altered cellular maintenance of the ECM which can contribute to the stiffening of fibers and tissues which in turn can cause a shift in the mean and range of fluctuations of homeostatically controlled parameters such as blood pressure.

8   Summary and Implications

It is now well documented that many cell types are sensitive to physical forces as well as the physical properties of the microenvironment and their signaling response influences many regulatory functions in the body. In the laboratory, such mechanotransduction is invariably studied using monotonous mechanical stimuli. Cells in the body, however, are exposed to irregular mechanical stimuli which are in general time varying but often exhibit spatial variations such as inhomogeneity and anisotropy of ECM stiffness. Cells sense and probe their mechanical microenvironment via adhesion molecules linked to the cytoskeleton and many cellular processes are sensitive to both the mean and the fluctuations of the mechanical boundary conditions they are exposed to. It is tempting to postulate that evolutionary forces must favor structures, both intra- and extracellular, that not only respond to the average but can also adapt to and take advantage of the fluctuations. As a consequence, cellular responses to fluctuations must be evolutionarily built into genetic mechanisms that eventually control long-term homeostasis within acceptable limits leading to both downward and upward causations. By accepting mechanical energy from the fluctuating boundary conditions via cytoskeletal and mitochondrial remodeling, vascular cells are capable of exploiting such mechanical fluctuations via increased ATP production. Several lung cell types appear to behave similarly. Since ATP is required for many basic cell functions, FDM leads to selective activation of signal transduction pathways. If the increase in ATP availability during FDM is also confirmed in other cells and organs, FDM may turn out to be a general phenomenon implying that active networks may be capable of harnessing energy from mechanical fluctuations in their microenvironment.

Many challenges and open questions remain however. At the most general level, we argued that FDM is a dynamic non-equilibrium process that never reaches a steady state. Hence, generalizing from results obtained under steady state MS conditions is not possible even when a range of amplitudes and frequencies are sequentially included in the steady state analysis. The primary difficulty is that no general thermodynamic description has been developed to describe the behavior of nonlinear dissipative systems driven far from equilibrium. There are several additional factors to consider. For example, fluctuations are not random because over long time scales, heart rate and breathing pattern show long-range correlations. Whether and how such time correlations in mechanical stimuli affect FDM is not known. Since cells and the ECM are viscoelastic implying mechanical memory, it is thus likely that FDM is also sensitive to the correlated nature of fluctuations. Physiological rates such as heart rate and respiratory rate also fluctuate which in turn result in variability in the strain rate boundary conditions for cells. Fluctuations in local strain rate lead to fluctuations in mechanical forces on proteins which, due to the viscoelastic nature of biological materials, can result in folding and unfolding patterns and bond dissociations that are different than in ideal elastic materials. The effects of the non-equilibrium unsteady conditions at the level of single proteins and bonds within the cell adhesion complex is not well understood. However, the general organization and the dynamics of the cytoskeleton, an active network under non-equilibrium conditions, may respond to FDM in a way that is nearly independent of cell type because all adherent cells have a subcortical actin network. Whether such generalization can be made should be experimentally tested. Additionally, nothing is known about the nuclear responses to FDM, which also need to be elucidated because the genetic response to fluctuations must be involved in the long-term homeostatic control of physiological parameters. Finally, the breakdown of a component within a hierarchical and bi-directional homeostatic control mechanism can amplify or attenuate mechanical fluctuations which in turn regulate FDM with aberrant intra-cellular signaling that can contribute to the pathogenesis and progression of diseases. In summary, uncovering how cells deal with physiological variabilities may also help unravel how cells regulate homeostatic control in the organs and tissues of the body with possible impact on better understanding vascular and respiratory diseases as well as neuro-degenerative and metabolic disorders and aging.

References

1. Cannon, W.R.: Organization for physiological homeostasis. Physiol. Rev. **9**, 399–431 (1929)
2. Atlas, S.A., Laragh, J.H.: Atrial natriuretic peptide: a new factor in hormonal control of blood pressure and electrolyte homeostasis. Annu. Rev. Med. **37**, 397–414 (1986). https://doi.org/10.1146/annurev.me.37.020186.002145
3. Kohan, D.E., Rossi, N.F., Inscho, E.W., Pollock, D.M.: Regulation of blood pressure and salt homeostasis by endothelin. Physiol. Rev. **91**(1), 1–77 (2011). https://doi.org/10.1152/physrev.00060.2009
4. Wiig, H., Luft, F.C., Titze, J.M.: The interstitium conducts extrarenal storage of sodium and represents a third compartment essential for extracellular volume and blood pressure homeostasis. Acta. Physiol. (Oxf.) **222**(3) (2018). https://doi.org/10.1111/apha.13006
5. Rodbard, S.: Vascular caliber. Cardiology **60**(1), 4–49 (1975). https://doi.org/10.1159/000169701
6. Murray, C.D.: The physiological principle of minimum work: I. The vascular system and the cost of blood volume. Proc. Natl. Acad. Sci. U S A. **12**(3), 207–214 (1926)
7. Zamir, M.: Shear forces and blood vessel radii in the cardiovascular system. J. Gen. Physiol. **69**(4), 449–461 (1977)
8. Humphrey, J.D., Dufresne, E.R., Schwartz, M.A.: Mechanotransduction and extracellular matrix homeostasis. Nat. Rev. Mol. Cell Biol. **15**(12), 802–812 (2014). https://doi.org/10.1038/nrm3896
9. Ingber, D.E.: Tensegrity: the architectural basis of cellular mechanotransduction. Annu. Rev. Physiol. **59**, 575–599 (1997). https://doi.org/10.1146/annurev.physiol.59.1.575
10. Chien, S.: Mechanotransduction and endothelial cell homeostasis: the wisdom of the cell. Am. J. Physiol. Heart Circulatory Physiol. **292**(3), H1209–H1224 (2007). https://doi.org/10.1152/ajpheart.01047.2006
11. Humphrey, J.D.: Vascular adaptation and mechanical homeostasis at tissue, cellular, and subcellular levels. Cell Biochem. Biophys. **50**(2), 53–78 (2008). https://doi.org/10.1007/s12013-007-9002-3
12. Canovic, E.P., Zollinger, A.J., Tam, S.N., Smith, M.L., Stamenovic, D.: Tensional homeostasis in endothelial cells is a multicellular phenomenon. Am. J. Physiol. Cell Physiol. **311**(3), C528–C535 (2016). https://doi.org/10.1152/ajpcell.00037.2016

13. Zollinger, A.J., Xu, H., Figueiredo, J., Paredes, J., Seruca, R., Stamenovic, D., Smith, M.L.: Dependence of tensional homeostasis on cell type and on cell-cell interactions. Cell Mol. Bioeng. **11**(3), 175–184 (2018). https://doi.org/10.1007/s12195-018-0527-x

14. Praetorius, H.A., Frokiaer, J., Leipziger, J.: Transepithelial pressure pulses induce nucleotide release in polarized MDCK cells. Am. J. Physiol. Renal. Physiol. **288**(1), F133–F141 (2005). https://doi.org/10.1152/ajprenal.00238.2004

15. Jacobs, C.R., Temiyasathit, S., Castillo, A.B.: Osteocyte mechanobiology and pericellular mechanics. Annu. Rev. Biomed. Eng. **12**, 369–400 (2010). https://doi.org/10.1146/annurev-bioeng-070909-105302

16. Shao, Y.Y., Wang, L., Welter, J.F., Ballock, R.T.: Primary cilia modulate Ihh signal transduction in response to hydrostatic loading of growth plate chondrocytes. Bone **50**(1), 79–84 (2012). https://doi.org/10.1016/j.bone.2011.08.033

17. Luo, N., Conwell, M.D., Chen, X., Kettenhofen, C.I., Westlake, C.J., Cantor, L.B., Wells, C.D., Weinreb, R.N., Corson, T.W., Spandau, D.F., Joos, K.M., Iomini, C., Obukhov, A.G., Sun, Y.: Primary cilia signaling mediates intraocular pressure sensation. Proc. Natl. Acad. Sci. U S A **111**(35), 12871–12876 (2014). https://doi.org/10.1073/pnas.1323292111

18. Martineau, L.C., Gardiner, P.F.: Insight into skeletal muscle mechanotransduction: MAPK activation is quantitatively related to tension. J. Appl. Physiol. **91**(2), 693–702 (2001)

19. Tock, Y., Ljubisavljevic, M., Thunberg, J., Windhorst, U., Inbar, G.F., Johansson, H.: Information-theoretic analysis of de-efferented single muscle spindles. Biol. Cybern. **87**(4), 241–248 (2002). https://doi.org/10.1007/s00422-002-0341-2

20. Fisher, A.B., Al-Mehdi, A.B., Manevich, Y.: Shear stress and endothelial cell activation. Crit. Care Med. **30**(5), S192–S197 (2002)

21. Osol, G.: Mechanotransduction by vascular smooth muscle. J. Vasc. Res. **32**(5), 275–292 (1995)

22. Suki, B., Ito, S., Stamenovic, D., Lutchen, K.R., Ingenito, E.P.: Biomechanics of the lung parenchyma: critical roles of collagen and mechanical forces. J. Appl. Physiol. **98**(5), 1892–1899 (2005)

23. Waters, C.M., Sporn, P.H., Liu, M., Fredberg, J.J.: Cellular biomechanics in the lung. Am. J. Physiol. Lung Cell Mol. Physiol. **283**(3), L503–509 (2002). https://doi.org/10.1152/ajplung.00141.2002

24. Wirtz, H.R., Dobbs, L.G.: Calcium mobilization and exocytosis after one mechanical stretch of lung epithelial cells. Science **250**(4985), 1266–1269 (1990)

25. Torday, J.S.: Homeostasis as the mechanism of evolution. Biol. (Basel) **4**(3), 573–590 (2015). https://doi.org/10.3390/biology4030573

26. Waldeck, M.R,, Lambert, M.I.. Heart rate during sleep: implications for monitoring training status. J. Sports Sci. Med. **2**(4), 133–138 (2003)

27. Peng, C.K., Buldyrev, S.V., Hausdorff, J.M., Havlin, S., Mietus, J.E., Simons, M., Stanley, H.E., Goldberger, A.L.: Non-equilibrium dynamics as an indispensable characteristic of a healthy biological system. Integr. Physiol. Behav. Sci. **29**(3), 283–293 (1994)

28. Peng, C.K., Havlin, S., Stanley, H.E., Goldberger, A.L.: Quantification of scaling exponents and crossover phenomena in nonstationary heartbeat time series. Chaos **5**(1), 82–87 (1995). https://doi.org/10.1063/1.166141

29. van de Borne, P., Schintgen, M., Niset, G., Schoenfeld, P., Nguyen, H., Degre, S., Degaute, J.P.: Does cardiac denervation affect the short-term blood pressure variability in humans? J. Hypertens. **12**(12), 1395–1403 (1994)

30. Dellaca, R.L., Aliverti, A., Lo Mauro, A., Lutchen, K.R., Pedotti, A., Suki, B.: Correlated variability in the breathing pattern and end-expiratory lung volumes in conscious humans. PLoS ONE **10**(3), e0116317 (2015). https://doi.org/10.1371/journal.pone.0116317

31. Hoffman, B.D., Grashoff, C., Schwartz, M.A.: Dynamic molecular processes mediate cellular mechanotransduction. Nature **475**(7356), 316–323 (2011). https://doi.org/10.1038/nature10316

32. Bartolak-Suki, E., Imsirovic, J., Parameswaran, H., Wellman, T.J., Martinez, N., Allen, P.G., Frey, U., Suki, B.: Fluctuation-driven mechanotransduction regulates mitochondrial-network structure and function. Nat. Mater. **14**(10), 1049–1057 (2015). https://doi.org/10.1038/nmat4358

33. Noble, D.: A theory of biological relativity: no privileged level of causation. Interface Focus **2**(1), 55–64 (2012). https://doi.org/10.1098/rsfs.2011.0067

34. Que, C.L., Kenyon, C.M., Olivenstein, R., Macklem, P.T., Maksym, G.N.: Homeokinesis and short-term variability of human airway caliber. J. Appl. Physiol. **91**(3), 1131–1141 (2001)

35. Suki, B., Parameswaran, H., Imsirovic, J., Bartolak-Suki, E.: Regulatory roles of fluctuation-driven mechanotransduction in cell function. Physiol. (Bethesda) **31**(5), 346–358 (2016). https://doi.org/10.1152/physiol.00051.2015

36. Alenghat, F.J., Ingber, D.E.: Mechanotransduction: all signals point to cytoskeleton, matrix, and integrins. Sci STKE. **2002**(119), pe6 (2002). https://doi.org/10.1126/stke.2002.119.pe6

37. Chatterjee, S., Fisher, A.B.: Mechanotransduction: forces, sensors, and redox signaling. Antioxid. Redox Signal. **20**(6), 868–871 (2014). https://doi.org/10.1089/ars.2013.5753

38. Gieni, R.S., Hendzel, M.J.: Mechanotransduction from the ECM to the genome: are the pieces now in place? J. Cell. Biochem. **104**(6), 1964–1987 (2008). https://doi.org/10.1002/jcb.21364

39. Ingber, D.E.: Cellular mechanotransduction: putting all the pieces together again. FASEB J. **20**(7), 811–827 (2006). https://doi.org/10.1096/fj.05-5424rev

40. Mammoto, T., Mammoto, A., Ingber, D.E.: Mechanobiology and developmental control. Ann. Rev. Cell Dev. Biol. **29**, 27–61 (2013). https://doi.org/10.1146/annurev-cellbio-101512-122340

41. Orr, A.W., Helmke, B.P., Blackman, B.R., Schwartz, M.A.: Mechanisms of mechanotransduction. Dev. Cell **10**(1), 11–20 (2006). https://doi.org/10.1016/j.devcel.2005.12.006

42. Schwartz, M.A., DeSimone, D.W.: Cell adhesion receptors in mechanotransduction. Curr. Opin. Cell Biol. **20**(5), 551–556 (2008). https://doi.org/10.1016/j.ceb.2008.05.005

43. Tschumperlin, D.J.: Mechanotransduction. Compr. Physiol. **1**(2), 1057–1073 (2011). https://doi.org/10.1002/cphy.c100016

44. Cavallaro, U., Christofori, G.: Cell adhesion and signalling by cadherins and Ig-CAMs in cancer. Nat. Rev. Cancer **4**(2), 118–132 (2004). https://doi.org/10.1038/nrc1276

45. Geiger, B., Spatz, J.P., Bershadsky, A.D.: Environmental sensing through focal adhesions. Nat. Rev. Mol. Cell Biol. **10**(1), 21–33 (2009). https://doi.org/10.1038/nrm2593

46. Juliano, R.L.: Signal transduction by cell adhesion receptors and the cytoskeleton: functions of integrins, cadherins, selectins, and immunoglobulin-superfamily members. Ann. Rev. Pharmacol. Toxicol. **42**, 283–323 (2002). https://doi.org/10.1146/annurev.pharmtox.42.090401.151133

47. Romer, L.H., Birukov, K.G., Garcia, J.G.: Focal adhesions: paradigm for a signaling nexus. Circ. Res. **98**(5), 606–616 (2006). https://doi.org/10.1161/01.RES.0000207408.31270.db

48. Hynes, R.O.: Integrins: bidirectional, allosteric signaling machines. Cell **110**(6), 673–687 (2002)

49. Hynes, R.O.: Integrins: versatility, modulation, and signaling in cell adhesion. Cell **69**(1), 11–25 (1992)

50. Ruoslahti, E.: Integrins. J. Clin. Invest. **87**(1), 1–5 (1991). https://doi.org/10.1172/JCI114957

51. Loftus, J.C., Liddington, R.C.: New insights into integrin-ligand interaction. J. Clin. Invest. **100**(11), S77–S81 (1997)

52. Hughes, P.E., Renshaw, M.W., Pfaff, M., Forsyth, J., Keivens, V.M., Schwartz, M.A., Ginsberg, M.H.: Suppression of integrin activation: a novel function of a Ras/Raf-initiated MAP kinase pathway. Cell **88**(4), 521–530 (1997)

53. Smyth, S.S., Joneckis, C.C., Parise, L.V.: Regulation of vascular integrins. Blood **81**(11), 2827–2843 (1993)

54. Schwartz, M.A.: Integrin signaling revisited. Trends Cell Biol. **11**(12), 466–470 (2001)

55. Aplin, A.E., Howe, A., Alahari, S.K., Juliano, R.L.: Signal transduction and signal modulation by cell adhesion receptors: the role of integrins, cadherins, immunoglobulin-cell adhesion molecules, and selectins. Pharmacol. Rev. **50**(2), 197–263 (1998)

56. Brancaccio, M., Hirsch, E., Notte, A., Selvetella, G., Lembo, G., Tarone, G.: Integrin signalling: the tug-of-war in heart hypertrophy. Cardiovasc. Res. **70**(3), 422–433 (2006). https://doi.org/10.1016/j.cardiores.2005.12.015

57. Iqbal, J., Zaidi, M.: Molecular regulation of mechanotransduction. Biochem. Biophys. Res. Commun. **328**(3), 751–755 (2005). https://doi.org/10.1016/j.bbrc.2004.12.087

58. Zhang, W., Liu, H.T.: MAPK signal pathways in the regulation of cell proliferation in mammalian cells. Cell Res. **12**(1), 9–18 (2002). https://doi.org/10.1038/sj.cr.7290105

59. Klemke, R.L., Cai, S., Giannini, A.L., Gallagher, P.J., de Lanerolle, P., Cheresh, D.A.: Regulation of cell motility by mitogen-activated protein kinase. J. Cell Biol. **137**(2), 481–492 (1997)

60. Morishima-Kawashima, M., Kosik, K.S.: The pool of map kinase associated with microtubules is small but constitutively active. Mol. Biol. Cell **7**(6), 893–905 (1996)

61. Clark, E.A., King, W.G., Brugge, J.S., Symons, M., Hynes, R.O.: Integrin-mediated signals regulated by members of the rho family of GTPases. J. Cell Biol. **142**(2), 573–586 (1998)

62. Amano, M., Ito, M., Kimura, K., Fukata, Y., Chihara, K., Nakano, T., Matsuura, Y., Kaibuchi, K.: Phosphorylation and activation of myosin by Rho-associated kinase (Rho-kinase). J. Biol. Chem. **271**(34), 20246–20249 (1996)

63. Roan. E., Waters, C.M.: What do we know about mechanical strain in lung alveoli. Am. J. Physiol. Lung Cell. Mol. Physiol. **301**(5), L625–L635 (2011). https://doi.org/10.1152/ajplung.00105.2011

64. Tschumperlin, D.J., Margulies, S.S.: Alveolar epithelial surface area-volume relationship in isolated rat lungs. J. Appl. Physiol. **86**(6), 2026–2033 (1999)

65. LaPrad, A.S., Lutchen, K.R., Suki, B.: A mechanical design principle for tissue structure and function in the airway tree. PLoS Comput. Biol. **9**(5), e1003083 (2013). https://doi.org/10.1371/journal.pcbi.1003083

66. Sinclair, S.E., Molthen, R.C., Haworth, S.T., Dawson, C.A., Waters, C.M.: Airway strain during mechanical ventilation in an intact animal model. Am. J. Respir. Crit. Care Med. **176**(8), 786–794 (2007). https://doi.org/10.1164/rccm.200701-088OC

67. Liu, H., Yambe, T., Sasada, H., Nanka, S., Tanaka, A., Nagatomi, R., Nitta, S.: Comparison of heart rate variability and stroke volume variability. Auton. Neurosci. **116**(1–2), 69–75 (2004). https://doi.org/10.1016/j.autneu.2004.09.003

68. Peng, C.K., Mietus, J., Hausdorff, J.M., Havlin, S., Stanley, H.E., Goldberger, A.L.: Long-range anticorrelations and non-Gaussian behavior of the heartbeat. Phys. Rev. Lett. **70**(9), 1343–1346 (1993). https://doi.org/10.1103/PhysRevLett.70.1343

69. Jelinek, H.F., Imam, H.M., Al-Aubaidy, H., Khandoker, A.H.: Association of cardiovascular risk using non-linear heart rate variability measures with the framingham risk score in a rural population. Front. Physiol. **4**. ARTN 186 (2013). https://doi.org/10.3389/fphys.2013.00186

70. Cheng, C., Helderman, F., Tempel, D., Segers, D., Hierck, B., Poelmann, R., van Tol, A., Duncker, D.J., Robbers-Visser, D., Ursem, N.T., van Haperen, R., Wentzel, J.J., Gijsen, F., van der Steen, A.F., de Crom, R., Krams, R.: Large variations in absolute wall shear stress levels within one species and between species. Atherosclerosis **195**(2), 225–235 (2007). https://doi.org/10.1016/j.atherosclerosis.2006.11.019

71. Karau, K.L., Krenz, G.S., Dawson, C.A.: Branching exponent heterogeneity and wall shear stress distribution in vascular trees. Am. J. Physiol. Heart Circ. Physiol. **280**(3), H1256–H1263 (2001)

72. Reneman, R.S., Hoeks, A.P.: Wall shear stress as measured in vivo: consequences for the design of the arterial system. Med. Biol. Eng. Comput. **46**(5), 499–507 (2008). https://doi.org/10.1007/s11517-008-0330-2

73. Uzarski, J.S., Scott, E.W., McFetridge, P.S.: Adaptation of endothelial cells to physiologically-modeled, variable shear stress. PLoS ONE **8**(2), e57004 (2013). https://doi.org/10.1371/journal.pone.0057004

74. Mortola, J.P., Lanthier, C.: Scaling the amplitudes of the circadian pattern of resting oxygen consumption, body temperature and heart rate in mammals. Comp. Biochem. Physiol. A: Mol. Integr. Physiol. **139**(1), 83–95 (2004). https://doi.org/10.1016/j.cbpb.2004.07.007

75. West, G.B., Brown, J.H., Enquist, B.J.: A general model for the origin of allometric scaling laws in biology. Science **276**(5309), 122–126 (1997)
76. Vogel, V.: Mechanotransduction involving multimodular proteins: converting force into biochemical signals. Ann. Rev. Biophys. Biomol. Struct. **35**, 459–488 (2006). https://doi.org/10.1146/annurev.biophys.35.040405.102013
77. Evans, E.A., Calderwood, D.A.: Forces and bond dynamics in cell adhesion. Science **316**(5828), 1148–1153 (2007). https://doi.org/10.1126/science.1137592
78. Liu, F., Ou-Yang, Z.C.: Force modulating dynamic disorder: a physical model of catch-slip bond transitions in receptor-ligand forced dissociation experiments. Phys. Rev. E: Stat., Nonlin, Soft Matter Phys. **74**(5 Pt 1), 051904 (2006). https://doi.org/10.1103/PhysRevE.74.051904
79. Imsirovic, J., Derricks, K., Buczek-Thomas, J.A., Rich, C.B., Nugent, M.A., Suki, B.: A novel device to stretch multiple tissue samples with variable patterns: application for mRNA regulation in tissue-engineered constructs. Biomatter. **3**(3) (2013). https://doi.org/10.4161/biom.24650
80. Eldib, M., Dean, D.A.: Cyclic stretch of alveolar epithelial cells alters cytoskeletal micromechanics. Biotechnol. Bioeng. **108**(2), 446–453 (2011). https://doi.org/10.1002/bit.22941
81. Parameswaran, H., Lutchen, K.R., Suki, B.: A computational model of the response of adherent cells to stretch and changes in substrate stiffness. J. Appl. Physiol. **116**(7), 825–834 (2014). https://doi.org/10.1152/japplphysiol.00962.2013
82. Kovacs, M., Thirumurugan, K., Knight, P.J., Sellers, J.R.: Load-dependent mechanism of nonmuscle myosin 2. Proc. Natl. Acad. Sci. U S A **104**(24), 9994–9999 (2007). https://doi.org/10.1073/pnas.0701181104
83. Norstrom, M.F., Smithback, P.A., Rock, R.S.: Unconventional processive mechanics of nonmuscle myosin IIB. J. Biol. Chem. **285**(34), 26326–26334 (2010). https://doi.org/10.1074/jbc.M110.123851
84. Wang, N., Tolic-Norrelykke, I.M., Chen, J., Mijailovich, S.M., Butler, J.P., Fredberg, J.J., Stamenovic, D.: Cell prestress. I. Stiffness and prestress are closely associated in adherent contractile cells. Am. J. Physiol. Cell Physiol. **282**(3), C606–C616 (2002). https://doi.org/10.1152/ajpcell.00269.2001
85. Gao, Y.Z., Saphirstein, R.J., Yamin, R., Suki, B., Morgan, K.G.: Aging impairs smooth muscle-mediated regulation of aortic stiffness: a defect in shock absorption function? Am. J. Physiol. Heart Circ. Physiol. **307**(8), H1252–H1261 (2014). https://doi.org/10.1152/ajpheart.00392.2014
86. Ehrenberg, B., Montana, V., Wei, M.D., Wuskell, J.P., Loew, L.M.: Membrane potential can be determined in individual cells from the nernstian distribution of cationic dyes. Biophys. J. **53**(5), 785–794 (1988). https://doi.org/10.1016/S0006-3495(88)83158-8
87. Kadenbach, B., Ramzan, R., Wen, L., Vogt, S.: New extension of the Mitchell Theory for oxidative phosphorylation in mitochondria of living organisms. Biochim. Biophys. Acta 1800 **3**, 205–212 (2010). https://doi.org/10.1016/j.bbagen.2009.04.019
88. Otera, H., Ishihara, N., Mihara, K.: New insights into the function and regulation of mitochondrial fission. Biochim. Biophys. Acta 1833 **5**, 1256–1268 (2013). https://doi.org/10.1016/j.bbamcr.2013.02.002
89. Bartolak-Suki, E., Imsirovic, J., Nishibori, Y., Krishnan, R., Suki, B.: Regulation of mitochondrial structure and dynamics by the cytoskeleton and mechanical factors. Int. J. Mol. Sci. **18**(8), 1812 (2017). https://doi.org/10.3390/ijms18081812
90. Edwards, Y.S., Sutherland, L.M., Power, J.H., Nicholas, T.E., Murray, A.W.: Cyclic stretch induces both apoptosis and secretion in rat alveolar type II cells. FEBS Lett. **448**(1), 127–130 (1999)
91. Arold, S.P., Bartolak-Suki, E., Suki, B.: Variable stretch pattern enhances surfactant secretion in alveolar type II cells in culture. Am. J. Physiol. Lung Cell Mol. Physiol. **296**(4), L574–581. pii: 90454.2008 (2009). https://doi.org/10.1152/ajplung.90454.2008
92. Majumdar, A., Arold, S.P., Bartolak-Suki, E., Parameswaran, H., Suki, B.: Jamming dynamics of stretch-induced surfactant release by alveolar type II cells. J. Appl. Physiol. **112**(5), 824–831 (2012). https://doi.org/10.1152/japplphysiol.00975.2010

93. Rose, F., Kurth-Landwehr, C., Sibelius, U., Reuner, K.H., Aktories, K., Seeger, W., Grimminger, F.: Role of actin depolymerization in the surfactant secretory response of alveolar epithelial type II cells. Am. J. Respir. Crit. Care Med. **159**(1), 206–212 (1999). https://doi.org/10.1164/ajrccm.159.1.9801106

94. Abonyo, B.O., Gou, D., Wang, P., Narasaraju, T., Wang, Z., Liu, L.: Syntaxin 2 and SNAP-23 are required for regulated surfactant secretion. Biochemistry **43**(12), 3499–3506 (2004). https://doi.org/10.1021/bi036338y

95. Sieber, J.J., Willig, K.I., Kutzner, C., Gerding-Reimers, C., Harke, B., Donnert, G., Rammner, B., Eggeling, C., Hell, S.W., Grubmuller, H., Lang, T.: Anatomy and dynamics of a supramolecular membrane protein cluster. Science **317**(5841), 1072–1076 (2007). https://doi.org/10.1126/science.1141727

96. Naruse, K., Sokabe, M.: Involvement of stretch-activated ion channels in Ca2+ mobilization to mechanical stretch in endothelial cells. Am. J. Physiol. **264**(4), C1037–C1044 (1993)

97. Bezrukov, S.M., Vodyanoy, I.: Noise-induced enhancement of signal transduction across voltage-dependent ion channels. Nature **378**(6555), 362–364 (1995)

98. Schurch, S., Bachofen, H., Goerke, J., Green, F.: Surface properties of rat pulmonary surfactant studied with the captive bubble method: adsorption, hysteresis, stability. Biochim. Biophys. Acta. **1103**(1), 127–136 (1992)

99. Tarbell, J.M., Simon, S.I., Curry, F.R.: Mechanosensing at the vascular interface. Ann. Rev. Biomed. Eng. **16**, 505–532 (2014). https://doi.org/10.1146/annurev-bioeng-071813-104908

100. Mazzag, B., Barakat, A.I.: The effect of noisy flow on endothelial cell mechanotransduction: a computational study. Ann. Biomed. Eng. **39**(2), 911–921 (2011). https://doi.org/10.1007/s10439-010-0181-5

101. Gouverneur, M., Berg, B., Nieuwdorp, M., Stroes, E., Vink, H.: Vasculoprotective properties of the endothelial glycocalyx: effects of fluid shear stress. J. Int. Med. **259**(4), 393–400 (2006). https://doi.org/10.1111/j.1365-2796.2006.01625.x

102. Kuchan, M.J., Jo, H., Frangos, J.A.: Role of G proteins in shear stress-mediated nitric oxide production by endothelial cells. Am. J. Physiol. **267**(3 Pt 1), C753–C758 (1994)

103. Morgan, K.Y., Black, D.L.: Creation of a bioreactor for the application of variable amplitude mechanical stimulation of fibrin gel-based engineered cardiac tissue. In: III MRaLDB. (ed.) Cardiac Tissue Engineering: Methods and Protocols, Vol. 1181, pp. 177–187. Springer Science + Business Media, New York (2014a). https://doi.org/10.1007/978-1-4939-1047-2_16

104. Morgan, K.Y., Black, L.D., III.: Investigation into the effects of varying frequency of mechanical stimulation in a cycle-by-cycle manner on engineered cardiac construct function. J. Tissue Eng. Regen. Med. (2014b). https://doi.org/10.1002/term.1915

105. Gupta, M., Doss, B., Lim, C.T., Voituriez, R., Ladoux, B.: Single cell rigidity sensing: a complex relationship between focal adhesion dynamics and large-scale actin cytoskeleton remodeling. Cell Adhes. Migr. **10**(5), 554–567 (2016). https://doi.org/10.1080/19336918.2016.1173800

106. Alves, C., Araujo, A.D., Oliveira, C.L., Imsirovic, J., Bartolak-Suki, E., Andrade, J.S., Suki, B.: Homeostatic maintenance via degradation and repair of elastic fibers under tension. Sci. Rep. **6**, 27474 (2016). https://doi.org/10.1038/srep27474

107. Nissen, R., Cardinale, G.J., Udenfriend, S.: Increased turnover of arterial collagen in hypertensive rats. Proc. Natl. Acad. Sci. U S A **75**(1), 451–453 (1978)

108. Rucklidge, G.J., Milne, G., McGaw, B.A., Milne, E., Robins, S.P.: Turnover rates of different collagen types measured by isotope ratio mass spectrometry. Biochim. Biophys. Acta **1156**(1), 57–61 (1992)

109. Kjaer, M., Langberg, H., Miller, B.F., Boushel, R., Crameri, R., Koskinen, S., Heinemeier, K., Olesen, J.L., Dossing, S., Hansen, M., Pedersen, S.G., Rennie, M.J., Magnusson, P.: Metabolic activity and collagen turnover in human tendon in response to physical activity. J. Musculoskelet. Neuronal. Interact. **5**(1), 41–52 (2005)

110. Heinemeier, K.M., Schjerling, P., Heinemeier, J., Magnusson, S.P., Kjaer, M.: Lack of tissue renewal in human adult Achilles tendon is revealed by nuclear bomb ^{14}C. FASEB J. **27**(5), 2074–2079 (2013). https://doi.org/10.1096/fj.12-225599

111. Goh, K.L., Holmes, D.F., Lu, Y., Purslow, P.P., Kadler, K.E., Bechet, D., Wess, T.J.: Bimodal collagen fibril diameter distributions direct age-related variations in tendon resilience and resistance to rupture. J. Appl. Physiol. **113**(6), 878–888 (2012). https://doi.org/10.1152/japplphysiol.00258.2012

112. Austad, S.N.: Comparative aging and life histories in mammals. Exp. Gerontol. **32**(1–2), 23–38 (1997)

113. Branchet, M.C., Boisnic, S., Frances, C., Lesty, C., Robert, L.: Morphometric analysis of dermal collagen fibers in normal human skin as a function of age. Arch. Gerontol. Geriatr. **13**(1), 1–14 (1991)

114. Araujo, A.D., Majumdar, A., Parameswaran, H., Yi, E., Spencer, J.L., Nugent, M.A., Suki, B.: Dynamics of enzymatic digestion of elastic fibers and networks under tension. Proc. Natl. Acad. Sci. U S A **108**(23), 9414–9419 (2011). https://doi.org/10.1073/pnas.1019188108

115. Gupta, M., Sarangi, B.R., Deschamps, J., Nematbakhsh, Y., Callan-Jones, A., Margadant, F., Mege, R.M., Lim, C.T., Voituriez, R., Ladoux, B.: Adaptive rheology and ordering of cell cytoskeleton govern matrix rigidity sensing. Nat. Commun. **6**, 7525 (2015). https://doi.org/10.1038/ncomms8525

116. Jesudason, R., Sato, S., Parameswaran, H., Araujo, A.D., Majumdar, A., Allen, P.G., Bartolak-Suki, E., Suki, B. Mechanical forces regulate elastase activity and binding site availability in lung elastin. Biophys. J. **99**(9), 3076–3083, pii: S0006-3495(10)01164-1 (2010). https://doi.org/10.1016/j.bpj.2010.09.018

117. Orgel, J.P., San Antonio, J.D., Antipova, O.: Molecular and structural mapping of collagen fibril interactions. Connect. Tissue Res. **52**(1), 2–17 (2011). https://doi.org/10.3109/03008207.2010.511353

118. Minary-Jolandan, M., Yu, M.F.: Nanomechanical heterogeneity in the gap and overlap regions of type I collagen fibrils with implications for bone heterogeneity. Biomacromolecules **10**(9), 2565–2570 (2009). https://doi.org/10.1021/bm900519v

119. Christiansen, D.L., Huang, E.K., Silver, F.H.: Assembly of type I collagen: fusion of fibril subunits and the influence of fibril diameter on mechanical properties. Matrix Biol. **19**(5), 409–420 (2000)

120. Bartolak-Suki, E., Suki, B.: Variability in cyclic stretch accelerates collagen secretion through cross talk with integrin Beta1 in cultured mouse fibroblasts. Am. J. Respir. Cell Mol. Biol. **5**, A3499 (2011)

121. Imsirovic, J., Bartolak-Suki, E., Jawde, S.B., Parameswaran, H., Suki, B.: Blood pressure-induced physiological strain variability modulates wall structure and function in aorta rings. Physiol. Meas. **39**(10), 105014 (2018). https://doi.org/10.1088/1361-6579/aae65f

122. Lavagnino, M., Brooks, A.E., Oslapas, A.N., Gardner, K.L., Arnoczky, S.P.: Crimp length decreases in lax tendons due to cytoskeletal tension, but is restored with tensional homeostasis. J. Orthop. Res. **35**(3), 573–579 (2017). https://doi.org/10.1002/jor.23489

123. Tobin, M.J., Mador, M.J., Guenther, S.M., Lodato, R.F., Sackner, M.A.: Variability of resting respiratory drive and timing in healthy subjects. J. Appl. Physiol. **65**(1), 309–317 (1988)

124. Kuratomi, Y., Okazaki, N., Ishihara, T., Arai, T., Kira, S.: Variability of breath-by-breath tidal volume and its characteristics in normal and diseased subjects: ventilatory monitoring with electrical impedance pneumography. Jpn. J. Med. **24**(2), 141–149 (1985)

125. Brack, T., Jubran, A., Tobin, M.J.: Effect of elastic loading on variational activity of breathing. Am. J. Respir. Crit. Care Med. **155**(4), 1341–1348 (1997). https://doi.org/10.1164/ajrccm.155.4.9105077

126. Brack, T., Jubran, A., Tobin, M.J.: Dyspnea and decreased variability of breathing in patients with restrictive lung disease. Am. J. Respir. Crit. Care Med. **165**(9), 1260–1264 (2002). https://doi.org/10.1164/rccm.2201018

127. Boker, A., Graham, M.R., Walley, K.R., McManus, B.M., Girling, L.G., Walker, E., Lefevre, G.R., Mutch, W.A.: Improved arterial oxygenation with biologically variable or fractal ventilation using low tidal volumes in a porcine model of acute respiratory distress syndrome. Am. J. Respir. Crit. Care Med. **165**(4), 456–462 (2002). https://doi.org/10.1164/ajrccm.165.4.2108006

128. Thammanomai, A., Hamakawa, H., Bartolak-Suki, E., Suki, B.: Combined effects of ventilation mode and positive end-expiratory pressure on mechanics, gas exchange and the epithelium in mice with acute lung injury. PLoS ONE **8**(1), e53934 (2013). https://doi.org/10.1371/journal.pone.0053934

129. Thammanomai, A., Hueser, L.E., Majumdar, A., Bartolak-Suki, E., Suki, B.: Design of a new variable-ventilation method optimized for lung recruitment in mice. J. Appl. Physiol. **104**(5), 1329–1340 (2008). https://doi.org/10.1152/japplphysiol.01002.2007

130. Bartolák-Suki, E., Noble, P.B., Bou Jawde, S., Pillow, J.J., Suki, B.: Optimization of variable ventilation for physiology, immune response and surfactant enhancement in preterm lambs. Front. Physiol. **8**, 425 (2017). https://doi.org/10.3389/fphys.2017.00425

131. Glass, L.: Synchronization and rhythmic processes in physiology. Nature **410**(6825), 277–284 (2001)

132. Mancia, G., Parati, G., Hennig, M., Flatau, B., Omboni, S., Glavina, F., Costa, B., Scherz, R., Bond, G., Zanchetti, A., Investigators, E.: Relation between blood pressure variability and carotid artery damage in hypertension: baseline data from the European Lacidipine Study on Atherosclerosis (ELSA). J. Hypertens. **19**(11), 1981–1989 (2001)

133. Su, D.F., Miao, C.Y.: Blood pressure variability and organ damage. Clin. Exp. Pharmacol. Physiol. **28**(9), 709–715 (2001)

134. Parati, G., Pomidossi, G., Albini, F., Malaspina, D., Mancia, G.: Relationship of 24-hour blood pressure mean and variability to severity of target-organ damage in hypertension. J. Hypertens. **5**(1), 93–98 (1987)

135. Rothwell, P.M., Howard, S.C., Dolan, E., O'Brien, E., Dobson, J.E., Dahlof, B., Sever, P.S., Poulter, N.R.: Prognostic significance of visit-to-visit variability, maximum systolic blood pressure, and episodic hypertension. Lancet **375**(9718), 895–905 (2010). https://doi.org/10.1016/S0140-6736(10)60308-X

136. Miao, C.Y., Su, D.F.: The importance of blood pressure variability in rat aortic and left ventricular hypertrophy produced by sinoaortic denervation. J. Hypertens. **20**(9), 1865–1872 (2002)

137. Xie, H.H., Shen, F.M., Cao, Y.B., Li, H.L., Su, D.F.: Effects of low-dose ketanserin on blood pressure variability, baroreflex sensitivity and end-organ damage in spontaneously hypertensive rats. Clin. Sci. (Lond.) **108**(6), 547–552 (2005). https://doi.org/10.1042/CS20040310

138. Schillaci, G., Bilo, G., Pucci, G., Laurent, S., Macquin-Mavier, I., Boutouyrie, P., Battista, F., Settimi, L., Desamericq, G., Dolbeau, G., Faini, A., Salvi, P., Mannarino, E., Parati, G.: Relationship between short-term blood pressure variability and large-artery stiffness in human hypertension: findings from 2 large databases. Hypertension **60**(2), 369–377 (2012). https://doi.org/10.1161/hypertensionaha.112.197491

139. Mitchell, G.F., Guo, C.Y., Benjamin, E.J., Larson, M.G., Keyes, M.J., Vita, J.A., Vasan, R.S., Levy, D.: Cross-sectional correlates of increased aortic stiffness in the community: the Framingham Heart Study. Circulation **115**(20), 2628–2636 (2007). https://doi.org/10.1161/CIRCULATIONAHA.106.667733

140. Zhang, Y., Agnoletti, D., Safar, M.E., Blacher, J.: Effect of antihypertensive agents on blood pressure variability: the Natrilix SR versus candesartan and amlodipine in the reduction of systolic blood pressure in hypertensive patients (X-CELLENT) study. Hypertension **58**(2), 155–160 (2011). https://doi.org/10.1161/hypertensionaha.111.174383

141. Constant, I., Laude, D., Elghozi, J.L., Murat, I.: Assessment of autonomic cardiovascular changes associated with recovery from anaesthesia in children: a study using spectral analysis of blood pressure and heart rate variability. Paediatr. Anaesth. **10**(6), 653–660 (2000)

142. Conci, F., Di Rienzo, M., Castiglioni, P.: Blood pressure and heart rate variability and baroreflex sensitivity before and after brain death. J. Neurol. Neurosurg. Psychiatry **71**(5), 621–631 (2001)

143. Calderwood, S.K., Murshid, A., Prince, T.: The shock of aging: molecular chaperones and the heat shock response in longevity and aging–a mini-review. Gerontology **55**(5), 550–558 (2009). https://doi.org/10.1159/000225957

144. Engelhardt, M., Martens, U.M.: The implication of telomerase activity and telomere stability for replicative aging and cellular immortality (review). Oncol. Rep. **5**(5), 1043–1052 (1998)
145. Lee, H.C., Wei, Y.H.: Mitochondrial alterations, cellular response to oxidative stress and defective degradation of proteins in aging. Biogerontology **2**(4), 231–244 (2001)
146. Avery, N.C., Bailey, A.J.: Enzymic and non-enzymic cross-linking mechanisms in relation to turnover of collagen: relevance to aging and exercise. Scand. J. Med. Sci. Sports **15**(4), 231–240 (2005)
147. Reiser, K.M., Hennessy, S.M., Last, J.A.: Analysis of age-associated changes in collagen crosslinking in the skin and lung in monkeys and rats. Biochem. Biophys. Acta. **926**(3), 339–348 (1987)
148. Liu, F., Mih, J.D., Shea, B.S., Kho, A.T., Sharif, A.S., Tager, A.M., Tschumperlin, D.J.: Feedback amplification of fibrosis through matrix stiffening and COX-2 suppression. J. Cell Biol. **190**(4), 693–706. pii: jcb.201004082 (2010). https://doi.org/10.1083/jcb.201004082
149. Phillip, J.M., Aifuwa, I., Walston, J., Wirtz, D.: The mechanobiology of aging. Ann. Rev. Biomed. Eng. **17**, 113–141 (2015). https://doi.org/10.1146/annurev-bioeng-071114-040829
150. Vafaie, F., Yin, H., O'Neil, C., Nong, Z., Watson, A., Arpino, J.M., Chu, M.W., Wayne Holdsworth, D., Gros, R., Pickering, J.G.: Collagenase-resistant collagen promotes mouse aging and vascular cell senescence. Aging Cell **13**(1), 121–130 (2014). https://doi.org/10.1111/acel.12155
151. Harman, D.: The biologic clock: the mitochondria? J. Am. Geriatr. Soc. **20**(4), 145–147 (1972)
152. Bratic, A., Larsson, N.G.: The role of mitochondria in aging. J. Clin. Invest. **123**(3), 951–957 (2013). https://doi.org/10.1172/JCI64125
153. Copeland, J.M., Cho, J., Lo Jr., T., Hur, J.H., Bahadorani, S., Arabyan, T., Rabie, J., Soh, J., Walker, D.W.: Extension of Drosophila life span by RNAi of the mitochondrial respiratory chain. Curr. Biol. **19**(19), 1591–1598 (2009). https://doi.org/10.1016/j.cub.2009.08.016
154. Acuna-Castroviejo, D., Carretero, M., Doerrier, C., Lopez, L.C., Garcia-Corzo, L., Tresguerres, J.A., Escames, G.: Melatonin protects lung mitochondria from aging. Age (Dordr) **34**(3), 681–692 (2012). https://doi.org/10.1007/s11357-011-9267-8
155. Quiros, P.M., Mottis, A., Auwerx, J.: Mitonuclear communication in homeostasis and stress. Nat. Rev. Mol. Cell Biol. **17**(4), 213–226 (2016). https://doi.org/10.1038/nrm.2016.23
156. Gomes, A.P., Price, N.L., Ling, A.J., Moslehi, J.J., Montgomery, M.K., Rajman, L., White, J.P., Teodoro, J.S., Wrann, C.D., Hubbard, B.P., Mercken, E.M., Palmeira, C.M., de Cabo, R., Rolo, A.P., Turner, N., Bell, E.L., Sinclair, D.A.: Declining NAD(+) induces a pseudohypoxic state disrupting nuclear-mitochondrial communication during aging. Cell **155**(7), 1624–1638 (2013). https://doi.org/10.1016/j.cell.2013.11.037
157. Labat-Robert, J., Robert, L.: Aging of the extracellular matrix and its pathology. Exp. Gerontol. **23**(1), 5–18 (1988)
158. Mackay, E.H., Banks, J., Sykes, B., Lee, G.: Structural basis for the changing physical properties of human pulmonary vessels with age. Thorax **33**(3), 335–344 (1978)
159. Sobin, S.S., Fung, Y.C., Tremer, H.M.: Collagen and elastin fibers in human pulmonary alveolar walls. J. Appl. Physiol. **64**(4), 1659–1675 (1988)
160. Mays, P.K., McAnulty, R.J., Campa, J.S., Laurent, G.J.: Age-related changes in collagen synthesis and degradation in rat tissues: importance of degradation of newly synthesized collagen in regulating collagen production. Biochem. J. **276**(2), 307–313 (1991)
161. Pierce, J.A., Resnick, H., Henry, P.H.: Collagen and elastin metabolism in the lungs, skin, and bones of adult rats. J. Lab. Clin. Med. **69**(3), 485–493 (1967)
162. Poole, A., Myllyla, R., Wagner, J.C., Brown, R.C.: Collagen biosynthesis enzymes in lung tissue and serum of rats with experimental silicosis. Br. J. Exp. Pathol. **66**(5), 567–575 (1985)
163. Cerami, A.: Hypothesis: glucose as a mediator of aging. J. Am. Geriatr. Soc. **33**(9), 626–634 (1985)
164. Bellmunt, M.J., Portero, M., Pamplona, R., Cosso, L., Odetti, P., Prat, J.: Evidence for the Maillard reaction in rat lung collagen and its relationship with solubility and age. Biochem. Biophys. Acta. **1272**(1), 53–60 (1995)

165. Miyata, T., Ishikawa, N., van Ypersele de Strihou, C. Carbonyl stress and diabetic complications. Clin. Chem. Lab. Med. CCLM/FESCC. **41**(9), 1150–1158 (2003). https://doi.org/10.1515/cclm.2003.178

166. Monnier, V.M.: Nonenzymatic glycosylation, the Maillard reaction and the aging process. J. Gerontol **45**(4), B105–B111 (1990)

167. Janson, I.A., Putnam, A.J. Extracellular matrix elasticity and topography: material-based cues that affect cell function via conserved mechanisms. J. Biomed. Mater. Res. Part A (2014). https://doi.org/10.1002/jbm.a.35254

168. Bereiter-Hahn, J., Voth, M.: Dynamics of mitochondria in living cells: shape changes, dislocations, fusion, and fission of mitochondria. Microsc. Res. Tech. **27**(3), 198–219 (1994). https://doi.org/10.1002/jemt.1070270303

169. Bach, D., Pich, S., Soriano, F.X., Vega, N., Baumgartner, B., Oriola, J., Daugaard, J.R., Lloberas, J., Camps, M., Zierath, J.R., Rabasa-Lhoret, R., Wallberg-Henriksson, H., Laville, M., Palacin, M., Vidal, H., Rivera, F., Brand, M., Zorzano, A.: Mitofusin-2 determines mitochondrial network architecture and mitochondrial metabolism. A novel regulatory mechanism altered in obesity. J Biol Chem **278**(19), 17190–17197 (2003). https://doi.org/10.1074/jbc.m212754200

170. Anesti, V., Scorrano, L.: The relationship between mitochondrial shape and function and the cytoskeleton. Biochim. Biophys. Acta **1757**(5–6), 692–699 (2006). https://doi.org/10.1016/j.bbabio.2006.04.013

171. Mancia, G., Bombelli, M., Facchetti, R., Madotto, F., Corrao, G., Trevano, F.Q., Grassi, G., Sega, R.: Long-term prognostic value of blood pressure variability in the general population: results of the Pressioni Arteriose Monitorate e Loro Associazioni Study. Hypertension **49**(6), 1265–1270 (2007). https://doi.org/10.1161/HYPERTENSIONAHA.107.088708

172. Wu, Z., Puigserver, P., Andersson, U., Zhang, C., Adelmant, G., Mootha, V., Troy, A., Cinti, S., Lowell, B., Scarpulla, R.C., Spiegelman, B.M.: Mechanisms controlling mitochondrial biogenesis and respiration through the thermogenic coactivator PGC-1. Cell **98**(1), 115–124 (1999). https://doi.org/10.1016/S0092-8674(00)80611-X

173. Veerman, D.P., Imholz, B.P., Wieling, W., Karemaker, J.M., van Montfrans, G.A.: Effects of aging on blood pressure variability in resting conditions. Hypertension **24**(1), 120–130 (1994)

174. Wu, C., Shlipak, M.G., Stawski, R.S., Peralta, C.A., Psaty, B.M., Harris, T.B., Satterfield, S., Shiroma, E.J., Newman, A.B., Odden, M.C., Health, A.B.C.S.: Visit-to-visit blood pressure variability and mortality and cardiovascular outcomes among older adults: the health, aging, and body composition study. Am. J. Hypertens. **30**(2), 151–158 (2017). https://doi.org/10.1093/ajh/hpw106

175. Bartolák-Suki, E., Suki, B.: Variability in stretch amplitude partially restores age-related decline in mitochondrial structure and function in lung fibroblasts. Am. J. Respir. Crit. Care Med. **195**, A2659 (2017)

Experimental Characterization of Adventitial Collagen Fiber Kinematics Using Second-Harmonic Generation Imaging Microscopy: Similarities and Differences Across Arteries, Species and Testing Conditions

Cristina Cavinato, Pierre Badel, Witold Krasny, Stéphane Avril and Claire Morin

Abstract Fibrous collagen networks are well known to play a central role in the passive biomechanical response of soft connective tissues to applied loads. In the current chapter we focus on vascular tissues and share our extensive experience in coupling mechanical loading and multi-photon imaging to investigate, across arteries, species and testing conditions, how collagen fibers move in response to mechanical loading. More specifically, we assess the deformations of collagen networks in rabbit, porcine or human arteries under different loading scenarios: uniaxial tension on flat samples, tension-inflation on tubular samples, bulge inflation on flat samples. We always observe that collagen fibers exhibit a wavy or crimped shape in load-free conditions, and tend to uncrimp when loads are applied, engaging sequentially to become the main load-bearing component. This sequential engagement, which is responsible for the nonlinear mechanical behaviour, is essential for an artery to function normally and appears to be less pronounced for arteries in elderly and aneurysmal patients. Although uncrimping of collagen fibers is a universal mechanism, we also observe large fiber rotations specific to tensile loading, with significant realignment along the loading axis. A unified approach is proposed to compare observations and quantitative analyses as the type of image processing may affect significantly the estimation of collagen fiber deformations. In summary, this chapter makes an important review of the basic roles of arterial microstructure and its deformations on the global

C. Cavinato · P. Badel · W. Krasny · S. Avril (✉) · C. Morin
Mines Saint-Etienne, Univ Lyon, Univ Jean Monnet, INSERM,
U 1059, Sainbiose, Centre CIS, 42023 Saint-Etienne, France
e-mail: avril@emse.fr

C. Cavinato
e-mail: Cristina.cavinato@emse.fr

P. Badel
e-mail: badel@emse.fr

C. Morin
e-mail: claire.morin@emse.fr

© Springer Nature Switzerland AG 2020
Y. Zhang (ed.), *Multi-scale Extracellular Matrix Mechanics and Mechanobiology*,
Studies in Mechanobiology, Tissue Engineering and Biomaterials 23,
https://doi.org/10.1007/978-3-030-20182-1_5

mechanical response. Eventually, directions for future studies combining mechanical loading and multi-photon imaging are suggested, with the aim of addressing open questions related to tissue adaptation and rupture.

1 Introduction

The passive biomechanical response of soft connective tissues to applied loads is well known to depend on their structure as well as on their composition [1–7]. Various proteins, like elastin and collagen, typically assemble together to form fibrils, fibers, and bundles of these fibers, thus creating a structural support for the extracellular matrix (ECM) of soft tissue [8]. The architecture and particular composition of this structural support is believed to respond to specific mechanical demands of each tissue [9–12].

Among such tissues, arterial tissue has been very much focused on due its essential role in human physiology. Arteries are made up of three concentric layers, with the medial one being the most important in healthy and normal physiological conditions [13–16]. The adventitia, the outermost layer, plays a crucial role in over-distention events or in pathological and aged tissues where its main component, collagen, is organized to act as a mechanically protective barrier [17, 18]. More precisely, adventitial collagen comes in the form of thick fiber bundles, organized to form a net. Those bundles are undulated at no or very low loads (the load being then born by medial tissue), and progressive uncrimping and possible re-orientation put them at play at increasing loads (a process generally referred to as fiber recruitment or fiber engagement) [19–24]. This sequential engagement of collagen becoming the main load-bearing component explains the typical non-linear stress-strain curve observed in these tissues [25–28]. Though this qualitative structural response of collagen network is understood and commonly accepted, structure-to-mechanics relationships in arterial collagenous networks remain to be fully elucidated and quantified as they are involved in biological processes like fluid transport throughout the ECM [29], mechanical stimuli provided from their environment to cells [30–32], and they may help in many clinically-relevant situations where mechanical function is impaired and requires accurate diagnosis for improved medical management.

For this reason, microstructural observations are needed, with the specific requirement of dynamically imaging loaded tissues. The latter prevents from using the most common techniques based on staining and/or histology, which are destructive or affect the mechanics of the tissue [33–37]. Though several other techniques may be appropriate for specific purposes, like OCT (optical coherence tomography) and X-ray micro-tomography [38, 39] for instance, one of the most suitable to perform studies at the scale of collagenous structures in arteries is second-harmonic generation (SHG) imaging [20, 40–45] which is very specific of collagen [46]. SHG is a complex physical phenomenon arising from the interactions of photons with matter. This phenomenon occurs in SHG-capable components (i.e. being a non-centrosymmetric medium), when two photons with the same energy simultaneously interact with the matter and combine into a single photon re-emitted at exactly double energy. To practically reach the high-intensity conditions required for SHG at a given point in a volume, a pulsed laser source, like a femtosecond laser, has to be focused at that point [40, 46–48]. As a consequence, a complete 3D image of a volume requires a scanning procedure. For this reason, SHG microscopy is intrinsically a technique that allows sectioning the imaged volume, which may explain why it is sometimes referred to as confocal microscopy [49] by mistake. SHG microscopy provides excellent in-plane resolution and allows through-depth imaging typically reaching one to two hundred microns in arterial tissue. Though this is a very attractive property, it is important to note that this depth is not sufficient to scan the whole thickness of human large arteries. It is also worth mentioning that the axial resolution of this technique is much smaller than its in-plane resolution, which prevents accurate out-of-plane geometrical measurements [49].

Irradiation with such a laser source will also generate two-photon fluorescence in sensitive components (called the fluorophores), which is the property used in two-photon microscopy [42, 50, 51]. However the light is re-emitted at a different wavelength and can be easily separated from the SHG signal; for instance, for an excitation wavelength of 830 nm, the SHG signal was emitted at a wavelength of 415 nm, whereas the autofluorescence signal was collected with a filter between 560 and 700 nm. In the present context, the advantage of the technique is that collagen is a structure that is intrinsically capable of SHG, unlike the other components of the ECM, which allows specific and non-destructive observation of this component within the arterial wall.

SHG microscopic imaging has focused for at least 2 decades on visualizing collagen fibers in a variety of connective tissues (e.g., skin [4, 52], bone [46, 53], tendon [5, 47, 54], blood vessels [19, 20, 44, 50, 55–58], and cornea [59]) and internal organs (e.g., cervix [60], liver [61], kidney [62], and lung [63, 64]), see [40] for a detailed review on this topic. This has permitted unprecedented imaging for the detection of collagenous fibrosis (with possible application for medical diagnosis) or for the quantification of extracellular matrix remodeling by comparing the arrangement of collagen at different physiological stages of a tissue or an organ. More recently, since 2012 and several pioneering studies, SHG microscopic imaging has been used to capture almost real time changes to the collagen structure during uniaxial or biaxial deformation in soft tissues [20, 21, 24, 66–69]. This has provided new quantitative evidence on the sequential engagement of collagen bundles and collagen fibers in response to mechanical loading.

In the current chapter we share our extensive experience in coupling mechanical loading and SHG imaging to demonstrate, across arteries, species, and testing conditions, the interesting sequential engagement of collagen fibers in response to mechanical loading, which is essential for an artery to function normally. More specifically, we analyze morphological changes of the microstructure of rabbit carotid arteries under loading, and investigate how the underlying microscopic mechanisms governing fiber reorientation compare with the same mechanisms in aortic tissues from porcine and human origin (healthy or aneurysmal). A unified analysis approach is proposed in order to compare observations and quantitative analyses in different loading scenarios: uniaxial tension on flat samples, tension-inflation on tubular samples, bulge-inflation on flat samples. Analysis and discussion focus on the kinematics, especially re-orientation, of adventitial collagen fibers and differences arising from the type of arteries and loading scenarios.

2 Materials and Methods

For a detailed and concise view on all realized tests, the nature, origin, and all relevant data of each tested sample are summarized in Table 1.

2.1 Tissue

2.1.1 Origin

In a first testing campaign [66, 67], carotid arteries were harvested from healthy male New Zealand White rabbits (R), weighing three kg approximately (as shown on Fig. 1a). These arterial samples were used for uniaxial and tension-inflation tests, as detailed in the following sections. In a second campaign dedicated to bulge-inflation experiments on aortic tissue [68], three types of tissue were collected, namely: (i) porcine whole aortae (P), at a local slaughterhouse from domestic pigs aged 6–12 months (as shown on Fig. 1b); (ii) fresh non-aneurysmal human aortae (H), through the French voluntary body donation program from the Department of Anatomy of the University of Saint-Etienne (France) (see Fig. 1c); and (iii) unruptured human ascending thoracic aortic aneurysms (ATAA), from patients undergoing elective surgery to replace the diseased segment with a graft (see Fig. 1d). Collections and experiments on all samples from human specimens were done in accordance with a protocol approved by the Institutional Review Board of the University Hospital Center of Saint Etienne. All aortic samples were classified as segments of ascending thoracic part (AT) or descending thoracic part (DT), and their external diameter was measured in different zones using a digital caliper.

2.1.2 Sample Storage and Preparation

Within 12 h after excision, each sample underwent careful removal of surrounding tissue by a trained operator, and was conserved in a PBS (phosphate buffered saline) solution at 5 °C until testing within the next 12–36 h maximum, except the samples dedicated to uniaxial tension, which were kept frozen after preparation. From the collected tissue samples, several samples were prepared, according to the following protocols:

Table 1 Characteristics of all tested samples analyzed in this chapter

# sample	Tissue type	Anatomical position	Sex/age	Ex vivo segment diameter (mm)	In vivo axial stretch	Ex vivo thickness (mm)	Load direction	Load levels for imaging
1	P	DT	NA	25.9	NA	1.55	Bulge inflation	0/200 (mmHg)
2	P	DT	NA	20.5	NA	2.01	Bulge inflation	0/200 (mmHg)
3	P	DT	NA	21.7	NA	1.81	Bulge inflation	0/200/450 (mmHg)
4	P	DT	NA	25.6	NA	1.95	Bulge inflation	0/200 (mmHg)
5	P	DT	NA	20.3	NA	1.62	Bulge inflation	0/200/450 (mmHg)
6	P	DT	NA	28.5	NA	1.97	Bulge inflation	0/200/450 (mmHg)
7	P	DT	NA	19.7	NA	1.48	Bulge inflation	0/200/450 (mmHg)
8	P	DT	NA	23.9	NA	2.21	Bulge inflation	0/200/450 (mmHg)
9	H	DT	F/92	22.2	NA	2.27	Bulge inflation	0/200/450 (mmHg)
10	H	DT	M/71	21.0	NA	2.23	Bulge inflation	0/200 (mmHg)
11				19.7		2.08	Bulge inflation	0/200/450 (mmHg)

(continued)

Table 1 (continued)

# sample	Tissue type	Anatomical position	Sex/age	Ex vivo segment diameter (mm)	In vivo axial stretch	Ex vivo thickness (mm)	Load direction	Load levels for imaging
12	ATAA	AT	M/69	52.1	NA	2.11	Bulge inflation	0/200/450 (mmHg)
13	ATAA	AT	M/51	57.6	NA	1.94	Bulge inflation	0/200/450 (mmHg)
14	ATAA	AT	F/55	55.7	NA	1.91	Bulge inflation	0/200/450 (mmHg)
15	ATAA	AT	M/44	57.1	NA	1.95	Bulge inflation	0/200 (mmHg)
16	ATAA	AT	M/48	49.2	NA	2.57	Bulge inflation	0/200 (mmHg)
17	ATAA	AT	F/61	51.2	NA	1.81	Bulge inflation	0/200 (mmHg)
18	ATAA	AT	M/65	50.0	NA	3.20	Bulge inflation	0/200 (mmHg)
19	ATAA	AT	M/69	44.3	NA	2.87	Bulge inflation	0/200 (mmHg)
20	ATAA	AT	M/76	57.7	NA	3.56	Bulge inflation	0/200 (mmHg)
21	ATAA	AT	M/50	57.3	NA	3.40	Bulge inflation	0/200 (mmHg)
22	ATAA	AT	M/79	54.3	NA	2.84	Bulge inflation	0/200 (mmHg)

(continued)

Table 1 (continued)

# sample	Tissue type	Anatomical position	Sex/age	Ex vivo segment diameter (mm)	In vivo axial stretch	Ex vivo thickness (mm)	Load direction	Load levels for imaging
23	R	C	R #1. C #1	NA	1.6	0.16	1D-circ	0/0.2/0.6/1.2/2.0 (MPa)
24							1D-long	0/0.3/0.9 (MPa)
25							1D-diag	0/0.9 (MPa)
26	R	C	R #2. C #1	NA	1.7	0.18	1D-circ	0/0.7 (MPa)
27							1D-long	0/0.1/0.3/0.9 (MPa)
28							1D-diag	0/1.7 (MPa)
29	R	C	R #3. C #1	NA	1.6	0.18	1D-circ	0/1.2 (MPa)
30							1D-long	0/1.8 (MPa)
31							1D-diag	0/0.9 (MPa)
32	R	C	R #4. C #1	NA	1.6	0.18	1D-circ	0/1.0 (MPa)
33	R	C	R #3. C #2	NA	1.7	0.19	1D-long	0/1.0 (MPa)
34							1D-diag	0/1.0 (MPa)
35	R	C	NA	NA	1.5	0.25	Tension under a pressure of 100 mmHg	1.3/1.6/1.8 stretch
36							Inflation under a 1.6 axial stretch	20/100/140 (mmHg)

(continued)

Table 1 (continued)

# sample	Tissue type	Anatomical position	Sex/age	Ex vivo segment diameter (mm)	In vivo axial stretch	Ex vivo thickness (mm)	Load direction	Load levels for imaging
37	R	C	NA	NA	1.5	0.13	Tension under a pressure of 100 mmHg	1.3/1.6/1.8 stretch
38							Inflation under a 1.6 axial stretch	20/100/140 (mmHg)
39	R	C	NA	NA	NA	NA	Inflation under a 1.3 axial stretch	20/100/140 (mmHg)
40							Inflation under a 1.6 axial stretch	20/100/140 (mmHg)
41							Inflation under a 1.8 axial stretch	20/100/140 (mmHg)

(continued)

Table 1 (continued)

# sample	Tissue type	Anatomical position	Sex/age	Ex vivo segment diameter (mm)	In vivo axial stretch	Ex vivo thickness (mm)	Load direction	Load levels for imaging
42	R	C	NA	NA	NA	NA	Tension under a pressure of 20 mmHg	1.3/1.6/1.8 stretch
43							Tension under a pressure of 100 mmHG	1.3/1.6/1.8 stretch
44							Tension under a pressure of 140 mmHg	1.3/1.6/1.8 stretch
45							Inflation under a 1.6 axial stretch	20/100/140 (mmHg)

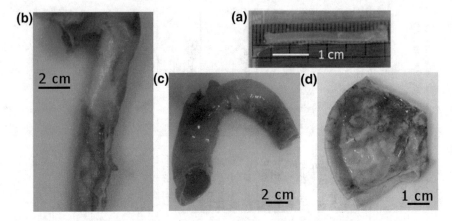

Fig. 1 **a** Excised carotid artery from a New Zealand White rabbit; **b** porcine aorta in its descending thoracic segment; **c** non-aneurysmal human thoracic aorta; **d** unruptured human ascending thoracic aortic aneurysm

- for uniaxial tension, 10-mm-long cylindrical segments were cut from the carotids and longitudinally cut open; rectangular strips were then cut into dogbone shapes [24] aligned along the three following in-plane directions: circumferential (circ.), longitudinal (long.), and along the first bisector of the circumferential-longitudinal plane, referred to in the sequel as the diagonal (diag.) direction;
- for tension-inflation, a 15 ± 1-mm-long cylindrical segment was cut from each carotid sample;
- for bulge-inflation, aortae were cut open along the longitudinal direction, following the line of the intercostal arteries; 45-mm^2 samples were cut, with edges parallel to the longitudinal and circumferential directions of the aortic tube.

2.1.3 Sample Measurement in the Reference Configuration

For stress and stretch measurement, the reference configuration was defined as the configuration of the sample after preconditioning (see Sect. 2.2 for a more precise description of the preconditioning). Besides, the thickness of each sample was measured in different positions with either a digital caliper and two rigid plastic plates for large samples (bulge-inflation testing) or using a macro objective camera (Nikon® D7200 equipped with Nikon® AF-S VR Micro-Nikkor optical 105 mm f/2.8G IF-ED lens) for small ring samples (tension and tension-inflation). Noticeably, in tension-inflation tests, the measurement of the arterial diameter showed a limited variability along the sample length (standard deviations being as low as 2% of the measured mean diameters).

2.2 Mechanical Testing Benches and Protocols

All mechanical tests were performed under quasi-static conditions, at an ambient temperature of 20–24 °C with samples immerged in PBS. For all tests, five cycles of preconditioning were performed prior to the mechanical test itself; in uniaxial tensile tests and tension-inflation tests, the preconditioning was performed up to the maximum load applied during the test itself, whereas in the bulge-inflation tests, preconditionning was stopped at a pressure of 80 mmHg. The mechanical tests consisted in increasing the mechanical load until predefined thresholds. At these thresholds, a stack of second-harmonic generation microscopy images was acquired. In the following section, we give details about the characteristics of each of these mechanical tests.

2.2.1 Uniaxial Tension and Tension-Inflation Testing

A screw-driven high precision tensile machine (Deben® Microtest tension/compression stage) with a 0.01 N precision load cell was used for uniaxial tensile loading. Both crossheads of the tensile machine moved in opposite directions, allowing the operator to identify and keep track of a motionless central region of interest to be imaged under the microscope.

Inflation loading was applied by a syringe pump (Harvard Apparatus®) equipped with a ±300 mmHg pressure transducer (FISO® optical fiber connected to the fluid network at the sample inlet) which infused PBS into the needles which were cannulated onto the sample. The setup enabled pressure control under constant or cyclic conditions, and was connected to the previously described tensile machine [66, 70].

The following testing protocols were used:

- For uniaxial tensile tests, a displacement was imposed at a rate of 0.5 mm.min^{-1}, corresponding to a stretch rate of 0.2 min^{-1}. Images were acquired at fixed loads, the maximum force being equal to 1 N (i.e. a maximum stress of about 2 MPa) (see Fig. 2, left).
- For tension-inflation tests, two loading scenarii were considered in agreement with well-established tension-inflation protocols for arteries [70, 71], namely: axial tension under a 100 mmHg pressure (corresponding to the rabbit average carotid pressure) and inflation under a 1.6 axial stretch (in vivo axial stretch of rabbit carotids). Few tests were also performed by maintaining either a 20 or a 140 mmHg pressure or a 1.3 or 1.8 axial stretch (see samples #39, #41, #42, and #44). Inflation was applied by steps of 20 mmHg every 30 s, and the applied pressure was maintained during image acquisition. Axial tension was applied at a rate of 2 mm.min^{-1} (i.e. an axial strain rate of 0.2 min^{-1}) and the applied axial stretch was maintained constant for image acquisition (see Fig. 2, middle). Note that for each test, after preconditioning, a first cycle was performed to acquire macroscopic data using the optical camera, followed by a cycle to acquire microscopic images.

For both uniaxial tension and tension-inflation, the samples were not fully unloaded, a small (but non-zero) loading was maintained to avoid buckling of the sample. For the exact characteristics of these tests, please refer to [66, 67].

2.2.2 Bulge-Inflation Testing

Aortic samples were clamped by gluing the adventitial side to a 30-mm-diameter PVC (Polyvinyl chloride) support and the intimal side to a second PVC support. The adventitial surface always faced outwards, while a hermetic closure was ensured on the intimal surface. An automatic water pumping system (WPI®, NE-501 Multi-Phaser) injected PBS at a constant rate of 2 mL.min^{-1} to inflate the sample. Pressure values were recorded simultaneously by a pressure transducer (Omega®). For image acquisition, inflation was stopped, and the volume was kept constant (see Fig. 2, right). For the exact characteristics of the bulge-inflation tests, please refer to [68].

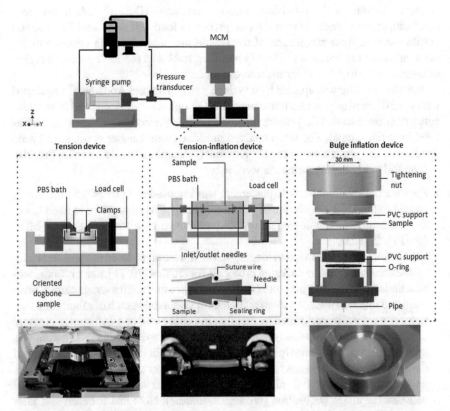

Fig. 2 Schematic views and pictures of the 3 test benches coupled with the two multi-photon microscopy setups. From left to right: uniaxial tension, tension-inflation and bulge inflation devices

2.3   Image Acquisition

2.3.1   Description of the Microscopes

Two multi-photon microscopy setups were used in these studies to image the collagen networks. A multiphoton microscope (NIKON, A1R MP PLUS®) of the IVTV platform (Engineering and Ageing of Living Tissues Platform, ANR-10-EQPX-06-01) was used for rabbit samples. A LEICA TCS SP2 upright microscope (HCX APO L UVI ×20 NA0.5 with a Ti:Sapphire femtosecond laser source Chameleon Vision I from COHERENT, Inc) equipped with a water immersion objective was used for the other samples. The scanning volumes were respectively of $512 \times 512 \ \mu m^2$ and $750 \times 750 \ \mu m^2$ and depth of view ranged up to 90 μm for rabbit samples or up to 200 μm for human samples, depending on sample transparency. All technical details of microscope settings and specific adjustments for acquisition are available in [66–68].

All microscopic investigations were performed such that the same region of interest was tracked during each test.

2.3.2   Choice of Mechanical Configurations for Microstructure Imaging

During uniaxial tensile tests, two to four microscopic configurations were acquired, corresponding to the zero-stress and the maximum imposed stress levels as well as stresses in the zone of large changes of the stress-stretch slope. The latter zone is the closest to the in vivo configuration.

During tensile tests under constant pressure, three configurations were acquired, at stretches of 1.3, 1.6, and 1.8, while during inflation tests at constant axial stretch, the three configurations were acquired at pressures of 20, 100, and 140 mmHg. In both cases, the three configurations correspond respectively to a low stress state with a sufficient decrimping degree, the in vivo axial stretch of rabbit carotids, and a stress state beyond the physiological range.

Finally, during bulge-inflation tests, three configurations were acquired: a zero pressure level was first imaged, corresponding to the reference configuration; then, a pressure level of 200 mmHg was chosen as it induced a wall stress close to diastolic-systolic loading conditions (equivalent to average in vivo pressure of ~120 mmHg [72]); finally, a third pressure level of 450 mmHg state was sometimes reached as it induced a stress corresponding to over-pressurization beyond the physiological range. The conversion between in vivo pressure levels and applied pressures assumed hemispherical membranes for both the in situ sample [73] and the in vivo aneurysm (or cylindrical shape for healthy vessels) and the use of the Laplace law (see [70] for details on the conversion process).

All imaged configurations are summarized in the last column of Table 1.

2.4 Post-processing

For all tested samples, preliminary observations in the radial direction showed the nearly planar orientation in the longitudinal-circumferential plane of the adventitial collagen network, in good agreement with other studies [19, 20, 74]. It was also shown that the most relevant morphological changes in the adventitial collagen network occur in this same plane. As a consequence, only planar orientation distributions of collagen fibers in the axial-circumferential plane were estimated by means of several of the methods available in the literature (see Table 2).

Table 2 Review of different methods to process multiphoton micrograph for characterizing collagen fibers in arteries

	Quantification technique	Applicability	Software tool	References
Fiber angle densities	Structure tensor method	All fiber networks	Custom or OrientationJ (ImageJ)	[75, 76]
	Fast Fourier transform	All fiber networks	Custom (Matlab, Fortran); "Oval profile plot and directionality" plugin (ImageJ)	[35, 77]
Fiber waviness	Semi-automatic tracking	Wavy fibers (crimped collagen)	Custom, NeuronJ (ImageJ)	[21, 78]
Fiber diameter	Custom image skeletonization	All fiber networks	Custom	[78, 79]
Fiber lengths	Custom image skeletonization	All fiber networks	Custom	[19, 79]
Fiber volume fractions	Image thresholding	All fiber networks	Custom, ImageJ	[13, 80]
Fiber tortuosity	Custom image skeletonization	Medial elastin, medial collagen	Custom (Matlab)	[45]
Node connectivity	Custom image skeletonization	Medial elastin, medial collagen	Custom (Matlab)	[45, 79]
Density of transversely oriented segments	Custom image skeletonization	Medial elastin, medial collagen	Custom (Matlab)	[45]

2.4.1 Extraction of Collagen Fiber Orientations

Before further analysis and discussion, it is here proposed to compare two commonly used methods, namely a structure tensor-based approach (OrientationJ, plugin available in ImageJ software) and a Fast Fourier Transform approach.

On the one hand, the structure tensor-based approach is based on the computation of intensity gradients and their related weighted 2D structure tensors at each pixel of a given slice, as described in [19]. In the present post-processing method, inspired from [68], the structure tensor was computed with a user-specified observation weighting window of 3 pixels using a Gaussian gradient interpolation. Two isotropic properties resulting from the structure tensor, namely energy and coherency, were used to separate significant and negligible oriented area. Such a distinction was made by considering pixels which had at least 3 and 15% of normalized energy and coherency, respectively, which provided a good trade-off between adequate number of usable pixels and elimination of insignificant information in the images. A weighted orientation histogram was built for each slice, discarding pixels below these thresholds and the average histogram within the stack was calculated for the subsequent analysis.

On the other hand, the Fast Fourier Transform method consists in performing a 2D discrete Fourier transform of the projection of the stack. We here compare two different options, namely: orthogonal projection of the whole stack onto a single image using a maximum intensity projection algorithm or a mean intensity projection algorithm. The 2D Fourier transforms are then converted to a power spectrum, eventually integrated by means of a wedge-shape sum approach, to create a fiber orientation distribution of the fiber network [35, 77, 81]. The whole procedure is shown in Fig. 3.

In the sequel, all resultant histograms were normalized with respect to their total area and only a qualitative comparison between the different methods is proposed.

2.4.2 Evaluation of the Load-Induced Changes in Fiber Orientation

Upon mechanical loading, the originally crimped and variously oriented collagen fibers are progressively stretched and possibly reoriented. First, principal orientations of the fibers were defined as follows: each histogram of fiber orientation averaged over a stack was fitted by a sum of four independent Gaussian functions y_i, $i \in \{1, \ldots, 4\}$ (with peak height y_{pi}, mean orientation θ_{mi}, and standard deviation σ_{mi}, $i \in \{1, \ldots, 4\}$) and a constant function y_b, corresponding to the base value (see Fig. 4).

Each Gaussian function y_i defines a family of fiber orientations; the resulting four families of orientations were first ranked by calculating their respective contribution to the global distribution, defined by the ratio between the area under the respective Gaussian curve and the area under the whole histogram: the principal fiber orientations within the observed region were defined as the mean orientation of the families whose area outweighed 25% of the cumulative area.

Extraction of angle densities
(example of adv. collagen sublayer)

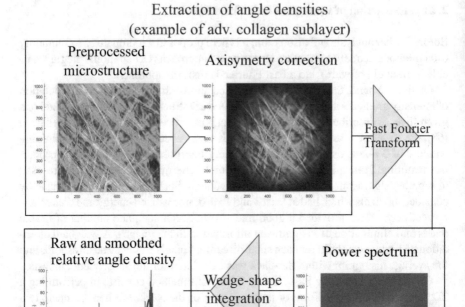

Fig. 3 Fast Fourier Transform approach. Flowchart and illustration of the different steps to get the fiber orientations from the raw multiphoton images

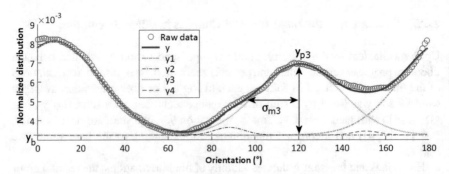

Fig. 4 Example of a normalized histogram of fiber orientations and of their approximation with a sum of four independent Gaussian distributions y_i, $i \in \{1, \ldots, 4\}$. y_b stands for a constant function representing the base value

Then, an alignment index and a dispersion index were defined to separately quantify the decrimping and reorientation mechanisms. They were computed as:

– The alignment index (AI), which characterizes the straightness of fibers, considers the dispersion of fiber orientations around their mean values:

$$AI = \sum_{i=1}^{4} \frac{y_{pi}}{\sigma_{mi}} \tag{1}$$

The narrower the orientation peaks, the higher the alignment index.

– The dispersion index (DI), which denotes the contribution of the base value to the total distribution, hence the extent to which the fibers are evenly spread over the orientation range, was computed as:

$$DI = \frac{180 y_b}{\int_0^{180} y(\theta)d\theta} \tag{2}$$

For each family, principal orientations, alignment and dispersion indexes, were calculated in each different configuration of the mechanical tests at which images were acquired, and their variations due to the specific loading were finally analyzed.

2.4.3   Kinematical Affinity Assessment

The affine or non-affine nature of kinematical transformations observed in soft tissues has seldom been questioned [66]. This aspect was addressed using the uniaxial tension and tension-inflation data because re-orientations were too small in bulge-inflation tests to be assessed accurately.

In previously published papers [66, 67], we analyzed the transformations of the collagen network by developing a custom algorithm for the extraction of local density maxima with their related dispersions. The corresponding orientations were used in a theoretical calculation yielding their affine re-orientation. More precisely, local density maxima were determined by a standard peak detection algorithm and the related dispersions were evaluated by determining the angles at which the density threshold that represented 80% of total fiber angles extracted from the image, was reached. We here propose an extended methodology that focuses not only on the changes of the peaks but also of the whole histogram. Adventitial collagen fibers were highly crimped in the unloaded configuration, which led to a difficult determination of global fiber orientations. Accordingly:

– For uniaxial tensile tests, we chose to apply the affine kinematics calculation taking the diagonally-loaded configuration as reference and applying the loading in two consecutive steps: first unloading along the diagonal direction with a diagonal stretch of $\lambda_{diag} = 1/1.9$, and with a transverse stretch of $\lambda_{tr} = 1/0.92$, then loading along either circumferential or the longitudinal direction, with, respectively, $\lambda_{circ} = 1.7$ and $\lambda_{long} = 2.1$, and, for each load direction, a transverse stretch of $\lambda_{tr} = 0.9$ (all numerical values correspond to experiments). For tensile tests under constant inner pressure, the chosen reference configuration was a partially loaded configuration, defined as: $\lambda_{long} = 1.3$ and $p = 100$ mmHg; the longitudinal applied stretch amounts to $\lambda_{long} = 1.8/1.3$, while the inner pressure leads to a transverse contraction evaluated to $\lambda_{circ} = 1 - 0.65.\lambda_{long}$.

– For inflation at a fixed axial stretch, the reference configuration corresponds to $\lambda_{long} = 1.6$ and $p = 20$ mmHg; the maximum applied pressure corresponds to a circumferentially measured stretch of $\lambda_{circ} = 1.37$.

For quantitative comparison between the affinely-transformed histograms and the experimentally-determined histograms, we compared the peak locations. To this aim, the different experimentally-determined histograms were fitted by the sum of a constant function and two Gaussian functions, similarly to the methodology described in Sect. 2.4.2. The mean angle of each Gaussian function θ_{exp} was then compared to the affinely-transformed mean angle θ_{aff} of the reference configuration (while θ_{ref} stands for the mean angle of the reference configuration); the error between the affine prediction and the experimental observations was computed as:

$$e = \frac{\theta_{aff} - \theta_{exp}}{\theta_{ref} - \theta_{exp}} \tag{3}$$

For all cases where only one peak was visible, only the mean angles of the highest peaks were compared.

3   Results

3.1   Comparison of Image Processing Methods to Derive Fiber Orientations

The comparison of image processing methods was performed on one sample of each type of specimen origin and of mechanical test. Figure 5 reports the comparison between the structure tensor approach and the mean and max intensity projection algorithms.

The choice of a mean or max intensity projection algorithm leads to almost the same orientation histograms, whether the fiber network was highly crimped (as it is the case for the lowest curves in each plot) or progressively stretched, as well as whatever the species or the healthy/aneurysmal state. In particular, the peaks were always located at the same angles. The comparison between the intensity projection algorithms and the structure tensor approach was more delicate: in general, when the tissue was under mechanical loading, i.e. when the fibers are (at least partially)

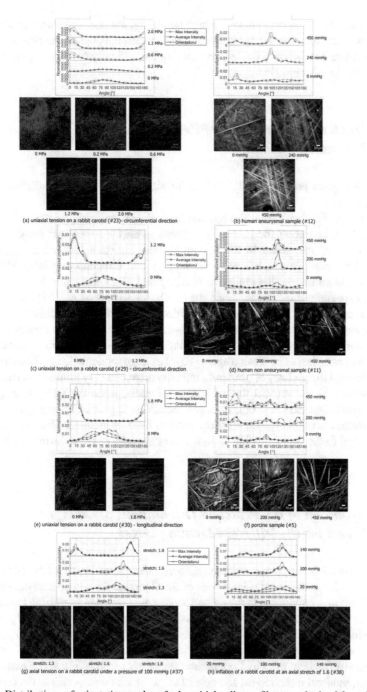

Fig. 5 Distributions of orientation angles of adventitial collagen fibers as obtained from 2D FFT (red circles and black squares) or from OrientationJ (blue crosses) on different types of tissues and for different types of tests, together with the corresponding projected images

uncrimped, the histograms showed a high similarity, and the peaks were at the same angles with comparable widths; in the crimped situation, however, the peak locations may be located at different angles, underlining the difficulty to assess a precise fiber orientation in this case.

3.2 Qualitative Approach to Fiber Reorientation Under Mechanical Loading

Figure 6 reports projected SHG images of adventitial collagen in different load configurations.

First, Fig. 6a shows how the adventitial collagen network, as observed in the unloaded configuration (bottom left), evolved when the tissue was subjected to a uniaxial tensile test either in the circumferential direction (bottom right), in the diagonal direction (top right), or in the longitudinal direction (top left). The fiber network has an impressive ability to decrimp and align along any load direction. This is especially remarkable, since the results of Krasny et al. [67] showed that both adventitial and medial elastin networks, along with medial collagen, do not display such reorientation abilities. However, a different scheme is observed when subjecting the arterial tissue to tension-inflation, as shown in Fig. 6b: in the case of cylindrical samples, axial tension again provokes large reorientations (as observed on the horizontal line of images), but inflation only provokes very limited reorientations (as observed on the vertical line of images). Finally, when the artery is subjected to bulge-inflation (see Fig. 6c), fiber reorientations appear to be negligible for all tested tissues (as observed on all three vertical lines of images). The next sections are devoted to the quantification of how fiber networks evolved during mechanical loading.

3.3 Evolution of the Principal Orientations of Collagen Fibers for the Different Loading Cases

We here use the principal orientations of fibers, as extracted by the method described in Sect. 2.4.2, i.e. the mean orientations of the Gaussian fits whose area outweighed 25% of the cumulative area. In the unloaded configuration, it is interesting to note that whatever the tissue under consideration, and although the tissue has been subjected to preconditioning cycles before acquiring images, the orientation peaks are distributed over the entire range of angles, as shown in Fig. 7.

The most frequent principal orientation for rabbit carotids, pig aorta, and ATAA is the longitudinal direction (at an angle of 90°), which corresponds to the principal orientation of the adventitial elastin network [67]. Although non-aneurysmal human samples do not display the same trend, one should underline the small number of such

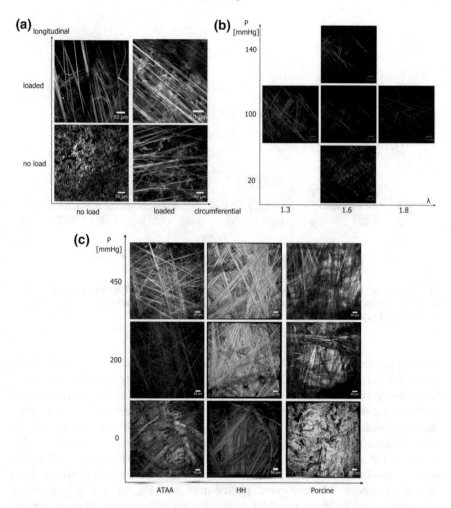

Fig. 6 Multiphoton images obtained during: **a** uniaxial tension in the circumferential, longitudinal and diagonal directions on rabbit carotids (samples #26–28); **b** tension-inflation on rabbit carotids (samples #35–36); **c** bulge inflation on porcine (P, #6), human non-aneurysmal (H, #11) and ascending thoracic aortic aneurysmal (ATAA, #14) samples

samples. Globally, there is a large dispersion of fiber orientations when the tissue is unloaded. This can be explained by the fact that the reported principal directions are actually local orientations of the highly crimped fibers which only straighten when the tissue is subjected to mechanical load.

We now analyze how these highly crimped fibers progressively decrimp and reorient, as displayed in Fig. 8.

Starting with uniaxial tensile tests, Fig. 8a shows the evolution of the principal orientations with respect to the stress state for samples #23–#30. The principal orien-

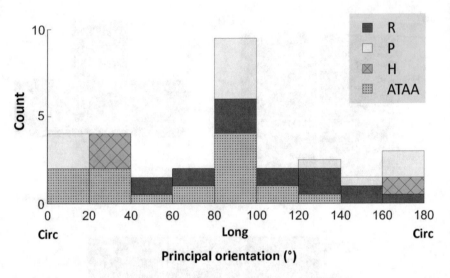

Fig. 7 Principal orientation distribution of collagen fibers at zero load for all analyzed samples, sorted by species. Circumferential and longitudinal directions are indicated by the labels Circ and Long respectively

tations were always evaluated by setting the 0 angle along the tension axis. We first notice that the principal orientations strongly vary with the applied load, confirming the dramatic reorganization that was qualitatively observed in the microstructural stack of images of Fig. 6a. Interestingly, whatever the load direction, the decrimping and reorientation processes occurred at low stresses and led to large stretches in the tissue. The consequence is a cluster of points grouped along the abscissa axis. Finally, when most fibers are recruited, they all align along the load direction (i.e. along the ordinate axis) and the tissue becomes much stiffer. Data are best described by a decreasing exponential trend curve, plotted in grey on Fig. 8a. Similarly, dramatic reorientations were also observed for the tensile tests under constant internal pressure (see Fig. 8b, which results from the postprocessing of samples #42–#44). Whatever the applied internal pressure (i.e. 20, 100, or 140 mmHg), the axial stretch provoked a progressive shift of the principal orientation towards the axial direction (at 90° here). The response was however quite different when a pressure was applied onto the arterial sample subjected to a constant axial stretch. The principal orientations showed only limited reorientations, as proved by the proximity of the dots on each line of the Fig. 8c (which results from the post-processing of samples #39–#41). Also, the larger the applied axial stretch, the more limited was the reorientation consecutive to inflation.

Finally, Fig. 9 reveals that the bulge-inflation tests also led to limited reorientations. Firstly, during the first part of the inflation process (i.e. between 0 and 200 mmHg), the decrimping process took place, which only slightly modifies the graph of principal orientations shown in Fig. 7 for the zero load configuration: peaks

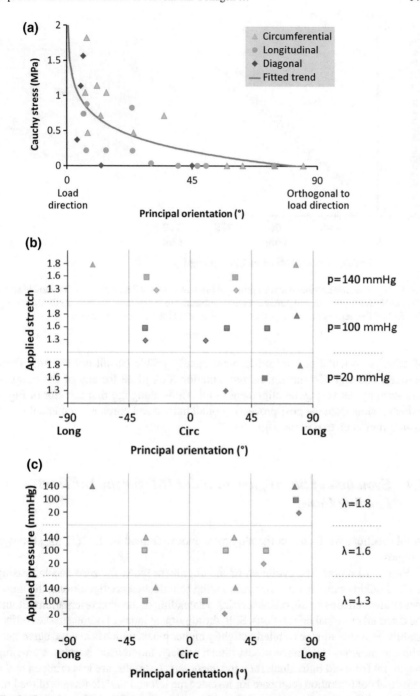

Fig. 8 Evolution of the principal orientations of the rabbit carotid collagen fibers as a function of the applied loading. **a** Uniaxial tensile tests in all three tested directions. The solid line corresponds to the exponential fitted trend curve **b** Axial tensile tests of cylindrical samples subjected to a constant inner pressure; **c** Inflation tests of cylindrical samples subjected to a constant axial stretch

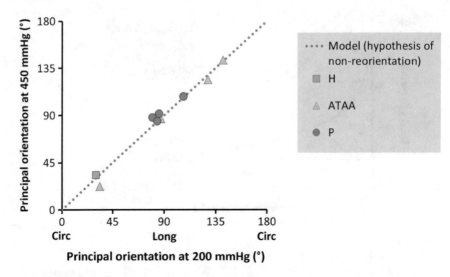

Fig. 9 Bivariate relationship between principal orientations of collagen fibers at 200 mmHg and at 450 mmHg for all human and porcine samples subjected to bulge-inflation tests up to 450 mmHg. The dashed line represents the hypothesis of non-reorientation of collagen fibers

of principal orientations are made more clearly visible (result not shown). Then, between 200 and 450 mmHg, no reorientation took place for any of the samples, as shown by the very good alignment of all points along the first bisector of Fig. 9 (which results from the post-processing of all human and porcine samples subjected to an inflation up to 450 mmHg).

3.4 Evolution of the Alignment Index (AI) for the Different Loading Cases

In this section, we focus on the alignment index, defined by Eq. (1) for all tested samples.

Figure 10 reports the evolution of the alignment index for zero and increasing loads. Looking only at values corresponding to unloaded configurations, the alignment index remains small, characterizing a spreading of the fiber orientations around the detected principal orientations. Still, the alignment index of human tissues (either healthy or aneurysmal) exhibited a slightly higher index, which is in good agreement with the previous qualitative observations: namely, the human samples, stemming in general from old individuals, and the aneurysmal samples are less crimped in the zero-load configuration (compare for instance the left and middle images of the bottom line of Fig. 6c to the bottom right image of Fig. 6c and to the bottom left image of Fig. 6a). Regarding the evolution of the alignment index for increasing loads,

Fig. 10a focuses on the samples subjected to uniaxial tension (with microstructure imaged for more than two loaded configurations, i.e. samples #23, #24, #25, #27), and with a normalized stress-stretch curve. Interestingly, AI followed the same profile as the stress-stretch curve, i.e. a slight increase at low stretches (i.e. until reaching the equivalent physiological state), followed by a much larger increase at high stretches. Regarding tensile tests under a fixed internal pressure (see Fig. 10b), for a low internal pressure of 20 mmHg, the observations are very similar to the previous results, i.e. stable AI up to a stretch of 1.6, followed by a rapid increase between 1.6 and 1.8 stretch. For higher internal pressures (of 100 and 140 mmHg), the initial AI was higher, reflecting fiber stretching due to the internal pressure; as a result, only very limited variations of AI were observed during axial tension of the sample.

For all (bulge-) inflation tests (see Fig. 10c, d), the major increase of AI takes place during the first part of the load, i.e. for mechanical states being lower than the in vivo, physiological state (corresponding to 200 mmHg for human and porcine samples, and to 100 mmHg for rabbit samples), while over-pressurizing the sample does not lead

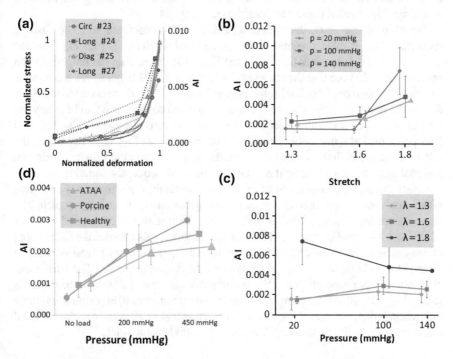

Fig. 10 Evolution of the alignment index (AI) with the applied load. **a** AI as a function of the applied uniaxial stretch for four of the tested samples. Superimposed to it, the evolution of the normalized stress with respect to the normalized applied uniaxial stretch for the same samples (solid lines); **b** AI as a function of the applied longitudinal stretch for different applied inner pressure on samples #42–#44; **c** AI as a function of the applied inner pressure for different applied longitudinal stretch on samples #39–#41; **d** AI as a function of the applied pressure for all porcine, healthy and aneurysmal human samples pressurized up to 450 mmHg

to a further large increase of AI. Concerning the bulge inflation tests (Fig. 10d), this phenomenon is very much pronounced for healthy and aneurysmal human samples, but is reduced for porcine samples. Two hypotheses may here be discussed: either the degradation of the sample microstructure (due to ageing or pathology) or the absence of a preferred fiber orientation. A very particular effect occurs for the inflation tests on rabbit sample (Fig. 10c): applying first an axial stretch of 1.8 induces a large alignment of the fiber, which the inflation test tends to disorganize, as revealed by the decreasing AI index, a phenomenon which was not observed for smaller axial stretches.

3.5 Evolution of the Dispersion Index (DI) for the Different Loading Cases

Now we analyze the evolution of the dispersion index (DI) during the different mechanical load cases under consideration (see Fig. 11).

For all analyzed tests, the dispersion index tends to decrease while the load is increased. This is clearly visible from the uniaxial tensile tests (see Fig. 11a), as well as from the bulge-inflation tests (see Fig. 11d). Regarding the bulge-inflation tests, one should that the strongest variation in the DI happens for the porcine samples. Going more into details, the largest changes in the dispersion index occurs during the first part of the test, between the zero-load configuration and the physiological one, when decrimping occurs. The same was also observed during inflation tests performed at a low or physiological axial stretch (see Fig. 11c). Regarding the inflation test at a high level of axial stretch, the large axial tension already uncrimped and realigned the fibers along the axial direction; consequently, inflating the sample leads to a small reorientation of the fibers towards the circumferential direction, which tends to slightly increase the dispersion index all along the loading path. The same conclusion can also be drawn for the tensile test at a fixed internal pressure of 100 and 140 mmHg (see Fig. 11b): these pressure levels allowed the decrimping of the collagen fibers as well as their reorientation, and the axial tension tends to disorient the fibers, leading to the increase of the dispersion index. The tensile test at an internal pressure of 20 mmHg exhibits the same trend as the uniaxial tensile test of Fig. 11a: fibers are crimped in the initial configuration, explaining the large dispersion index, and their progressive decrimping and reorientation along the axial direction leads to the final strong decrease of the dispersion index.

3.6 Assessment of Kinematic Affinity

The comparison between affine predictions and experimental observations is shown in Fig. 12 for uniaxial tension, Fig. 13 left column for tensile tests at a fixed inner

pressure, and Fig. 13 right column for inflation tests at a fixed axial stretch. For all
tests, the error between the affine prediction and the experimental observations, as
computed by Eq. (3) is reported in Table 3.

For all samples subjected to uniaxial tensile tests, experimental observations show
that collagen fibers undergo a strong reorientation, with fibers being almost parallel
to the load direction at a stretch larger than 1.7 (see Fig. 6 for qualitative observations
and the circle curves of Fig. 12, with peaks close to 0/180°, 90°, and 45/135° for tests
performed in the longitudinal, circumferential, and diagonal directions respectively).
The affine rule would require a much larger stretch to reach such a reorientation, and
consequently it strongly underpredicts the observed fiber rotations, with an averaged
error of 59% (compare the location of the solid line and of the circles with same
color on each plot of Fig. 12, and see Table 3 for quantitative details). Very similar
results were obtained for the samples subjected to tensile tests on cylindrical samples
subjected to constant pressure, confirming the previous results (see left column of
Fig. 13, with initial experimental peaks located close to the circumferential direction
and moving towards the longitudinal direction during axial tension). Once again, the

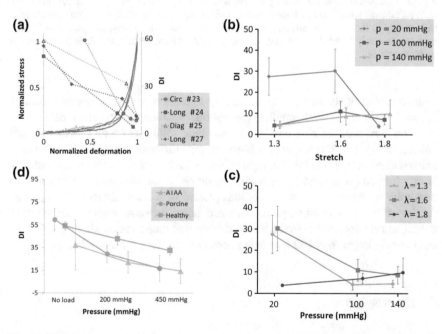

Fig. 11 Evolution of the dispersion index with the applied load. **a** DI as a function of the applied
uniaxial stretch for four of the tested samples. Superimposed to it, the evolution of the normalized
stress with respect to the normalized applied uniaxial stress with respect to the normalized applied
uniaxial stretch for the same samples (solid lines); **b** DI as a function of the applied longitudinal
stretch for different applied inner pressure on samples #42–#44; **c** DI as a function of the applied
inner pressure for different applied longitudinal stretch on samples #39-#41; **d** DI as a function
of the applied pressure for all porcine, healthy and aneurysmal human samples pressurized up to
450 mmHg

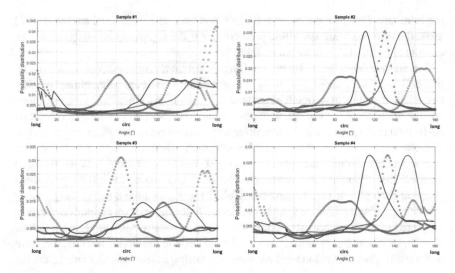

Fig. 12 Comparison between affine predictions (solid lines) and experimental observations (circles) for uniaxial tension tests for the four tested samples. Curves in black, blue, and magenta refer respectively to the diagonal, circumferential, and longitudinal directions. Samples #1–#4 correspond respectively to the three directions of samples #23–#34 of Table 1, see Table 3 for the exact correspondence

collagen network showed a dramatic ability to reorient under mechanical loading. Again, the affine rule underpredicts the real fiber reorientation by about 61% (compare, for instance, the blue circles with the blue solid line on all plots of the left column in Fig. 13). The response is however quite different during the inflation tests at fixed axial stretch (see Fig. 13, right column). In these tests, the fiber reorientation is very limited (as noticed by the superimposition of the two circle curves and their proximity to the reference configurations plotted with black circles), and although the deviation between affine predictions and actual measurements is reduced here, the affine rule does not reproduce well this limited reorientation either (as seen by errors being larger than 100% or being negative), see [66] for more details.

4 Discussion

This chapter gathers observations and analyses made from in situ multiphoton microscopic imaging during several mechanical tests performed on samples from several different species. It shows that collagen fiber networks in arterial walls share common patterns but one has to be cautious as the type of loading or even the image processing method may affect significantly the observations.

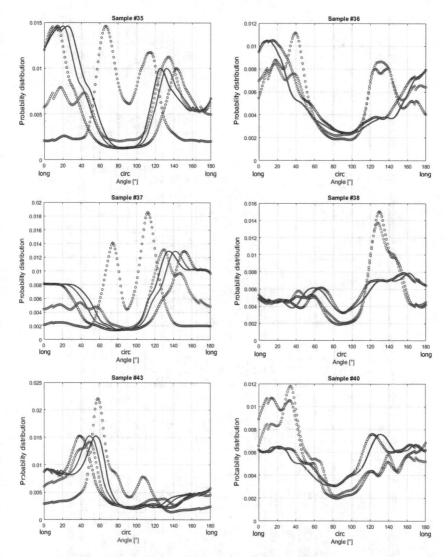

Fig. 13 Comparison between affine predictions and experimental observations for tension inflation tests: left column: axial tension at fixed internal pressure; right column: inflation at fixed axial stretch. The different configurations are displayed with different colors: black: reference, red: intermediate, blue: final; circles: experimental observations, solid lines: affine prediction of the intermediate and final states

Table 3 Peak angle as measured experimentally (Exp.), as predicted by affine transformation (Aff.), and as measured in the reference configuration (Ref.), and the associated error

			Peak #1				Peak #2			
			Exp.	Aff.	Ref.	Error (%)	Exp.	Aff.	Ref.	Error (%)
Uniaxial tension	S #1	Circ. #23	82.3	139.3	145.7	89.9				
		Long. #24	175.4	168.5	145.7	23.2				
	S #2	Circ. #26	89.6	110.6	130	52.0				
		Long. #27	165	147.7	130	49.4				
	S #3	Circ. #29	81	96.1	117.2	41.7				
		Long. #30	176	114.2	117.2	105.1				
	S #4	Circ. #32	85.4	113.5	133	59.0				
		Long. #33	171.9	151.4	133	52.7				
Tension-inflation	S #35	tension	110.5	125.7	143.9	45.5	67.1	26.3	15.6	79.2
	S #37	tension	114.7	140.6	157	61.2	72.9	39.4	24.8	69.6
	S #43	tension	58.3	27.9	17	73.6	67	57.2	40.2	36.6
Tension-inflation	S #36	inflation	134	170.5	167.6	108.6	27.1	2.9	4.3	106.1
	S #38	Inflation	33.5	29.5	38.2	−85.1	135.6	122.8	NA	
	S #40	inflation	131.7	130.8	122.7	10.0	17.4	5.8	8.2	126.1

For uniaxial tensile test, only one peak was visible, while two peaks were measured in the tension-inflation tests. For uniaxial test, the sample number given in the second column corresponds to the one of Fig. 12, while the one given in the third column as well as the ones of the tension-inflation tests refer to the numbering of Table 1

4.1 Influence of the Post-processing Method

Two methods, OrientationJ—a structure tensor-based method [19, 75, 76]—and 2D FFT-based method [35, 82–85], were compared on a large dataset of second-harmonic generation microscopy images obtained from various arterial tissues. The nature of these images is such that automatic analysis and extraction of relevant data is most often limited. The two methods of interest in this chapter are currently among the most commonly used in previous research about arterial microstructure characterization focusing on fiber orientations, see [21, 22, 28, 86–88] and references therein, justifying the need for a detailed comparative assessment.

Overall, both methods are comparable and provide close results. This is especially true when analyzing non-unloaded tissue, where correlations between both methods' results (peaks in fiber angle distribution and their widths) are excellent. The case of unloaded tissue is more complex and the results presented in this contribution demonstrate the difficulty of analyzing global fiber orientations in such configuration which is, however, the easiest one from an experimental point of view. In particular, peaks may be found at different angles in those complex images. Indeed loaded tissues show mainly linear fiber bundles which are well handled by any algorithm, whereas the configuration of unloaded tissues reveals highly entangled and crimped structures. In that case, fiber orientation is characterized by two scales, which may distort the results of the presented algorithms, as also noted in [21, 67]. The largest scale is that of the whole fiber or bundle (at least its visible portion in the image), and the smallest scale is that of undulations of these fibers/bundles. The presence of these two scales in the images renders the analysis inaccurate when using local algorithms as those presented here. This should be kept in consideration and carefully managed. Currently, this issue raises the need for more robust methods based on either non-local algorithms (such as fiber tracking method, which follows the evolution of the global orientation of the fibers during the loading [19, 20, 89]) or algorithms combining various approaches. An example could be to use inverse methods in which the morphological parameters of synthetic images would be sought to match the true image with highest confidence.

4.2 Comparison with Existing State of the Art on Arterial Microstructure

Despite these limitations in analysis of fibrous orientations, several trends could be observed and quantified with confidence along this work. In the following, the authors highlight and discuss the main findings about the microstructural characterization of collagen networks and microstructure-to-mechanics relationships that could be evidenced thanks to these observations in various adventitial tissues.

First, the microstructural arrangement of collagen fiber bundles was observed and analyzed in unloaded tissues. Overall, adventitial collagen networks are characterized by a very low alignment of fibers (i.e. they are widely spread around preferred directions, if any) and a large dispersion of preferred orientations, in good agreement with, e.g., [12, 19, 57]. This translates into very smooth orientation histograms as shown in the present results, for all examined tissues and species. Similarly to [12, 22, 90], a slightly higher occurrence of the 90° (longitudinal) orientation was observed (see Fig. 8) and should be mentioned, but no definitive conclusion about the orientations of collagen bundles in adventitial tissue could be drawn from this study. This high-dispersion trend was especially marked in animal (healthy and young) tissues. Looking more into details, an interesting difference could be observed from human aneurysmal tissues which showed more aligned fibers and less dispersed orientation distributions. It is hypothesized that this observable trend results from tissue remodeling in which collagen bundle arrangements may be affected [91]. Nevertheless, it is important to note that such remodeling could be the result of either normal aging or pathological remodeling. Unfortunately access to healthy tissue is very difficult and limited, which prevents from drawing any conclusions on that specific aspect. Also, aged and/or aneurysmal tissues are known to be stiffer and less extensible [11, 92–97], suggesting a reduced stretch difference between load and no load states compared to young healthy tissue. This is especially true for calcified tissue, and could explain that fibers are less crimped, almost straight and oriented like in loaded configurations, even at no load.

The second focus of the present chapter was to compare re-orientation of fibers along with various loading scenarii. Obviously, re-orientation depends on the loading scenario, but it was interestingly observed that the nature of the load influenced the extent or inclination of fiber bundles to re-orient. First, uniaxial tension was the loading case which induced most remarkable reorientation. The different loading directions (i.e. longitudinal or circumferential) behaved similarly and the general trend of orienting towards the loading direction as stress increases was strongly marked in those uniaxial tests; these observations made on rabbit carotids were similar to the ones made on porcine coronary adventitia in [22]. In this phenomenon, it is important to mention that the collagen reorientation response seems to be different in the media, as detailed in [22, 67, 98] related to both rabbit carotids and porcine coronaries, underlining the likely effect of neighboring components, morphological arrangements and interactions with them. In addition to this global behavior of reorienting towards the loading direction, we showed that the proposed fiber alignment index followed a non-linear evolution with imposed stretch, remarkably similar to the evolution of stress. This confirms that the characteristic J-curve response of such tissue is due to underlying mechanisms of progressive straightening of fiber bundles which govern their recruitment for load bearing [74]. This hypothesis was formulated long time ago [20, 35, 37, 74] and was recently confirmed in other similar experimental studies on soft connective tissues like skin for instance [4, 52]. Our results do not only confirm this uncrimping phenomenon, but they also specify that fibers and/or bundles tend to be more and more parallel to each other during this process. When considering combined longitudinal tension and inflation of tubular vessels, other observations could be made which emphasize the complexity of re-orientation mechanisms. Two distinct responses were indeed evidenced. In tests where pressure is maintained at a fixed value, the longitudinal tension produced reorientations characterized by an increase, respectively decrease, in indexes of alignment and dispersion of fiber bundles. This response is similar to that observed in uniaxial tension tests with, however, slight differences to be noticed depending on the initial pressure load. Indeed, the pressure level dictates an initial state of the micro-structure from which the dynamics of re-orientation may slightly vary (see also [23] for similar results). Specifically, at low pressure, the response is similar to that observed in uniaxial tests with marked changes in dispersion and alignment indexes, along with large re-orientation angles. At higher pressures, alignment and dispersion indexes were already respectively at higher and lower than relaxed levels before applying tension. For this reason, their variations were smaller when applying tension, which also combined with smaller reorientation angles. The case of inflation at a given stretch value revealed different responses, mainly characterized by limited changes of alignment and dispersion metrics (as also noted in [20], where only uncrimping occurs but no reorientation). In that testing configuration, the competition between longitudinal tension effects and inflation-induced circumferential stretch seems to explain the observations. The higher the initial stretch, the lower the influence of the pressure load, up to a point where its effect would invert. Interestingly, at such high imposed stretch, the effect of the additional pressure load was rather to slightly disperse and de-align fibers. That effect was marginal and looked rather like a perturbation of the already installed tension-driven morphology. Of course, the influence of boundary conditions during these tests has to be discussed. In fixed pressure tensile tests, a circumferential Poisson's effect is allowed which is in favor of fiber rotation towards the longitudinal traction direction. On the opposite, pressurization at fixed longitudinal stretch, or tensile tests at fixed circumferential stretch (as imposed by [23]) does not allow longitudinal Poisson's effect. In addition, the pressure boundary condition, unlike displacement boundary conditions, did not enforce as large circumferential stretch due to the stiffening of the tissue. These reasons are likely to explain the limited effects of pressurization observed in our experiments.

Last, we addressed bulge-inflation testing. In this test, when material anisotropy is low or negligible, tissue inflates in a rotational symmetry shape, inducing an equibiaxial stretch test at the top of the bulged test (where our image acquisitions were made). In such conditions, it is sensible to expect low or null reorientation. Our observations and associated analyses fully confirm that response of the microstructure. The relationship between the main orientations of bundles at 200 mmHg and their corresponding orientation at 450 mmHg remarkably supports this conclusion (Fig. 9). We also quantified fiber parallel alignment and orientation dispersion. This analysis revealed clear trends in which alignment increased and dispersion of orientation decreased (see also [99]). This observation is valid for all species, with an interesting distinction that could be evidenced between human (elderly) specimens and (young) pig specimens, with the latter exhibiting much higher variations in both indexes. This observation is probably related to higher decrimping arising from differences in the tissue itself as observed from unloaded images (human fibers were already at a higher alignment that pig ones). This also suggests a possibly higher fiber bundle mobility. Unfortunately, this specific aspect cannot be individually assessed from these tests, but it would deserve further investigations as it may constitute one key effects of aging.

4.3 Structure-to-Mechanics and Implications

The previous discussion emphasizes the role of the micro-structure and its kinematic response in the global mechanical response of arterial tissue. In addition to the previous discussion which detailed the progressive uncrimping and alignment of fibers/bundles, additional elements were brought to the description of kinematics in these fibrous structures thanks to the analysis testing the hypothesis of affine kinematics.

These results confirm that the kinematics of fiber bundles does not obey affine transformation rules [67, 100]. This means that, under given boundary conditions, these fibrous structures do not behave like a simple network of wires would. Interactions of the fibrous components with their neighboring components like other solid components, cells, or liquids impact their kinematic behavior, hence the global mechanical response. A complete understanding of these coupling phenomena remains to be provided but, already, hypotheses have been made to explain them. In this context, multi-scale homogenization models taking into account the micro-structural content and morphology succeeded at reproducing this non-affine response [52]. In these models, the non-affinity is due to shear interactions between fibers and their surrounding matrix, model of the interstitial fluid. This supports the importance of mechanical interactions between fibers and interstitial fluid as a first order mechanism in the deformation of these soft tissues. Furthermore, a very relevant study by Ehret et al. [29] recently demonstrated the essential role of interstitial fluid mobility in the mechanical properties of soft tissues, governed by so-called inverse poro-elasticity. Inverse poro-elasticity induces chemo-mechanical coupling, such that tensile forces are modulated by the chemical potential of the interstitial fluid, so being coupled to the osmotic pressure. They concluded that water mobility would determine the tissue's ability to adapt to deformation through compaction and dilation of the collagen fiber network.

These conclusions confirm that mechanical and chemo-mechanical interactions between the fibrous structure and its neighborhood have significant role in the mechanics of soft collagenous tissues. Accordingly, they raise a number of questions related to cellular sensing and interactions coming into play when tissue undergoes deformation. Without going into the details of such interactions, the present work also suggests that these interactions would dramatically change with aging of tissues. We showed that aged and diseased tissues presented very limited fiber movements, including uncrimping and alignment phenomena, compared to young tissue. It is sensible to hypothesize that this loss of deformation-induced mobility in the fibrous structure would impact interactions with the environment, limiting water mobility for instance, and affecting cell mechano-sensing. As a cause or a consequence, remodeling of tissues, within the processes of either aging or disease, would be impacted as well. Similarly, fiber volume fractions have direct mechanical impact on these interactions. Differences in fiber volume fraction were visible in our different samples, and raise the question of possible evolution with aging. This phenomenon is also probably an additional element involved in the cascade of tissue degeneration and/or remodeling. Our analyses did not make it possible to accurately estimate fiber volume fraction, but this should be investigated in future work as another piece of the complex understanding of soft tissue mechanics and its evolution.

5 Conclusion and Future Work

In the current chapter we shared our extensive experience in coupling mechanical loading and multi-photon imaging to demonstrate, across arteries, species and testing conditions, the interesting sequential engagement of collagen fibers in response to mechanical loading, which is essential for an artery to function normally. More specifically, we gathered observations and analyses to show that collagen fiber networks in arterial walls share common patterns: applying a mechanical loading on an arterial sample provokes large movements of the adventitial collagen network, which, first, uncrimps and stretches, and eventually, depending on the principal strain directions, reorient to stiffen the arterial wall in these directions and prevent overdistension. As a consequence, while the uncrimping process is observed for all species and all tested samples, the reorientation process is more pronounced in uniaxial tension, which generally produces larger stretches than inflation, while all planar directions are equivalently principal strain directions during bulge inflation test, leading to no reorientation. Regarding the fiber orientation distribution, the uncrimping process leads to a less dispersed fiber orientation, i.e. a lowering of the base value of the orientation profile, while the reorientation of the fibers, if any, leads to more pronounced, i.e. narrower and higher peaks. This seems to be true whatever the investigated species and samples. Despite the very limited number of human healthy samples, aneurysmal and healthy but old human samples seemed to exhibit almost no crimping, with fibers already stretched and with less mobility as compared to young porcine or rabbit samples. This study also evidences the particular role of the nearly physiological configuration, which is generally situated at the slope change of the mechanical response, but which also corresponds to the point where the majority of the fibers are uncrimped and where reorientation starts to play a major role. This was clearly visible in the change of variation of both the alignment and dispersion indexes. However, our study also has severe limitations: first, one has to be cautious as the type of loading or even the image processing method may affect significantly the observations; second more tests should be made to be able to really compare the fibrous structure among the different species and tissues, in the in vivo configuration as well as during the application of the load; last but not least, our study does not allow to differentiate between samples stemming from old individuals and aneurysmal samples: a study dedicated to how fiber kinematics is affected by disease should be undertaken. Still, all these results emphasize the role of the microstructure and its kinematical response in the global mechanical response of arterial tissue.

Regarding future works, more focus should be put on the evolution of the arterial microstructure during ageing and during the development of pathologies. Among the different mechanisms potentially involved, chemo-mechanical interactions between the fibrous structure and its neighborhood seem to deserve major investigations (e.g., role of hydration) to develop multi-scale mechanobiological models of arterial walls. Also, the role of the interactions between the collagen and the elastin networks (e.g., through the different crosslinks) should be investigated, since the latter is known to be strongly damaged in the elderly as well as in pathological cases, changing the morphology of the collagen network and its possible mobility. All these future works would improve the understanding of rupture mechanisms and pathology evolution.

Acknowledgements This research was supported by the European Research Council, Starting Grant No. 638804, AArteMIS, as well as by the ARC 2 "Bien-être et vieillissement" research program of the Auvergne-Rhône-Alpes region (FR). The authors thank Dr. Hélène Magoariec (Ecole Centrale Lyon, Université de Lyon, FR), Prof. Eric Viguier (VetAgro Sup, Université de Lyon, FR), Dr. Caroline Boulocher (VetAgro Sup, Université de Lyon, FR) and Mr. Fabrice Desplanches (Centre Lago, Vonnas, FR) for their help in the experimental studies on rabbit samples, as well as Dr. Ambroise Duprey for his help in provision of human arterial specimen. Authors also thank the IVTV (ANR-10-EQPX-06-01) team and the Hubert Curien laboratory (University Jean Monnet, Saint-Etienne, France) for their support during the imaging process.

References

1. Abrahams, M.: Mechanical behaviour of tendon in vitro; a preliminary report. Med. Biol. Engng. **5**, 433–443 (1967)
2. Thorpe, C.T., Birch, H.L., Clegg, P.D., Screen, H.R.C.: The role of the non-collagenous matrix in tendon function. Int. J. Exp. Pathol. **94**, 248–259 (2013). https://doi.org/10.1111/iep.12027
3. Wells, S.M., Langille, B.L., Lee, J.M., Adamson, S.L.: Determinants of mechanical properties in the developing avine thoracic aorta. Am. J. Physiol. Heart Circ. Physiol. **277**, H1385–H1391 (1999)
4. Lynch, B., Bancelin, S., Bonod-Bidaud, C., Gueusquin, J.-B., Ruggiero, F., Schanne-Klein, M.-C., Allain, J.-M.: A novel microstructural interpretation for the biomechanics of mouse skin derived from multiscale characterization. Acta Biomater. **50**, 302–311 (2017)
5. Goulam Houssen, Y., Gusachenko, I., Schanne-Klein, M.-C., Allain, J.-M.: Monitoring micrometer-scale collagen organization in rat-tail tendon upon mechanical strain using second harmonic microscopy. J. Biomech. **44**, 2047–2052 (2011). https://doi.org/10.1016/j.jbiomech.2011.05.009
6. Morin, C., Krasny, W., Avril, S.: Multiscale mechanical behavior of large arteries. Encycl. Biomed. Eng. (2017) (Elsevier)
7. O'Rourke, M.: Mechanical principles in arterial disease. Hypertension **26**, 2–9 (1995). https://doi.org/10.1161/01.HYP.26.1.2
8. Brodsky, B., Eikenberry, E.F.: Characterization of fibrous forms of collagen. Methods Enzymol. **82**, 127–174 (1982)
9. Akhtar, R., Sherratt, M.J., Cruickshank, J.K., Derby, B.: Characterizing the elastic properties of tissues. Mater. Today **14**, 96–105 (2011). https://doi.org/10.1016/S1369-7021(11)70059-1
10. Wagenseil, J.E., Mecham, R.P.: Vascular extracellular matrix and arterial mechanics. Physiol. Rev. **89**, 957–989 (2010). https://doi.org/10.1152/physrev.00041.2008.

11. Gasser, T.C., Gallinetti, S., Xing, X., Forsell, C., Swedenborg, J., Roy, J.: Spatial orientation of collagen fibers in the abdominal aortic aneurysm's wall and its relation to wall mechanics. Acta Biomater. **8**, 3091–3103 (2012)

12. Niestrawska, J.A., Viertler, C., Regitnig, P., Cohnert, T.U., Sommer, G., Holzapfel, G.A.: Microstructure and mechanics of healthy and aneurysmatic abdominal aortas: experimental analysis and modelling. J. R. Soc. Interface. **13** (2016). https://doi.org/10.1098/rsif.2016.0620

13. O'Connell, M.K., Murthy, S., Phan, S., Xu, C., Buchanan, J., Spilker, R., Dalman, R.L., Zarins, C.K., Denk, W., Taylor, C.A.: The three-dimensional micro- and nanostructure of the aortic medial lamellar unit measured using 3D confocal and electron microscopy imaging. Matrix Biol. **27**, 171–181 (2008)

14. Dingemans, K.P., Teeling, P., Lagendijk, J.H., Becker, A.E.: Extracellular matrix of the human aortic media: an ultrastructural histochemical and immunohistochemical study of the adult aortic media. Anat. Rec. **258**, 1–14 (2000)

15. Clark, J.M., Glagov, S.: Transmural organization of the arterial media. The lamellar unit revisited. Arterioscler. Thromb. Vasc. Biol. **5**, 19–34 (1985). https://doi.org/10.1161/01.atv.5.1.19

16. Ratz, P.H.: Mechanics of vascular smooth muscle. Compr. Physiol. (2014)

17. Humphrey, J.D., Holzapfel, G.A.: Mechanics, mechanobiology, and modeling of human abdominal aorta and aneurysms. J. Biomech. **45**, 805–814 (2012). https://doi.org/10.1016/j.jbiomech.2011.11.021

18. Rizzo, R.J., McCarthy, W.J., Dixit, S.N., Lilly, M.P., Shively, V.P., Flinn, W.R., Yao, J.S.T.: Collagen types and matrix protein content in human abdominal aortic aneurysms. J. Vasc. Surg. **10**, 365–373 (1989). https://doi.org/10.1016/0741-5214(89)90409-6

19. Rezakhaniha, R., Agianniotis, A., Schrauwen, J.T.C., Griffa, A., Sage, D., Bouten, C.V.C., Van De Vosse, F.N., Unser, M., Stergiopulos, N.: Experimental investigation of collagen waviness and orientation in the arterial adventitia using confocal laser scanning microscopy. Biomech. Model. Mechanobiol. **11**, 461–473 (2012). https://doi.org/10.1007/s10237-011-0325-z

20. Schrauwen, J.T.C., Vilanova, A., Rezakhaniha, R., Stergiopulos, N., van de Vosse, F.N., Bovendeerd, P.H.M.: A method for the quantification of the pressure dependent 3D collagen configuration in the arterial adventitia. J. Struct. Biol. **180**, 335–342 (2012). https://doi.org/10.1016/j.jsb.2012.06.007

21. Chow, M.-J., Turcotte, R., Lin, C.P., Zhang, Y.: Arterial extracellular matrix: a mechanobiological study of the contributions and interactions of elastin and collagen. Biophys. J. **106**, 2684–2692 (2014). https://doi.org/10.1016/j.bpj.2014.05.014

22. Chen, H., Liu, Y., Slipchenko, M.N., Zhao, X., Cheng, J.-X.X., Kassab, G.S.: The layered structure of coronary adventitia under mechanical load. Biophys. J. **101**, 2555–2562 (2011). https://doi.org/10.1016/j.bpj.2011.10.043

23. Chen, H., Slipchenko, M.N., Liu, Y., Zhao, X., Cheng, J.-X., Lanir, Y., Kassab, G.S.: Biaxial deformation of collagen and elastin fibers in coronary adventitia. J. Appl. Physiol. **115**, 1683–1693 (2013). https://doi.org/10.1152/japplphysiol.00601.2013

24. Hill, M.R., Duan, X., Gibson, G.A., Watkins, S., Robertson, A.M.: A theoretical and non-destructive experimental approach for direct inclusion of measured collagen orientation and recruitment into mechanical models of the artery wall. J. Biomech. **45**, 762–771 (2012). https://doi.org/10.1016/j.jbiomech.2011.11.016

25. Tower, T.T., Neidert, M.R., Tranquillo, R.T.: Fiber alignment imaging during mechanical testing of soft tissues. Ann. Biomed. Eng. **30**, 1221–1233 (2002). https://doi.org/10.1114/1.1527047

26. Sutton, M.A., Ke, X., Lessner, S.M., Goldbach, M., Yost, M., Zhao, F., Schreier, H.W.: Strain field measurements on mouse carotid arteries using microscopic three-dimensional digital image correlation. J. Biomed. Mater. Res., Part A **84**, 178–190 (2008)

27. Genovese, K., Lee, Y.-U., Lee, A.Y., Humphrey, J.D.: An improved panoramic digital image correlation method for vascular strain analysis and material characterization. J. Mech. Behav. Biomed. Mater. **27**, 132–142 (2013). https://doi.org/10.1016/j.jmbbm.2012.11.015

28. Wang, R., Brewster, L.P., Gleason Jr., R.L.: In-situ characterization of the uncrimping process of arterial collagen fibers using two-photon confocal microscopy and digital image correlation. J. Biomech. **46**, 2726–2729 (2013). https://doi.org/10.1016/j.jbiomech.2013.08.001
29. Ehret, A.E., Bircher, K., Stracuzzi, A., Marina, V., Zündel, M., Mazza, E.: Inverse poroelasticity as a fundamental mechanism in biomechanics and mechanobiology. Nat. Commun. **8**, 1002 (2017). https://doi.org/10.1038/s41467-017-00801-3
30. Faury, G.: Function–structure relationship of elastic arteries in evolution: from microfibrils to elastin and elastic fibres. Pathol. Biol. **49**, 310–325 (2001). https://doi.org/10.1016/S0369-8114(01)00147-X
31. Faury, G., Pezet, M., Knutsen, R.H., Boyle, W.A., Heximer, S.P., McLean, S.E., Minkes, R.K., Blumer, K.J., Kovacs, A., Kelly, D.P., Li, D.Y., Starcher, B., Mecham, R.P.: Developmental adaptation of the mouse cardiovascular system to elastin haploinsufficiency. J. Clin. Invest. **112**, 1419–1428 (2003). https://doi.org/10.1172/JCI19028
32. Humphrey, J.D.: Mechanisms of arterial remodeling in hypertension. Hypertension **52**, 195–200 (2008)
33. Canham, P.B., Finlay, H.M., Dixon, J.A.N.G., Boughner, D.R., Chen, A.: Measurements from light and polarised light microscopy of human coronary arteries fixed at distending pressure. Cardiovasc. Res. **23**, 973–982 (1989)
34. Sáez, P., García, A., Peña, E., Gasser, T.C., Martínez, M.A.: Microstructural quantification of collagen fiber orientations and its integration in constitutive modeling of the porcine carotid artery. Acta Biomater. **33**, 183–193 (2016). https://doi.org/10.1016/j.actbio.2016.01.030
35. Schriefl, A.J., Reinisch, A.J., Sankaran, S., Pierce, D.M., Holzapfel, G.A.: Quantitative assessment of collagen fibre orientations from two-dimensional images of soft biological tissues. J. R. Soc. Interface. **9**, 3081–3093 (2012)
36. Dahl, S.L.M., Rhim, C., Song, Y.C., Niklason, L.E.: Mechanical properties and compositions of tissue engineered and native arteries. Ann. Biomed. Eng. **35**, 348–355 (2007). https://doi.org/10.1007/s10439-006-9226-1
37. Wolinsky, H., Glagov, S.: Structural basis for the static mechanical properties of the aortic media. Circ. Res. **14**, 400–413 (1964). https://doi.org/10.1161/01.RES.14.5.400
38. Fujimoto, J.G., Boppart, S.A., Tearney, G.J., Bouma, B.E., Pitris, C., Brezinski, M.E.: High resolution in vivo intra-arterial imaging with optical coherence tomography. Heart **82**, 128–133 (1999)
39. Acosta, V., Flechas Garcia, M., Molimard, J., Avril, S.: Three-dimensional full-field strain measurements across a whole porcine aorta subjected to tensile loading using optical coherence tomography–digital volume correlation. Front. Mech. Eng. (2018). https://doi.org/10.3389/fmech.2018.00003
40. Campagnola, P.J., Dong, C.Y.: Second harmonic generation microscopy: principles and applications to disease diagnosis. Laser Photonics Rev. **5**, 13–26 (2011). https://doi.org/10.1002/lpor.200910024
41. Rouède, D., Schaub, E., Bellanger, J.J., Ezan, F., Scimeca, J.C., Baffet, G., Tiaho, F.: Determination of extracellular matrix collagen fibril architectures and pathological remodeling by polarization dependent second harmonic microscopy. Sci. Rep. **7** (2017). https://doi.org/10.1038/s41598-017-12398-0
42. Zipfel, W.R., Williams, R.M., Christie, R., Nikitin, A.Y., Hyman, B.T., Webb, W.W.: Live tissue intrinsic emission microscopy using multiphoton-excited native fluorescence and second harmonic generation. Proc. Natl. Acad. Sci. **100**, 7075–7080 (2003). https://doi.org/10.1073/pnas.0832308100
43. Williams, R.M., Zipfel, W.R., Webb, W.W.: Interpreting second-harmonic generation images of collagen I fibrils. Biophys. J. **88**, 1377–1386 (2005). https://doi.org/10.1529/biophysj.104.047308
44. Zoumi, A., Lu, X., Kassab, G.S., Tromberg, B.J.: Imaging coronary artery microstructure using second-harmonic and two-photon fluorescence microscopy. Biophys. J. **87**, 2778–2786 (2004). https://doi.org/10.1529/biophysj.104.042887

45. Koch, R.G., Tsamis, A., D'Amore, A., Wagner, W.R., Watkins, S.C., Gleason, T.G., Vorp, D.A.: A custom image-based analysis tool for quantifying elastin and collagen micro-architecture in the wall of the human aorta from multi-photon microscopy. J. Biomech. **47**, 935–943 (2014). https://doi.org/10.1016/j.jbiomech.2014.01.027

46. Chen, X., Nadiarynkh, O., Plotnikov, S., Campagnola, P.J.: Second harmonic generation microscopy for quantitative analysis of collagen fibrillar structure. Nat. Protoc. **7**, 654–669 (2015). https://doi.org/10.1038/nprot.2012.009.Second

47. Freund, I., Deutsch, M., Sprecher, A.: Optical second-harmonic microscopy, crossed-beam summation, and small-angle scattering in rat-tail tendon. Biophys. J. **50**, 693–712 (1986)

48. Gannaway, J.N., Sheppard, C.J.R.: Second-harmonic imaging in the scanning optical microscope. Opt. Quantum Electron. **10**, 435–439 (1978). https://doi.org/10.1007/BF00620308

49. Semwogerere, D., Weeks, E.R.: Confocal microscopy. Encycl. Biomater. Biomed. Eng. 1–10 (2005). https://doi.org/10.1081/e-ebbe-120024153

50. Zoumi, A., Yeh, A., Tromberg, B.J.: Imaging cells and extracellular matrix in vivo by using second-harmonic generation and two-photon excited fluorescence. Proc. Natl. Acad. Sci. **99**, 11014–11019 (2002). https://doi.org/10.1073/pnas.172368799

51. Denk, W., Strickler, J., Webb, W.W.: Two-photon laser scanning fluorescence microscopy. Science (80) **248**, 73–76 (1990)

52. Bancelin, S., Lynch, B., Bonod-Bidaud, C., Ducourthial, G., Psilodimitrakopoulos, S., Dokládal, P., Allain, J.-M., Schanne-Klein, M.-C., Ruggiero, F.: Ex vivo multiscale quantitation of skin biomechanics in wild-type and genetically-modified mice using multiphoton microscopy. Sci. Rep. **5**, 17635 (2015)

53. Ambekar, R., Chittenden, M., Jasiuk, I., Toussaint, K.C.: Quantitative second-harmonic generation microscopy for imaging porcine cortical bone: Comparison to SEM and its potential to investigate age-related changes. Bone **50**, 643–650 (2012). https://doi.org/10.1016/j.bone.2011.11.013

54. Gusachenko, I., Tran, V., Goulam Houssen, Y., Allain, J.-M., Schanne-Klein, M.-C.: Polarization-resolved second-harmonic generation in tendon upon mechanical stretching. Biophys. J. **102**, 2220–2229 (2012). https://doi.org/10.1016/j.bpj.2012.03.068

55. Morin, C., Avril, S.: Inverse problems in the mechanical characterization of elastic arteries. MRS Bull. **40** (2015). https://doi.org/10.1557/mrs.2015.63

56. Zeinali-Davarani, S., Chow, M.-J., Turcotte, R., Zhang, Y.: Characterization of biaxial mechanical behavior of porcine aorta under gradual elastin degradation. Ann. Biomed. Eng. **41**, 1528–1538 (2013). https://doi.org/10.1007/s10439-012-0733-y

57. Zeinali-Davarani, S., Wang, Y., Chow, M. J., Turcotte, R., Zhang, Y.: Contribution of collagen fiber undulation to regional biomechanical properties along porcine thoracic aorta. J. Biomech. Eng. **137**, 51001 (2015)

58. Chow, M.-J., Zhang, Y.: Changes in the mechanical and biochemical properties of aortic tissue due to cold storage. J. Surg. Res. **171**, 434–442 (2011). https://doi.org/10.1016/j.jss.2010.04.007

59. Park, C.Y., Lee, J.K., Chuck, R.S.: Second harmonic generation imaging analysis of collagen arrangement in human cornea. Investig. Ophthalmol. Vis. Sci. **56**, 5622–5629 (2015). https://doi.org/10.1167/iovs.15-17129

60. Narice, B.F., Green, N.H., Macneil, S., Anumba, D.: Second Harmonic Generation microscopy reveals collagen fibres are more organised in the cervix of postmenopausal women. Reprod. Biol. Endocrinol. **14**, 70–78 (2016). https://doi.org/10.1186/s12958-016-0204-7

61. Jayyosi, C., Coret, M., Bruyere-Garnier, K.: Characterizing liver capsule microstructure via in situ bulge test coupled with multiphoton imaging. J. Mech. Behav. Biomed. Mater. **54**, 229–243 (2016)

62. Olson, E., Levene, M.J., Torres, R.: Multiphoton microscopy with clearing for three dimensional histology of kidney biopsies. Biomed. Opt. Express. **7**, 3089 (2016). https://doi.org/10.1364/BOE.7.003089

63. Pena, A.-M., Fabre, A., Débarre, D., Marchal-Somme, J., Crestani, B., Martin, J.-L., Beaurepaire, E., Schanne-Klein, M.-C.: Three-dimensional investigation and scoring of extracellular

matrix remodeling during lung fibrosis using multiphoton microscopy. Microsc. Res. Tech. **70**, 162–170 (2007). https://doi.org/10.1002/jemt

64. Wang, C.-C., Li, F.-C., Wu, R.-J., Hovhannisyan, V.A., Lin, W.-C., Lin, S.-J., So, P.T.C., Dong, C.-Y.: Differentiation of normal and cancerous lung tissues by multiphoton imaging. J. Biomed. Opt. **14**, 044034 (2009). https://doi.org/10.1117/1.3210768

65. Campa, J.S., Greenhalgh, R.M., Powell, J.T.: Elastin degradation in abdominal aortic aneurysms. Atherosclerosis **65**, 13–21 (1987). https://doi.org/10.1016/0021-9150(87)90003-7

66. Krasny, W., Magoariec, H., Morin, C., Avril, S.: Kinematics of collagen fibers in carotid arteries under tension-inflation loading. J. Mech. Behav. Biomed. Mater. **77**, 718–726 (2018). https://doi.org/10.1016/j.jmbbm.2017.08.014

67. Krasny, W., Morin, C., Magoariec, H., Avril, S.: A comprehensive study of layer-specific morphological changes in the microstructure of carotid arteries under uniaxial load. Acta Biomater. **57**, 342–351 (2017). https://doi.org/10.1016/j.actbio.2017.04.033

68. Cavinato, C., Helfenstein-Didier, C., Olivier, T., du Roscoat, S.R., Laroche, N., Badel, P.: Biaxial loading of arterial tissues with 3D in situ observations of adventitia fibrous microstructure: a method coupling multi-photon confocal microscopy and bulge inflation test. J. Mech. Behav. Biomed. Mater. **74**, 488–498 (2017). https://doi.org/10.1016/j.jmbbm.2017.07.022

69. Jayyosi, C., Affagard, J.-S., Ducourthial, G., Bonod-Bidaud, C., Lynch, B., Bancelin, S., Ruggiero, F., Schanne-Klein, M.-C., Allain, J.-M., Bruyère-Garnier, K., et al.: Affine kinematics in planar fibrous connective tissues: an experimental investigation. Biomech. Model. Mechanobiol. **16**(4), 1459–1473 (2017)

70. Humphrey, J.D.D.: Cardiovascular Solid Mechanics: Cells, Tissues, and Organs. Springer, New-York (2002)

71. Keyes, J.T., Haskett, D.G., Utzinger, U., Azhar, M., Vande Geest, J.P.: Adaptation of a planar microbiaxial optomechanical device for the tubular biaxial microstructural and macroscopic characterization of small vascular tissues. J. Biomech. Eng. **133**, 075001 (2011). https://doi.org/10.1115/1.4004495

72. Länne, T., Sonesson, B., Bergqvist, D., Bengtsson, H., Gustafsson, D.: Diameter and compliance in the male human abdominal aorta: influence of age and aortic aneurysm. Eur. J. Vasc. Endovasc. Surg. **6**, 178–184 (1992)

73. Romo, A., Badel, P., Duprey, A., Favre, J.-P., Avril, S.: In vitro analysis of localized aneurysm rupture. J. Biomech. **47**, 607–616 (2014)

74. Roy, S., Boss, C., Rezakhaniha, R., Stergiopulos, N.: Experimental characterization of the distribution of collagen fiber recruitment. J. Biorheol. **24**, 84–93 (2010). https://doi.org/10.1007/s12573-011-0027-2

75. Jähne, B.: Spatio-Temporal Image Processing: Theory and Scientific Applications. Springer Science & Business Media (1993)

76. Bigun, J., Bigun, T., Nilsson, K.: Recognition by symmetry derivatives and the generalized structure tensor. IEEE Trans. Pattern Anal. Mach. Intell. **26**, 1590–1605 (2004)

77. Ayres, C., Jha, B.S., Meredith, H., Bowman, J.R., Bowlin, G.L., Henderson, S.C., Simpson, D.G.: Measuring fiber alignment in electrospun scaffolds: a user's guide to the 2D Fast Fourier Transform approach. J. Biomater. Sci. Polym. Ed. **19**, 603–621 (2008)

78. Phillippi, J.A., Green, B.R., Eskay, M.A., Kotlarczyk, M.P., Hill, M.R., Robertson, A.M., Watkins, S.C., Vorp, D.A., Gleason, T.G.: Mechanism of aortic medial matrix remodeling is distinct in patients with bicuspid aortic valve. J. Thorac. Cardiovasc. Surg. **147**, 1056–1064 (2014). https://doi.org/10.1016/j.jtcvs.2013.04.028

79. D'Amore, A., Stella, J.A., Wagner, W.R., Sacks, M.S., D'Amore, A., Stella, J.A., Wagner, W.R., Sacks, M.S.: Characterization of the complete fiber network topology of planar fibrous tissues and scaffolds. Biomaterials **31**, 5345–5354 (2010)

80. Tonar, Z., Nemecek, S., Holota, R., Kocova, J., Treska, V., Molacek, J., Kohoutek, T., Hadravska, S.: Microscopic image analysis of elastin network in samples of normal, atherosclerotic and aneurysmatic abdominal aorta and its biomechanical implications. J. Appl. Biomed. **1**, 149–160 (2003)

81. Schriefl, A.J., Wolinski, H., Regitnig, P., Kohlwein, S.D., Holzapfel, G.A.: An automated approach for three-dimensional quantification of fibrillar structures in optically cleared soft biological tissues. J. R. Soc. Interface. **10**, 20120760 (2013)
82. Polzer, S., Gasser, T.C., Forsell, C., Druckmüllerova, H., Tichy, M., Staffa, R., Vlachovsky, R., Bursa, J.: Automatic identification and validation of planar collagen organization in the aorta wall with application to abdominal aortic aneurysm. Microsc. Microanal. **19**, 1395–1404 (2013)
83. Ayres, C., Bowlin, G.L., Henderson, S.C., Taylor, L., Shultz, J., Alexander, J., Telemeco, T.A., Simpson, D.G.: Modulation of anisotropy in electrospun tissue-engineering scaffolds: analysis of fiber alignment by the fast Fourier transform. Biomaterials **27**, 5524–5534 (2006)
84. Morrill, E.E., Tulepbergenov, A.N., Stender, C.J., Lamichhane, R., Brown, R.J., Lujan, T.J.: A validated software application to measure fiber organization in soft tissue. Biomech. Model. Mechanobiol. **15**, 1467–1478 (2016)
85. Sander, E.A., Barocas, V.H.: Comparison of 2D fiber network orientation measurement methods. J. Biomed. Mater. Res. Part A. **88**, 322–331 (2009)
86. Morin, C., Krasny, W., Avril, S.: Multiscale mechanical behavior of large arteries. Encycl. Biomed. Eng. **2**. Elsevier (2019)
87. Keyes, J.T., Lockwood, D.R., Utzinger, U., Montilla, L.G., Witte, R.S., Vande Vande Geest, J.P., Geest, J.P.: Comparisons of planar and tubular biaxial tensile testing protocols of the same porcine coronary arteries. Ann. Biomed. Eng. **41**, 1579–1591 (2013). https://doi.org/10.1007/s10439-012-0679-0
88. Wan, W., Dixon, J.B., Gleason Jr., R.L.: Constitutive modeling of mouse carotid arteries using experimentally measured microstructural parameters. Biophys. J. **102**, 2916–2925 (2012). https://doi.org/10.1016/j.bpj.2012.04.035
89. Pourdeyhimi, B.: Imaging and Image Analysis Applications for Plastics. William Andrew (1999)
90. Chen, Q., Pugno, N.M.: Bio-mimetic mechanisms of natural hierarchical materials: a review. J. Mech. Behav. Biomed. Mater. **19**, 3–33 (2013). https://doi.org/10.1016/j.jmbbm.2012.10.012
91. Tsamis, A., Krawiec, J.T., Vorp, D.A.: Elastin and collagen fibre microstructure of the human aorta in ageing and disease: a review. J. R. Soc. Interface. **10** (2013). https://doi.org/10.1098/rsif.2012.1004
92. Brüel, A., Oxlund, H.: Changes in biomechanical properties, composition of collagen and elastin, and advanced glycation endproducts of the rat aorta in relation to age. Atherosclerosis **127**, 155–165 (1996). https://doi.org/10.1016/S0021-9150(96)05947-3
93. Cox, R.H.: Age-related changes in arterial wall mechanics and composition of NIA Fischer rats. Mech. Ageing Dev. **23**, 21–36 (1983). https://doi.org/10.1016/0047-6374(83)90096-9
94. Fornieri, C., Quaglino, D., Mori, G.: Role of the extracellular matrix in age-related modifications of the rat aorta. Ultrastructural, morphometric, and enzymatic evaluations. Arterioscler. Thromb. Vasc. Biol. **12**, 1008–1016 (1992). https://doi.org/10.1161/01.atv.12.9.1008
95. Valenta, J., Vitek, K., Cihak, R., Konvickova, S., Sochor, M., Horny, L.: Age related constitutive laws and stress distribution in human main coronary arteries with reference to residual strain. Biomed. Mater. Eng. **12**, 121–134 (2002)
96. Duprey, A., Trabelsi, O., Vola, M., Favre, J.P., Avril, S.: Biaxial rupture properties of ascending thoracic aortic aneurysms. Acta Biomater. **42**, 273–285 (2016). https://doi.org/10.1016/j.actbio.2016.06.028
97. Farzaneh, S., Trabelsi, O., Avril, S.: Inverse identification of local stiffness across ascending thoracic aortic aneurysms. Biomech. Model. Mechanobiol. **18**(1), 137–153 (2019). https://doi.org/10.1007/s10237-018-1073-0
98. Sokolis, D.P., Kefaloyannis, E.M., Kouloukoussa, M., Marinos, E., Boudoulas, H., Karayannacos, P.E.: A structural basis for the aortic stress-strain relation in uniaxial tension. J. Biomech. **39**, 1651–1662 (2006). https://doi.org/10.1016/j.jbiomech.2005.05.003
99. Sugita, S., Matsumoto, T.: Multiphoton microscopy observations of 3D elastin and collagen fiber microstructure changes during pressurization in aortic media. Biomech. Model. Mechanobiol. 1–11 (2016). https://doi.org/10.1007/s10237-016-0851-9

100. Morin, C., Avril, S., Hellmich, C.: Non-affine fiber kinematics in arterial mechanics: a continuum micromechanical investigation. ZAMM - J. Appl. Math. Mech. / Zeitschrift für Angew. Math. und Mech. 1–21 (2018). https://doi.org/10.1002/zamm.201700360

Intrinsic Optical Imaging of ECM Mechanics

Raphaël Turcotte and Yanhang Zhang

Abstract This chapter provides a comprehensive introduction to optical imaging in the context of extracellular matrix (ECM) mechanics. The goal is to address the critical aspects of optical microscopy so that researchers in the biomechanics research field are able to take full advantage of optical imaging technologies in their studies. Optics is of particular interest to answering questions of multiscale mechanics because of its multiscale nature; information can be obtained at the tissue, fiber, and even molecular levels. Revealing the microstructure of the load-bearing ECM constituents, elastin and collagen, and the interplay between ECM structure and mechanical loading is probably the area where optical microscopy has contributed the most to our understanding of ECM mechanics in recent years. The discussion is therefore confined to optical imaging technologies able to resolve ECM fibers using signals from molecules endogenous to tissue and emphasizes the importance of unbiased imaging and image analysis. Descriptions of polarimetric multiphoton microscopy and adaptive optics are also provided because of their potential for enabling discoveries in ECM mechanics. Although arteries are used here as an exemplar tissue, all concepts covered in this chapter are expected to be generalizable to other organs and tissues.

1 Introduction

Elastic and collagen fibers are the major ECM components in blood vessels (Fig. 1). Elastic fibers endow blood vessels with critical mechanical properties such as flexibility and extensibility. They are essential for accommodating deformations encountered during physiological function of arteries, which undergo repeated cycles of extension and recoil. In the medial layer of elastic arteries, elastic fibers form thick

R. Turcotte (✉)
University of Oxford, Oxford, UK
e-mail: raphael.turcotte@pharm.ox.ac.uk

Y. Zhang
Boston University, Boston, MA, USA

© Springer Nature Switzerland AG 2020   165
Y. Zhang (ed.), *Multi-scale Extracellular Matrix Mechanics and Mechanobiology*,
Studies in Mechanobiology, Tissue Engineering and Biomaterials 23,
https://doi.org/10.1007/978-3-030-20182-1_6

concentric fenestrated layers of elastic lamella. Together with alternating layers of smooth muscle cells and medial collagen, these elastic lamella form a functional unit of the arterial wall (Fig. 1a–c) [21]. Adventitial collagen, commonly described as a meshwork of helically woven fibers layered around the vessel wall [40], provides structural integrity at higher strains (Fig. 1d).

Determining how the ECM components contribute to the mechanical behavior of an artery is essential for understanding the mechanisms of vascular remodeling and disease progressions. Optical imaging plays an important role in the study of the effect of ECM structural changes on vascular function. Information obtained through improved medical imaging and biomechanical testing methods would be useful in the understanding of vascular remodeling associated with diseases. Several methods, including optical coherence tomography and Brillouin microscopy, can be used to directly measure mechanical properties [45, 75, 79]. Other optical technologies such as point-scanning microscopy and micro-optical coherence tomography are able to provide information about the ECM organization within biological tissue [19, 117]. In particular, the last decade has seen a surge in the application of multiphoton microscopy to ECM mechanics. Multiphoton microscopy enables visualizing the microstructure of the two main ECM constituents, collagen and elastin fibers, in three dimensions (Fig. 1a–d) [80] and imaging in intact tissue, even in living organisms. The ability to image intact tissue is of paramount significance as it implies that imaging can be coupled with mechanical loading (Fig. 1e). The opportunity to

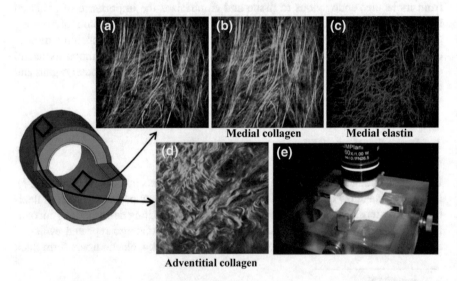

Fig. 1 Optical imaging of ECM in porcine thoracic aorta with multiphoton microscopy. **a–c** Images from the media are shown with **a** both collagen (green) and elastin (magenta), **b** only collagen, and **c** only elastin (image width: 300 μm). **d** Image of collagen from the adventitia (image width: 200 μm). **e** Combining imaging and mechanical loading that allows the sample to be imaged while subjected to biaxial stretching [17]

link the microstructure of elastin and collagen to the tissue mechanical properties provides important insights into the complex multiscale nature of tissue mechanics. The mechanical roles of ECM constituents that cannot be optically measured can also be studied indirectly through their impact on the elastin and collagen fibers microstructure and mechanics [60, 61]. Finally, it is also possible to widen the multiscale window offered by multiphoton microscopy when considering polarization to include the molecular organization of ECM fibers [34, 91].

2 Fundamentals of Optical Imaging

Before diving into a discussion on aspects of imaging directly related to multiscale ECM mechanics, it is necessary to briefly introduce some fundamental concepts of optical imaging. Here, we attempt to provide an intuitive picture sufficient for fully understanding the core ideas of this chapter.

2.1 Optical Images

An optical imaging system is an instrument that uses light to obtain spatially-defined information about a sample [6, 27, 63]. The general working principle of optical imaging is that light is shined onto a sample, altered by the sample through a light-matter interaction, and then detected by a device converting the optical signal to an electrical one. The type of information acquired does not only depend on the light-matter interactions occurring in the sample, but on the nature of the illumination and light-detection systems as well. The strength, or magnitude, of the light-matter interaction being space-variant, the physical parameter containing the desired quantitative information about the sample is usually represented in a structure-preserving map: the *image*.

The above definition of optical imaging is broad, and it aims to encompass the wide variety of existing experimental systems. Essential to the definition of imaging is the fact that a physical property should vary in space; otherwise, the generated uniform image is equivalent to a point measurement. This leads to an alternative definition of (scientific) images that is more closely related to the experimental configuration: a collection of point measurements in an n-dimensional space arranged to conserve the relative dimensional organization. It should be noted that imaging technologies do not perfectly map the object information. For example, the signal at one location may be "contaminated" with signal from other locations [64] or optical aberrations might alter the mapping [27]. Also, different sources of noise can contribute to the imperfect mapping of information in optical images [63].

2.2 Contrast and Spatial Resolution

How much the signal varies between "adjacent" points within the observed volume is often referred to as the *contrast* (or modulation) and is a useful metric to characterize the performance of an imaging system or modality, although it can be dependent on the imaged sample. Contrast is expressed in term of the signal intensity (I) and is quantified as [37]:

$$C = \frac{I_{max} - I_{min}}{I_{max} + I_{min}} \tag{1}$$

where I_{max} corresponds to the brightest pixel and I_{min} to the dimmest one. It can intuitively be understood as our ability to visualize objects (white), to distinguish them from the background (black, Fig. 2) [7]. The contrast is ideally high and close to its maximal value of 1.

Another metric to characterize the performance of an imaging system, the *spatial resolution,* addresses how the intensity varies spatially (Fig. 3a, b). Here, it is essential to understand that an infinitesimal object will appear to have a larger size in the image (Fig. 3c). Indeed, filtering of the optical signal occurs in such a way that all objects (and borders) will have a minimal size determined by the physical parameters of the microscope. Great importance is (rightfully) placed on spatial resolution because it dictates the length scale of objects that can be resolved with a given microscopy technique. More formally, spatial resolution is commonly defined in two ways: (1) the width of the impulse response of a system (Fig. 3) [63], or (2) the minimal distance required to resolve, or identify the presence of, two adjacent point objects [103]. Each of these two definitions comes with its own criteria on how to quantify resolution. Metrics for characterizing spatial resolution are usually presented as scalars for simplicity. It is nevertheless important to keep in mind that spatial resolution may be anisotropic. Often, the resolution is constant within the imaging plane (lateral axes), but lower in the direction orthogonal to the imaging plane (axial axis) [63].

Fig. 2 Contrast. Images of point objects with, from left to right, increasing contrast

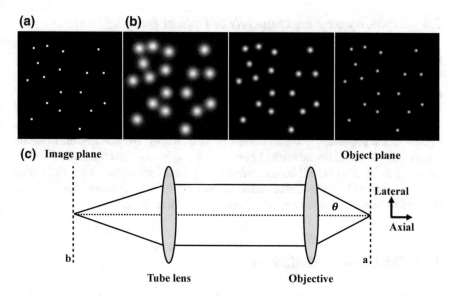

Fig. 3 Spatial resolution. **a** Point objects distribution. **b** Images of point objects with, from left to right, increasing spatial resolution. **c** Schematic of an imaging system

2.3 Temporal Resolution

Resolution can also be temporal. The *temporal resolution* quantifies the minimal time between two sequential images, e.g. the highest achievable imaging rate. It can be limited by the instrumentation because some operations have a finite duration (i.e., a camera takes time to read and send data to a computer). The brightness of the sample must also be considered because enough photons must be detected to obtain an image of sufficient quality and this means that the duration of the illumination period may have to be increased for achieving high contrast in dim samples [65].

When thinking about ECM constituents such as elastin and collagen and their mechanical roles, knowing their spatial arrangement is essential. To determine their microstructure, it is necessary to perform volumetric, or three-dimensional (3D), structural imaging. How long it takes to acquire 3D data is determined by the imaging rate. This parameter will often pose a practical limit on the observable volume fraction (spatial sampling) within an experiment. Imaging can also be dynamic: the same region is imaged repetitively. This would enable, for example, seeing structural changes during viscoelastic relaxation [98]. Dynamic imaging is also necessary for functional imaging where the goal is to measure molecular transients in cells or any other marker of functions. For instance, it might be of relevance to look at calcium transients in smooth muscle cells (functional imaging), when cellular contraction changes the structure of the soft tissue, and describe how the ECM around those cells is changed during a single event (dynamic imaging) [39]. In short, the temporal resolution dictates the time scale of observable processes.

2.4 Summary on Fundamentals of Optical Imaging

Image, contrast, spatial resolution, and temporal resolution are concepts that have to be considered when selecting an imaging modality and system. Contrast and spatial resolution are closely related, but we would argue that they need to be independent to be meaningful. Indeed, some definitions convolute contrast with resolution and some information is then lost about the imaging system. This section was not meant to present the physical principles behind those concepts. For more information on these topics, we recommend reading *Fundamentals of Biomedical Optics* by Caroline Boudoux as its content will be accessible to experts in non-optics fields (including biomechanics) [7]. For a formal introduction to optical microscopy, the reader is directed to *Introduction to Optical Microscopy* by Mertz [63].

3 ECM Imaging Modalities

The development of optical imaging technologies is closely related to their usage in biological applications. Several systems have been applied to visualize the main constituents of the ECM, collagen and elastin, or to measure their mechanical properties [79, 116, 117]. Nowadays, optical microscopes are ubiquitous in research institutions and their operation is becoming more user-friendly. It nevertheless remains essential to understand how signal is being generated and how tissue is altering this signal. In this section, we will present ECM imaging modalities based on different contrast mechanisms, which are light-matter interactions modulating the light field and thus making the contrast, as defined in Eq. 1, non-null.

We will focus our discussion around intrinsic contrast mechanisms. Such mechanisms are defined by the fact that the light-matter interaction takes place with molecules endogenous to the tissue of interest [113]. This is in contrast to exogenous contrast mechanisms which require the introduction of an additional molecule into the sample through some labelling strategies (molecular targeting or genetic modification). Intrinsic, or endogenous, contrast mechanisms possess a series of properties making them advantageous for imaging in the context of ECM mechanics:

1. Issues related to labelling uniformity are avoided. It is important that the detected signal quantitatively represents the distribution (including the concentration) of the molecular species of interest in the tissue. Nonuniformities associated with extrinsic labeling will lead to nonlinearities in the mapping between the sample and the image, thus limiting the usefulness of quantitative information.
2. The signal directly informs on the structure of interest. Directly probing the molecules of interest not only allows accurately mapping structures but also enables characterizing other properties [10]. With exogenous contrast, such measurements, when at all possible, required specialized labeling agents.
3. Tissue processing is not needed. Beyond saving time, the elimination of tissue processing ensures that the tissue's mechanical properties are not altered by the

introduction of exogenous agents forming chemical bonds with the tissue and potentially altering the molecular environment (viscosity, hydration, etc.).

4. Compatible with in vivo (intravital) experiments. Intrinsic imaging techniques developed for excised tissue samples can readily be employed in live animals as no additional sample processing is needed other than the required surgery.

We should note here that the use of exogenous signals also has some important advantages over intrinsic ones. First, specific molecules can be targeted in a fashion that is not dependent on the molecular integrity of the structure of interest. This ability makes it possible to visualize altered ECM constituents in a pathological context. Also, not all ECM constituents can emit light endogenously, but exogenous labels can make them visible. Finally, the signal strength of exogenous agents and their spectral emission bandwidth can be modulated according to experimental needs, resulting in a more flexible system.

3.1 Scattering and Fluorescence Microscopy

In this section, we address **linear, or one-photon, interactions, which occur when one photon interacts with the tissue to generate a different photon.** As photons travel through tissue, they have a probability (termed "cross-section") to interact with molecules. A large portion of the light will be absorbed or diffused by the tissue and will not produce a useful contrast. This absorption and diffusion lead to an exponential decrease of the illumination power as a function of the propagation path length and limits the imaging depth. Other interactions generate a signal suitable for ECM imaging: fluorescence and scattering, both of which can be understood through energy level diagrams (Fig. 4) [7].

Elastin and collagen can both be visualized by fluorescence (Fig. 5) [53, 111]. For fluorescence, a photon is absorbed by a molecule in the ground state $|S_0\rangle$ (Fig. 4a). The photon energy E_{exc} brings the molecule into an excited state $|S_1\rangle$. When the molecule relaxes, on a time scale of nanosecond, a photon of lower energy E_{em}, or equivalently of longer wavelength, is emitted. This wavelength shift between excitation and emission is known as the Stokes shift (Fig. 6a). The Stokes shift originates from the relaxation of vibrational states v_i, which occurs on a picosecond timescale. The Stokes shift facilitates the spectral separation between the excitation and emission. In the case of elastin and collagen, they can both be excited in the ultra-violet region, with a peak at ~335 nm (Fig. 6a) [94]. Fortunately, elastin undergoes a larger shift than collagen, enabling the spectral separation of their endogenous fluorescence emission [86]. Endogenous fluorescence can also be referred to as autofluorescence. This "self" fluorescence is often considered detrimental in most exogenous fluorescence imaging applications based on extrinsic labels as it can reduce the specificity of the detected signal.

Linear fluorescence can be generated at any location along the light path. Even if the object of interest is properly positioned with respect to the objective lens and its

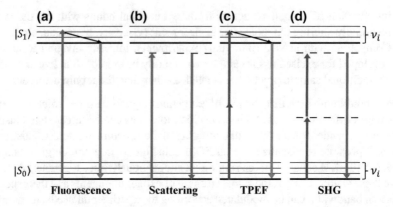

Fig. 4 Energy level diagrams for **a** one-photon fluorescence, **b** linear scattering, **c** two-photon excited fluorescence (TPEF), and **d** second-harmonic generation (SHG). An energy level diagram, also known as a Jablonksi diagram, is a representation of the electronic states for molecule (horizontal lines) and the transition routes between those states (vertical arrows). The ground state $|S_0\rangle$, singlet excited state $|S_1\rangle$, and their related vibrational states v_i are typically included. Virtual states can also be included (horizontal dashed lines). The vertical axis represents an energy scale. The horizontal axis facilitates the display of event sequences but is not strictly a temporal axis

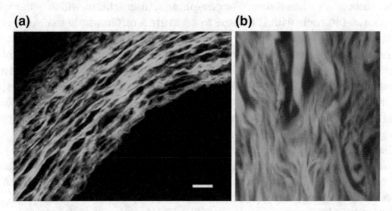

Fig. 5 Linear fluorescence images of **a** elastin in a circumferential cross-section of a porcine carotid artery (scale bar: 50 μm) and **b** collagen in human skin. Reproduced with permission from [53] and [111], respectively

signal correctly transmitted to the optical detector in the image plane, out-of-focus fluorescence from objects located above or below the object plane can contaminate the image [64]. As determining the 3D organization of ECM fibers requires knowing the axial plane from which the signal originate, a strategy for rejection of the out-of-focus signal is needed. Confocal microscopy is a solution to this problem. In confocal microscopy, the illumination light is focused to a single point by the objective and the fluorescence output is filtered in the image plane by a pinhole [70]. The position of the pinhole is such that only the in-focus fluorescence (light from the axial plane

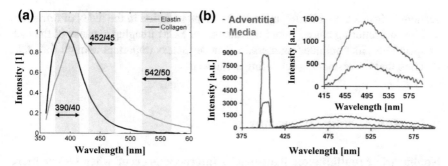

Fig. 6 Emission spectra. **a** One-photon autofluorescence emission spectra from elastin and collagen and three spectral regions (gray, numbers indicate central wavelength/bandwidth) for fiber identification through ratiometric analysis [86]. **b** Combined second-harmonic generation (SHG) and two-photon excited fluorescence (TPEF) from arterial tissue at different excitation wavelength [116]. Reproduced with permission

where the focus occurs in the object, optical sectioning) will go through and reach a point photo-detector (e.g. photomultiplier tube and avalanche photodiode). An image is formed by translating the focal point laterally and sequentially acquire the signal for each pixel. Even when hundreds of thousands of pixels must successively be recorded to form a complete image, the imaging rate can reach 30 frames per second [93].

Achieving optical sectioning is also possible with widefield fluorescence imaging techniques, in which the entire field of view is illuminated at once and the image formed onto a camera. These techniques have in common that (*i*) multiple widefield images need to be acquired with some variation in the illumination pattern and (*ii*) the optically sectioned image is produced in post-processing by combining the multiple raw images. For example, optical sectioning structure illumination microscopy (OS-SIM) requires at least three images recorded with a sinusoidal illumination pattern that is phase shifted between the images [64, 68, 104]. HiLo microscopy requires two images: one with a uniform illumination and a second with non-uniform illumination containing high spatial frequency variations [54, 55]. OS-SIM and HiLo microscopy have not been widely applied to ECM imaging, but would be suitable when a high imaging speed is required. Indeed, although multiple raw images must be acquired to produce a single optically-sectioned image, widefield imaging can reach hundreds and even thousands of frame per second [20, 57, 84].

Scattering is another important light-matter interaction for imaging. During a scattering event, the photon is not absorbed by a molecule, but causes an oscillation of the electronic cloud, which brings the molecule to a virtual state [7]. At the end of the oscillation, a photon of equal energy to the incident one leaves the system, with a different direction (Fig. 4b). The concept of confocal point-scanning microscopy also applies to scattering. Scattering differs from fluorescence in that it doesn't offer chemical specificity, i.e. the signal doesn't originate from a specific molecular component allowing one to distinguish collagen from elastin or any other scatterers. Instead, a scattering microscope maps variation in the index of refraction and linear scattering has a limited usefulness for ECM imaging because of the lack of specificity. Its nonlinear counterpart is nevertheless important enough for ECM imaging that understanding scattering is relevant.

3.2 Multiphoton Microscopy

Nonlinear, or multiphoton, light-matter interactions occur when two or more photons interact with a material and generate another photon [9]. Two-photon excited fluorescence (TPEF) is the nonlinear counterpart to one-photon fluorescence and occurs when a molecule absorbs two photons simultaneously and reaches an excited state (Fig. 4c) [114]. The following processes of molecular relaxation and emission of a fluorescence photon are then equivalent between linear and nonlinear excitation. Because two photons contribute to the excitation, their energy only needs to be about half of that required for single photon excitation. Therefore, lasers operating at longer wavelengths, in the near infrared between 700 nm and 900 nm, are employed for multiphoton ECM imaging [117]. The use of longer wavelengths increases the penetration depth by reducing the scattering [43, 47], but also by reducing absorption from water and blood [81]. This is a substantial benefit for imaging in thick tissue, especially considering the short linear excitation wavelength of elastin and collagen (<400 nm).

The optimal illumination wavelength for TPEF of elastin is at 730 nm [117]. The emission spectrum is broad with a peak at ~495 nm (Fig. 6b). The large wavelength shift (~80 nm) between the one- and two-photon emission spectra is indicative that different endogenous molecules are at the origin of elastin autofluorescence [100]. Indeed, not all molecules generating a strong linear fluorescence signal will generate an equivalently strong nonlinear signal, and reciprocally, some molecules with a relatively strong nonlinear signal might yield a weak linear fluorescence [23]. In fact, the TPEF from collagen is weak to the point that a different multiphoton contrast mechanism is used to visualize collagen: second-harmonic generation (SHG).

SHG is the nonlinear counterpart to scattering (Fig. 4d) and requires two photons to cause oscillations of the electronic cloud on the same molecule for a single photon at exactly half the wavelength to be produced. The SHG emission spectrum has nonetheless a non-zero width but this width remains very narrow compared to fluorescence (Fig. 6b). The optimal excitation wavelength for SHG from collagen is at 800 nm, which results in an SHG emission at 400 nm [117]. This SHG emission does not overlap with TPEF from elastin, which facilitate the spectral separation of signals in comparison to linear fluorescence. SHG can only occur in non-centrosymmetric media (not possessing an inversion symmetry) [9]. There are few biological structures having permanent dipole moments that are sufficiently ordered and non-centrosymmetric. However, collagen is a highly ordered fibrillar structure and can produce strong SHG which effectively provide signal specificity [11, 14]. As a consequence, SHG was rapidly adopted to image collagen in arteries, skin, tendon,

(a) **(b)**

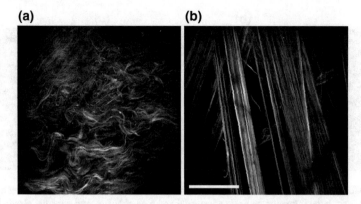

Fig. 7 Multiphoton imaging in porcine aorta. **a, b** SHG images from adventitial collagen **a** with no strain and **b** with a 40% strain (scale bar: 180 μm). Reproduced with permission [91]

and other tissues where the ECM plays an important mechanical role [8, 114, 116, 117]. It should be noted that different types of collagen do not generate the same amount of second-harmonic signal. Fibrillar collagen of type I and II produce the strongest SHG signal, whereas non-fibrillar type IV and V provided very limited SHG contrast. Collagen of type III yields weak SHG signal even though it has a fibrillar arrangement [72]. Exemplar images of adventitial collagen in porcine aorta during a mechanical loading experiment are shown in Fig. 7. Of note, the molecular alignment of collagen fibril creates a polarization effect that can be deleterious for structural imaging applications as the signal will be dependent on the fiber orientation in the laboratory frame and the incident polarization [85]. Fortunately, this effect can be alleviated by using a circularly polarized incident light [14].

The probability that two photons arrive at the same location and at the same time, within the extent of the Heisenberg uncertainty principle, is low and thus so is the cross-section for TPEF and SHG [114]. For this reason, multiphoton microscopy is generally implemented with spatially focused excitation (point-scanning) from a pulsed laser source. The use of laser pulses effectively compresses in time the arrival of photons at the focus. These two strategies suffice to generate enough photons for high-quality images to be formed. The requirement for a high photon density for multiphoton events to occur has the benefit that the out-of-focus background is negligible. Optical sectioning is therefore an intrinsic property of multiphoton microscopy (Fig. 8); no confocal pinhole is necessary.

3.3 Summary of ECM Imaging Modalities

In recent years, there have been a very large number of studies in which collagen and elastin were imaged with SHG and TPEF microscopy, respectively, to study the

(a)   **(b)**   **(c)**

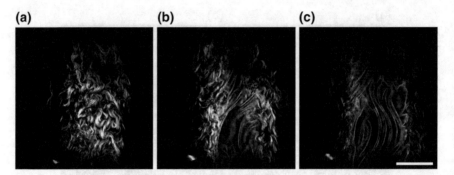

Fig. 8 Multiphoton imaging in murine carotid artery. Images at different depths showing **a** adventitial collagen, **b** the transition between adventitial and medial layers, and **c** medial elastin (TPEF: green; SHG: white; scale bar: 100 μm). Reproduced with permission from [108]

relation between tissue mechanics and microstructure. Constitutive modeling of soft biological tissue is a challenging field of research due to the complexity of the tissue behavior, hierarchical structures in the tissue, and the ability of biological tissue to remodel in response to stimuli. Advances in optical methods and image processing techniques have made possible exploring/quantifying the architecture of ECM structural components of soft tissues at different scales. Using multiphoton microscopy, arterial elastin and collagen has been simultaneously captured in previous studies [16, 17, 26, 110]. Constitutive formulations that are motivated by the tissue structure has been proposed for biological tissues, along with many studies attempted to incorporate some of the structural information into the structural-based constitutive models [38, 95, 99, 109, 110]. In sum, SHG and TPEF are powerful modalities for ECM imaging because intrinsic signals are available to visualize collagen and elastin, respectively, in 3D at a length-scale appropriate for mapping the fiber microstructure.

4 Prospective ECM Imaging Technologies

This chapter on optical imaging of the ECM would not be complete without discussing more advanced technologies. Some of these technologies can be implemented as add-ons to standard microscopes. Beyond instrumentation, advanced imaging techniques can also refer to the status of the sample. In this section, we will cover the following technologies: polarimetric multiphoton microscopy and adaptive optics. These technologies have yet to be exploited to their full potential to study ECM mechanics.

4.1 Polarimetric Multiphoton Microscopy of ECM Molecular Order

So far, we have focused on techniques quantifying intensity, not phase or polarization. Intensity provides information on the microscopic scale and is ideal to map the micro-architecture of the ECM. The multiscale relation between microstructure and tissue mechanics can be extended to also include molecular organization of ECM fibers. The molecular order can be probed by polarimetric multiphoton microscopy [14, 25, 69, 85]. This technique is particularly relevant for SHG collagen imaging applications, but the appropriate extrinsic labelling agent could enable similar measurements for elastin [10, 52].

At the core of polarimetric multiphoton microscopy is a *polarization dependence* effect. The polarization dependence describes the modulation of the detected (SHG) intensity as a function of the 3D angular relation between the molecular dipole axis of scatterers and the spatial orientation of the electric field oscillation of the illumination light. This dependence can be mathematically expressed by a tensorial quantity, the nonlinear susceptibility [9]. By controlling the polarization of the incident light, it is therefore possible to interrogate the molecular dipole arrangement by determining the nonlinear susceptibility tensor or related quantities.

Polarimetric multiphoton microscopy is usually performed with linearly polarized light, for which the electric field oscillates within a single plane. For collagen fibers residing in the object plane, the SHG intensity will vary as a function of the polar angle between the polarization axis and average molecular orientation within each focal volume (Fig. 9a). By recording multiple images with different orientation of the incident linear polarization, the polarization dependence of the SHG signal can be plotted and fitted to an analytical expression having nonlinear susceptibility tensor elements as free parameters (Fig. 9b). The exact form of the analytical expression depends on the sample and typically involves many geometrical and symmetry assumptions [3, 18]. The intensity variation is large for ECM collagen fibers as they have a highly symmetric molecular structure. If the molecular dipoles had a random distribution, their average orientation would be isotropic and there would be no polarization dependence.

By performing the fitting at every pixel in the image, it is possible to produce maps of the nonlinear susceptibility tensor components (Fig. 10a, b) [13, 18, 62]. These are particularly useful as they show differential molecular organization, which may not be visible with intensity measurements alone or by analyzing intensity variations for the entire image. The existence of different types of collagen is further evidence by the bimodality of the tensor element histogram (Fig. 10c, d).

Several other optical properties related to polarization can be mapped in this way. Of particular relevance for ECM mechanics is the mapping of the principal orientation per pixels [89]. This mapping is spatially more resolved than the results from FFT analysis, as the results from the latter depends on surrounding pixels, whereas adjacent pixels are independent in the polarization analysis.

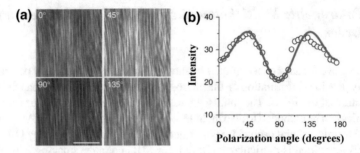

Fig. 9 Polarization dependence. **a** SHG images of collagen recorded at one location with different orientation of the incident linear polarization (scale bar: 50 μm). **b** Plot of the SHG intensity as a function of the polarization angle. Reproduced with permission from [14]

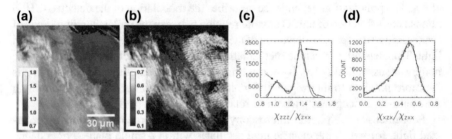

Fig. 10 Molecular imaging at a tendon-muscle junction. **a, b** Nonlinear susceptibility tensor element χ_{ijk} ratio mapping of **a** χ_{zzz}/χ_{zxx} and **b** χ_{xzx}/χ_{zxx}. **c, d** Histogram distribution of the tensor ratio **c** χ_{zzz}/χ_{zxx} and **d** χ_{xzx}/χ_{zxx} fitted with the sum of two Gaussian functions. Reproduced with permission from [13]

It was mentioned in Sect. 3.2 that the use of a circularly polarized illumination alleviated the effect of polarization dependence and that this was essential for structural imaging of collagen with SHG. It should nevertheless be noted that the nonlinear susceptibility tensor can be characterized using circularly polarized light [24] and that there is a distinction between an image taken with circularly polarized light and an image of the amplitude calculated from fitting the analytical expression for polarization dependence [91].

Polarimetric multiphoton microscopy has been used extensively to study the properties of collagen, but often not enough consideration was given to how the mechanical state of the tissue can affect the measurement. Reciprocally, very few studies have taken advantage of the additional length-scale provide by polarization analysis and this approach has yet to provide new insight into the biomechanics of the ECM. It is expected that as fibers are being recruited to carry a load their molecular organization will change to become more uniformly aligned. Consequentially, the SHG signal measured with circular polarization should increase and the intensity modulation form the polarization dependence should be more pronounced (Fig. 11a, b). Two studies have reported preliminary observation supporting these predictions. The first study reported changes in the SHG signal from adventitial collagen during equal

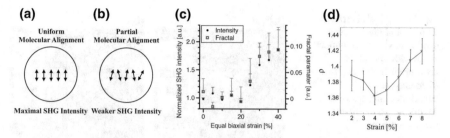

Fig. 11 Molecular ECM mechanics. **a**, **b** The SHG intensity is dependent on molecular alignment for any incident polarization state [91]. **c** Fractal analysis and polarization-dependent intensity measurements during equal biaxial loading of SHG images from adventitial porcine aorta [91]. **d** Optical parameters ρ from rat-tendon obtained from polarization SHG data as a function of strain [34]. Reproduced with permission

biaxial loading in porcine aorta and was measured with circularly polarized light. The observed variation corresponded closely with previous results obtained from fractal analysis—a measure of the degree of self-similarity of an image at different scales. Here changes in the fibrous network was quantified with a parameter corresponding to the normalized variation in fractal number to the initial experimental condition—(Fig. 11c) [91]. A second study characterized different optical parameters in rat tail tendon under mechanical loading [34]. In particular, it is shown how the anisotropy, a parameter related to the nonlinear susceptibility tensor, varies with strain (Fig. 11d), but the authors link this result to fibril orientation. Much remains to be done to link the molecular information to the fiber and tissue levels in order to yield new insight into the multiscale ECM mechanics.

Several factors contribute to the alteration of the polarization state during propagation in biological tissue: birefringence [32, 33], geometrical parameters [88], and scattering [22]. They will not be discussed in detail, but it should be understood that polarimetric multiphoton microscopy is challenging to perform at depth and will often require empirical corrections in order to be accurate. Another important limitation comes from the fact that the structure to be probed does not necessarily lies in the imaging plane, i.e. the azimuthal angle between the linear polarization and the fiber orientation may not be zero. In such common cases, a complete characterization of the nonlinear susceptibility tensor would require probing with axially oscillating polarization states, which requires more advanced optical tools to be generated.

4.2 Adaptive Optics

To assess the mechanical properties of the ECM, it is necessary to image the ECM constituents in situ as they interact together microscopically to give tissue level properties. This multiscale shift precludes relevant mechanical and microstructural information from being obtained by imaging in thin tissue sections or along cuts made

Fig. 12 Optical aberrations. Two-photon fluorescence excitation volume **a** without and **b** with optical aberrations introduced by a tilted glass surface (X: lateral; Z: axial; scale bar: 2 μm). Reproduced with permission from [90]

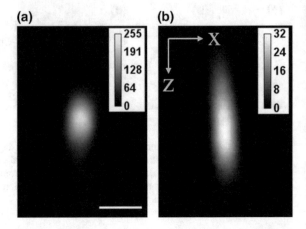

on excise volumes. As previously mentioned, imaging deep in biological tissue is limited by absorption and scattering, which cause an exponential attenuation of the illumination power. The imaging depth is further decreased by optical aberrations. Optical aberrations are characterized by an alteration of the illumination wavefront,[1] which results in an intensity and spatial resolution losses (Fig. 12) [4, 44]. Therefore, not only is the imaging depth reduced, but the image quality at the accessible depths is degraded. Optical aberrations caused by the sample can have multiple origins. They can originate from a mismatch in index of refraction between the immersion milieu for the optics and the sample [5]. They can also arise from the interface between the sample and immersion milieu being tilted [90] or curved [59]. These situations will result in wavefronts with predictable conformation such as spherical and comatic aberrations.

More challenging to predict are the aberrations generated by the heterogenous index of refraction distribution in biological tissue. Indeed, biological tissues do not, in general, possess uniform optical properties and this can have severe effects on image quality. Methods for wavefront control are therefore needed. Adaptive optics refers to a group of technologies for wavefront control [4, 44]. As there exists many variations in implementation of adaptive optics, we will not review them here. Instead, the two core principles, common to all adaptive optics methods, will be presented: sensing and correction. Sensing refers to the determination of the aberrated wavefront. The measurement of aberrations can either be performed directly or indirectly. Direct wavefront sensing involves measuring the wavefront itself and requires additional optical components to be integrated into the imaging system [2, 96, 97]. In general, direct wavefront sensing has the advantage of being fast, as only a single measurement is needed, but has the disadvantages of having increased instru-

---

[1] A wavefront is a surface orthogonal to the local propagation vectors on which the phase of the light wave is uniform. For a plane wave, aberrations are causing the wavefront to no longer reside in a single plane, but instead to form a complex 3D surface, i.e. the phases in the plane are not uniform.

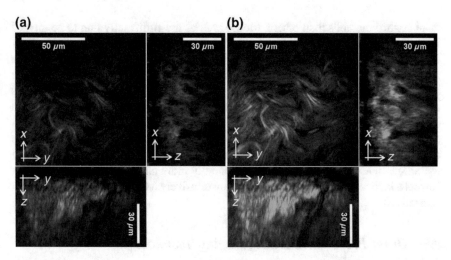

Fig. 13 Adaptive optics. Ex vivo murine skin images (xy) and axial-lateral reconstructions (xz and yz) **a** without and **b** with aberration correction. Green: SHG from collagen. Reproduced with permission from [67]

mental complexity and requiring calibration procedures to be routinely performed for performance to be maintained.

Indirect sensing methods do not directly measure the wavefront, but instead use the signal of interest, either in the form of a point or as an image, as a beacon for an optimization procedure. For instance, the optimization could aim at increasing the contrast in an image or the brightness of a point. In opposition to direct sensing, indirect sensing requires a large number of measurements and is thus slower [44]. It possesses the advantage of not necessitating modification of the microscope, the standard imaging detector being a sufficient sensor. The selection between indirect and direct sensing should therefore be based on a comparison between time constraints (which are likely low for structural ECM imaging) and the availability of support to operate more advanced imaging systems.

Correction of the aberrations is done following a single approach for multiphoton imaging: pre-compensation. The pre-compensation consists in imparting to an unaberrated incident wavefront the same amplitude of phase error as will be encountered in the sample but of the opposite sign. As the shaped wavefront travels through the tissue, the pre-compensation will cancel out the effect of tissue-induced aberrations. Wavefront shaping is achieved by positioning a segmented device which can control the local slopes within a wavefront (either a deformable mirror or a liquid-crystal spatial light modulator) in the appropriate Fourier-conjugate plane. Adaptive optics has been implemented most frequently in bio-imaging for neuroscience applications using multiphoton point-scanning fluorescence microscopy [44], but a few examples of collagen imaging with SHG can be found (Fig. 13) [67].

Increasing the imaging depth and improving image quality with adaptive optics would benefit all ECM imaging applications, but some would profit more from wavefront control methods than others. Mouse arteries are sufficiently thin to be imaged through their entire thickness with multiphoton microscopy. Unfortunately, their large curvature on the scale of the wavefront from their cylindrical shapes generate a substantial amount of aberrations, making it challenging to visualize medial elastin and to reconstruct elastic lamellae [107]. Collagen bundles in the adventitial layers are also highly aberrating and cause underlying structures to appear dim and smeared (Fig. 22). As will be further discussed, the local shadowing effect from collagen bundles can make quantitative analysis challenging. Unlike the undesirable effect from tissue curvature, collagen bundles cause aberrations that are likely spatially-variant within the field-of-view and are consequentially more challenging to correct. Most current adaptive optics methods are suitable exclusively for spatially-invariant optical aberrations.

4.3 Other Prospective ECM Imaging Technologies

In this chapter, we focus on only two technologies (polarimetric multiphoton microscopy and adaptive optics) to preserve generality. Several other approaches, such as super-resolution imaging, expansion microscopy, and optical clearing, could potentially advance the field of ECM mechanics as well. We will mention them here briefly. Super-resolution microscopy has been demonstrated for ECM imaging in the brain [92]. Its main benefit is the extended spatial resolution it provides, going beyond the diffraction-limit, which could enable detailed mapping of individual fibers in a complex meshwork [78]. Increased spatial resolution is also achieved with expansion microscopy by chemically processing the sample for it to be physically and homo-morphically expanded [12]. Unlike super-resolution imaging, this approach is not compatible with mechanical testing, but provides increased light penetration depth by making the sample more transparent. In optical clearing methods, the chemical treatment exclusively aims to make the sample optically transparent without expansion. Here again the goal is to increase the imaging depth and this method has been used together with SHG to visualize skeletal muscle [71].

5 Unbiased Imaging

It is often tempting to immediately place a sample under the microscope and acquire images. Before doing so, there is nevertheless a critical, yet oft-omitted step: designing the imaging study. Even imaging experts will frequently not perform an imaging study and instead report a proof-of-principle that a certain measurement can be performed with their latest device. Unfortunately, this is insufficient to answer questions of ECM mechanics. More likely is that imaging at many (but not all) locations in several samples will be necessary. Going beyond making single measurements is essential but establishing the optimal sampling strategy is not trivial. In this section, we will explore how to think about sample and sampling in the context of ECM mechanics.

5.1 Stereology for Biomechanics

Stereology is the science of analyzing 3D materials such as biological tissue from spatially sampled data [66, 77, 101]. The goal of stereology can be understood as creating a set of approaches and tools to evaluate quantitative parameters—volume, surface area, length, and number, which will be referred to as the first-order stereological parameters—for a sample, and the population it belongs to, when only a fraction of the sample is probed [66]. Stereology can thus be divided into two parts: sampling and analysis. Here, we will look at the sampling; the analysis will be addressed in a later section.

Historically, the spatial sampling consisted of discontinuous cross-sections from fixed and cut tissue because it was not possible to image and analyze the entirety of a sample volume [101]. When imaging is discontinuous, it is not possible to determine with certainty the structures located between two adjacent planes [66]. The main challenge therefore involved extending the information obtained from 2D planes to the full 3D tissue. This extension is not as simple as one might think and failure to account for the difference in dimensionality leads to significant inaccuracies and imprecisions. Most early stereological tools are related to the "2D-to-3D" correspondence and therefore suitable for optical imaging data, which is often intrinsically 2D.

Thanks to advances in automation and imaging speed, it is now common to image samples continuously along the light propagation axis. For the imaging to be continuous implies that the axial sampling is sufficiently high with respect to the spatial resolution such that the volumetric information can be reconstructed between adjacent planes; the uncertainty about structures in between planes is eliminated. As previously mentioned, performing volumetric imaging is critical when assessing the organization of ECM structure, which are located in multiple axial planes due to their waviness, their bundle arrangement, and the tissue curvature. For instance, medial elastic lamellae in mouse carotid arteries might appear as large wavy fibers under two photon fluorescence microscopy if the axial sampling is not sufficient (Fig. 8b, c), whereas a volumetric reconstruction will reveal its lamellar nature [107]. Volumetric reconstructions are also necessary when assessing relations between two objects, such as the distance between two elastic lamellae. Even if a lamella rests exactly in a single imaging plane, tissue deformation will change the absolute position of this plane with respect to the imaging device such that under some mechanical conditions the lamella might no longer be visible if the axial sampling (the distance between two imaging planes) is insufficient.

For smaller vessels, it is possible that nearly the full thickness can be imaged from the adventitial side (Fig. 14b). For larger vessels, the entire thickness of the sample is rarely imaged (Fig. 14a). The reasons for this are two-fold. First, light is scattered as it propagates through tissue, thus posing a limitation on the imaging depth (as discussed in Sect. 3.1). Second, it is more efficient to keep in line with the stereological perspective, using several 3D sub-volumes distributed throughout a sample to probe the 3D structure of the whole sample. However, in tissues that are thick in comparison to the optical penetration depth, the number of accessible

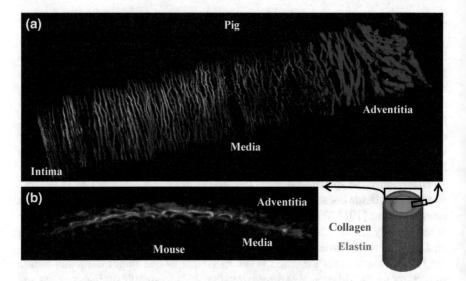

Fig. 14 Multiphoton images of collagen (blue) and elastin (green) acquired in **a** a circumferential cross-section from a pig aorta (width: ~1200 μm) and **b** an intact mouse carotid artery reconstructed to show a circumferential cross-section view (width: 350 μm)

sub-volumes is limited. It is thus questionable whether the ECM structure obtained from a small depth is representative of the ECM network structure in the arterial wall. Transmural variation in elastin fiber distribution was demonstrated by acquiring images from 3D slabs across the thickness (Fig. 15) [108]. It was concluded that the transmural variation in fiber orientation distribution is important in characterizing the anisotropic mechanical behavior of ECM network and should be considered in constitutive modeling of tissue mechanics.

An important question is "how much of the tissue should be sampled?" To answer this question, we should be reminded that the goal of most scientific measurements is to test the information that will be obtained against a multitude of hypotheses and for the conclusions to be extrapolatable to a population. In order to achieve this, it is not only necessary to accurately determine the value of stereological parameters, but also their variation. Therefore, a sufficient fraction of the tissue should be imaged such that this fraction encompasses most of the biological variation in a parameter within a sample, most of the intra-sample biological heterogeneity. A corollary is that enough samples should also be imaged to assess the inter-sample variability. Quantification of both the intra- and inter-sample biological variations are essential. Imaging a large number of samples but only at one location will yield inaccurate and imprecise data. Imaging a single sample in its entirety doesn't provide information about the population it belongs to. In brief, stereology dictates rule which guarantee that spatially sampled imaging data is suitable for statistical inferences [66].

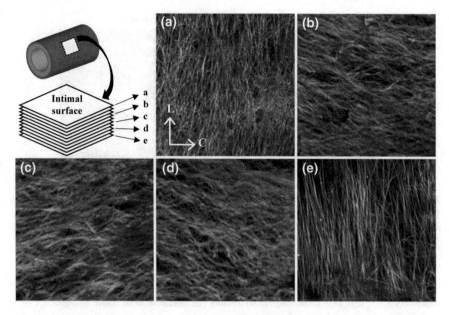

Fig. 15 Multiphoton images showing the distribution of elastin fibers in **a** the inner media, **b–d** middle media, and **e** outer media. Images width 250 μm; L: longitude; and C: circumference. Adapted from [108]

5.2 Reference Space

In order to compare different samples, it is essential that the probed regions be consistent. Otherwise the accuracy of the subsequent analysis may be worsened, and the inter-sample variation will be increased. It will then be impossible to separate the variation due to this sampling error and the real biological variation, thus limiting the performance and adequacy of statistical inference methods [42]. It is well known that the content and architecture of elastin and collagen and, hence, the mechanics of aorta vary with anatomical locations throughout the aortic tree [29, 46, 51, 74, 76, 115]. For instance, elastin content [15, 35, 36, 82] and the number of lamellar units [83, 105] appear to drop markedly from proximal thoracic to distal regions of aorta. Therefore, it is essential to establish a *reference space* that can be consistently imaged in different subjects [66].

The structure of the ECM largely dictates the properties of the organs and tissues in which they are located. The goal of imaging in this context is to inform the relation between the ECM and tissue mechanical properties. Consequentially, the reference space should be defined in a way that reflects biological functions [66]. Ideally, the borders of the reference space will correspond exactly with the borders of an organ or biological structure. This definition of the reference space is the most robust one for repeatability. It also enables normalization of volumes to create standardized tissue maps. In addition, volumetric changes/alterations due to growth and remodeling are

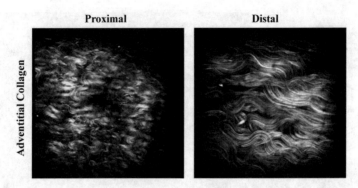

Fig. 16 Representative SHG images of collagen in adventitial collagen of proximal (left) and distal (right) regions of thoracic aortas. Adventitial collagen fibers are highly undulated in the proximal region compared to the distal region while there is no obvious difference in medial elastin and collagen waviness between proximal and distal region (Images width: 360 μm)

common when comparing experimental groups of different age, gender, strain, or disease status, and this definition also allows for correction of these biases.

In large organs, taking a fraction (a sub-region) of a biological structure as the reference space may be beneficial. This is particularly true when results can be sample-matched between ECM imaging and mechanical testing, i.e. the same sample is used for both measurements. The requirement for the reference space to be deterministic remains regardless of the fact that it does not incorporate an entire function or structural biological unit. Indeed, if cuts are made in a tissue, such cuts should be made at morphologically defined positions, and a unique position should be used for a study. For example, the ECM microstructures was found to vary longitudinally in porcine aorta, in accordance to an equivalent mechanical variation (Fig. 16), and such gradient could not have been observed with a random selection of samples within aortas [110]. More importantly, such gradient would generate undesirable inter-sample variation and potentially mask subtle biomechanical phenomena. This illustrates the benefit of using sub-regions as reference space.

We will conclude the discussion on reference spaces by distinguishing them from regions of interest (ROIs). While reference spaces are deterministic, ROIs will only be selected after the imaging and according to the distribution of interesting features within the dataset. The goal of selecting ROIs is to identify groups of pixels on which further analysis should be performed, which contrast with the physiological nature of reference spaces. It is usually challenging to establish formal criteria to define ROIs. As a consequence, it is frequent for different individuals to obtain different quantitative results. Nonetheless, the same qualitative observations should be made. Otherwise, it might be that the ROIs were poorly chosen.

5.3   Systematic Random Sampling

Once the reference space is defined, the next step consists in establishing how it will be sampled. In order to make such a determination two questions must be answered: (1) what fraction of the volume should be imaged and (2) how should the probes (imaging area) be spatially distributed? The volume fraction that should be imaged is determined by the intra-sample variation. The aim is to image enough of the sample to capture its biological heterogeneity. Of course, if the entire sample is imaged, the variance of the imaged volume fraction will correspond exactly to the variance of the sample. However, acquiring such a large amount of data is often technically unrealistic. In the study of ECM mechanics, it is often more informative to image a sample under a number of mechanical conditions. We routinely determine properties of the ECM in porcine aorta by imaging only 10% of the optically accessible volume fraction [17, 60].

The optimal volume fraction can be found by performing a pilot study [1, 66, 102]. This pilot study would consist in imaging a large fraction of a single sample and evaluate the statistical estimates for different volume fractions artificially generated in post-processing. By plotting the statistical estimates as a function of volume fraction, one can then find the smallest volume fraction at which the statistical estimates have converged to the sample values. As independence is needed between samples for statistical inference, the decision of including the sample in which the pilot study was conducted should be taken carefully; independence is lost if a sample included in the experimental group affects the selection of the other samples.

The spatial distribution of the imaging areas should be random for the imaging and sequential quantitative analysis to be unbiased. It is indeed essential for the reference space to be probed randomly for the results to be unbiased. One would correctly expect a non-random and random distribution to appear as shown in Figs. 17a and b, respectively. The reason Fig. 17a represents a non-random probe distribution is because the reference space in partially probed, not because of the systematic arrangement of the probes. It is indeed the case that a partial sampling of the reference space is necessarily not random, as exemplified in Fig. 17c. In fact, the distribution in Fig. 17a is a limiting case of the one in Fig. 17c, where 100% of the partial reference space is probed.

Optimal probe distributions for ECM imaging are shown in Fig. 18 [31]. Two key features are apparent: the probes have a uniform distribution and the distribution is based on the geometry of the reference space. Unbiased stereology requires that no mathematical models be used to inform the probe distribution and that no assumptions be made about the geometry of the reference space [66], the rationale being that biological objects do not follow geometrical rules, and the assumptions and models are thus imperfect. The distributions shown in Fig. 18 are termed *systematic random sampling* if the origin is randomly assigned and the spacing between probes is uniform (systematic) and preserved from sample to sample [31]. The systematic arrangement of probes does not affect the randomness as long as the starting point is randomly assigned for each sample. In previous studies, we performed such assign-

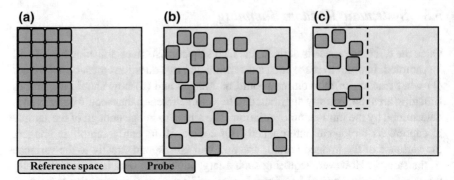

Fig. 17 Sampling distributions. **a** Non-random probe distribution. **b** Random probe distribution. **c** Partial covering of the reference space; non-random probe distribution

Fig. 18 Systematic random sampling. **a** Equidistant probe separation forms a systematic sampling. The origin must be determined randomly. **b** The same distribution is used for different samples but the origin is randomly changed

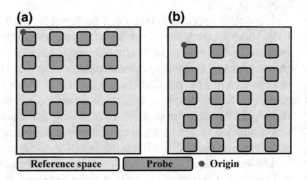

ment by arbitrarily positioning the samples, and thus the reference space, under the microscope. Another option would have been to position each sample very exactly under the microscope and use computer generated origin coordinates for each sample. Systematic sampling possesses the important advantage of being more efficient than non-systematic sampling [66]. Less data is required with systematic than non-systematic sampling for the sample estimates to converge. In addition, the uniform spacing makes the data easier to acquire manually or through automation. It should nonetheless be noted that systematic and non-systematic random sampling will provide data of equivalent accuracy.

5.4   Other Experimental Considerations

The last step before beginning data acquisition is to determine a key experimental parameter, the illumination power. The illumination power is the number of photons per unit of time (in watts) incident onto the specimen. For most biological applications, it is desirable to keep this power low because too much light may cause damage, thus altering the biology of interest, and engender photobleaching, a decrease of the signal due to the inability of molecules to be excited multiple times. Collagen and elastin are less vulnerable to photodamage than cells and minimally susceptible to photobleaching [10, 30, 41, 48]. These properties allow for large illumination powers to be used for ECM imaging, although one should be mindful of collagen cross-linking; collagen cross-linking can alter the mechanical properties of the sample and spectrally contaminate the multiphoton fluorescence signal of elastin with new fluorescence signal from collagen [50, 73]. No absolute power values are given here because photobleaching, photodamage, and collagen cross-linking depend not only on the illumination power, but also on other physical parameters such as pixel dwell time and imaging rate.

Using a large illumination power has several advantages. First, it will improve the image quality by increasing the signal-to-noise ratio. Another way of increasing the signal-to-noise ratio consists in acquiring multiple images sequentially and averaging them, which requires more time but is nonetheless advisable (averaging just a few frames can substantially improve image quality). Second, a larger dynamic range is made available, making features which are less bright visible. Third, the imaging depth is increased. The effective illumination power decreases exponentially through tissue due to absorption and scattering, thus limiting the depth at which imaging can be performed. In the past, our strategy has been to find the saturation point at the surface of the samples and use a fixed laser power. Alternatively, the power could be increased to compensate for the attenuation. Adjusting the power of the light source can usually easily be done either through software control or by rotating a wave plate located in front of a prism. In any case, the power should be measured directly with a power meter calibrated for the wavelengths and power range of the source, kept constant for all samples, and saturation absolutely avoided. Due to inter-sample variation in optical properties, it is possible that there will be saturation appearing between the preliminary and main studies or between samples within the main study. It is therefore a good practice to sacrifice some dynamic range by decreasing the laser power by a few percent to avoid saturation. The illumination power should always be reported in scientific publications, as should the reference space and sampling parameters.

5.5   Unbiased Imaging Summary

By performing systematic random sampling of the reference space in a sufficiently large number of samples, a dataset will be generated that is suited for statistical inference. Abiding by stereological rules also means that the experiment will not have to be repeated and can serve as a reference. For instance, we conducted an imaging study of the ECM in porcine aorta to characterize the engagement of collagen and elastin during mechanical loading [17]. Because of the use of unbiased imaging, we were able to use the same data as a baseline to assess the specific role of glycosaminoglycans [60]. Of course, only the imaging strategies is unbiased. A different set of considerations might be necessary to make the mechanical testing more quantitatively robust. Finally, we would like to note that the terminology used and the selection of concepts to be introduced was based on *Principles and practices of unbiased stereology* by Mouton [66]. Readers should nevertheless note that we depart significantly from the content of that book.

6 Methods for Image Analysis

Image analysis is an important part of imaging, as is the accompanying post-processing. There certainly are many valid ways of analyzing a single dataset, and ideally, each of these should bring us to the same conclusion. Therefore, guidelines for image analysis and reporting of imaging data are presented here that can be generalizable to any chosen analysis method. Quantitative analysis will also be placed into a stereological context. For ECM imaging, it is important to characterize the complex networks at different structural hierarchies; an emphasis will thus be put on how to describe geometric objects in biological tissue efficiently and accurately.

6.1 Standard Approach for Image Analysis

6.1.1 Quantitative Image Analysis

When deploying an imaging modality for a new application, such as imaging an organ from a new experimental animal model, the first step of image analysis should consist in verifying that the optical signal originates from the structure of interest. For example, in early imaging of elastin in the descending aorta of mice, multiphoton fluorescence images were presented with images from histological methods (Fig. 19). It is quite striking in Fig. 19b which elastic lamellae is different from the others, and one might be tempted to stop the analysis at this point. We argue here that imaging analysis should always be *quantitative*, even if the interpretation of the results will be qualitative. Unbiased quantitative analysis is necessary for meaningful reporting of conclusions from statistical inference testing, i.e. significance of the observation should be demonstrated.

Quantifying the imaging data will also help to select the images to display for reporting. Displayed images should be *representative* of the populations studied. An unbiased way of performing this selection is to systematically pick the image for which the quantitative metric is the closest to the average value of the sample. For instance, the image in Fig. 19b of the experimental condition in transgenic (KO) mice should have a "fiber dispersion" value close to 34° as suggested by Fig. 19c. Of course, displaying extreme cases would be more evocative and may thus facilitate the communication of the results, which is an aspect that should not be neglected. It

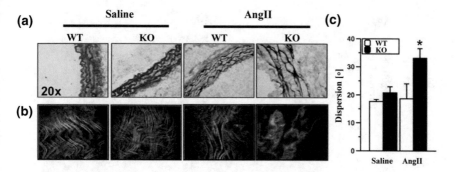

Fig. 19 Exemplar imaging analysis figure. Representative images of **a** elastin-stained aortic sections and **b** elastic lamellae by multiphoton fluorescence microscopy (green: elastin, blue: collagen; image width: 306 μm). **c** Quantification of the elastin signal from multiphoton data (1-sided Wilcoxon rank-sum test, *P < 0.05; n = 4 for each saline conditions and n = 5 for each experimental condition, AngII) based on the elastin fiber dispersion in the descending aorta. WT: wild-type, KO: transgenic model. Reproduced from [28]

is our opinion that representative imaging data should necessarily be presented, and can, if desired, be supplemented with more drastic examples. In any cases, the two should be differentiated explicitly.

6.1.2 Direct Metrics

The second step in image analysis is to choose a metric for quantification of a physical property of the tissue. The metric can directly capture the properties of interest. This is the case when stereological parameters are measured (number, length, surface area, and volume). Some ECM relevant examples are *i*) counting the number of elastic lamellae in the medial layer of an artery and *ii*) quantifying the volume occupied by collagen bundles in the adventitial layer of porcine arteries. Direct metrics are straightforward to interpret but can nonetheless be challenging to obtain. For instance, assessing the waviness of collagen bundles in adventitial collagen of porcine aorta by measuring the end-to-end distance and the contour length of individual fibers and then taking their ratio as the straightness parameter involves tedious manual tracing (Fig. 20a).

6.1.3 Indirect Metrics

An alternative solution is to select an indirect metric serving as a proxy for a physical parameter. Such indirect metrics are often more complex to interpret as they do not necessarily have a physical definition in the stereological sense and their variations can have multiple phenomenological origins. This complexity can lead to erroneous conclusion, and interpretations should be validated when needed. Figure 20b shows

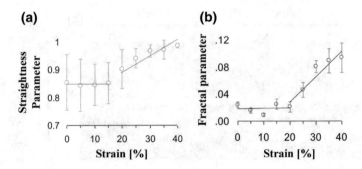

Fig. 20 Comparison between **a** a manual quantification of the waviness through direct length measurements and **b** automatic quantification with the proxy metric (fractal analysis) to characterize waviness in porcine adventitial collagen bundles during mechanical loading imaged with SHG. Reproduced with permission from [17]

the results of a fractal analysis as a function of strain of adventitial collagen from porcine aorta. For this specific case, the fractal analysis appears to be equivalent to the manual waviness analysis and can thus serve as a proxy for waviness (Fig. 20a was originally made to validate the interpretation of Fig. 20b). A clear advantage here is that the fractal analysis is computer automated. The absence of a need for manual inputs makes the analysis more consistent among users. Another important advantage is that the analysis can easily be adapted to other tissue types. Of course, when it is possible to automate the quantification of a direct metric, this option is preferable.

It was mentioned that indirect metrics variations can have a multitude of origins. To explain this fact, we will look again at the fractal analysis, but this time consider both medial collagen and elastin in porcine aorta during mechanical loading. The images reveal that waviness would have only a partial contribution in characterizing changes in the complex network of intertwined fibers (Fig. 1b) [17]. Yet, the fractal analysis revealed progressive changes in fiber organization and sequential recruitment of the different fiber families. It turns out that the fractal analysis metric is highly sensitive to any structural changes, including but not limited to waviness. It was therefore concluded that in the context of mechanical testing variations in the fractal number represent global changes in fiber engagement. Here, by making the interpretation more general, it also becomes more exact. This case also evidences another benefit of automated quantification of indirect metrics as they are a unique solution to quantitatively characterize tissue on which manual measurements cannot realistically be performed.

6.2   Dimensionality

Technical advances now make it possible to image 3D volumes in biological tissue with submicrometer spatial resolution. This capability should substantially circumvent the need to implement stereological methods of analysis of 2D images that inform on 3D parameters. However, 3D data requires analysis methods of the same dimensionality. The transition from 2D to 3D analysis is not trivial. Any manual measurements become much more time consuming. More importantly, automated methods that can perform essential tasks, such as following a fiber in multiple planes (for length measurements) or segmenting fiber bundles and lamellae for 3D reconstruction (for surface area and volume measurements), still need to be developed and their usage made robust for end-users.

In the meantime, a common approach in ECM imaging has consisted in compressing the 3D data to a 2D plane by performing intensity projections (Fig. 21). This approach allows one to use 2D analysis tools, which are extensively developed, for 3D data. The cost of this convenience is a loss of information. The point here is not to discourage the use of dimensional compression. After all, several important new observations about the ECM and its mechanical functions have been made with this simplistic strategy [17, 60]. However, it is important to be aware of the existing biases.

In projections, fiber length and waviness measurements are proxies for the actual fiber length and waviness in the 3D space; they are not stereological parameters. Lengths will appear shorter in projections if fibers do not reside in a single imaging plane. This bias is further pronounced during mechanical loading because as fibers are getting engaged, they will tend to rest increasingly within the imaging plane, i.e. the bias may change within a measurement. This applies for biaxial or uniaxial loading when the imaging plane contains one of the stretch axes. Another source of bias in characterizing fiber distribution originates from the use of the surface fraction, instead of volumetric fraction, occupied by a fiber group within a projected image as the weighting, or normalization, factor. Alleviating this bias is crucial as

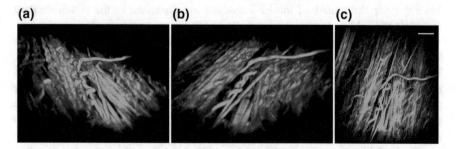

(a)   **(b)**   **(c)**

Fig. 21 Space compression. SHG signal from adventitial collagen in porcine aorta under mechanical loading shown as **a, b** 3D reconstructions from different angles and **c** a maximum intensity projection image along the axial dimension (scale bar: 50 μm)

(a)   **(b)**

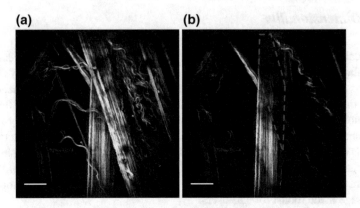

Fig. 22 Shadowing effect. SHG images from adventitial collagen in porcine aorta under mechanical loading. **a** Maximum intensity projection image (MIP) of the axial image stack. **b** MIP of the deeper half of the axial image stack. The red contour line indicates the shadowed area caused by the overlying collagen fibers. Scale bars: 50 μm

the distribution of fiber orientations is a central metric in mechanical modeling and to evaluate fiber recruitment [17, 49, 108, 109].

A common method to characterize the distribution of fiber orientation is the fast Fourier transform (FFT) analysis [17, 49, 106, 112]. During an FFT analysis [56], a projected image is divided into a matrix of small square regions. Then, each of these sub-regions is binarized independently before being Fourier transformed into the spatial frequency domain. Finally, the principal orientation is found for every sub-region through fitting an ellipse on the transform output. Typically, a histogram of the principal orientation (fiber angle) is generated to illustrate the distribution.

In addition to being susceptible to the bias caused by weighting, an optical shadowing effect further biases the results. The shadowing effect occurs when light propagating through a fiber bundle is altered to the extent the optical signal from any underlying structure is substantially degraded (Fig. 22). As a consequence, the results of FFT analysis are more heavily weighted toward bright and superficial structures. Several other limitations of the FFT analysis are discussed in the supplementary materials of [17].

To minimize the inherent bias associated with 2D image analysis, some effort has been made toward the development of 3D reconstruction tools for ECM images. Recently, the 3D surfaces of elastic lamellae in the mouse carotid artery were segmented (Fig. 23). Reconstruction is nevertheless only one of the necessary steps for a complete analysis. Due to the lack of quantification tools suitable to process 3D surfaces, the main discovery of this study, which provided the microstructural explanation for transmurally uniform lamellar stretching through a transmural gradient in lamellar unfolding, was still obtained from 2D circumferential cross-sections (but without dimensional compression) [107].

The distance between elastic lamellae was a key parameter in the above study (Fig. 23). A possible unbiased stereological strategy to quantify this parameter would

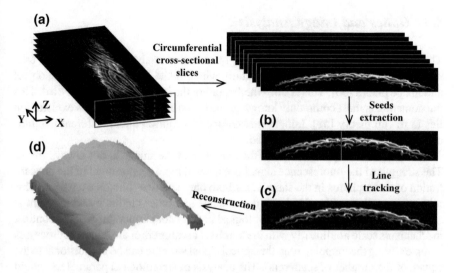

Fig. 23 3D image reconstruction. **a** The axial image stack images (TPEF from elastin in murine carotid arteries) are resliced before applying an algorithm to reconstruct the lamellar layers, which primarily involved **b** extraction of seed points and **c** line tracking from the seed points. **d** Reconstructed 3D elastic lamella. Reproduced with permission from [107]

Fig. 24 Cylindrical geometry, Schematic of concentric elastic lamellar layers in the arterial wall. Assuming a cylindrical geometry substantially simplifies the evaluation of the interlamellar distances (d_i). Adapted from [107]

have been to evaluate the volumes between two lamella and to divide this value by the surface area corrected for curvature. The last step adds quite a lot of complexity, but this approach would ensure that no geometrical assumptions are made in the analysis and its outcome is hence unbiased. A way to simplify the quantification of interlamellar distances is to assume that the tissue possesses a cylindrical symmetry. Indeed, fitting circular segments onto lamellar profiles in circumferential cross-section was amply sufficient to measure precisely the distance between layers (Fig. 24). Selecting a geometry to which the biological samples are never a perfect match will cause bias. Nonetheless, this choice can be suitable when model-free stereological approaches are too complex for implementation.

6.3   Optics and Image Analysis

The optical system can also impact the quantitative analysis. Firstly, length estimates are dependent on the magnification, which is the scaling between the object and image planes [66]. This is true whether or not the imaging system is optical. This phenomenon is most commonly known as the *Coastline Paradox* and its explanation lies in fractal theory [58]. Length measurements acquired under different magnifications cannot therefore be compared.

Secondly, the mapping between the sample and the image is not always linear. The strength of the fluorescence signal is expected to be proportional to the concentration of fluorophores in the sample, for both one- and two- photon excitation. This relation is modulated by the depth dependence and other factors by which light is altered when propagating through biological tissue. More importantly, some contrast mechanisms scale nonlinearly with the number of scatterers or are coherent processes complexifying the mapping (e.g. the optical signal strength can be proportional to the square of the number of scatterers). The analysis of stereological parameters should be insensitive to the scaling (beyond how the scaling affect thresholding), but when using an indirect metric caution should be used.

Thirdly, optical images are a spatial convolution of the object, which occurs as light propagates through the instrument. This impulse response of optical systems is known as the *point spread function* (PSF) (Fig. 12a) and related to the spatial resolution (Fig. 3). For point-scanning systems, the axial extent is usually larger than the lateral one and this gives the PSF an ellipsoidal shape. The convolution will not affect distance measurements if they are performed between geometric centers of objects, but it will lead to a systematic overestimation of surfaces and volumes. If the PSF is known, a deconvolution can be numerically performed to limit the bias it introduces [87]. Even with deconvolution, one still doesn't get an infinite spatial frequency spectrum, but an improved image. It should also be noted that the PSF may vary as a function of depth, and sometimes lateral position, in biological tissue (see Sect. 4.2) and that applying deconvolution might not always be possible.

In summary, methods for image analysis should be validated, quantitative, and unbiased. All steps of image processing and analysis should be described when reporting results. It is also a good practice to take sufficient notes when performing manual measurements such that the analysis can be exactly repeated.

7   Conclusion and Future Directions

In this chapter, we presented fundamental concepts related to optical imaging, contrast mechanisms, imaging strategies, and image analysis. These concepts were positioned within the context of ECM imaging in an attempt to provide integrative guidelines for experimental design. We would like to reiterate that even if the discussion was focused on imaging for arteries, most, if not all, concepts addressed apply to other types of biological tissues as well. Imaging data and statistical inferences can together provide powerful means for understanding complex multi-scale tissue mechanics. However, this can only be achieved by adopting an imaging method suitable for the length-scale of the structure of interest and also possessing the required signal specificity. Multiphoton microscopy is currently the optimal method for visualizing the microstructure of elastin and collagen ECM in intact tissue and we expect that this fundamental approach will yield many more discoveries in multiscale biomechanics in the years to come. The restricted imaging depth is the most important limitation of multiphoton microscopy to ECM imaging and advances on this front using a variety of strategies (e.g.: adaptive optics and optical clearing) would be highly beneficial through increased axial sampling. Increasing the imaging speed such that imaging in living animals is enabled would also constitute a substantial gain because it would then be possible to couple in vivo tissue deformation, ECM microstructure, and cellular dynamics in a unified experimental model.

Acknowledgements We would like to thank Matthew A. Mandelkern for proofreading. RT is very grateful to Prof. Charles P. Lin, who kindly provided a microscope over many years to perform mechanical ECM imaging studies. YZ gratefully acknowledges funding support from the National Institute of Health (2R01HL098028) and the National Science Foundation (CMMI 1463390 and CAREER 0954828).

References

1. Andersen, B.B., Korbo, L., Pakkenberg, B.: A quantitative study of the human cerebellum with unbiased stereological techniques. J. Comp. Neurol. **326**, 549–560 (1992)
2. Aviles-Espinosa, R., Andilla, J., Porcar-Guezenec, R., Olarte, O.E., Nieto, M., Levecq, X., Artigas, D., Loza-Alvarez, P.: Measurement and correction of in vivo sample aberrations employing a nonlinear guide-star in two-photon excited fluorescence microscopy. Biomed Opt. Express. **2**, 3135–3149 (2011)
3. Bélanger, E., Turcotte, R., Daradich, A., Sadetsky, G., Gravel, P., Bachand, K., De Koninck, Y., Côté, D.C.: Maintaining polarization in polarimetric multiphoton microscopy. J. Biophotonics **8**, 884–888 (2015)
4. Booth, M.J.: Adaptive optical microscopy: the ongoing quest for a perfect image. Light Sci Appl. **3**, e165 (2014)
5. Booth, M.J., Neil, M., Wilson, T.: Aberration correction for confocal imaging in refractive-index-mismatched media. J. Microsc. **192**, 90–98 (1998)
6. Born, M., Wolf, E.: Principle of Optics. Cambridge University Press (1999)
7. Boudoux, C.: Fundamentals of Biomedical Optics. Blurb (2017)
8. Boulesteix, T., Pena, A.M., Pagès, N., Godeau, G., Sauviat, M.P., Beaurepaire, E., Schanne-Klein, M.C.: Micrometer scale ex vivo multiphoton imaging of unstained arterial wall structure. Cytometry A. **69**, 20–26 (2006)
9. Boyd, R.W.: Nonlinear Optics. Elsevier (2008)
10. Campagnola, P.: Second harmonic generation imaging microscopy: applications to diseases diagnostics. Anal. Chem. **83**, 3224–3231 (2011)
11. Campagnola, P.J., Loew, L.M.: Second-harmonic imaging microscopy for visualizing biomolecular arrays in cells, tissues and organisms. Nat. Biotechnol. **21**, 1356–1360 (2003)
12. Chen, F., Tillberg, P.W., Boyden, E.S.: Expansion microscopy. Science **347**, 543–548 (2015)

13. Chen, W.L., Li, T.H., Su, P.J., Chou, C.K., Fwu, P.T., Lin, S.J., Kim, D., Dong, C.Y.: Second harmonic generation χ tensor microscopy for tissue imaging. Appl. Phys. Lett. **94**, 183902 (2009)
14. Chen, X., Nadiarynkh, O., Plotnikov, S., Campagnola, P.J.: Second harmonic generation microscopy for quantitative analysis of collagen fibrillar structure. Nat. Protoc. **7**, 654–669 (2012)
15. Cheuk, B.L.Y., Cheng, S.W.K.: Expression of integrin α5β1 and the relationship to collagen and elastin content in human suprarenal and infrarenal aortas. Vasc Endovascular Surg. **39**, 245–251 (2005)
16. Chow, M.-J., Choi, M., Yun, S.H., Zhang, Y.: The effect of static stretch on elastin degradation in arteries. PLoS ONE **8**, e81951 (2013)
17. Chow, M.-J., Turcotte, R., Lin, C.P., Zhang, Y.: Arterial extracellular matrix: a mechanobiological study of the contributions and interactions of elastin and collagen. Biophys. J. **106**, 2684–2692 (2014)
18. Chu, S.-W., Chen, S.-Y., Chern, G.-W., Tsai, T.-H., Chen, Y.-C., Lin, B.-L., Sun, C.-K.: Studies of χ(2)/χ(3) tensors in submicron-scaled bio-tissues by polarization harmonics optical microscopy. Biophys. J. **86**, 3914–3922 (2004)
19. Curatolo, A., Villiger, M., Lorenser, D., Wijesinghe, P., Fritz, A., Kennedy, B.F., Sampson, D.D.: Ultrahigh-resolution optical coherence elastography. Opt. Lett. **41**, 21–24 (2016)
20. Dan, D., Lei, M., Yao, B., Wang, W., Winterhalder, M., Zumbusch, A., Qi, Y., Xia, L., Yan, S., Yang, Y., Gao, P., Ye, T., Zhao, W.: DMD-based LED-illumination super-resolution and optical sectioning microscopy. Sci. Reports. **3**, 1116 (2013)
21. Davis, E.C.: Smooth muscle cell to elastic lamina connections in developing mouse aorta. Role in aortic medial organization. Lab. Invest. **68**, 89–99 (1993)
22. de Aguiar, H.B., Gasecka, P., Brasselet, S.: Quantitative analysis of light scattering in polarization-resolved nonlinear microscopy. Opt. Express **23**, 8960–8973 (2015)
23. Drobizhev, M., Makarov, N.S., Tillo, S.E., Hughes, T.E., Rebane, A.: Two-photon absorption properties of fluorescent proteins. Nat. Methods **8**, 393–399 (2011)
24. Duboisset, J., Rigneault, H., Brasselet, S.: Filtering of matter symmetry properties by circularly polarized nonlinear optics. Phys. Rev. A. **90**, 063827 (2014)
25. Empedocles, S.A., Neuhauser, R., Bawendi, M.G.: Three-dimensional orientation measurements of symmetric single chromophores using polarization microscopy. Nature **399**, 126–130 (1999)
26. Fata, B., Carruthers, C.A., Gibson, G., Watkins, S.C., Gottlieb, D., Mayer, J.E., Sacks, M.S.: Regional structural and biomechanical alterations of the ovine main pulmonary artery during postnatal growth. J. Biomech. Eng. **135**, 021022 (2013)
27. Fellgett, P.B., Linfoot, E.H.: On the assessment of optical images. Philos. Trans. R. Soc. Lond. **247**, 369–407 (1955)
28. Fry, J.L., Shiraishi, Y., Turcotte, R., Yu, X., Gao, Y.Z., Akiki, R., Bachschmid, M., Zhang, Y., Morgan, K.G., Cohen, R.A., Seta, F.: Vascular Smooth Muscle Sirtuin-1 Protects Against Aortic Dissection During Angiotensin II-Induced Hypertension. J. Am. Heart Assoc. **4**, e002384 (2015)
29. García-Herrera, C.M., Celentano, D.J., Cruchaga, M.A., Rojo, F.J., Atienza, J.M., Guinea, G.V., Goicolea, J.M.: Mechanical characterisation of the human thoracic descending aorta: Experiments and modelling. Comput. Methods Biomech. Biomed. Engin. **15**, 185–193 (2012)
30. Gerson, C.J., Goldstein, S., Heacox, A.E.: Retained structural integrity of collagen and elastin within cryopreserved human heart valve tissue as detected by two-photon laser scanning confocal microscopy. Cryobiology **59**, 171–179 (2009)
31. Gundersen, H.J.G., Jensen, E.B.: The efficiency of systematic sampling in stereology and its prediction. J. Microsc. **147**, 229–263 (1987)
32. Gusachenko, I., Latour, G., Schanne-Klein, M.-C.: Polarization-resolved Second Harmonic microscopy in anisotropic thick tissues. Opt. Express **18**, 19339–19352 (2010)
33. Gusachenko, I., Schanne-Klein, M.-C.: Numerical simulation of polarization-resolved second-harmonic microscopy in birefringent media. Phys. Rev. A **88**, 101–115 (2013)

34. Gusachenko, I., Tran, V., Goulam Houssen, Y., Allain, J.-M., Schanne-Klein, M.-C.: Polarization-resolved second-harmonic generation in tendon upon mechanical stretching. Biophys. J. **102**, 2220–2229 (2012)

35. Halloran, B.G., Davis, V.A., McManus, B.M., Lynch, T.G., Baxter, B.T.: Localization of aortic disease is associated with intrisinc differences in aortic structure. J. Surg. Res. **59**, 17–22 (1995)

36. Harkness, M.L.R., Harkness, R.D., McDonald, D.A.: The collagen and elastin content of the arterial wall in the dog. Proc. R. Soc. Lond., B, Biol. Sci. **146**, 541–551 (1957)

37. Hecht, E.: Optique (fr). Pearson (2005)

38. Hill, M.R., Duan, X., Gibson, G.A., Watkins, S., Robertson, A.M.: A theoretical and non-destructive experimental approach for direct inclusion of measured collagen orientation and recruitment into mechanical models of the artery wall. J. Biomech. **45**, 762–771 (2012)

39. Hill-Eubanks, D.C., Werner, M.E., Heppner, T.J., Nelson, M.T.: Calcium signaling in smooth muscle. Cold Spring Harb. Perspect. Biol. **3**, a004549 (2011)

40. Holzapfel, G.A., Gasser, T.C., Ogden, R.W.: A new constitutive framework for arterial wall mechanics and a comparative study of material models. J. Elast. **61**, 1–48 (2000)

41. Hovhannisyan, V., Ghazaryan, A., Chen, Y.-F., Chen, S.-J., Dong, C.Y.: Photophysical mechanisms of collagen modification by 80 MHz femtosecond laser. Opt. Express **18**, 24037–24047 (2010)

42. Hunziker, E.B., Cruz-Orive, L.M.: Consistent and efficient delineation of reference spaces for light microscopical stereology using a laser microbeam system. J. Microsc. **142**, 95–99 (1986)

43. Jacques, S.L.: Optical properties of biological tissues: a review. Phys. Med. Biol. **58**, R37–61 (2013)

44. Ji, N.: Adaptive optical fluorescence microscopy. Nat. Methods **14**, 374–380 (2017)

45. Kennedy, B.F., McLaughlin, R.A., Kennedy, K.M., Chin, L., Curatolo, A., Tien, A., Latham, B., Saunders, C.M., Sampson, D.D.: Optical coherence micro-elastography: mechanical-contrast imaging of tissue microstructure. Biomed. Opt. Express. **5**, 2113–2124 (2014)

46. Kim, J., Hong, J.-W., Baek, S.: Longitudinal differences in the mechanical properties of the thoracic aorta depend on circumferential regions. J Biomed Mater Res A. **101**, 1525–1529 (2013)

47. Kobat, D., Durst, M.E., Nishimura, N., Wong, A.W., Schaffer, C.B., Xu, C.: Deep tissue multiphoton microscopy using longer wavelength excitation. Opt. Express **17**, 13354–13364 (2009)

48. König, K., So, P.T., Mantulin, W.W., Gratton, E.: Cellular response to near-infrared femtosecond laser pulses in two-photon microscopes. Opt. Lett. **22**, 135–136 (1997)

49. Krasny, W., Morin, C., Magoariec, H., Avril, S.: A comprehensive study of layer-specific morphological changes in the microstructure of carotid arteries under uniaxial load. Acta Biomater. **57**, 342–351 (2017)

50. Kwok, S.J.J., Kuznetsov, I.A., Kim, M., Choi, M., Scarcelli, G., Yun, S.H.: Selective two-photon collagen crosslinking in situ measured by Brillouin microscopy. Optica. **3**, 469–472 (2016)

51. Labrosse, M.R., Beller, C.J., Mesana, T., Veinot, J.P.: Mechanical behavior of human aortas: Experiments, material constants and 3-D finite element modeling including residual stress. J. Biomech. **42**, 996–1004 (2009)

52. Le Floc'h, V., Brasselet, S., Roch, J.-F., Zyss, J.: Monitoring of orientation in molecular ensembles by polarization sensitive nonlinear microscopy. J. Phys. Chem. B. **107**, 12403–12410 (2003)

53. Lee, K.-W., Stolz, D.B., Wang, Y.: Substantial expression of mature elastin in arterial constructs. Proc. Natl. Acad. Sci. U.S.A. **108**, 2705–2710 (2011)

54. Lim, D., Chu, K.K., Mertz, J.: Wide-field fluorescence sectioning with hybrid speckle and uniform-illumination microscopy. Opt. Lett. **33**, 1819–1821 (2008)

55. Lim, D., Ford, T.N., Chu, K.K., Mertz, J.: Optically sectioned in vivo imaging with speckle illumination HiLo microscopy. J. Biomed. Opt. **16**, 016014 (2011)

56. Liu, Z.Q.: Scale space approach to directional analysis of images. Appl. Opt. **30**, 1369–1373 (1991)
57. Lu-Walther, H.-W., Kielhorn, M., Förster, R., Jost, A., Wicker, K., Heintzmann, R.: fastSIM: a practical implementation of fast structured illumination microscopy. Methods Appl. Fluoresc. **3**, 014001 (2015)
58. Mandelbrot, B.: How long is the coast of britain? Statistical self-similarity and fractional dimension. Science. **156**, 636–638 (1967)
59. Matsumoto, N., Konno, A., Inoue, T., Okazaki, S.: Aberration correction considering curved sample surface shape for non-contact two-photon excitation microscopy with spatial light modulator. Sci. Rep. **8**, 9252 (2018)
60. Mattson, J.M., Turcotte, R., Zhang, Y.: Glycosaminoglycans contribute to extracellular matrix fiber recruitment and arterial wall mechanics. Biomech. Model. Mechanobiol. **16**, 213–225 (2017)
61. Mattson, J.M., Wang, Y., Zhang, Y.: Contributions of glycosaminoglycans to collagen fiber recruitment in constitutive modeling of arterial mechanics. J. Biomech. **82**, 211–219 (2019)
62. Mazumder, N., Deka, G., Wu, W.-W., Gogoi, A., Zhuo, G.-Y., Kao, F.-J.: Polarization resolved second harmonic microscopy. Methods **128**, 105–118 (2017)
63. Mertz, J.: Introduction to optical microscopy. Roberts and Company Publishers (2010).
64. Mertz, J.: Optical sectioning microscopy with planar or structured illumination. Nat. Methods **8**, 811–819 (2011)
65. Mondal, P.P.: Temporal resolution in fluorescence imaging. Front Mol Biosci. **1**, 11 (2014)
66. Mouton, P.R.: Principles and practices of unbiased stereology. The Johns Hopkins University Press (2002)
67. Müllenbroich, C.M., McGhee, E.J., Wright, A.J., Anderson, K.I., Mathieson, K.: Strategies to overcome photobleaching in algorithm-based adaptive optics for nonlinear in-vivo imaging. J. Biomed. Opt. **19**, 16021 (2014)
68. Neil, M., Juskaitis, R., Wilson, T.: Method of obtaining optical sectioning by using structured light in a conventional microscope. Opt. Lett. **22**, 1905–1907 (1997)
69. Ostroverkhov, V., Singer, K.D., Petschek, R.G.: Second-harmonic generation in nonpolar chiral materials: relationship between molecular and macroscopic properties. J. Opt. Soc. Am. B. **18**, 1858–1865 (2001)
70. Pawley, J.: Handbook of Biological Confocal Microscopy. Springer (2006)
71. Plotnikov, S., Juneja, V., Isaacson, A.B., Mohler, W.A., Campagnola, P.J.: Optical clearing for improved contrast in second harmonic generation imaging of skeletal muscle. Biophys. J. **90**, 328–339 (2006)
72. Ranjit, S., Dvornikov, A., Stakic, M., Hong, S.-H., Levi, M., Evans, R.M., Gratton, E.: Imaging fibrosis and separating collagens using second harmonic generation and phasor approach to fluorescence lifetime imaging. Sci. Rep. **5**, 13378 (2015)
73. Raub, C.B., Suresh, V., Krasieva, T., Lyubovitsky, J., Mih, J.D., Putnam, A.J., Tromberg, B.J., George, S.C.: Noninvasive assessment of collagen gel microstructure and mechanics using multiphoton microscopy. Biophys. J. **92**, 2212–2222 (2007)
74. Roccabianca, S., Figueroa, C.A., Tellides, G., Humphrey, J.D.: Quantification of regional differences in aortic stiffness in the aging human. J. Mech. Behav. Biomed. Mater. **29**, 618–634 (2014)
75. Rogowska, J., Patel, N., Plummer, S., Brezinski, M.E.: Quantitative optical coherence tomographic elastography: method for assessing arterial mechanical properties. Br. J. Radiol. **79**, 707–711 (2006)
76. Rouleau, L., Tremblay, D., Cartier, R., Mongrain, R., Leask, R.L.: Regional variations in canine descending aortic tissue mechanical properties change with formalin fixation. Cardiovasc. Pathol. **21**, 390–397 (2012)
77. Russ, J.C., Dehoff, R.T., Deh: Practical stereology. Springer (2012)
78. Sahl, S.J., Hell, S.W., Jakobs, S.: Fluorescence nanoscopy in cell biology. Nat. Rev. Mol. Cell Biol. **18**, 685–701 (2017)

79. Scarcelli, G., Yun, S.H.: Confocal Brillouin microscopy for three-dimensional mechanical imaging. Nat. Photonics **2**, 39–43 (2007)
80. Schenke-Layland, K.: Non-invasive multiphoton imaging of extracellular matrix structures. J. Biophotonics **1**, 451–462 (2008)
81. Smith, A.M., Mancini, M.C., Nie, S.: Bioimaging: second window for in vivo imaging. Nat. Nanotechnol. **4**, 710–711 (2009)
82. Sokolis, D.P.: Passive mechanical properties and structure of the aorta: segmental analysis. Acta Physiol. **190**, 277–289 (2007)
83. Sokolis, D.P., Boudoulas, H., Kavantzas, N.G., Kostomitsopoulos, N., Agapitos, E.V., Karayannacos, P.E.: A morphometric study of the structural characteristics of the aorta in pigs using an image analysis method. Anat. Histol. Embryol. **31**, 21–30 (2002)
84. Song, L., Lu-Walther, H.-W., Förster, R., Jost, A., Kielhorn, M., Zhou, J., Heintzmann, R.: Fast structured illumination microscopy using rolling shutter cameras. Meas. Sci. Technol. **27**, 055401 (2016)
85. Stoller, P., Reiser, K.M., Celliers, P.M., Rubenchik, A.M.: Polarization-modulated second harmonic generation in collagen. Biophys. J. **82**, 3330–3342 (2002)
86. Sun, Y., Sun, Y., Stephens, D., Xie, H., Phipps, J., Saroufeem, R., Southard, J., Elson, D.S., Marcu, L.: Dynamic tissue analysis using time- and wavelength-resolved fluorescence spectroscopy for atherosclerosis diagnosis. Opt. Express **19**, 3890–3901 (2011)
87. Swedlow, J.R., Sedat, J.W., Agard, D.A.: Deconvolution in optical microscopy. In: Deconvolution of images and spectra. pp. 284–309, Orlando, FL, USA (1997)
88. Teulon, C., Gusachenko, I., Latour, G., Schanne-Klein, M.-C.: Theoretical, numerical and experimental study of geometrical parameters that affect anisotropy measurements in polarization-resolved SHG microscopy. Opt. Express **23**, 9313–9328 (2015)
89. Tuer, A.E., Akens, M.K., Krouglov, S., Sandkuijl, D., Wilson, B.C., Whyne, C.M., Barzda, V.: Hierarchical Model of fibrillar collagen organization for interpreting the second-order susceptibility tensors in biological tissue. Biophys. J. **103**, 2093–2105 (2012)
90. Turcotte, R., Liang, Y., Ji, N.: Adaptive optical versus spherical aberration corrections for in vivo brain imaging. Biomed. Opt. Express. **8**, 3891–3902 (2017)
91. Turcotte, R., Mattson, J.M., Wu, J.W., Zhang, Y., Lin, C.P.: Molecular order of arterial collagen using circular polarization second-harmonic generation imaging. Biophys. J. **110**, 530–533 (2016)
92. Tønnesen, J., Inavalli, V.V.G.K., Nägerl, U.V.: Super-resolution imaging of the extracellular space in living brain tissue. Cell **172**, 1108–1121 (2018)
93. Veilleux, I., Spencer, J.A., Bliss, D.P., Côté, D., Lin, C.P.: In vivo cell tracking with video rate multimodality laser scanning microscopy. IEEE J. Sel. Top. Quantum Electron. 14, 10–18 (2008)
94. Wagnieres, G.A., Star, W.M., Wilson, B.C.: In vivo fluorescence spectroscopy and imaging for oncological applications. Photochem. Photobiol. **68**, 603–632 (1998)
95. Wan, W., Dixon, J.B., Gleason, R.L.: Constitutive modeling of mouse carotid arteries using experimentally measured microstructural parameters. Biophys. J. **102**, 2916–2925 (2012)
96. Wang, K., Milkie, D.E., Saxena, A., Engerer, P., Misgeld, T., Bronner, M.E., Mumm, J., Betzig, E.: Rapid adaptive optical recovery of optimal resolution over large volumes. Nat. Methods **11**, 625–628 (2014)
97. Wang, K., Sun, W., Richie, C.T., Harvey, B.K., Betzig, E., Ji, N.: Direct wavefront sensing for high-resolution in vivo imaging in scattering tissue. Nat. Commun. **6**, 7276 (2015)
98. Wang, Y., Li, H., Zhang, Y.: Understanding the viscoelastic behavior of arterial elastin in glucose via relaxation time distribution spectrum. J. Mech. Behav. Biomed. Mater. **77**, 634–641 (2018)
99. Wang, Y., Zeinali-Davarani, S., Zhang, Y.: Arterial mechanics considering the structural and mechanical contributions of ECM constituents. J. Biomech. **49**, 2358–2365 (2016)
100. Webb, W.W., Xu, C.: Measurement of two-photon excitation cross sections of molecular fluorophores with data from 690 to 1050 nm. J. Opt. Soc. Am. B. **13**, 481–491 (1996)

101. West, M.J.: Basic Stereology for Biologists and Neuroscientists. Cold Spring Harbor Press (2012)
102. West, M.J., Gundersen, H.J.G.: Unbiased Stereological Estimation of the number of neurons in the human hippocampus. J. Comp. Neurol. **296**, 1–22 (1990)
103. Wicker, K., Heintzmann, R.: Resolving a misconception about structured illumination. Nat. Photonics **8**, 342–344 (2014)
104. Wicker, K., Mandula, O., Best, G., Fiolka, R., Heintzmann, R.: Phase optimisation for structured illumination microscopy. Opt. Express **21**, 2032–2049 (2013)
105. Wolinsky, H., Glagov, S.: Comparison of abdominal and thoracic aortic medial structure in mammals. Circ. Res. **25**, 677–686 (1969)
106. Wu, S., Li, H., Yang, H., Zhang, X., Li, Z., Xu, S.: Quantitative analysis on collagen morphology in aging skin based on multiphoton microscopy. J. Biomed. Opt. **16**, 040502 (2011)
107. Yu, X., Turcotte, R., Seta, F., Zhang, Y.: Micromechanics of elastic lamellae: unravelling the role of structural inhomogeneity in multi-scale arterial mechanics. J. R. Soc. Interface. **15**, 20180492 (2018)
108. Yu, X., Wang, Y., Zhang, Y.: Transmural variation in elastin fiber orientation distribution in the arterial wall. J. Mech. Behav. Biomed. Mater. **77**, 745–753 (2018)
109. Zeinali-Davarani, S., Chow, M.-J., Turcotte, R., Zhang, Y.: Characterization of biaxial mechanical behavior of porcine aorta under gradual elastin degradation. Ann Biomed Eng. **41**, 1528–1538 (2013)
110. Zeinali-Davarani, S., Wang, Y., Chow, M.-J., Turcotte, R., Zhang, Y.: Contribution of collagen fiber undulation to regional biomechanical properties along porcine thoracic aorta. J. Biomech. Eng. **137**, 051001 (2015)
111. Zhao, H.L., Zhang, C.P., Zhu, H., Jiang, Y.F., Fu, X.B.: Autofluorescence of collagen fibres in scar. Skin. Res. Technol. **23**, 588–592 (2017)
112. Zhu, X., Zhuo, S., Zheng, L., Lu, K., Jiang, X., Chen, J., Lin, B.: Quantified characterization of human cutaneous normal scar using multiphoton microscopy. J. Biophotonics. **3**, 108–116 (2010)
113. Zipfel, W.R., Williams, R.M., Christie, R., Nikitin, A.Y., Hyman, B.T., Webb, W.W.: Live tissue intrinsic emission microscopy using multiphoton-excited native fluorescence and second harmonic generation. Proc. Natl. Acad. Sci. USA **100**, 7075–7080 (2003)
114. Zipfel, W.R., Williams, R.M., Webb, W.W.: Nonlinear magic: multiphoton microscopy in the biosciences. Nat. Biotechnol. **21**, 1369–1377 (2003)
115. Zou, Y., Zhang, Y.: An experimental and theoretical study on the anisotropy of elastin network. Ann. Biomed. Eng. **37**, 1572–1583 (2009)
116. Zoumi, A., Lu, X., Kassab, G.S., Tromberg, B.J.: Imaging coronary artery microstructure using second-harmonic and two-photon fluorescence microscopy. Biophys. J. **87**, 2778–2786 (2004)
117. Zoumi, A., Yeh, A., Tromberg, B.J.: Imaging cells and extracellular matrix in vivo by using second-harmonic generation and two-photon excited fluorescence. Proc. Natl. Acad. Sci. USA **99**, 11014–11019 (2002)

Collagen Self-assembly: Biophysics and Biosignaling for Advanced Tissue Generation

David O. Sohutskay, Theodore J. Puls and Sherry L. Voytik-Harbin

Abstract Type I collagen is the predominant protein in the body and the extracellular matrix, where it gives rise to the vast diversity of tissue form and function. Within the extracellular matrix, this natural polymer exists as the fibrillar scaffolding that not only dictates tissue-specific structure and mechanical properties but also interacts with cells and other biomolecules to orchestrate complex processes associated with tissue development, homeostasis, and repair. For this reason, the hierarchical self-assembly of collagen molecules and their inherent biochemical and biophysical signaling capacity have been a long-standing subject of study across multiple disciplines, including structural biochemistry, biomechanics, biomaterials and tissue engineering, computational modeling, and medicine. This review works to capture some of the major discoveries and innovative technologies related to the supramolecular assembly of collagen in vivo and in vitro, with a focus on motivating their integration and application for advanced tissue fabrication and regenerative medicine therapies.

D. O. Sohutskay · T. J. Puls · S. L. Voytik-Harbin (✉)
Weldon School of Biomedical Engineering, College of Engineering, Purdue University, West Lafayette, IN, USA
e-mail: harbins@purdue.edu

D. O. Sohutskay
e-mail: dsohutsk@purdue.edu

T. J. Puls
e-mail: tpuls@purdue.edu

S. L. Voytik-Harbin
Department of Basic Medical Sciences, College of Veterinary Medicine, Purdue University, West Lafayette, IN, USA

D. O. Sohutskay
Indiana University School of Medicine, Indianapolis, IN, USA

© Springer Nature Switzerland AG 2020
Y. Zhang (ed.), *Multi-scale Extracellular Matrix Mechanics and Mechanobiology*,
Studies in Mechanobiology, Tissue Engineering and Biomaterials 23,
https://doi.org/10.1007/978-3-030-20182-1_7

1 Tissue Engineering and Regenerative Medicine: The Goal and Challenge

The fields of tissue engineering and regenerative medicine, which operate at the interface of engineering and life sciences, have evolved over the last three decades with the goal of restoring damaged or dysfunctional tissues and organs through the development of biological substitutes and/or the promotion of tissue regeneration. Miniaturized in vitro human tissue systems are also highly sought after as an alternative to animals for cosmetic and chemical toxicity testing, high-throughput/high-content drug screening, and basic research. One foundational element of such efforts has been development of biomaterials that recreate the extracellular matrix (ECM) component of tissues. The ECM constitutes non-living material produced and secreted by cells within which they are distributed and organized. It represents a composite material, largely composed of an insoluble collagen-fibril scaffold surrounded by an interstitial fluid phase, giving tissues both poroelastic and viscoelastic properties [84]. More specifically, applied deformation to the composite will intrinsically lead to fluid flow that homogenizes scaffold pore pressure. At the same time, the composite will undergo viscoelastic deformation, exhibiting both viscous (liquid) and elastic (solid) characteristics [137]. ECM is found in all tissues and organs, providing not only the essential physical structure that organizes and supports cellular constituents but also crucial biochemical and biomechanical signaling required for tissue morphogenesis, homeostasis, and remodeling. In fact, a dynamic and reciprocal dialogue exists between cells and their surrounding ECM, such that multi-scale tissue architecture and function are integrated [216]. As such, the ability to recapitulate this natural scaffold and dynamic cell-ECM interactions has been a focused effort of tissue engineering and regenerative medicine even prior to the formal definition of these fields.

When tissue engineering emerged as a new field in the early 1990s, emphasis was placed on the use of synthetic polymers for development of porous scaffolds to mimic the structural features of ECM [115]. Synthetic materials received preference over natural polymers, such as collagen, largely owing to advantages associated with cost, batch-to-batch reproducibility, mechanical stability, as well as amenability to customization, processing, and scale-up manufacturing. Furthermore, at the time, medical devices containing candidate synthetic materials had already received FDA-approval, documenting their biocompatibility and paving the way for translation into the clinic. To date, extensive effort has been invested in the design and manufacturing of synthetic biomaterials that are biocompatible (non-toxic to cells) and possess the structural and mechanical properties of a target tissue. Another fundamental design criteria was that the biomaterial should be biodegradable, allowing host cells to progressively deposit site-appropriate replacement tissue over time [64, 107, 201]. However, in recent times, concerns have been raised regarding the immune-mediated foreign-body responses elicited by synthetic materials [46, 98] as well as their lack of biological signaling capacity [71, 88]. As a result, design criteria for next-generation biomaterials are changing, moving away from merely providing bulk structure and mechanical properties to strategies that guide biological processes underlying tissue regeneration [3, 71, 128].

Despite this initial focus on synthetics, others targeted the use of natural materials, including intact ECMs prepared from various tissues and their component molecules (e.g., collagen, fibrin, glycosaminoglycans). Here, the goal was to capitalize on the biological signaling capacity inherent to these molecules and their assemblies for purposes of inducing site-appropriate tissue regeneration. Interestingly, evaluation of the present-day tissue engineering and regenerative medicine market, shows that biologically-derived materials (e.g., decellularized tissues) and natural polymers, specifically type I collagen, account for the majority of translated technologies [8]. Within this context, this chapter focuses on type I collagen and its use for tissue engineering and regenerative medicine applications. We start by providing a historical overview of milestone discoveries related to collagen biochemistry and collagen-based biomaterials, highlighting their impact on research and medicine. The next section describes what is known regarding the biosynthesis and hierarchical self-assembly of type I collagen as it occurs within the body. This is followed by a brief description of collagen biomechanics and the more recent discovery of collagen's participation in mechanobiology signaling, which collectively have contributed new and important design criteria for cell-instructive biomaterials. We then rigorously define and compare various collagen preparations, lending support to the notion that "all collagens or collagen-containing materials are not alike." We then hone in on collagen advancements and applications that support next-generation, multi-scale design and custom fabrication of collagen scaffolds and tissues. Finally, we conclude with a look to the future, where this natural polymer interfaces with other tissue engineering and regeneration advancements, including stem cells (adult, induced pluripotent), computational modeling, and advanced manufacturing, to address today's challenges and unmet clinical needs.

2 Collagen Biomaterials: The History

Scientific inquiry and applications of collagen as a tool and in medicine date back millennia. Figure 1 provides a timeline, outlining some of the major milestones in the development and application of collagen biomaterials. The word collagen is Greek, from the roots "κόλλα" (glue) and "γέν" (to make), so called because the first application of denatured collagen (gelatin) was as an adhesive for wood furnishings [63]. The first medical application of collagen as an implantable biomaterial was likely "catgut" suture, which was documented as early as 150 A.D. by Galen of Pergamon [41, 124]. Despite the moniker, these collagenous threads were typically formed from decellularized sheep intestine, not cats. Although catgut sutures were used for centuries, it wasn't until the late 19th century that their production was perfected, with the development of chromic acid-based sterilization procedures by Lister and MacEwen [67, 121]. Catgut persisted into modern use, though largely supplanted by resorbable synthetic products due to their ease of manufacturing and

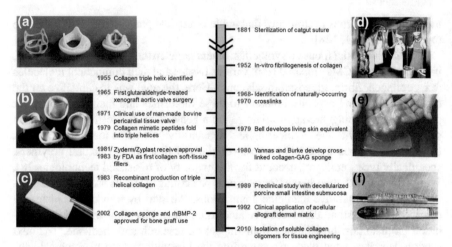

Fig. 1 Timeline of key developments in the history of collagen biomaterials. **a** Assembly and reinforcement of glutaraldehyde-treated aortic valve xenograft onto supports (reprinted with permission from Zudhi et al. [226]). **b** Man-made bioprosthetic valves prepared from glutaraldehyde-treated bovine pericardial tissue (reprinted with permission from Society for Cardiothoracic Surgery in Great Britain and Ireland). **c** Freeze-dried collagen-glycosaminoglycan sponge. **d** Processing and sterilization of catgut sutures using the Kuhn procedure (reprinted with permission from Dietz et al. [56]). **e** Living-skin equivalent prepared from fibroblast-contracted collagen matrix (reprinted with permission from Bell et al. [15]). **f** Vascular graft fashioned from decellularized small intestine submucosa (reprinted with permission from Badylak et al. [10])

sterilization. Despite the common usage of collagen over this early time period, it's unique structure as a semiflexible, triple helical rod was not determined until the 1950s. Ramachandran and others used fiber diffraction analysis and model building, together with early amino acid composition and sequence data, to elucidate that the three component polypeptide chains, each in an extended left-handed polyproline II-helix conformation, were supercoiled in a right-handed manner about a common axis [166].

In the mid-1960s, another historical milestone was reached for collagen biomaterials—the use of biological tissue valves derived from porcine or bovine sources. The very first xenograft (porcine) aortic valve replacement in a human patient was performed in 1965 by Carpentier and his team [18]. It was later discovered that stent reinforcements and treatment of these valves with exogenous glutaraldehyde crosslinking reduced their antigenicity and degradation, dramatically improving clinical success rates [127, 226]. The first clinical use of an "engineered" or man-made heart valve followed in 1971, when Marian Ion Ionescu introduced the novel concept of constructing heart valves by attaching glutaraldehyde-treated bovine pericardium to a support frame [90]. This application of a replenishable collagen tissue source for valve design and manufacturing has contributed significantly to the evolution of the heart valve industry. Today, innovative, non-invasive trans-catheter approaches involving stented pericardial tissue are paving the way for expanded valve applications and patient populations, including children [187].

More widespread use of collagen for tissue-engineered medical products came with the isolation and decellularization of porcine small intestine submucosa (SIS), developed at Purdue University in the late 1980s [10]. Here, the design strategy was to remove all cellular components while maintaining the complex molecular composition, architecture, mechanical properties, and biological activity inherent to the naturally-occurring ECM. With a focus on inducing tissue regeneration, SIS became one of the first major tissue engineering industry success stories [122, 123], with Cook Biotech continuing to expand its portfolio of wound management and surgical reconstruction products based on this technology. Today, a number of decellularized tissue products populate the market, including those derived from multiple animal tissue sources (porcine and bovine small intestine, dermis, and urinary bladder) as well as human tissue sources (dermis and placenta). It is notable that AlloDerm, produced by LifeCell, was the first decellularized human dermal tissue on the market, receiving initial FDA approval in 1992 for treatment of burns [204].

As an alternative to these top-down approaches to tissue design, others have applied bottom-up strategies, focused on applications of purified collagen in both insoluble fibrillar and soluble, fibril-forming (self-assembling) formats. Improvements in biotechnology and development of scalable extraction procedures, such as those developed by Miller and Rhodes [134], facilitated large-scale production of high-purity collagens, paving the way for their use in tissue engineering and medicine. One of the first and most successful products created from insoluble fibrillar collagen was the "collagen-glycosaminoglycan membrane," which was initially developed by Yannas and Burke for management of skin wounds [51, 219, 220]. These scaffolds were created by freeze drying a viscous slurry of purified bovine hide particulate and chondroitin 6-sulfate from shark cartilage followed by chemical crosslinking. Design criteria including pore size, mechanical properties, and degradation (resorption) rate were modulated, with the goal of retarding wound contraction while carefully controlling host cell infiltration and tissue deposition. This technology was acquired by Integra, which successfully entered the burn market with the first dermal regeneration template in 1995. Integra's tissue-engineered products have become a significant commercial success with many applications, including burns, diabetic ulcers, and dental wounds. One might argue this is, in large part, owing to the design control afforded by their fabrication process.

Insoluble fibrillar collagen also served as the starting material for injectable soft tissue fillers products that reached popularity for cosmetic applications in the late twentieth century [188]. More specifically, Zyderm and its chemically crosslinked counterpart Zyplast consisted of insoluble bovine dermal collagen dispersed in phosphate-buffered saline, which contained lidocaine as a local anesthetic. Because these injectable collagens required multiple injections and chemical crosslinking to enhance their stability in vivo, they are no longer on the market and have been superseded by hyaluronic acid products [99]. Lyophilized collagen sponges, again which comprise insoluble fibrillar collagen, have also been used as drug or growth factor carriers. One particular example of a mainstay collagen-based drug delivery device is

InFuse bone graft, which received approval in the early 2000s. This product involves the application of recombinant human bone morphogenetic protein-2 (rhBMP-2) to a lyophilized collagen sponge prior to implantation into bone defects [33, 65].

Some of the first descriptions of in vitro collagen self-assembly, also referred to as fibrillogenesis or polymerization, came in 1952 by Gross and Schmitt as well as Jackson and Fessler in 1955 [73, 92]. Collagen self-assembly refers to the spontaneous and precise multi-scale aggregation of collagen molecules to form longitudinal staggered arrays, giving rise to insoluble fibrous networks with a characteristic banding pattern. Additional details regarding this process as it occurs in vivo and in vitro can be found in Sects. 3 and 5, respectively.

Although some earlier studies identified the ability of cells to interface with collagen, it was Bell and co-workers, in 1979, who reported that human dermal fibroblasts encapsulated within a reconstituted collagen matrix reorganized the fibrous scaffold into a "dermal equivalent" following culture in vitro [13]. This landmark discovery, which came at the infancy of tissue engineering, eventually gave rise to Apligraf, the first "living" dermal-epidermal skin product [15]. Apligraf was produced by culturing human keratinocytes on the surface of the contracted collagen-fibroblast dermal layer. It received initial FDA approval in 1998 and remains on the market to date with indications for venous leg and diabetic foot ulcers that are not responding to conventional therapy.

Although self-assembling collagens have received considerable attention for development of 3D in vitro tissue systems, tissue-engineered constructs, and drug delivery vehicles, translation into medically useful products has been limited to date. There have been and continue to be numerous commercial products consisting of acid-soluble collagen in lyophilized or solution format for research or cell culture applications. These formulations represent single collagen molecules (monomeric collagen) extracted and purified from various tissue sources; however, little focus is given to self-assembly as a functional and standardizable collagen property [112]. As a result, significant product-to-product and lot-to-lot variation exists in the time required for collagen self-assembly (polymerization kinetics) as well as the physical properties (microstructure and mechanical properties) of self-assembled construct [5, 112]. Other persistent challenges of monomeric collagens include long polymerization times (>30 min), low mechanical integrity of formed constructs, and rapid degradation following culture in vitro and/or implantation in vivo [97].

Increased attention on self-assembling collagens came in the late 1990s and early 2000s, with the emergence of recombinant collagens, collagen mimetic peptides, and oligomeric collagen. Advancements in recombinant technology and peptide synthesis facilitated the pursuit of recombinant human collagen (rhCOL) and synthetic collagen-mimetic peptides (CMPs) as potential alternative collagen sources [4]. Today, rhCOL has been produced in plant, insect, yeast, and bacterial systems which co-express the necessary enzymes to create stable collagen triple helices; however, only a subset of these can self-assemble into fibrils [4, 177]. The first report of tissue-derived oligomeric collagen for tissue engineering applications came in 2010 [47, 112]. Unlike monomeric collagens, oligomers represented aggregates of collagen molecules (e.g., trimers) that retained their natural intermolecular crosslinks [12].

Published work shows that oligomers overcome many of the limitations of conventional monomeric formulations, with rapid polymerization, dramatically improved mechanical integrity, and resistance to proteolytic degradation both in vitro and in vivo (see Sect. 5 for specific details).

Collagen has a storied history as the preeminent biomaterial of the body and medicine. The current landscape has led to a variety of collagen formats and formulations, which are routinely categorized as crosslinked tissues, decellularized ECMs (dECMs), insoluble fibrillar collagens, and self-assembling collagens. There exists great promise and potential at the interface of self-assembling collagens, bioinspired multi-scale tissue design, and scalable manufacturing processes for advanced tissue design and fabrication. Additionally, unraveling the mechanisms by which this natural polymer guides fundamental cell behaviors through biochemical and biophysical signaling will continue to inspire approaches to promote tissue regeneration.

3 Hierarchical Design of Collagen In Vivo

Understanding the unique hierarchical organization of type I collagen and its associated physical properties, interactions with other biomolecules, and metabolism (turnover) is fundamental to its use in the fabrication of next-generation biomaterials and tissue-engineered medical products. As the most prevalent protein, collagen is widely distributed throughout the body, where it is found in load-bearing tissues (e.g., skin, bone, tendon, cartilage, and blood vessels), organs (e.g., bladder, stomach, and intestine), and other connective tissues (e.g., pericardium, fat, and placenta). Collagen molecules are produced by cells and deposited within the extracellular space, where they self-assemble in a multi-scale fashion to give rise to the fibrillar scaffold of the tissue ECM. As shown in Fig. 2, this supramolecular assembly involves several aggregation steps: first from single polypeptide chains to a stable triple helical molecule, then to microfibrils, fibrils, and fibers, and finally to macro-scale tissues. Although its primary sequence is identical across tissues, post-translational modifications and formation of intermolecular crosslinks contribute to diversification of collagen building blocks, ECM collagen-fibril networks, and therefore tissue-specific form and function [61, 109].

3.1 Biosynthesis of Collagen

The biosynthesis and folding of collagen as it occurs within the cell has been the topic of extensive study since the 1950s. It represents a highly complex process involving various post-translation events, including hydroxylation, glycosylation, trimerization, and crosslinking, so only the fundamentals are covered here. For more comprehensive coverage, the reader is referred to recent reviews [36, 100, 183].

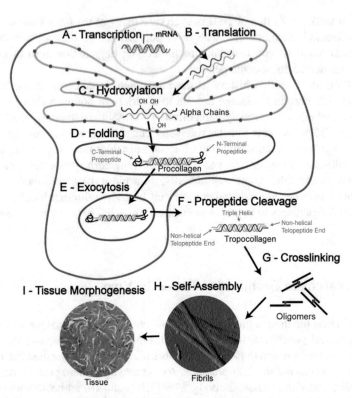

Fig. 2 Multi-scale synthesis and assembly of collagen as occurs in vivo. (A) Collagen genes are transcribed from DNA into RNA. (B) Translation of component polypeptide alpha chains by ribosomes and translocation into the rough endoplasmic reticulum. (C) Hydroxylation of alpha chains by lysyl hydroxylases. (D) Folding of trimeric procollagen molecule. (E) Transfer of procollagen to Golgi for additional post-translational modification and packaging for exocytosis. (F) Enzymatic cleavage of propeptide ends yielding tropocollagen molecules. (G) Crosslinking of tropocollagen molecules by lysyl oxidases to form oligomers. (H) Self-assembly of collagen molecules into D-banded fibrils. (I) Fibrils merge to form fibers and networks, giving rise to complex tissue architecture

Type I collagen is a trimeric protein composed of two α1 and one α2 polypeptide chains. Each of these chains contains the hallmark Gly-X-Y repeat, where X and Y can be any amino acid but are usually proline and hydroxyproline, respectively. This repeating sequence results in the formation of left-handed helices by component polypeptide chains, the interaction of which results in an overall right-handed triple helical structure. The full-length, processed tropocollagen molecule, which represents the fundamental building block of tissues, is approximately 300 nm in length and about 1.5 nm in diameter. Mutations in any of the component α chains, particularly ones that cause problems with folding and crosslinking, have significant consequences on tissue architecture and function, such as the heritable disease osteogenesis imperfecta (OI), a potentially lethal brittle bone disease [141].

As outlined in Fig. 2, synthesis begins with transcription and translation of individual soluble protocollagen chains. Within the endoplasmic reticulum, protocollagen α chains are strategically hydroxylated on proline and lysine residues by specific hydroxylase enzymes. These hydroxylation reactions are important not only for protein folding, but also for downstream intra- and inter-molecular crosslinking. Processed polypeptides then fold and assemble into the procollagen molecule, which contains a central triple-helical region, flanked by non-helical telopeptide and propeptide domains on each end. Terminal propeptides, most notably the one found at the carboxy terminus, and the Gly-X-Y repeats, are critical to proper protein folding [36, 118]. This folding and trimerization process is further assisted by molecular chaperones and enzymes [144, 212]. Additional post-translational processing of procollagen molecules includes the addition of carbohydrate moieties prior to translocation to the Golgi apparatus, where modification of N-linked oligosaccharides is known to occur.

Secretion of procollagen from cells is similar to that of other extracellular proteins, where molecules passing through the Golgi are packaged into secretory vesicles prior to moving to the cell surface for release by exocytosis. After secretion, amino- and carboxy-terminal propeptides are cleaved by multiple C– and N– terminal proteinases. This conversion is critical for proper self-assembly of fibrils, since tropocollagen has a drastically decreased critical aggregation concentration [36]. In fact, defects in N-terminal proteinase ADAMTS2 (a disintegrin and a metalloproteinase with thrombospondin repeats 2) have been shown to lead to the dermatosparaxis variant of Ehlers-Danlos syndrome, which is characterized by fragile, hyperextensible soft tissues [141].

3.2   In Vivo Collagen Self-assembly and Crosslinking

In contrast to the intracellular biosynthetic pathways described above, the precise mechanisms underlying collagen fibril assembly and tissue-specific organization are less well defined. Various models have been proposed to describe the progressive assembly of micro-fibrils, fibrils, fibers, and fiber bundles; however, significant mechanistic gaps that lack corroborating experimental evidence remain. There is, however, strong support suggesting that molecular aggregation begins within secretory vesicles, with the rest of the assembly process occurring exterior to the cell [36]. An important element of collagen assembly and stabilization is the formation of crosslinks, catalyzed by members of the lysyl oxidase (LOX) family. It is here where divergent theories exist, with lysyl oxidase often portrayed as a "welding" mechanism for already assembled collagen fibers. However, the isolation and properties of soluble collagen oligomers, representing stable crosslinked collagen molecules (e.g., trimers), together with what appears as a strategic tissue-specific distribution of crosslink chemistries (Fig. 3) challenges this notion [61]. Furthermore, it has been documented experimentally, that LOX is unable to penetrate the fibril surface, despite the presence of crosslinking throughout the fibril [48]. Based upon these find-

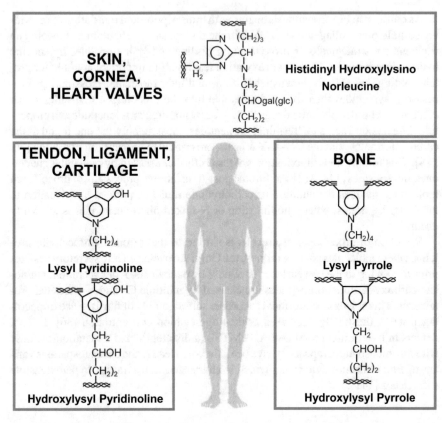

Fig. 3 Structures of mature, trivalent collagen intermolecular crosslinks and their associated tissue-specific distribution (based on Eyre and Wu [61])

ings and our experience with collagen oligomers, we are a proponent of the theory where collagen assembles as prefibrillar aggregates of staggered monomers, with LOX binding and catalyzing the formation of oligomers [48]. In turn, these early oligomer precursors serve as nucleation sites and direct the progressive molecular packing and assembly that ultimately gives rise to tissue-specific ECMs.

Naturally-occurring intra- and inter-molecular collagen crosslinks, which impart mechanical strength to collagen assemblies, have been extensively studied since the 1960s. These bonds form not only between collagen molecules of the same type in homopolymeric fibrils but also between different types of collagen molecules that give rise to heteropolymeric structures [61]. The significant contribution of different crosslinks chemistries in tissue-specific structure and function can be gleaned by analyzing their distribution (Fig. 3), where crosslink number and type appear to be associated with mechanical loading and collagen turnover [60, 61]. Furthermore, these crosslink chemistries, like the primary sequence of collagen, are well conserved across species. Finally, evidence that crosslink content is a critical determinant of collagen fibril ultrastructure, ECM microstructure, and tissue mechanical properties is derived from numerous hereditable diseases as well as in vitro and in vivo studies where specific crosslinking enzymes (e.g., lysyl oxidase or lysyl hydroxylase) are selectively inhibited or genetically knocked out [80, 125, 156]. Our own in vitro work with purified soluble oligomers shows the profound effect of these crosslinked collagen building blocks on the supramolecular assembly, including assembly kinetics, fibril-fibril associations, scaffold mechanical properties and persistence (resistance to proteolytic degradation), and modulation of fundamental cellular processes, such as vessel morphogenesis and tumor cell invasion.

The bulk of research defining the basic pathways of collagen crosslinking was performed over 3 decades ago, with the identification of new crosslink chemistries and their implications continuing today. For detailed reviews, see [7, 61, 174]. In brief, major collagen crosslinks are derived from the oxidative deamination of ε-amino groups of specific lysine and hydroxylysines by LOX within non-triple helical telopeptides regions of the molecules. In turn, the resulting aldehydes react with lysine or hydroxylysine residues within the central triple-helical region of adjacent molecules to form intermediate divalent crosslinks of the aldol, hydroxyaldol, or ketoimmine varieties. Upon maturation, these divalent crosslinks convert into more stable trivalent crosslinks such as the histidine derivative histidinyl-hydroxylysinonorleucine (HHL) which is prominent in skin and hydroxylysyl pyrrole which is prevalent in bone.

3.3 Supramolecular Collagen Assemblies

The supramolecular assembly of collagen is not random but ordered, and much of the process is inherent to the post-translationally modified molecule itself. While residual propeptides have an inhibitory effect on fibril formation, telopeptides are required for proper molecular registry and alignment [184]. The generally accepted Petruska model of collagen fibril structure is a repeated lattice, where collagen molecules are present in a head-to-tail quarter staggered array generating a characteristic banding pattern with 67 nm D-spacing (Fig. 2). While this general value of D-spacing is most commonly found in the literature, there is ample evidence suggesting that a distribution of values occurs throughout tissues, and may vary with age as observed with estrogen depletion in osteoporosis [62, 205]. Additionally, atomic force microscopy, x-ray diffraction, and crystallography studies have elucidated more complex 3D structures within fibrils, including polar ends, tilted or twisted molecules, and crystalline and disordered regions [87, 102, 152]. Oligomers may serve to nucleate formation of branch sites or connections between fibrils during self-assembly, providing an additional mechanism of stabilization [209]. In turn, these fibrils and their networks merge as well as entangle with each other to form larger composite structures such as fibers, bundles or fibrils, and lamellae. Tissues contain an array of higher-order collagen network structures that might be recreated in tissue engineering to give rise to improved functional outcomes. For example, skin is well-known for its anisotropic basket weave structure that contributes to its multiaxial tensile strength [29, 116, 117]. Tendons are composed of criss-crossing fibers densely bundled in parallel, making them ideally suited for their load-bearing function [16]. Other unique structures include the orthogonal lattice of the cornea [133] and the parallel lamellae in osteons of bone [59].

The high conservation of collagen primary sequence and crosslink chemistries across species illustrate their importance as determinants of tissue form and function [61]. Collagen molecules also contain many critical functional domains (motifs) that allow adhesion of cells, binding to other ECM molecules and growth factors, and control of proteolytic degradation. In fact, one fundamental reason why synthetic polymers have failed to displace collagen as a leading tissue engineering material is because of the immense biological activity held in its multifunctional domains. A comprehensive summary and diagram of these various domains has been provided by Sweeney and co-workers [195]. Reciprocal binding interactions between collagen, growth factors, heparin, fibronectin, and other matrix components lends further stability, fluid retention, and biological signaling capacity to the ECM [113]. Collagen is recognized by several cell surface receptors including integrins, discoidin domain receptor (DDR) receptor tyrosine kinases, glycoprotein VI for platelet adhesion, and the immunomodulatory leukocyte-associated immunoglobulin-like receptor 1 (LAIR1) [119]. Of these, integrins are exquisitely mechanosensitive and a prime target for tissue engineering and regeneration design.

4 Biomechanics and Mechanobiology of Collagen

Energy storage, transmission, and dissipation are some of the key mechanical functions provided by ECMs, contributing to bulk tissue mechanical properties as well as guiding cellular behavior through mechanochemical transduction. The hierarchical structure of collagen lends itself to both experimental and computational approaches for deciphering structure-function relationships at the various size scales as well as determining how forces are transmitted between the matrix and resident cells.

4.1 Scaffold and Tissue Biomechanics

To date, measurements of mechanical properties have been made on single molecules, individual collagen fibrils, collagen fibers, as well as native and engineered collagen tissues, with atomic force microscopy (AFM) serving as an important tool at the smaller size scales [23, 173, 208]. From these efforts, the elastic modulus, which provides a measure of rigidity or stiffness, and the fracture strength for a single tropocollagen molecule, has been estimated at 6–7 and 11 GPa, respectively, supporting its role as a "rigid rod" [31]. As we move up size scales, the mechanical properties of fibrils, fibers, and tissues are somewhat less and largely a function of their nano- and micro-structural organization.

The diversity of tissue mechanical properties is a manifestation and optimization of collagen structure at its various size scales. In general, collagenous tissues exhibit a non-linear stress-strain behavior with characteristic strain-stiffening, where the network becomes more rigid with increased deformation [136]. The small strain region, also known as the toe region, corresponds to removal of crimp, both at the molecular and fibrillar levels. The following phase of mechanical testing is a linear region, where the stiffness of collagen fibrils increases considerably with extension. This region has been associated with stretching of collagen triple helices or of the crosslinks between helices, implying a side-by-side gliding of neighboring molecules. Finally, at failure, a disruption of component fibrils occurs. It is well established that initial loading curves for collagenous tissues are different from subsequent loadings, therefore conventionally tissues are "preconditioned" via application of several loading and unloading cycles prior to measurement of mechanical properties. Preconditioning assists in reducing the contributions weak bonds/entanglements and the subsequent reorientation of component fibrils [83, 194]. The stress-strain response is also sensitive to strain rate, a characteristic of viscoelastic materials. Other behaviors exhibited by tissues and other viscoelastic materials include hysteresis—time-based dependence of a material's output on its history, stress-relaxation—decrease in stress in response to a persistent strain (deformation), and creep—tendency to deform in response to a persistent stress [138, 145, 146].

Experimental studies on intact tissues and engineered collagen-fibril constructs as well as computational simulations indicate that key determinants of tissue viscoelastic and poroelastic properties include intrinsic stiffness of the constituent fibrils, interfiber connectivity (branching, bundling), fibril/fiber dimensions (length and diameter), and interactions between insoluble collagen fibrils, other ECM components, and the surrounding interstitial fluid. For example, when fibrils are aligned in parallel to the applied force, constructs fail at lower strain but higher stress values than those with more random fibril organizations. With aligned fibrils, low levels of deformation are required for their recruitment and reorganization in the axis of extension, where they are able to bear load. By contrast, with randomly organized fibrils, higher deformation levels are required for fibril reorganization and not all fibrils are positioned to bear load due to bending or buckling. In addition, while fibril diameter and length certainly contribute to bulk mechanical properties, fibril connectivity is likely the most important determinant, with native and engineered tissues with increased fibril connectivity (branching) and stronger fibril-fibril associations (bundling) able to store increased elastic energy. Supporting this notion we find Young's modulus values for tendon, where fibrils and fibers are parallel aligned are 43–1600 MPa, while reported values are 21–39 MPa for dermis with its basket weave construction and 0.6–3.5 MPa for artery and vein with their layered laminae [131]. The high tear-resistance of skin also has been attributed to unique features of collagen networks, namely fibril straightening and reorientation, elastic stretching and interfibrillar sliding, which redistribute stresses and do not allow tear propagation [218]. While these molecular level events associated with preyield deformation of tissues are fairly well established, those that occur from the yield point to tissue failure (post yield) are less well defined. A number of studies on the overloading of tendon have documented fibril dissociation into their fine subfibrillar components [108, 200], while others report events associated with molecular unfolding [202, 203].

4.2 Mechanobiology and Functional Tissue Engineering

Since the early days of tissue engineering, significant focus has been placed creating constructs that matched the physical characteristics of natural tissues, such as geometry and structure, or the mechanical measures, such as Young's modulus (stiffness) or failure strength. However, with the advent of mechanobiology, it is now recognized that cells can sense and respond to mechanical cues at the molecular and micro-scale levels, just as easily as they do chemical ones. Now, tissue design has shifted from simply mimicking the physical properties (e.g., architecture, mechanical properties) of tissue to focusing on creating biomaterials that provide the correct mechanochemical signals to direct cell phenotype and function as well as tissue morphogenesis [76]. This viewpoint was formalized as "functional tissue engineering" in 2000 by a United States National Committee on Biomechanics subcommittee. Their main goal was to increase awareness of the importance of engineered tissue biomechanics by identifying criteria for mechanical requirements and encouraging tissue engineers to incorporate biomechanics into their design process [34]. This encourages a more multi-scale design approach to tissue engineering and regeneration strategies, which is more focused on guiding the cell response, including therapeutic cell populations within the construct as well as host cells. This perspective is further bolstered by advancements in the stem cell area, where plentiful numbers of multi-potential cell populations can be harvested directly from tissues (e.g., fat, bone marrow, blood) or developed from induced pluripotent stem cells, which are created by reprogramming skin or blood cells into an embryonic-like pluripotent state.

When approaching tissue fabrication, whether in the body or man-made, it is important to understand how hierarchical collagen construction contributes to not only tissue-level mechanical properties but also transmission of loads across size scales to cells and vice versa. Biophysical cues such as those originating from the ECM microstructure and mechanical properties are now recognized as major signaling sources, regulating growth and differentiation of cells [126]. It's important to note that this transmission of biophysical signals is a two-way street, evoked by the contractile machinery of resident cells or by loads applied externally. This exchange of biophysical information is further facilitated by the physical connectivity between cells and collagen fibrils, which in large part is mediated through specific cell surface receptor proteins known as integrins. It was Donald Ingber that first depicted the dynamic force balance that exists between cells and their ECM using the popular tensegrity model, where cytoskeleton and ECM form a single, tensionally integrated structural system [89]. It is at this interface where specific design criteria and constraints for advanced tissue fabrication continue to emerge. While certainly a difficult task, sophisticated methods designed to probe biophysical and biomolecular responses of living cells within tissues continue to assist in elucidation of the mechanochemical signaling that occurs from tissue level through the ECM to the cell nucleus.

5 Collagens as a Natural Polymer for Custom Tissue Fabrication

Because type I collagen is one of most commonly used biomaterials in both research and clinical settings, there exists a wide variety of formulations, as alluded to in Sects. 1 and 2. Most collagen-based products used clinically represent processed intact tissues (e.g., dECMs) or insoluble fibrillar collagenin various formats (e.g., sponge, particulate), with only a few products prepared from self-assembling collagens. This section focuses on advancements related to self-assembling collagen formulations and their potential for multi-scale tissue design. We begin with molecular and micro-level design control, identifying how specific collagen building blocks, assembly conditions, and exogenous crosslinking affect the microstructure of engineered biomaterials and tissues. This is followed by a description of higher-level fabrication and manufacturing techniques for controlling macro-scale properties, including 3D geometry and physical properties (e.g., mechanical strength and stiffness). Special emphasis is placed on the cellular response, whether in vitro or in vivo, documenting its dependence upon multiple size scale features, extending from molecular to macroscopic.

5.1 Micro-scale Design Control

The first fundamental level of design control for collagen materials resides at the molecular level. Molecular level features largely determine the achievable range of chemical, biological, and physical attributes of resulting scaffolds and tissue constructs; however, user control at this level is often overlooked.

5.1.1 Molecular Building Blocks

The molecular make-up, structure, and self-assembly capacity of various collagen building blocks are summarized in Table 1, where extraction, processing and reconstitution techniques are known to be a source of variation. Insoluble fibrillar collagen, which is the starting material for many freeze-dried collagen and collagen-glycosaminoglycan sponge products, represents a particulate of undissociated collagen fibers isolated and purified from comminuted tissues. While this form of collagen does not self-assemble or offer molecular and fibril-microstructure control, it does aggregate to form a viscous gel or slurry when swollen in acid or hydrated in phosphate buffered saline, which has proven useful for various medical applications. As documented by Yannas and Burke and others, insoluble fibrillar collagen supports cell adhesion and offers design control of larger scale material features such as particulate content, porosity, and resorption rate [51, 219, 220].

Unlike fibrillar collagen, other collagen building blocks do have the capacity to self-assemble or form fibrils in vitro, providing control over molecular and fibril microstructure features. The ability of relatively pure collagen molecules to spontaneously form fibrils when brought to physiologic conditions (pH and ionic strength) and warmed was first reported by Gross in the 1950s and has since been the subject of extensive research [73, 74]. Collagen is routinely extracted and purified from various tissue sources (rat tail tendon or calf skin) using either dilute acid or enzymatic digestion (pepsin), yielding a solution composed predominantly of single molecules (monomers) [5]. Historically, crosslinked oligomers and insoluble molecular aggregates that accompanied monomers were viewed as undesirable by-products, especially for studies focused on collagen molecule structure and fibril assembly [143]. In fact, enzymatic digestion, secondary purification strategies, or young or lathrytic animals were routinely used to minimize or eliminate these components [39, 132, 134, 142]. Acetic acid extraction followed by salt precipitation is one of the most common approaches used to generate telocollagen, which represents full length tropocollagen molecules with telopeptide regions intact [25]. The addition of pepsin to the extraction mixture increases yield but causes cleavage of telopeptide regions, giving rise to atelocollagen [179]. More recently, a sodium citrate extraction process was applied to porcine dermis, generating a high fraction of soluble oligomeric collagen for biomaterials development [12, 112]. Oligomers represent aggregates of individual collagen molecules (e.g., trimers) that retain their natural intermolecular crosslinks.

Monomeric collagens, specifically telocollagen and atelocollagen, continue to be the most commonly used self-assembling collagens because of their relatively facile extraction and commercial availability. However, the shortcomings of these preparations are well established and commonly cited by users, including lot-to-lot variability in purity and self-assembly capacity, long polymerization times (often >30 min), lack of user control, low mechanical strength, and poor stability in vitro and in vivo [1, 5, 112]. When comparing telocollagen and atelocollagen, it has been shown that telopeptide preservation is important for the thermal stability of the collagen triple helix and the organized arrangement of collagen molecules into fibrils [82, 183]. The loss of the telopeptide regions in atelocollagen significantly hinders and slows assembly kinetics, resulting in less organized fibrils that vary in size and lack natural D-banding pattern [21, 79, 179]. This difference in molecular chemistry and fibril microstructure also affects matrix physical properties and proteolytic resistance, with atelocollagen generating weaker (i.e., Young's modulus and yield strength) constructs that are more prone to rapid dissolution and proteolytic degradation [81, 112].

Table 1 Summary of collagen building block characteristics and their associated level of design control for engineered biomaterials and tissues

Building block	Molecular composition, structure, and self-assembly capacity	Design control
Fibrillar collagen	• Insoluble • Collagen fiber particulate processed and purified from comminuted tissues • Does not exhibit self-assembly	• Macrolevel
Atelocollagen	• Soluble • Tropocollagen molecule devoid of telopeptide ends • Exhibits fibril assembly with modified or no D-banding • Contains collagen functional domains	• Microlevel • Macrolevel
Telocollagen	• Soluble • Telopeptide ends allow for formation of D-banded fibrils • Contains collagen functional domains	• Microlevel • Macrolevel
Oligomer	• Soluble • Aggregates of tropocollagen molecules (e.g., trimers) that retain natural intermolecular crosslink • Exhibits fibril and suprafibrillar assembly with D-banding and high fibril-fibril connectivity or branching • Contains collagen functional domains	• Microlevel • Macrolevel
Recombinant	• Soluble • Recombinant human procollagen • Post-translational modification requires co-expression of relevant enzymes • Endopeptidase treatment yields self-assembling atelocollagen • Contains collagen functional domains	• Molecular • Microlevel • Macrolevel
Collagen mimetic peptides	• Soluble • Peptides (~30 to 60 amino acids) containing repeats of helical region sequences • Self-assembly into helices and fibers largely driven by electrostatic interactions • Lack collagen functional domains	• Molecular • Microlevel • Macrolevel

Oligomers, a more recently-discovered collagen building block, appear to play a critical role in collagen self-assembly, both in vitro and in vivo. Over the past decade, oligomer preparations have proven to be quite robust and reproducible, exhibiting rapid polymerization (<1 min at 37 °C) and generating distinct fibril microstructures compared to telocollagen and atelocollagen formulations [12, 112, 190]. Since oligomers retain intermolecular crosslinks, they exhibit a higher average molecular weight compared to monomers and a distinct protein and peptide banding pattern [12, 112]. In addition, the presence of crosslinked oligomers induces fibrillar as well as suprafibrillar assembly, resulting in networks with high fibril-fibril connectivity and branching. These higher-order assembly properties support formation of collagen scaffolds that not only retain their shape but exhibit a much broader range of physical properties and slow turnover (Fig. 4) [21, 112, 190]. In particular, collagen oligomer scaffolds demonstrate significantly increased shear, tensile, and compressive moduli compared to their monomeric counterparts (Fig. 4c–e). Since these parameters increase linearly or quadratically with oligomer concentration, differences become even greater at high concentration. The improved stability and mechanical integrity exhibited by oligomer effectively eliminates the need for exogenous crosslinking, which is routinely applied to constructs produced from monomeric collagens [77, 136].

Other approaches for generating purified collagen molecule preparations, especially human, include recombinant technology or peptide synthesis. Production of collagen molecules and peptides via these techniques supports design control at the molecular level (Fig. 5a), which is especially useful for elucidating relationships between specific molecular motifs/domains and functional properties [26]. To date, researchers have successfully genetically modified mammalian, bacterial, and plant systems to produce recombinant human procollagen, from which self-assembling collagen formulations can be derived [96, 155, 176, 197]. One of the challenges associated with recombinant collagen production has been the ability to introduce and co-express various genes involved in collagen post-translational modifications, including prolyl-4-hydroxylase and lysyl hydroxylase, which are necessary for triple-helix stabilization [217]. To date, a number of groups have overcome this obstacle, successfully generating stable procollagen triple helices [111, 165, 167, 170, 189] Since procollagen molecules are unable to undergo self-assembly due to the presence of propeptide ends, endopeptidase treatment (e.g., pepsin, ficin) is routinely applied to yield fibril-forming recombinant human atelocollagens [11, 26]. At present, atelocollagen produced recombinantly yields thinner fibrils with less mechanical integrity than their tissue-derived counterparts (Fig. 5b) [211]. Researchers focused on recombinant collagen development for biomedical applications continue to work to scale their processes to support more cost-effective, large-scale production [26, 167].

Collagen mimetic peptides (CMPs) produced using synthetic chemistry methods are another means of achieving molecular-level design control [191]. Relatively short sequences, roughly 30 amino acids in length, are synthesized with the goal of forming homo- or hetero-trimeric collagen helices, which in turn self-assemble into fibrils. The majority of sequences consist of amino acid triplet repeats found within the helical region of collagen, capitalizing on electrostatic forces to drive molecular

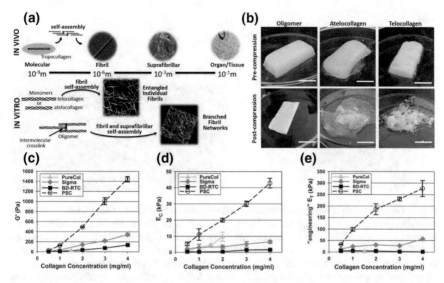

Fig. 4 Self-assembly of collagen and comparison of different collagen formulations. **a** Hierarchical, multi-scale assembly of type I collagen as occurs in vivo and in vitro with polymerizable monomer (atelocollagen and telocollagen) and oligomer formulations (reprinted with permission from Blum et al. [21]). **b** Representative images of oligomer, atelocollagen, and telocollagen constructs before (3.5 mg/mL) and after (24.5 mg/mL) confined compression (86% strain or 7×), demonstrating differences in shape retention and mechanical properties. Scale bars = 2 mm (reprinted with permission from Blum et al. [21]). **c–e** Comparison of mechanical properties, including (**c**) shear storage modulus (G'), unconfined compressive modulus (E_C), and tensile modulus (E_T), for oligomer (PSC) and commercial atelocollagen (PureCol) and telocollagen (Sigma, BD-RTC) formulations (reprinted with permission from Kreger et al. [112])

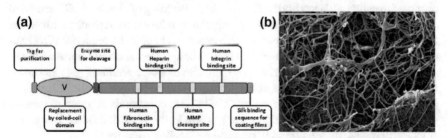

Fig. 5 Examples of the utility of recombinant collagens. **a** Schematic representation of recombinant bacterial collagen construct, showing examples of possible sequence manipulations (reprinted with permission from Brodsky and Ramshaw [26]). **b** Scanning electron microscopy of fibrils formed from purified recombinant human atelocollagen produced in tobacco plants. Scale bar = 5μm (reprinted with permission from Stein et al. [189])

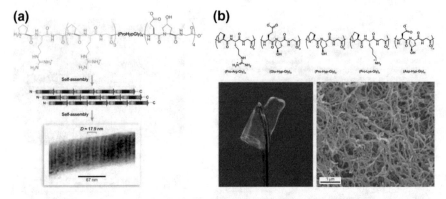

Fig. 6 Notable examples of self-assembling CMPs. **a** Schematic of CMP and associated Coulombic forces between cationic and anionic blocks that yield self-assembled fibrils. Transmission electron microscopy image of CMP fibril shows D-periodicity with D = 17.9 nm. Natural type I collagen has D ≈ 67 nm (reprinted with permission from Shoulders and Raines [183]). **b** Chemical structure of the common amino-acid triplets used to generate CMPs. Photo and scanning electron microscopy image show shape-retaining fibrillar gel (1%) formed following self-assembly of CMPs consisting of (Pro-Lys-Gly)$_4$(Pro-Hyp-Gly)$_4$(Asp-Hyp-Gly)$_4$ (reprinted with permission from O'Leary et al. [154])

assembly. While a number of groups have created CMPs that generate fibrils, creating peptides that mimic the various levels of collagen supramolecular assembly, including staggered alignment, has been a challenge [111]. In 2009, Chaikof, Conticello and co-workers reported a CMP that, in part, formed fibrils with a regular D-spacing pattern; however, the periodicity was about 18 nm rather than the characteristic 67 nm observed in native collagen fibrils (Fig. 6a) [170]. Building off this work, O'Leary and colleagues prepared a new CMP, where arginine residues were replaced with lysine and glutamate residues were replaced with aspartate, to give the sequence (Pro-Lys-Gly)$_4$(Pro-Hyp-Gly)$_4$(Asp-Hyp-Gly)$_4$ [154]. These CMPs showed improved fibril- and hydrogel-forming characteristics, giving rise to shape-retaining gels as shown in Fig. 6b. Finally, although functional domains, such as integrin binding sequences (e.g. GFOGER), can be engineered into CMPs, size constraints inherent to peptide synthesis (about 60 amino acids or less) preclude the inclusion of all functional collagen domains, thereby limiting overall biosignaling capacity [70].

5.1.2 Polymerization Conditions

In addition to the various collagen building blocks described above, there are a wide variety of external means by which collagen self-assembly can be modulated to create hydrogels, matrices, and scaffolds with distinct structural and physical properties. This section summarizes various conditions, such as concentration, pH, temperature, and ionic strength, that have been used to modulate collagen assembly kinetics and outcomes. These conditions can be carefully controlled to modulate fibril density, fiber length, fibril diameter, fibril-fibril associations (e.g., branching), and pore size, all of which, in turn, determine functional physical properties, including strength, stiffness, fluid and mass transport, and proteolytic degradation. When cells are encapsulated in these self-assembled collagen matrices, they quickly adhere to the collagen fibrils, sensing and responding to differences in microstructure mechanical properties [72, 160, 171]. Through this mechanotransduction signaling, fundamental cell behavior is modulated, including cell-induced matrix contraction and remodeling, morphogenesis, proliferation, migration, and differentiation. Additionally, these microstructure features dictate how external mechanical loads are transmitted from the construct or macro-level to resident cells [14].

A landmark study by Wood and Keech in 1960 showed that increasing collagen concentration or temperature, and decreasing pH and ionic strength, accelerated the formation of individual collagen fibrils [213]. Additionally, they reported that higher temperatures, increased pH, and lower ionic strengths yielded thinner fibrils; however, no correlation was made between structural and mechanical properties of resulting fibrils was provided [213]. The effects of temperature, pH, and ionic strength on fibril assembly of telocollagen have been confirmed by many, and studies have expanded to include more detailed mechanical characterization [2, 6, 168, 175, 178]. In addition to the effects on self-assembly kinetics, increasing the temperature or pH of the reaction results in decreased pore size and fibril diameter, which have been shown to increase compressive, tensile, and shear storage moduli [2, 6, 168, 175]. The effect of ionic strength on matrix mechanics seems to be dependent on pH and temperature, thus making distinctive trends difficult to decipher [2].

Collagen concentration is another primary means by which many researchers vary matrix mechanics, since increasing collagen concentration leads to increased fibril density which increases matrix stiffness (compressive, tensile and shear) [112, 160, 175]. In attempts to independently control collagen fibril density and matrix stiffness, many have created composite systems, sometimes termed interpenetrating networks, composed of mixtures such as gelatin and collagen [17], alginate and collagen [50], polyethylene glycol and collagen [120]. Whittington and co-workers identified another approach for independently controlling fibril density and matrix stiffness which did not rely on non-collagenous agents. Here, the total content and ratio of type I collagen oligomers to monomers were used to independently vary fibril density and the extent of fibril-fibril branching, both of which are known determinants of in vivo ECM stiffness [209, 210].

Another way in which researchers have attempted to gain design control of collagen self-assembly is motivated by the fact that collagen fibrillogenesis and assembly in vivo is guided by other collagenous and non-collagenous proteins and proteoglycans of the ECM. For example, fibronectin and collagen assembly in vivo are known to be reciprocally dependent such that interruption of one decreases the other [101]. However, early experimental evidence from Brokaw and co-workers suggested that in vitro, the addition of fibronectin only affected collagen self-assembly kinetics, with no changes in the resulting microstructure [27]. On the other hand, it has also been shown that co-polymerization of fibronectin with collagen increases the tensile strength of formed matrices, supporting the notion that fibronectin affects collagen fibril organization and microstructure [68]. Type V collagen also affects in vivo collagen assembly, where it is thought to serve as a nucleation site. Loss-of-function mutations are embryonic lethal, characterized by lack of collagen fibril formation in the mesenchyme [101]. While type I collagen can self-assemble in vitro without type V collagen, Birk et al. showed that the presence of collagen V during in vitro self-assembly yielded heterotypic fibrils with decreased diameter and altered D-periodicity [19]. More recently, Piechocka et al. demonstrated that these relatively minor changes in microstructure caused drastic decreases in shear storage modulus [158]. These authors propose that this discrepancy between in vitro results and in vivo mechanisms may be due to the fact that the type V collagen used in vitro is pepsin treated and lacks the N-propeptide region which is present during in vivo ECM assembly. One final example demonstrating how other ECM components guides collagen assembly and mechanics involves glycosaminoglycans and proteoglycans, which consist of glycosaminoglycans attached to a core protein. Interestingly, polymerization of monomeric collagen in the presence of the glycosaminoglycan dermatan sulfate resulted in increased lateral fibril aggregation and decreased tensile properties compared to control matrices. On the other hand, co-polymerization of collagen and the dermatan sulfate containing proteoglycan decorin yielded highly interconnected, long, thin fibrils with increased tensile strength [66, 159, 169]. Collectively, these studies highlight the impact of other ECM components on the hierarchical organization of collagen. Discrepancies between in vivo and in vitro results, as well as between studies reveal the sensitive nature of these reactions and their dependence of specific molecular features and reaction conditions. Continued elucidation of mechanisms underlying supramolecular collagen assembly both in vivo and in vitro will continue to inspire tissue engineering and regeneration design strategies.

5.1.3 Exogenous Crosslinking

Mechanical integrity, metabolic turnover, and degradation resistance are properties afforded to in vivo collagen assemblies, in part, by the formation of natural intra- and inter-molecular crosslinks as described in Sect. 3.2. These natural crosslinks are controlled via post-translational modifications and enzymatic reactions that occur within and outside the cell, respectively, making them difficult to recreate in vitro [61]. The application of oligomeric collagen allows tissue-engineered constructs to capture some of the performance characteristics imparted by natural intermolecular crosslinks. However, for materials produced from insoluble fibrillar collagen or self-assembling monomeric collagens, the development and application of exogenous physical and chemical crosslinking is commonplace to improve mechanical properties and proteolytic resistance [78].

Glutaraldehyde is one of the most commonly employed chemical crosslinking agents [37]. It is well established that glutaraldehyde enhances collagenous material stiffness, strength, and resistance to proteolytic degradation through formation of intramolecular and intermolecular crosslinks by non-specifically reacting with lysine and hydroxylysine residues on collagen [52]. Despites its widespread use, glutaraldehyde is far from ideal as its crosslinks are transient and release of glutaraldehyde monomers over time is cytotoxic [78, 207]. Additionally, calcification of glutaraldehyde crosslinked tissues upon implantation remains a challenge [180, 181].

Dehydrothermal treatment (DHT) and ultraviolet (UV) radiation have been examined as alternatives to glutaraldehyde since the 1980s [135, 207]. DHT and UV crosslinking methods are thought to be advantageous because they do not introduce any exogenous toxic chemicals; however, these treatments can induce partial denaturation or fragmentation of collagen [78]. Carbodiimide treatment is another technique used to form amide-type bonds within collagen. Here, the only by-product is urea, which can be washed away after crosslinking [78]. The combination of 1-ethyl-3-(3-dimethylaminopropyl)-carbodiimide hydrochloride with N-hydroxysuccinimide (EDC/NHS) is the most commonly used strategy and has been applied both during and after self-assembly of monomeric collagen to enhance scaffold strength [224]. Interestingly, when EDC crosslinking was applied to scaffolds created from oligomeric collagen, it did not enhance the mechanical properties thus suggesting that the presence of the natural intermolecular crosslink outweighs the effect of these unnatural chemistries [150]. It is important to note that owing to their non-specificity and cytotoxicity, the majority of exogenous crosslinking strategies are incompatible with self-assembled collagen constructs formed in the presence of cells.

Enzymatically crosslinking collagen with LOX and transglutaminase or generating advanced glycation end-products (AGE) with sugars such as ribose are crosslinking strategies that are reported to be more compatible with cells. However, these methods only modestly improve mechanical strength and are reported to be cost-prohibitive for large/clinical scale applications [78, 192]. Despite being non-cytotoxic in the short term, non-enzymatic glycation, as occurs during ageing and pathological processes such as diabetes, has been linked to reactive oxygen species production and cellular inflammation via the receptor for advanced glycation end products (RAGE) pathways, suggesting this method of crosslinking may be suboptimal for many engineered products intended to for permanent tissue replacements [186]. Finally, genipin, a plant-derived chemical used in traditional Chinese medicine is another collagen crosslinker that has been shown to be cell-compatible at low concentrations [192, 193]. However, genipin crosslinking turns crosslinked collagenous materials blue and upon in vivo implantation induces inflammation and an associated foreign body response, although the extent is reduced compared to glutaraldehyde [40, 55].

Collectively, molecular and microscale features, including molecular composition, endogenous or exogenous crosslinking, and fibril ultrastructure and architecture, are important considerations when designing next-generation tissue engineering and regenerative medicine strategies. This is especially true when working to promote a regenerative phenotype since cells naturally interface with collagen at these levels and can readily detect and respond to changes at these size scales.

5.2 Meso- and Macro-scale Design Control

The ECM component of tissues has a complex construction with spatial gradients, anisotropies, and higher-order structures. By contrast, the majority of constructs formed by encapsulation of cells within self-assembling collagens in vitro represent isotropic random fibril networks, and are often limited in concentration or fibril density due to the solubility and phase behavior of collagen. For accurate recreation of tissues, the density and spatial organization of the collagen-fibril ECM is an important design consideration. Historically, these meso-scale features have been difficult to control, making the engineering of functional tissue replacements challenging. Recent years have seen the rise of process engineering and manufacturing techniques to address these challenges.

5.2.1 Compression

Initial efforts to convert polymerized collagen-fibril matrices into constructs with tissue-like histology and consistency relied on the remodeling properties of cells to densify or compact surrounding collagen fibrils. More specifically, collagen-fibril matrices seeded with fibroblasts and cultured up to 2 weeks yielded contracted or condensed dermal-like tissue equivalents [15]. Seeding keratinocytes on the surface of these dermal equivalents resulted in the formation of a multilayered epidermis, yielding a tissue-engineered living skin, which ultimately was produced by Organogenesis and gained FDA approval in 1998 for management of diabetic ulcers and hard-to-heal venous ulcers. Persistent drawbacks to this product include its costly manufacturing process, limited shelf-life (5–10 days) and the slight risk of disease transmission, all of which are due to the requirement of allogeneic cells to contract and further mature the ECM and finished product [182].

In 2005, Brown described a process designed to "engineer tissue-like constructs without cell participation." This "cell-independent" approach involved polymerization of monomeric collagen in the presence or absence of cells followed by plastic compression (PC) in an unconfined format and/or capillary fluid flow into absorbent layers to reduce the interstitial fluid content [30]. Here, low-loads (50–60 g or 1.1 kPa) are applied to the top surface of a collagen matrix to achieve significant fluid reduction (approximately 85–99.8% compressive strain) through a supporting nylon mesh (Fig. 7a). The resulting densified collagen sheets, which measure 20–200 μm in thickness are still fragile, requiring spiraling and multiple compressions to facilitate handling and further mechanical testing. Tensile strength and modulus values of 0.6 ± 0.11 MPa and 1.5 ± 0.36 MPa, respectively, have been supported with 85% viability of encapsulated cells [20]. Additional compression of spiraled constructs improves mechanical integrity but reduces cellular viability [20, 42]. This technology contributed to development of the RAFT 3D Cell Culture System by TAP Biosystems (now part of Sartorius Stedim Biotech Group), which applies their patented absorber

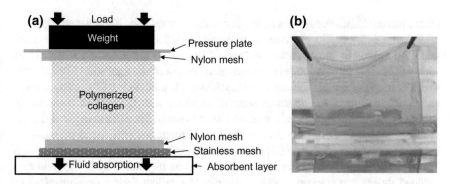

Fig. 7 Densification of collagen-fibril constructs through plastic deformation achieved with unconfined compression and absorption. **a** Plastic compression is achieved by applying known weights to low-density collagen-fibril matrices to achieve fluid reduction through a supporting nylon mesh into an absorbent layer (based on Brown et al. [30]). **b** Dermo-epidermal skin substitute produced by densification of monomeric type I collagen in the presence of human dermal fibroblasts. Seven days following culture the collagen-fibroblast construct was seeded with human keratinocytes (reprinted with permission from Braziulis et al. [24])

technology to monomeric rat tail collagen to create densified tissue constructs for research applications.

An adaptation of this PC technology was reported by Reichmann's group for generation of an autologous tissue-engineered skin. This work involved custom-fabrication of a large (7 × 7 cm) compression chamber, fashioned to support weights on top and absorbent filter paper on the bottom [24]. This device was used to cast square polymerized collagen matrices containing human dermal fibroblasts, which in turn were compressed to 0.5–0.6 mm thickness and then transferred to culture dishes. Following 7 days of culture and maturation in vitro, a high density of keratinocytes was applied and cultured for an additional 7 days. To date, analyses of histological outcomes as well as gene expression of relevant dermal and epidermal markers have been conducted [24, 151]; however, mechanical properties testing has yet to be reported. This tissue-engineered autologous dermo-epidermal skin graft, referred to as denovoSkin (Fig. 7b), has obtained orphan drug designation as a treatment for burns by Swissmedic, European Medicine Agency, and the FDA. Reports indicate that this product can be safely and conveniently handled by surgeons, and matures into high quality skin in animal models as well as recently performed clinical studies [151].

Expanded efforts on this front, include work by Voytik-Harbin and collaborators where scalable plastic compression processes have been applied to type I oligomeric collagen, providing increased versatility in product design and geometry as well as predictive meso-scale control [21, 149]. As mentioned previously, type I oligomeric collagen exhibits not only fibrillar but also suprafibrillar assembly, yielding highly interconnected collagen-fibril scaffolds with substantially improved proteolytic resistance and mechanical integrity compared to standard monomeric collagens. In this work, plastic compression was applied in a confined, rather than an unconfined, format to increase fibril density via controlled fluid removal (Fig. 8a). Interestingly, this approach was not applicable to monomeric matrices due to the inability of the resultant fibril microstructure to sustain or support associated compressive and fluid shear forces [21]. This fabrication process provided control of the final solid fibril content (fibril density) of the compressed construct through modulation of starting volume and concentration of the oligomer solution together with the applied compressive strain [149]. Additionally, strain rate was used to control steepness of fibril density gradient, and placement of porous polyethylene foam and associated porous-solid boundary conditions defined high-order spatial fibril organization (e.g., alignment). Finite element analysis confirmed this process to be dependent upon the fluid flow induced during compression, with steepness of gradient formation dependent on strain rate [149]. These early findings support the notion that controlled, plastic compression together with computational models can be used for predictive design and scalable manufacture of a diverse array of precision-tuned tissue constructs. To date, this fabrication method has been applied for the development of cartilage constructs for laryngeal reconstruction [28, 225], articular cartilage constructs with continuous fibril density gradients that recapitulate the different histological zones in native cartilage (Fig. 8c) [149], acellular and cellular dermal replacements (Fig. 8b) [149], as well as an in vitro model of cardiac fibrosis [215].

5.2.2 Electrospinning

Electrospinning is a fiber-forming process that applies a large electric field between a polymer solution reservoir and a collection plate to form polymer fibers with nanometer-scale diameters. More specifically, when a sufficiently high voltage is applied to a liquid polymer droplet, the body of liquid becomes charged, and electrostatic repulsion counteracts the surface tension and the droplet is stretched. At a critical point a stream of liquid erupts from the surface forming a fiber. This fiber elongates and thins, and the solvent evaporates, as it moves towards the grounded collector where it is deposited. Published work on the electrospinning of collagen dates back nearly two decades [85, 86, 130]. In this case, materials are designed to mimic the geometry (e.g., diameter) of collagen fibrils or fibers found in vivo within the extracellular matrix. Since that time a large number of design variables including solvents, molecular make-up of collagen, collagen concentration and viscosity, applied electric field, flow rates, collection distance, and collection strategies (plates, rotating mandrels) have been explored [35, 130]. At present, this technique has been used to generate collagen-based scaffolds of varying geometries (tubes, mats) and architectures (randomly oriented, aligned, high porosity, low porosity) for various tissue applications including bone [163], nerve [164], blood vessel [22, 196], and skin [162, 172]. For more details, the readers are encouraged to see DeFrates et al. [54]. A major limitation associated with present-day electrospinning is its requirement for volatile solvents (e.g., fluoroalcohols), which denature the native structure of collagen yielding gelatin. Furthermore, resulting materials lack collagen fiber ultrastructure (axial periodicity and D-banding) and therefore display

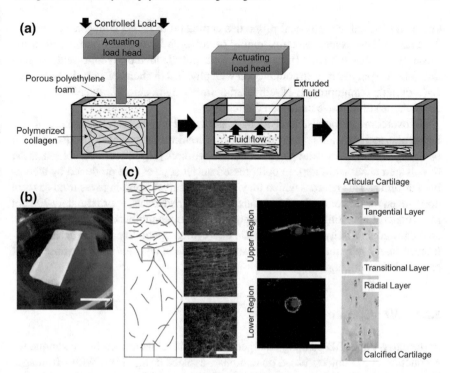

Fig. 8 Controlled confined compression for fabrication of acellular and cellular constructs with and without continuous structural gradients **a** Schematic depicting controlled confined compression process for densification of collagen-fibril constructs. A low-density collagen-fibril matrix is formed in a mold and then compressed at a controlled strain rate to achieve a specified strain. Fluid flow is directed through the porous boundary (adapted from Blum et al. [21]). **b** Densified sheet formed via controlled confined compression of type I oligomeric collagen. Scale bar = 2 mm (reprinted with permission by Blum et al. [21]). **c** Gradient densification of collagen-fibril matrices as achieved via controlled confined compression. Type I oligomer matrices were compressed with a porous platen, directing fluid flow through an upper porous boundary condition. Confocal reflection microscopy revealed a gradient in fibril density, with a high density of fibrils aligned parallel to the construct surface near the top progressing to a low-density region of randomly organized fibrils near the bottom. Scale bar = 100 μm (reprinted with permission from Novak et al. [149]). Encapsulated cells responded to their local microenvironment as a result of densification, as detected 1 week with confocal microscopy (green = F-actin; blue = nucleus). Cells in the high-density region developed a spindle shape and were oriented parallel to the fibrils, while cells in the low-density regions displayed a more rounded morphology. Scale bar = 10μm. Such gradients in collagen microstructure and cell morphology/phenotype are reminiscent of the gradient layers found in articular cartilage

altered biological and physical properties compared to native collagen assemblies. To address these issues, electrospinning of collagen is routinely performed in the presence of other synthetic [polycaprolactone, poly(lactic-co-glycolic acid)] or natural (elastin) polymers or in conjunction with physical or chemical crosslinking (e.g., N-(3-dimethylaminopropyl)-N'-ethylcarbodiimide hydrochloride and NHS, and glutaraldehyde) to improve mechanical integrity.

To overcome the persistent challenges associated with the electrospinning of collagen, alternative manufacturing processes are continued to be developed for creation of collagen fibers. For example, Polk and co-workers [161] described volatilization of collagen using a high-speed compressed air jet such as that produced by a common airbrush. This process which they termed pneumatospinning was used to form non-woven meshes of randomly organized and aligned fibrils, approximately 200 nm in diameter. Interestingly, pneumatospun and electrospun fibers formed from acetic acid showed similarity in size, strength, and cytocompatibility. However, like electrospun fibers, pneumatospun fibers were not stable in aqueous media in absence of chemical crosslinking.

5.2.3   3D Bioprinting

In the early 1990s, 3D printing emerged as an additive manufacturing technique for production of 3D objects based on computer-assisted design [49]. With advantages of mass production and fine tuning of spatial-dimensional properties, this process has been adapted for purposes of developing functional tissues and organ constructs. Such constructs are being fashioned for use as in vitro model systems for basic research or drug screening [110, 147], delivery of pharmaceutical agents (genes, drugs) or cells [93, 185], and tissue-engineered medical products for tissue replacement or reconstruction [153]. Bioprinting involves sequential layer-by-layer deposition of biomaterials in the presence and absence of specific cell populations in predetermined spatial-dimensional patterns with millimeter or nanometer scale resolution. In this way, porosity, permeability and mechanical properties, and cell-cell and cell-ECM associations within the construct may be controlled. Of the various 3D printing technologies, direct ink writing and inkjet printing have received the most widespread use for bioprinting applications. For direct ink writing, high viscosity hydrogels in the presence or absence of cells are extruded to obtain 3D structures with or without a carrier. By contrast, inkjet bioprinting applies low viscosity solutions or suspensions as droplets.

A critical component of bioprinting are the "bioinks", which typically are polymeric materials that are used to deposit cells and/or serve as the extracellular scaffold. Ideally, bioink materials need to exhibit (i) good printability, (ii) biocompatibility for maintaining cell viability without eliciting immune reactions, (iii) cell-friendly curability, (iv) mechanical stability with shape retainability, (v) predictive biodegradability including mechanism (hydrolysis or proteolytic degradation) and kinetics, and (vi) predictable material-cell interface with ability to promote fundamental cellular behaviors (adhesion and remodeling, migration, proliferation, differentiation) and processes (morphogenesis) [95]. While bioink materials used to date satisfy a subset of these design requirements, bioink development and characterization remains a high-priority activity, together with optimization of the bioink-bioprinter interface [139].

To date a number of synthetic, nature-derived, and natural biomaterials have been used for a myriad of bioprinting activities and have been the subject of recent comprehensive reviews [53, 95, 139]. Here, we focus on the application of various collagen-based formulations, especially those that exhibit self-assembly. As stated previously, the use of collagen is advantageous because of its inherent biocompatibility and biosignaling capacity. However, a persistent limitation with conventional monomeric collagens has been their poor mechanical properties and long polymerization times, contributing to poor shape retaining properties and printing resolution. To circumvent these problems, collagen and its denatured counterpart gelatin have been modified by introducing new functional groups or used in conjunction with other biomaterials.

Gelatin methacryloyl (GelMA) represents one of the most popular bioinks, offering fast polymerization, good biocompatibility as well as tunable mechanical properties (for recent reviews see [223]). GelMA is a chemically-modified version of gelatin that exhibits photopolymerization (gelation) upon exposure to light irradiation in the presence of photoinitiators. Gelatin is distinct from native collagen in that it represents a mixture of collagen peptides and single-stranded polypeptide chains produced by collagen hydrolysis and denaturation. Although gelatin retains arginine-glycine-aspartic acid (RGD) sequences that promote cell attachment as well as target sequences for matrix metalloproteases, it does not maintain collagen's native triple helical structure and therefore inherent fibril-forming capacity. Introduction of methacryloyl groups confers to gelatin the capacity to be photocrosslinked with the assistance of photoinitiators and exposure to light. Many physical parameters of GelMA hydrogels, such as mechanical properties, pore sizes, degradation rates, and swell ratio can be readily tailored by changing the degree of methacryloyl substitution, GelMA prepolymer concentration, initiator concentration, and light exposure time [43, 148].

More recently, tissue- and organ-derived dECMs that retain collagen's fibril-forming capacity have been gaining increased use as bioinks for 3D bioprinting applications [106]. While traditionally intact dECMs derived from various allogeneic and xenogeneic tissue sources have been used clinically (surgical mesh, wound management), recent studies have focused on adapting these materials for tissue-specific 3D bioprinting applications [94, 157]. Creation of dECM bioinks involves application of various decellularization methods to remove cells from tissues and organs. The resulting dECMs are then exposed to acid-treatment in the presence or absence of pepsin, yielding a complex mixture of self-assembling collagen as well as other ECM components (glycosaminoglycans, proteoglycans, growth factors, fibronectin). It is noteworthy that that dECM bioink composition can vary widely since it depends largely on the decellularization and solubilization protocols employed. Furthermore, removal of cells and their associated components is essential so to avoid elicitation of immune-mediated responses when used in vivo [9]. At present, most dECM bioinks form soft hydrogels, therefore the use of exogenous crosslinking in commonplace.

To date dECM bioinks have been derived from various tissues and organs including heart, liver, fat, cartilage, skeletal muscle, skin, and vascular tissue. For a more comprehensive review of dECM bioink use in 3D bioprinting see [45].

5.2.4   Extrusion, Electrochemical Processes, and Magnetic Fields

One of the first applications of flow to induce preferential alignment of collagen fibrils was provided by Elsdale and Bard in 1972 [58]. This method, referred to as the "draining method", involved pipetting polymerizable collagen into a dish and placing the dish at an incline to achieve gravity-induced flow and aligned bundles of fibrillar collagen. These findings have been extended to more scalable, industrial processes such as extrusion. Extrusion is formally defined as the act or process of shaping a material by forcing it through a die. In the late 1980s and early 1990s, Kato and co-workers described a scalable process for collagen fiber production that involved extrusion of acid-swollen dispersions of insoluble fibrillar collagen through polyethylene tubing into a phosphate-based buffer reservoir to induce gelation (Fig. 9a) [103, 105]. The resultant fibers were then transferred to isopropyl alcohol followed by air drying under tension. Chemical and physical crosslinking resulted in fibers with ultimate tensile strength values that were comparable to those for rat tail tendon fibers (24–66 MPa). When fashioned and implanted as tendon and ligaments, implants showed inflammatory reactions, degradation profiles, and neotissue formation that varied with type of crosslinking [57, 69, 104, 206]. More recently, a similar approach was applied to soluble telocollagen and atelocollagen formulations, yielding "strings" of flow-aligned collagen fibrils (Fig. 9b) [91, 114]. Although this process could be applied to yield a wide variety of geometries and patterns, including sheets, meshes, and tubes, functional mechanical properties for tissue engineering applications have yet to be achieved.

Taking advantage of the rapid polymerization and improved mechanical integrity of type I oligomeric collagen, Brookes and co-workers described methods of extruding self-assembling oligomer solutions in the presence of muscle progenitor cells for creating engineered skeletal muscle for laryngeal reconstruction (Fig. 9c). This process yielded mechanically stable constructs with aligned cells surrounded by highly aligned collagen fibrils [28]. Resident muscle progenitor cells readily fused, forming multi-nucleated myotubes upon culture in vitro. When used for laryngeal muscle reconstruction in a rat hemi-laryngectomy model, these tissue-engineered muscle constructs integrated with the surrounding host tissue in absence of a significant inflammatory response. Furthermore, functional muscle regeneration and maturation occurred over a 3 month period marked by progressive increases in striations, innervation, and functional motor unit activity [28]. Other recently reported methods for achieving aligned cellularized collagen constructs include the multi-step process referred to as gel aspiration-ejection [129]. Here, isotropic, densified collagen constructs are aspirated into a syringe and then ejected through capillary tubes, 0.3–0.9 mm in diameter. Initial in vitro studies showed that constructs formed by this process containing mesenchymal stem cells showed accelerated osteoblast

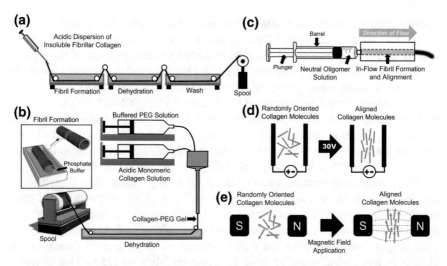

Fig. 9 Extrusion processes for production of collagen threads and aligned collagen. **a** Collagen fiber formation from acidic dispersions of insoluble fibrillar collagen (based on Kato et al. [103]). Collagen dispersions are extruded into a sodium phosphate based fiber formation buffer. Resulting threads are sequentially dehydrated in isopropyl alcohol, washed in water, and air dried prior to spooling. **b** Wet spinning of collagen fibers (adapted from Caves et al. [38]). An acidic solution of collagen monomers is aggregated into a gel-like fiber by mixing with a buffered PEG solution. The extruded fiber is dehydrated in ethanol prior to collection on a spool. Resulting threads are exposed to phosphate buffer to induce formation of D-banded collagen fibrils, rinsed, and then air dried prior to spooling. **c** In-flow collagen fibril formation and alignment (based on Brookes et al. [28]). Neutral solutions of oligomer collagen in the presence or absence of cells are extruded through a die, resulting in alignment of self-assembled collagen fibrils and resident cells. **d** Electrochemical aggregation and alignment of collagen. Soluble collagen molecules are placed within an electrochemical cell consisting of two parallel electrode wires. Isoelectric focusing occurs with application of DC voltage resulting in molecular accumulation into compacted threads. Formation of D-banded collagen fibrils occurs when resulting collagen thread is placed in phosphate buffered saline. **e** Magnetic alignment of collagen as occurs when neutralized collagen solutions are placed within high strength magnetic fields. Mechanical torque on molecules results in alignment orthogonal to the applied field

and neuronal differentiation when cultured in the appropriate differentiation medium formulations.

Methods other than extrusion have been used to create anisotropic collagen constructs. Specifically, collagen monomer solutions have been exposed to electrochemical processes, where isoelectric focusing is used to drive aggregation of collagen molecules (Fig. 9d). While this process does not produce the staggered arrangement of molecules observed in native collagen fibrils, D-spacing can be achieved with exposure to phosphate buffered saline [44]. Follow-up processing of these electrochemically aligned collagen threads by sequential treatment with genipin crosslinking, peracetic acid/ethanol exposure, and heparinization in EDC/NHS yields heparinized sutures that can be used for growth factors such as platelet derived growth factor [221, 222]. On the other hand, application of large magnetic fields to polymerizable collagens, which was first described in the 1980s [140, 199], orients collagen molecules and associated fibril-forming counterparts perpendicular to the applied field (Fig. 9e). This outcome is largely attributed to diamagnetism of the peptide bond [214]. Since that time, magnetic fields have been applied to generate anisotropic constructs for mechanistic studies of cell contact guidance [75] as well as generating tissue-engineered constructs cartilage [150], cornea [32, 198], and peripheral nerve replacement [32, 198] and regeneration. Notable findings from this work, was that orthogonal patterns of collagen fibrils, similar to those found in native cornea stroma, could be generated by polymerization-rotation-polymerization of sequential layers of collagen in the presence of a magnetic field [198]. Resident cells, whether grown in culture or infiltrating from surrounding tissue following implantation, align by contact guidance along the long axis of the fibrils. Interestingly, magnetically aligned constructs produced with atelocollagen and telocollagen showed improved handling and mechanical properties upon exogenous crosslinking [44]. However, magnetically aligned constructs produced with oligomeric collagen showed no significant change in mechanical integrity upon chemical crosslinking [150]. This observation was attributed to the fact that compared to monomers, oligomer produces more mechanically stable fibril microstructures with increased connectivity between fibrils (interfibril branching).

6 Conclusion and Future Directions

To date, tissue engineering and regenerative medicine approaches involving collagen-based scaffolds, cells and combinations thereof have led to a number of new, FDA-approved therapies. However, many would say that the field, in general, has still not lived up to promises and enthusiasm generated early on. The ability to replace or regenerate damaged or diseased tissues and organs remains one of the great challenges and unmet needs facing medicine and society. Continued translation and commercialization of next generation therapies must forge new pathways that interface biomolecules and cells, scalable manufacturing processes, and regulatory policy. Careful consideration of the scientific, regulatory, and business hurdles is paramount in streamlining translation and maximizing clinical impact. Integration of computational modeling for predictable, customizable design will facilitate precision medicine, applications, which works to account for the inevitable variability in health status and intrinsic healing/remodeling potential between patients. Translating these biomedical advances to medical successes will help fulfill the long-standing promise of tissue engineering and regenerative medicine to patients, clinicians, investors, and society.

References

1. Abraham, L.C., Zuena, E., Perez-Ramirez, B., Kaplan, D.L.: Guide to collagen characterization for biomaterial studies. J. Biomed. Mater. Res. B Appl. Biomater. **87**(1), 264–285 (2008)
2. Achilli, M., Mantovani, D.: Tailoring mechanical properties of collagen-based scaffolds for vascular tissue engineering: the effects of pH, temperature and ionic strength on gelation. Polymers **2**(4), 664–680 (2010)
3. Aguado, B.A., Grim, J.C., Rosales, A.M., Watson-Capps, J.J., Anseth, K.S.: Engineering precision biomaterials for personalized medicine. Sci. Transl. Med. **10**(424), eaam8645 (2018)
4. An, B., Kaplan, D.L., Brodsky, B.: Engineered recombinant collagen as an alternative collagen-based biomaterial for tissue engineering. Front. Chem. **2**, 40 (2014)
5. Antoine, E.E., Vlachos, P.P., Rylander, M.N.: Review of collagen I hydrogels for bioengineered tissue microenvironments: characterization of mechanics, structure, and transport. Tissue Eng. Part B Rev. **20**(6), 683–696 (2014)
6. Antoine, E.E., Vlachos, P.P., Rylander, M.N.: Tunable collagen I hydrogels for engineered physiological tissue micro-environments. PLoS ONE **10**(3), e0122500 (2015)
7. Avery, N.C., Bailey, A.J.: Restraining cross-links responsible for the mechanical properties of collagen fibers: natural and artificial. In: Fratzl, P. (ed.) Collagen Structure and Mechanics, pp. 81–110. Springer, New York (2008)
8. BCCResearch: Tissue Engineering and Regeneration: Technologies and Global Markets (2016)
9. Badylak, S.F., Gilbert, T.W.: Immune response to biologic scaffold materials. Semin. Immunol. **20**(2), 109–116 (2008)
10. Badylak, S.F., Lantz, G.C., Coffey, A., Geddes, L.A.: Small intestinal submucosa as a large diameter vascular graft in the dog. J. Surg. Res. **47**(1), 74–80 (1989)
11. Baez, J., Olsen, D., Polarek, J.W.: Recombinant microbial systems for the production of human collagen and gelatin. Appl. Microbiol. Biotechnol. **69**(3), 245–252 (2005)
12. Bailey, J.L., Critser, P.J., Whittington, C., Kuske, J.L., Yoder, M.C., Voytik-Harbin, S.L.: Collagen oligomers modulate physical and biological properties of three-dimensional self-assembled matrices. Biopolymers **95**(2), 77–93 (2011)
13. Bell, E., Ivarsson, B., Merrill, C.: Production of a tissue-like structure by contraction of collagen lattices by human fibroblasts of different proliferative potential in vitro. Proc. Natl. Acad. Sci. **76**(3), 1274–1278 (1979)
14. Bell, B.J., Nauman, E., Voytik-Harbin, S.L.: Multiscale strain analysis of tissue equivalents using a custom-designed biaxial testing device. Biophys. J. **102**(6), 1303–1312 (2012)
15. Bell, E., Sher, S., Hull, B., Merrill, C., Rosen, S., Chamson, A., Asselineau, D., Dubertret, L., Coulomb, B., Lapiere, C., Nusgens, B., Neveux, Y.: The reconstitution of living skin. J. Invest. Dermatol. **81**(1 Suppl), 2s–10s (1983)
16. Benjamin, M., Kaiser, E., Milz, S.: Structure-function relationships in tendons: a review. J. Anat. **212**(3), 211–228 (2008)
17. Berger, A.J., Linsmeier, K.M., Kreeger, P.K., Masters, K.S.: Decoupling the effects of stiffness and fiber density on cellular behaviors via an interpenetrating network of gelatin-methacrylate and collagen. Biomaterials **141**, 125–135 (2017)
18. Binet, J.P., Carpentier, A., Langlois, J., Duran, C., Colvez, P.: Implantation of heterogenic valves in the treatment of aortic cardiopathies. C. R. Acad. Sci. Hebd. Seances Acad. Sci. D **261**(25), 5733–5734 (1965)
19. Birk, D.E., Fitch, J.M., Babiarz, J.P., Doane, K.J., Linsenmayer, T.F.: Collagen fibrillogenesis in vitro: interaction of types I and V collagen regulates fibril diameter. J. Cell Sci. **95**(4), 649–657 (1990)
20. Bitar, M., Salih, V., Brown, R.A., Nazhat, S.N.: Effect of multiple unconfined compression on cellular dense collagen scaffolds for bone tissue engineering. J. Mater. Sci. Mater. Med. **18**(2), 237–244 (2007)

21. Blum, K.M., Novak, T., Watkins, L., Neu, C.P., Wallace, J.M., Bart, Z.R., Voytik-Harbin, S.L.: Acellular and cellular high-density, collagen-fibril constructs with suprafibrillar organization. Biomater. Sci. **4**(4), 711–723 (2016)

22. Boland, E.D., Matthews, J.A., Pawlowski, K.J., Simpson, D.G., Wnek, G.E., Bowlin, G.L.: Electrospinning collagen and elastin: preliminary vascular tissue engineering. Front. Biosci. **9**, 1422–1432 (2004)

23. Bozec, L., Horton, M.: Topography and mechanical properties of single molecules of type I collagen using atomic force microscopy. Biophys. J. **88**(6), 4223–4231 (2005)

24. Braziulis, E., Diezi, M., Biedermann, T., Pontiggia, L., Schmucki, M., Hartmann-Fritsch, F., Luginbuhl, J., Schiestl, C., Meuli, M., Reichmann, E.: Modified plastic compression of collagen hydrogels provides an ideal matrix for clinically applicable skin substitutes. Tissue Eng. Part C Methods **18**(6), 464–474 (2012)

25. Brennan, M., Davison, P.F.: Role of aldehydes in collagen fibrillogenesis in vitro. Biopolymers **19**(10), 1861–1873 (1980)

26. Brodsky, B., Ramshaw, J.A.: Bioengineered collagens. Fibrous Proteins: Structures and Mechanisms, pp. 601–629. Springer, Berlin (2017)

27. Brokaw, J.L., Doillon, C.J., Hahn, R.A., Birk, D.E., Berg, R.A., Silver, F.H.: Turbidimetric and morphological studies of type I collagen fibre self assembly in vitro and the influence of fibronectin. Int. J. Biol. Macromol. **7**(3), 135–140 (1985)

28. Brookes, S., Voytik-Harbin, S., Zhang, H., Halum, S.: Three-dimensional tissue-engineered skeletal muscle for laryngeal reconstruction. Laryngoscope **128**(3), 603–609 (2018)

29. Brown, I.A.: Scanning electron microscopy of human dermal fibrous tissue. J. Anat. **113**(Pt 2), 159–168 (1972)

30. Brown, R.A., Wiseman, M., Chuo, C.B., Cheema, U., Nazhat, S.N.: Ultrarapid engineering of biomimetic materials and tissues: fabrication of nano- and microstructures by plastic compression. Adv. Funct. Mater. **15**(11), 1762–1770 (2005)

31. Buehler, M.J.: Nature designs tough collagen: explaining the nanostructure of collagen fibrils. Proc. Natl. Acad. Sci. **103**(33), 12285–12290 (2006)

32. Builles, N., Janin-Manificat, H., Malbouyres, M., Justin, V., Rovere, M.R., Pellegrini, G., Torbet, J., Hulmes, D.J., Burillon, C., Damour, O., Ruggiero, F.: Use of magnetically oriented orthogonal collagen scaffolds for hemi-corneal reconstruction and regeneration. Biomaterials **31**(32), 8313–8322 (2010)

33. Burkus, J.K., Transfeldt, E.E., Kitchel, S.H., Watkins, R.G., Balderston, R.A.: Clinical and radiographic outcomes of anterior lumbar interbody fusion using recombinant human bone morphogenetic protein-2. Spine (Phila Pa 1976) **27**(21), 2396–2408 (2002)

34. Butler, D.L., Goldstein, S.A., Guilak, F.: Functional tissue engineering: the role of biomechanics. J. Biomech. Eng. **122**(6), 570–575 (2000)

35. Buttafoco, L., Kolkman, N.G., Engbers-Buijtenhuijs, P., Poot, A.A., Dijkstra, P.J., Vermes, I., Feijen, J.: Electrospinning of collagen and elastin for tissue engineering applications. Biomaterials **27**(5), 724–734 (2006)

36. Canty, E.G., Kadler, K.E.: Procollagen trafficking, processing and fibrillogenesis. J. Cell Sci. **118**(Pt 7), 1341–1353 (2005)

37. Carpentier, A., Lemaigre, G., Robert, L., Carpentier, S., Dubost, C.: Biological factors affecting long-term results of valvular heterografts. J. Thorac. Cardiovasc. Surg. **58**(4), 467–483 (1969)

38. Caves, J.M., Kumar, V.A., Wen, J., Cui, W., Martinez, A., Apkarian, R., Coats, J.E., Berland, K., Chaikof, E.L.: Fibrillogenesis in continuously spun synthetic collagen fiber. J. Biomed. Mater. Res. B Appl. Biomater. **93**(1), 7175–7182 (2009)

39. Chandrakasan, G., Torchia, D.A., Piez, K.A.: Preparation of intact monomeric collagen from rat tail tendon and skin and the structure of the nonhelical ends in solution. J. Biol. Chem. **251**(19), 6062–6067 (1976)

40. Chang, Y., Tsai, C.C., Liang, H.C., Sung, H.W.: In vivo evaluation of cellular and acellular bovine pericardia fixed with a naturally occurring crosslinking agent (genipin). Biomaterials **23**(12), 2447–2457 (2002)

41. Chattopadhyay, S., Raines, R.T.: Review collagen-based biomaterials for wound healing. Biopolymers **101**(8), 821–833 (2014)
42. Cheema, U., Brown, R.A.: Rapid fabrication of living tissue models by collagen plastic compression: understanding three-dimensional cell matrix repair in vitro. Adv. Wound Care **2**(4), 176–184 (2013)
43. Chen, Y.C., Lin, R.Z., Qi, H., Yang, Y., Bae, H., Melero-Martin, J.M., Khademhosseini, A.: Functional human vascular network generated in photocrosslinkable gelatin methacrylate hydrogels. Adv. Funct. Mater. **22**(10), 2027–2039 (2012)
44. Cheng, X., Gurkan, U.A., Dehen, C.J., Tate, M.P., Hillhouse, H.W., Simpson, G.J., Akkus, O.: An electrochemical fabrication process for the assembly of anisotropically oriented collagen bundles. Biomaterials **29**(22), 3278–3288 (2008)
45. Choudhury, D., Tun, H.W., Wang, T., Naing, M.W.: Organ-derived decellularized extracellular matrix: a game changer for bioink manufacturing? Trends Biotechnol. **36**(8), 787–805 (2018)
46. Chung, L., Maestas Jr., D.R., Housseau, F., Elisseeff, J.H.: Key players in the immune response to biomaterial scaffolds for regenerative medicine. Adv. Drug Del. Rev. **114**, 184–192 (2017)
47. Critser, P.J., Kreger, S.T., Voytik-Harbin, S.L., Yoder, M.C.: Collagen matrix physical properties modulate endothelial colony forming cell-derived vessels in vivo. Microvasc. Res. **80**(1), 23–30 (2010)
48. Cronlund, A.L., Smith, B.D., Kagan, H.M.: Binding of lysyl oxidase to fibrils of type I collagen. Connect. Tissue Res. **14**(2), 109–119 (1985)
49. Crump, S.S.: Apparatus and method for creating three-dimensional objects. 5,121,329, 9 June 1992
50. da Cunha, C.B., Klumpers, D.D., Li, W.A., Koshy, S.T., Weaver, J.C., Chaudhuri, O., Granja, P.L., Mooney, D.J.: Influence of the stiffness of three-dimensional alginate/collagen-I interpenetrating networks on fibroblast biology. Biomaterials **35**(32), 8927–8936 (2014)
51. Dagalakis, N., Flink, J., Stasikelis, P., Burke, J.F., Yannas, I.V.: Design of an artificial skin. Part III. Control of pore structure. J. Biomed. Mater. Res. **14**(4), 511–528 (1980)
52. Damink, L.O., Dijkstra, P.J., Van Luyn, M., Van Wachem, P., Nieuwenhuis, P., Feijen, J.: Glutaraldehyde as a crosslinking agent for collagen-based biomaterials. J. Mater. Sci. Mater. Med. **6**(8), 460–472 (1995)
53. Datta, P., Barui, A., Wu, Y., Ozbolat, V., Moncal, K.K., Ozbolat, I.T.: Essential steps in bioprinting: from pre- to post-bioprinting. Biotechnol. Adv. **36**(5), 1481–1504 (2018)
54. DeFrates, K.G., Moore, R., Borgesi, J., Lin, G., Mulderig, T., Beachley, V., Hu, X.: Protein-based fiber materials in medicine: a review. Nanomaterials (Basel) **8**(7), 457 (2018)
55. Delgado, L.M., Bayon, Y., Pandit, A., Zeugolis, D.I.: To cross-link or not to cross-link? Cross-linking associated foreign body response of collagen-based devices. Tissue Eng. Part B Rev. **21**(3), 298–313 (2015)
56. Dietz U.A., Kehl F., Hamelmann W., Weißer C.: On the 100th anniversary of sterile catgut Kuhn: franz Kuhn (1866–1929) and the epistemology of catgut sterilization. World J. Surg. **31**(12), 2275–2283 (2007)
57. Dunn, M.G., Tria, A.J., Kato, Y.P., Bechler, J.R., Ochner, R.S., Zawadsky, J.P., Silver, F.H.: Anterior cruciate ligament reconstruction using a composite collagenous prosthesis. Am. J. Sports Med. **20**(5), 507–515 (1992)
58. Elsdale, T., Bard, J.: Collagen substrata for studies on cell behavior. J. Cell Biol. **54**(3), 626–637 (1972)
59. Evans, F.G., Vincentelli, R.: Relation of collagen fiber orientation to some mechanical properties of human cortical bone. J. Biomech. **2**(1), 63–71 (1969)
60. Eyre, D.R., Paz, M.A., Gallop, P.M.: Cross-linking in collagen and elastin. Annu. Rev. Biochem. **53**, 717–748 (1984)
61. Eyre, D.R., Wu, J.J.: Collagen cross-links. In: Collagen, vol. 247, pp. 207–229 (2005)
62. Fang, M., Goldstein, E.L., Turner, A.S., Les, C.M., Orr, B.G., Fisher, G.J., Welch, K.B., Rothman, E.D., Holl, M.M.B.: Type I collagen D-spacing in fibril bundles of dermis, tendon, and bone: bridging between nano- and micro-level tissue hierarchy. A.C.S. Nano **6**(11), 9503–9514 (2012)

63. Fratzl, P.: Collagen: structure and mechanics, an introduction. Collagen, pp. 1–13. Springer, Berlin (2008)
64. Freed, L.E., Vunjak-Novakovic, G., Biron, R.J., Eagles, D.B., Lesnoy, D.C., Barlow, S.K., Langer, R.: Biodegradable polymer scaffolds for tissue engineering. Biotechnology (N. Y.) 12(7), 689–693 (1994)
65. Friedlaender, G.E., Perry, C.R., Cole, J.D., Cook, S.D., Cierny, G., Muschler, G.F., Zych, G.A., Calhoun, J.H., LaForte, A.J., Yin, S.: Osteogenic protein-1 (bone morphogenetic protein-7) in the treatment of tibial nonunions—A prospective, randomized clinical trial comparing rhOP-1 with fresh bone autograft. J. Bone Joint Surg. Am. 83a(Pt 2), S151–S158 (2001)
66. Garg, A.K., Berg, R.A., Silver, F.H., Garg, H.G.: Effect of proteoglycans on type I collagen fibre formation. Biomaterials 10(6), 413–419 (1989)
67. Gibson, T.: Evolution of catgut ligatures: the endeavours and success of Joseph Lister and William Macewen. Br. J. Surg. 77(7), 824–825 (1990)
68. Gildner, C.D., Lerner, A.L., Hocking, D.C.: Fibronectin matrix polymerization increases tensile strength of model tissue. Am. J. Physiol. Heart Circ. Physiol. 287(1), H46–H53 (2004)
69. Goldstein, J.D., Tria, A.J., Zawadsky, J.P., Kato, Y.P., Christiansen, D., Silver, F.H.: Development of a reconstituted collagen tendon prosthesis. A preliminary implantation study. J. Bone Joint Surg. Am. 71(8), 1183–1191 (1989)
70. Golser, A.V., Scheibel, T.: Routes towards novel collagen-like biomaterials. Fibers 6(2), 21 (2018)
71. Green, J.J., Elisseeff, J.H.: Mimicking biological functionality with polymers for biomedical applications. Nature 540(7633), 386–394 (2016)
72. Grinnell, F.: Fibroblast biology in three-dimensional collagen matrices. Trends Cell Biol. 13(5), 264–269 (2003)
73. Gross, J., Highberger, J.H., Schmitt, F.O.: Collagen structures considered as states of aggregation of a kinetic unit—the tropocollagen particle. Proc. Natl. Acad. Sci. 40(8), 679 (1954)
74. Gross, J., Kirk, D.: The heat precipitation of collagen from neutral salt solutions: some rate-regulating factors. J. Biol. Chem. 233(2), 355–360 (1958)
75. Guido, S., Tranquillo, R.T.: A methodology for the systematic and quantitative study of cell contact guidance in oriented collagen gels. Correlation of fibroblast orientation and gel birefringence. J. Cell Sci. 105(Pt 2), 317–331 (1993)
76. Guilak, F., Butler, D.L., Goldstein, S.A., Baaijens, F.P.: Biomechanics and mechanobiology in functional tissue engineering. J. Biomech. 47(9), 1933–1940 (2014)
77. Hall, M.S., Alisafaei, F., Ban, E., Feng, X., Hui, C.Y., Shenoy, V.B., Wu, M.: Fibrous nonlinear elasticity enables positive mechanical feedback between cells and ECMs. Proc. Natl. Acad. Sci. 113(49), 14043–14048 (2016)
78. Hapach, L.A., VanderBurgh, J.A., Miller, J.P., Reinhart-King, C.A.: Manipulation of in vitro collagen matrix architecture for scaffolds of improved physiological relevance. Phys. Biol. 12(6), 061002 (2015)
79. Helseth, D., Veis, A.: Collagen self-assembly in vitro. Differentiating specific telopeptide-dependent interactions using selective enzyme modification and the addition of free amino telopeptide. J. Biol. Chem. 256(14), 7118–7128 (1981)
80. Herchenhan, A., Uhlenbrock, F., Eliasson, P., Weis, M., Eyre, D., Kadler, K.E., Magnusson, S.P., Kjaer, M.: Lysyl oxidase activity is required for ordered collagen fibrillogenesis by tendon cells. J. Biol. Chem. 290(26), 16440–16450 (2015)
81. Herrera-Perez, M., Voytik-Harbin, S.L., Rickus, J.L.: Extracellular matrix properties regulate the migratory response of glioblastoma stem cells in three-dimensional culture. Tissue Eng. Part A 21(19–20), 2572–2582 (2015)
82. Holmes, R., Kirk, S., Tronci, G., Yang, X., Wood, D.: Influence of telopeptides on the structural and physical properties of polymeric and monomeric acid-soluble type I collagen. Mater. Sci. Eng. C Mater. Biol. Appl. 77, 823–827 (2017)
83. Hosseini, S.M., Wilson, W., Ito, K., van Donkelaar, C.C.: How preconditioning affects the measurement of poro-viscoelastic mechanical properties in biological tissues. Biomech. Model. Mechanbiol. 13(3), 503–513 (2014)

84. Hu, Y.H., Suo, Z.G.: Viscoelasticity and poroelasticity in elastomeric gels. Acta Mech. Solida Sin. **25**(5), 441–458 (2012)
85. Huang, L., Apkarian, R.P., Chaikof, E.L.: High-resolution analysis of engineered type I collagen nanofibers by electron microscopy. Scanning **23**(6), 372–375 (2001)
86. Huang, L., Nagapudi, K., Apkarian, R.P., Chaikof, E.L.: Engineered collagen-PEO nanofibers and fabrics. J. Biomater. Sci. Polym. Ed. **12**(9), 979–993 (2001)
87. Hulmes, D.J., Wess, T.J., Prockop, D.J., Fratzl, P.: Radial packing, order, and disorder in collagen fibrils. Biophys. J. **68**(5), 1661–1670 (1995)
88. Ingber, D.E., Mow, V.C., Butler, D., Niklason, L., Huard, J., Mao, J., Yannas, I., Kaplan, D., Vunjak-Novakovic, G.: Tissue engineering and developmental biology: going biomimetic. Tissue Eng. **12**(12), 3265–3283 (2006)
89. Ingber, D.E., Wang, N., Stamenovic, D.: Tensegrity, cellular biophysics, and the mechanics of living systems. Rep. Prog. Phys. **77**(4), 046603 (2014)
90. Ionescu, M.I., Pakrashi, B.C., Holden, M.P., Mary, D.A., Wooler, G.H.: Results of aortic valve replacement with frame-supported fascia lata and pericardial grafts. J. Thorac. Cardiovasc. Surg. **64**(3), 340–353 (1972)
91. Isobe, Y., Kosaka, T., Kuwahara, G., Mikami, H., Saku, T., Kodama, S.: Oriented collagen scaffolds for tissue engineering. Materials **5**(3), 501–511 (2012)
92. Jackson, D.S., Fessler, J.H.: Isolation and properties of a collagen soluble in salt solution at neutral pH. Nature **176**(4471), 69–70 (1955)
93. Jamroz, W., Szafraniec, J., Kurek, M., Jachowicz, R.: 3D printing in pharmaceutical and medical applications—Recent achievements and challenges. Pharm. Res. **35**(9), 176 (2018)
94. Jang, J., Park, H.J., Kim, S.W., Kim, H., Park, J.Y., Na, S.J., Kim, H.J., Park, M.N., Choi, S.H., Park, S.H., Kim, S.W., Kwon, S.M., Kim, P.J., Cho, D.W.: 3D printed complex tissue construct using stem cell-laden decellularized extracellular matrix bioinks for cardiac repair. Biomaterials **112**, 264–274 (2017)
95. Ji, S., Guvendiren, M.: Recent advances in bioink design for 3D bioprinting of tissues and organs. Front Bioeng. Biotechnol. **5**, 23 (2017)
96. John, D.C., Watson, R., Kind, A.J., Scott, A.R., Kadler, K.E., Bulleid, N.J.: Expression of an engineered form of recombinant procollagen in mouse milk. Nat. Biotechnol. **17**(4), 385–389 (1999)
97. Johnson, K.R., Leight, J.L., Weaver, V.M.: Demystifying the effects of a three-dimensional microenvironment in tissue morphogenesis. Methods Cell Biol. **83**, 547–583 (2007)
98. Julier, Z., Park, A.J., Briquez, P.S., Martino, M.M.: Promoting tissue regeneration by modulating the immune system. Acta Biomater. **53**, 13–28 (2017)
99. Kablik, J., Monheit, G.D., Yu, L., Chang, G., Gershkovich, J.: Comparative physical properties of hyaluronic acid dermal fillers. Dermatol. Surg. **35**(Suppl 1), 302–312 (2009)
100. Kadler, K.E.: Fell muir lecture: collagen fibril formation in vitro and in vivo. Int. J. Exp. Pathol. **98**(1), 4–16 (2017)
101. Kadler, K.E., Hill, A., Canty-Laird, E.G.: Collagen fibrillogenesis: fibronectin, integrins, and minor collagens as organizers and nucleators. Curr. Opin. Cell Biol. **20**(5), 495–501 (2008)
102. Kadler, K.E., Holmes, D.F., Graham, H., Starborg, T.: Tip-mediated fusion involving unipolar collagen fibrils accounts for rapid fibril elongation, the occurrence of fibrillar branched networks in skin and the paucity of collagen fibril ends in vertebrates. Matrix Biol. **19**(4), 359–365 (2000)
103. Kato, Y.P., Christiansen, D.L., Hahn, R.A., Shieh, S.J., Goldstein, J.D., Silver, F.H.: Mechanical properties of collagen fibres: a comparison of reconstituted and rat tail tendon fibres. Biomaterials **10**(1), 38–42 (1989)
104. Kato, Y.P., Dunn, M.G., Zawadsky, J.P., Tria, A.J., Silver, F.H.: Regeneration of Achilles tendon with a collagen tendon prosthesis. Results of a one-year implantation study. J. Bone Joint Surg. Am. **73**(4), 561–574 (1991)
105. Kato, Y.P., Silver, F.H.: Formation of continuous collagen fibres: evaluation of biocompatibility and mechanical properties. Biomaterials **11**(3), 169–175 (1990)

106. Kim, B.S., Kim, H., Gao, G., Jang, J., Cho, D.W.: Decellularized extracellular matrix: a step towards the next generation source for bioink manufacturing. Biofabrication **9**(3), 034104 (2017)
107. Kim, B.S., Mooney, D.J.: Development of biocompatible synthetic extracellular matrices for tissue engineering. Trends Biotechnol. **16**(5), 224–230 (1998)
108. Knorzer, E., Folkhard, W., Geercken, W., Boschert, C., Koch, M.H.J., Hilbert, B., Krahl, H., Mosler, E., Nemetschekgansler, H., Nemetschek, T.: New aspects of the etiology of tendon-rupture—an analysis of time-resolved dynamic-mechanical measurements using synchrotron radiation. Arch. Orthop. Trauma Surg. **105**(2), 113–120 (1986)
109. Knott, L., Bailey, A.J.: Collagen cross-links in mineralizing tissues: a review of their chemistry, function, and clinical relevance. Bone **22**(3), 181–187 (1998)
110. Knowlton, S., Onal, S., Yu, C.H., Zhao, J.J., Tasoglu, S.: Bioprinting for cancer research. Trends Biotechnol. **33**(9), 504–513 (2015)
111. Kotch, F.W., Raines, R.T.: Self-assembly of synthetic collagen triple helices. Proc. Natl. Acad. Sci. **103**(9), 3028–3033 (2006)
112. Kreger, S.T., Bell, B.J., Bailey, J., Stites, E., Kuske, J., Waisner, B., Voytik-Harbin, S.L.: Polymerization and matrix physical properties as important design considerations for soluble collagen formulations. Biopolymers **93**(8), 690–707 (2010)
113. Kubow, K.E., Vukmirovic, R., Zhe, L., Klotzsch, E., Smith, M.L., Gourdon, D., Luna, S., Vogel, V.: Mechanical forces regulate the interactions of fibronectin and collagen I in extracellular matrix. Nat. Commun. **6**, 8026 (2015)
114. Lai, E.S., Anderson, C.M., Fuller, G.G.: Designing a tubular matrix of oriented collagen fibrils for tissue engineering. Acta Biomater. **7**(6), 2448–2456 (2011)
115. Langer, R., Vacanti, J.P.: Tissue engineering. Science **260**(5110), 920–926 (1993)
116. Lanir, Y., Fung, Y.C.: Two-dimensional mechanical properties of rabbit skin. I. Experimental system. J. Biomech. **7**(1), 29–34 (1974)
117. Lanir, Y., Fung, Y.C.: Two-dimensional mechanical properties of rabbit skin. II. Experimental results. J. Biomech. **7**(2), 171–182 (1974)
118. Lees, J.F., Tasab, M., Bulleid, N.J.: Identification of the molecular recognition sequence which determines the type-specific assembly of procollagen. EMBO J. **16**(5), 908–916 (1997)
119. Leitinger, B., Hohenester, E.: Mammalian collagen receptors. Matrix Biol. **26**(3), 146–155 (2007)
120. Liang, Y., Jeong, J., DeVolder, R.J., Cha, C., Wang, F., Tong, Y.W., Kong, H.: A cell-instructive hydrogel to regulate malignancy of 3D tumor spheroids with matrix rigidity. Biomaterials **32**(35), 9308–9315 (2011)
121. Lister, J.: An address on the catgut ligature. Br. Med. J. **1**(1050), 219–221 (1881)
122. Lysaght, M.J., Nguy, N.A., Sullivan, K.: An economic survey of the emerging tissue engineering industry. Tissue Eng. **4**(3), 231–238 (1998)
123. Lysaght, M.J., Reyes, J.: The growth of tissue engineering. Tissue Eng. **7**(5), 485–493 (2001)
124. Mackenzie, D.: The history of sutures. Med. Hist. **17**(2), 158–168 (1973)
125. Maki, J.M., Sormunen, R., Lippo, S., Kaarteenaho-Wiik, R., Soininen, R., Myllyharju, J.: Lysyl oxidase is essential for normal development and function of the respiratory system and for the integrity of elastic and collagen fibers in various tissues. Am. J. Pathol. **167**(4), 927–936 (2005)
126. Mammoto, T., Mammoto, A., Ingber, D.E.: Mechanobiology and developmental control. Annu. Rev. Cell Dev. Biol. **29**, 27–61 (2013)
127. Manji, R.A., Lee, W., Cooper, D.K.C.: Xenograft bioprosthetic heart valves: past, present and future. Int. J. Surg. **23**(Pt B), 280–284 (2015)
128. Mao, A.S., Mooney, D.J.: Regenerative medicine: current therapies and future directions. Proc. Natl. Acad. Sci. **112**(47), 14452–14459 (2015)
129. Marelli, B., Ghezzi, C.E., James-Bhasin, M., Nazhat, S.N.: Fabrication of injectable, cellular, anisotropic collagen tissue equivalents with modular fibrillar densities. Biomaterials **37**, 183–193 (2015)

130. Matthews, J.A., Wnek, G.E., Simpson, D.G., Bowlin, G.L.: Electrospinning of collagen nanofibers. Biomacromolecules 3(2), 232–238 (2002)
131. McKee, C.T., Last, J.A., Russell, P., Murphy, C.J.: Indentation versus tensile measurements of Young's modulus for soft biological tissues. Tissue Eng. Part B Rev. 17(3), 155–164 (2011)
132. McPherson, J.M., Wallace, D.G., Sawamura, S.J., Conti, A., Condell, R.A., Wade, S., Piez, K.A.: Collagen fibrillogenesis in vitro: a characterization of fibril quality as a function of assembly conditions. Coll. Relat. Res. 5(2), 119–135 (1985)
133. Meek, K.M., Knupp, C.: Corneal structure and transparency. Prog. Retin. Eye Res. 49, 1–16 (2015)
134. Miller, E.J., Rhodes, R.K.: Preparation and characterization of the different types of collagen. Methods Enzymol. 82(Pt A), 33–64 (1982)
135. Ming-Che, W., Pins, G.D., Silver, F.H.: Collagen fibres with improved strength for the repair of soft tissue injuries. Biomaterials 15(7), 507–512 (1994)
136. Motte, S., Kaufman, L.J.: Strain stiffening in collagen I networks. Biopolymers 99(1), 35–46 (2013)
137. Muiznieks, L.D., Keeley, F.W.: Molecular assembly and mechanical properties of the extracellular matrix: a fibrous protein perspective. Biochim. Biophys. Acta 1832(7), 866–875 (2013)
138. Munster, S., Jawerth, L.M., Leslie, B.A., Weitz, J.I., Fabry, B., Weitz, D.A.: Strain history dependence of the nonlinear stress response of fibrin and collagen networks. Proc. Natl. Acad. Sci. 110(30), 12197–12202 (2013)
139. Murphy, S.V., Atala, A.: 3D bioprinting of tissues and organs. Nat. Biotechnol. 32(8), 773–785 (2014)
140. Murthy, N.S.: Liquid crystallinity in collagen solutions and magnetic orientation of collagen fibrils. Biopolymers 23(7), 1261–1267 (1984)
141. Myllyharju, J., Kivirikko, K.I.: Collagens and collagen-related diseases. Ann. Med. 33(1), 7–21 (2001)
142. Na, G.C., Butz, L.J., Bailey, D.G., Carroll, R.J.: In vitro collagen fibril assembly in glycerol solution: evidence for a helical cooperative mechanism involving microfibrils. Biochemistry 25(5), 958–966 (1986)
143. Na, G.C., Butz, L.J., Carroll, R.J.: Mechanism of in vitro collagen fibril assembly. Kinetic and morphological studies. J. Biol. Chem. 261(26), 12290–12299 (1986)
144. Nagata, K.: Hsp47: a collagen-specific molecular chaperone. Trends Biochem. Sci. 21(1), 22–26 (1996)
145. Nam, S., Hu, K.H., Butte, M.J., Chaudhuri, O.: Strain-enhanced stress relaxation impacts nonlinear elasticity in collagen gels. Proc. Natl. Acad. Sci. 113(20), 5492–5497 (2016)
146. Nam, S., Lee, J., Brownfield, D.G., Chaudhuri, O.: Viscoplasticity enables mechanical remodeling of matrix by cells. Biophys. J. 111(10), 2296–2308 (2016)
147. Nguyen, D.G., Pentoney Jr., S.L.: Bioprinted three dimensional human tissues for toxicology and disease modeling. Drug Discov. Today Technol. 23, 37–44 (2017)
148. Nichol, J.W., Koshy, S.T., Bae, H., Hwang, C.M., Yamanlar, S., Khademhosseini, A.: Cell-laden microengineered gelatin methacrylate hydrogels. Biomaterials 31(21), 5536–5544 (2010)
149. Novak, T., Seelbinder, B., Twitchell, C.M., van Donkelaar, C.C., Voytik-Harbin, S.L., Neu, C.P.: Mechanisms and microenvironment investigation of cellularized high density gradient collagen matrices via densification. Adv. Funct. Mater. 26(16), 2617–2628 (2016)
150. Novak, T., Voytik-Harbin, S.L., Neu, C.P.: Cell encapsulation in a magnetically aligned collagen-GAG copolymer microenvironment. Acta Biomater. 11, 274–282 (2015)
151. Oostendorp, C., Meyer, S., Sobrio, M., van Arendonk, J., Reichmann, E., Daamen, W.F., van Kuppevelt, T.H.: Evaluation of cultured human dermal- and dermo-epidermal substitutes focusing on extracellular matrix components: comparison of protein and RNA analysis. Burns 43(3), 520–530 (2017)
152. Orgel, J.P., Irving, T.C., Miller, A., Wess, T.J.: Microfibrillar structure of type I collagen in situ. Proc. Natl. Acad. Sci. 103(24), 9001–9005 (2006)

153. Ozbolat, I.T.: Bioprinting scale-up tissue and organ constructs for transplantation. Trends Biotechnol. **33**(7), 395–400 (2015)
154. O'Leary, L.E., Fallas, J.A., Bakota, E.L., Kang, M.K., Hartgerink, J.D.: Multi-hierarchical self-assembly of a collagen mimetic peptide from triple helix to nanofibre and hydrogel. Nat. Chem. **3**(10), 821–828 (2011)
155. Pakkanen, O., Hamalainen, E.R., Kivirikko, K.I., Myllyharju, J.: Assembly of stable human type I and III collagen molecules from hydroxylated recombinant chains in the yeast Pichia pastoris. J. Biol. Chem. **278**(34), 32478–32483 (2003)
156. Pasquali-Ronchetti, I., Baccarani-Contri, M., Young, R.D., Vogel, A., Steinmann, B., Royce, P.M.: Ultrastructural analysis of skin and aorta from a patient with Menkes disease. Exp. Mol. Pathol. **61**(1), 36–57 (1994)
157. Pati, F., Jang, J., Ha, D.H., Won Kim, S., Rhie, J.W., Shim, J.H., Kim, D.H., Cho, D.W.: Printing three-dimensional tissue analogues with decellularized extracellular matrix bioink. Nat. Commun. **5**, 3935 (2014)
158. Piechocka, I.K., van Oosten, A.S., Breuls, R.G., Koenderink, G.H.: Rheology of heterotypic collagen networks. Biomacromolecules **12**(7), 2797–2805 (2011)
159. Pins, G.D., Christiansen, D.L., Patel, R., Silver, F.H.: Self-assembly of collagen fibers. Influence of fibrillar alignment and decorin on mechanical properties. Biophys. J. **73**(4), 2164–2172 (1997)
160. Pizzo, A.M., Kokini, K., Vaughn, L.C., Waisner, B.Z., Voytik-Harbin, S.L.: Extracellular matrix (ECM) microstructural composition regulates local cell-ECM biomechanics and fundamental fibroblast behavior: a multidimensional perspective. J. Appl. Physiol. **98**(5), 1909–1921 (2005)
161. Polk, S., Sori, N., Thayer, N., Kemper, N., Maghdouri-White, Y., Bulysheva, A.A., Francis, M.P.: Pneumatospinning of collagen microfibers from benign solvents. Biofabrication **10**(4), 045004 (2018)
162. Powell, H.M., Boyce, S.T.: Engineered human skin fabricated using electrospun collagen-PCL blends: morphogenesis and mechanical properties. Tissue Eng. Part A **15**(8), 2177–2187 (2009)
163. Prabhakaran, M.P., Venugopal, J., Ramakrishna, S.: Electrospun nanostructured scaffolds for bone tissue engineering. Acta Biomater. **5**(8), 2884–2893 (2009)
164. Prabhakaran, M.P., Venugopal, J.R., Ramakrishna, S.: Mesenchymal stem cell differentiation to neuronal cells on electrospun nanofibrous substrates for nerve tissue engineering. Biomaterials **30**(28), 4996–5003 (2009)
165. Przybyla, D.E., Chmielewski, J.: Metal-triggered radial self-assembly of collagen peptide fibers. J. Am. Chem. Soc. **130**(38), 12610–12611 (2008)
166. Ramachandran, G.N., Kartha, G.: Structure of collagen. Nature **174**(4423), 269–270 (1954)
167. Ramshaw, J.A., Werkmeister, J.A., Dumsday, G.J.: Bioengineered collagens: emerging directions for biomedical materials. Bioengineered **5**(4), 227–233 (2014)
168. Raub, C.B., Suresh, V., Krasieva, T., Lyubovitsky, J., Mih, J.D., Putnam, A.J., Tromberg, B.J., George, S.C.: Noninvasive assessment of collagen gel microstructure and mechanics using multiphoton microscopy. Biophys. J. **92**(6), 2212–2222 (2007)
169. Reese, S.P., Underwood, C.J., Weiss, J.A.: Effects of decorin proteoglycan on fibrillogenesis, ultrastructure, and mechanics of type I collagen gels. Matrix Biol. **32**(7–8), 414–423 (2013)
170. Rele, S., Song, Y., Apkarian, R.P., Qu, Z., Conticello, V.P., Chaikof, E.L.: D-periodic collagen-mimetic microfibers. J. Am. Chem. Soc. **129**(47), 14780–14787 (2007)
171. Rhee, S., Grinnell, F.: Fibroblast mechanics in 3D collagen matrices. Adv. Drug Del. Rev. **59**(13), 1299–1305 (2007)
172. Rho, K.S., Jeong, L., Lee, G., Seo, B.M., Park, Y.J., Hong, S.D., Roh, S., Cho, J.J., Park, W.H., Min, B.M.: Electrospinning of collagen nanofibers: effects on the behavior of normal human keratinocytes and early-stage wound healing. Biomaterials **27**(8), 1452–1461 (2006)
173. van der Rijt, J.A., van der Werf, K.O., Bennink, M.L., Dijkstra, P.J., Feijen, J.: Micromechanical testing of individual collagen fibrils. Macromol. Biosci. **6**(9), 697–702 (2006)

174. Robins, S.P.: Biochemistry and functional significance of collagen cross-linking. Biochem. Soc. Trans. **35**(Pt 5), 849–852 (2007)
175. Roeder, B.A., Kokini, K., Sturgis, J.E., Robinson, J.P., Voytik-Harbin, S.L.: Tensile mechanical properties of three-dimensional type I collagen extracellular matrices with varied microstructure. J. Biomech. Eng. **124**(2), 214–222 (2002)
176. Ruggiero, F., Exposito, J.Y., Bournat, P., Gruber, V., Perret, S., Comte, J., Olagnier, B., Garrone, R., Theisen, M.: Triple helix assembly and processing of human collagen produced in transgenic tobacco plants. FEBS Lett. **469**(1), 132–136 (2000)
177. Rutschmann, C., Baumann, S., Cabalzar, J., Luther, K.B., Hennet, T.: Recombinant expression of hydroxylated human collagen in Escherichia coli. Appl. Microbiol. Biotechnol. **98**(10), 4445–4455 (2014)
178. Sapudom, J., Rubner, S., Martin, S., Kurth, T., Riedel, S., Mierke, C.T., Pompe, T.: The phenotype of cancer cell invasion controlled by fibril diameter and pore size of 3D collagen networks. Biomaterials **52**, 367–375 (2015)
179. Sato, K., Ebihara, T., Adachi, E., Kawashima, S., Hattori, S., Irie, S.: Possible involvement of aminotelopeptide in self-assembly and thermal stability of collagen I as revealed by its removal with proteases. J. Biol. Chem. **275**(33), 25870–25875 (2000)
180. Schmidt, C.E., Baier, J.M.: Acellular vascular tissues: natural biomaterials for tissue repair and tissue engineering. Biomaterials **21**(22), 2215–2231 (2000)
181. Schoen, F.J., Levy, R.J.: Calcification of tissue heart valve substitutes: progress toward understanding and prevention. Ann. Thorac. Surg. **79**(3), 1072–1080 (2005)
182. Shahrokhi, S., Arno, A., Jeschke, M.G.: The use of dermal substitutes in burn surgery: acute phase. Wound Repair Regen. **22**(1), 14–22 (2014)
183. Shoulders, M.D., Raines, R.T.: Collagen structure and stability. Annu. Rev. Biochem. **78**, 929–958 (2009)
184. Silver, F.H., Freeman, J.W., Seehra, G.P.: Collagen self-assembly and the development of tendon mechanical properties. J. Biomech. **36**(10), 1529–1553 (2003)
185. Skeldon, G., Lucendo-Villarin, B., Shu, W.: Three-dimensional bioprinting of stem-cell derived tissues for human regenerative medicine. Philos. Trans. R. Soc. Lond. B Biol. Sci. **373**(1750), 20170224 (2018)
186. Snedeker, J.G., Gautieri, A.: The role of collagen crosslinks in ageing and diabetes—the good, the bad, and the ugly. Muscles Ligaments Tendons J. **4**(3), 303–308 (2014)
187. Soares, J.S., Feaver, K.R., Zhang, W., Kamensky, D., Aggarwal, A., Sacks, M.S.: Biomechanical behavior of bioprosthetic heart valve heterograft tissues: characterization, simulation, and performance. Cardiovasc. Eng. Technol. **7**(4), 309–351 (2016)
188. Stegman, S.J., Tromovitch, T.A.: Implantation of collagen for depressed scars. J. Dermatol. Surg. Oncol. **6**(6), 450–453 (1980)
189. Stein, H., Wilensky, M., Tsafrir, Y., Rosenthal, M., Amir, R., Avraham, T., Ofir, K., Dgany, O., Yayon, A., Shoseyov, O.: Production of bioactive, post-translationally modified, heterotrimeric, human recombinant type-I collagen in transgenic tobacco. Biomacromolecules **10**(9), 2640–2645 (2009)
190. Stephens, C.H., Orr, K.S., Acton, A.J., Tersey, S.A., Mirmira, R.G., Considine, R.V., Voytik-Harbin, S.L.: In situ type I oligomeric collagen macroencapsulation promotes islet longevity and function in vitro and in vivo. Am. J. Physiol. Endocrinol. Metab. **315**(4), E650–E661 (2018)
191. Strauss, K., Chmielewski, J.: Advances in the design and higher-order assembly of collagen mimetic peptides for regenerative medicine. Curr. Opin. Biotechnol. **46**, 34–41 (2017)
192. Sundararaghavan, H.G., Monteiro, G.A., Lapin, N.A., Chabal, Y.J., Miksan, J.R., Shreiber, D.I.: Genipin-induced changes in collagen gels: correlation of mechanical properties to fluorescence. J. Biomed. Mater. Res. A **87**(2), 308–320 (2008)
193. Sung, H.W., Chang, W.H., Ma, C.Y., Lee, M.H.: Crosslinking of biological tissues using genipin and/or carbodiimide. J. Biomed. Mater. Res., Part A **64**(3), 427–438 (2003)
194. Susilo, M.E., Paten, J.A., Sander, E.A., Nguyen, T.D., Ruberti, J.W.: Collagen network strengthening following cyclic tensile loading. Interface Focus **6**(1), 20150088 (2016)

195. Sweeney, S.M., Orgel, J.P., Fertala, A., McAuliffe, J.D., Turner, K.R., Di Lullo, G.A., Chen, S., Antipova, O., Perumal, S., Ala-Kokko, L., Forlino, A., Cabral, W.A., Barnes, A.M., Marini, J.C., San Antonio, J.D.: Candidate cell and matrix interaction domains on the collagen fibril, the predominant protein of vertebrates. J. Biol. Chem. **283**(30), 21187–21197 (2008)
196. Tillman, B.W., Yazdani, S.K., Lee, S.J., Geary, R.L., Atala, A., Yoo, J.J.: The in vivo stability of electrospun polycaprolactone-collagen scaffolds in vascular reconstruction. Biomaterials **30**(4), 583–588 (2009)
197. Tomita, M., Kitajima, T., Yoshizato, K.: Formation of recombinant human procollagen I heterotrimers in a baculovirus expression system. J. Biochem. **121**(6), 1061–1069 (1997)
198. Torbet, J., Malbouyres, M., Builles, N., Justin, V., Roulet, M., Damour, O., Oldberg, A., Ruggiero, F., Hulmes, D.J.: Orthogonal scaffold of magnetically aligned collagen lamellae for corneal stroma reconstruction. Biomaterials **28**(29), 4268–4276 (2007)
199. Torbet, J., Ronziere, M.C.: Magnetic alignment of collagen during self-assembly. Biochem. J. **219**(3), 1057–1059 (1984)
200. Torp, S.: Effects of age and of mechanical deformation on the ultrastructure of tendon. In: Structure of Fibrous Biopolymers, pp. 223–250 (1975)
201. Vacanti, J.P., Langer, R.: Tissue engineering: the design and fabrication of living replacement devices for surgical reconstruction and transplantation. Lancet **354**(Suppl 1), SI32–34 (1999)
202. Veres, S.P., Harrison, J.M., Lee, J.M.: Repeated subrupture overload causes progression of nanoscaled discrete plasticity damage in tendon collagen fibrils. J. Orthop. Res. **31**(5), 731–737 (2013)
203. Veres, S.P., Lee, J.M.: Designed to fail: a novel mode of collagen fibril disruption and its relevance to tissue toughness. Biophys. J. **102**(12), 2876–2884 (2012)
204. Wainwright, D.J.: Use of an acellular allograft dermal matrix (Alloderm) in the management of full-thickness burns. Burns **21**(4), 243–248 (1995)
205. Wallace, J.M., Erickson, B., Les, C.M., Orr, B.G., Banaszak Holl, M.M.: Distribution of type I collagen morphologies in bone: relation to estrogen depletion. Bone **46**(5), 1349–1354 (2010)
206. Wasserman, A.J., Kato, Y.P., Christiansen, D., Dunn, M.G., Silver, F.H.: Achilles tendon replacement by a collagen fiber prosthesis: morphological evaluation of neotendon formation. Scanning Microsc. **3**(4), 1183–1197; discussion 1197–1200 (1989)
207. Weadock, K., Olson, R.M., Silver, F.H.: Evaluation of collagen crosslinking techniques. Biomater. Med. Devices Artif. Organs **11**(4), 293–318 (1983)
208. Wenger, M.P., Bozec, L., Horton, M.A., Mesquida, P.: Mechanical properties of collagen fibrils. Biophys. J. **93**(4), 1255–1263 (2007)
209. Whittington, C.F., Brandner, E., Teo, K.Y., Han, B., Nauman, E., Voytik-Harbin, S.L.: Oligomers modulate interfibril branching and mass transport properties of collagen matrices. Microsc. Microanal. **19**(5), 1323–1333 (2013)
210. Whittington, C.F., Yoder, M.C., Voytik-Harbin, S.L.: Collagen-polymer guidance of vessel network formation and stabilization by endothelial colony forming cells in vitro. Macromol. Biosci. **13**(9), 1135–1149 (2013)
211. Willard, J.J., Drexler, J.W., Das, A., Roy, S., Shilo, S., Shoseyov, O., Powell, H.M.: Plant-derived human collagen scaffolds for skin tissue engineering. Tissue Eng. Part A **19**(13–14), 1507–1518 (2013)
212. Wilson, R., Lees, J.F., Bulleid, N.J.: Protein disulfide isomerase acts as a molecular chaperone during the assembly of procollagen. J. Biol. Chem. **273**(16), 9637–9643 (1998)
213. Wood, G.C., Keech, M.K.: The formation of fibrils from collagen solutions. 1. The effect of experimental conditions: kinetic and electron-microscope studies. Biochem. J. **75**(3), 588–598 (1960)
214. Worcester, D.L.: Structural origins of diamagnetic anisotropy in proteins. Proc. Natl. Acad. Sci. **75**(11), 5475–5477 (1978)
215. Worke, L.J., Barthold, J.E., Seelbinder, B., Novak, T., Main, R.P., Harbin, S.L., Neu, C.P.: Densification of type I collagen matrices as a model for cardiac fibrosis. Adv. Healthc. Mater. **6**(22), 1700114 (2017)

216. Xu, R., Boudreau, A., Bissell, M.J.: Tissue architecture and function: dynamic reciprocity via extra- and intra-cellular matrices. Cancer Metastasis Rev. **28**(1–2), 167–176 (2009)
217. Yang, C., Hillas, P.J., Baez, J.A., Nokelainen, M., Balan, J., Tang, J., Spiro, R., Polarek, J.W.: The application of recombinant human collagen in tissue engineering. Biodrugs **18**(2), 103–119 (2004)
218. Yang, W., Sherman, V.R., Gludovatz, B., Schaible, E., Stewart, P., Ritchie, R.O., Meyers, M.A.: On the tear resistance of skin. Nat. Commun. **6**, 6649 (2015)
219. Yannas, I.V., Burke, J.F.: Design of an artificial skin. I. Basic design principles. J. Biomed. Mater. Res. **14**(1), 65–81 (1980)
220. Yannas, I.V., Burke, J.F., Gordon, P.L., Huang, C., Rubenstein, R.H.: Design of an artificial skin. II. Control of chemical composition. J. Biomed. Mater. Res. **14**(2), 107–132 (1980)
221. Younesi, M., Donmez, B.O., Islam, A., Akkus, O.: Heparinized collagen sutures for sustained delivery of PDGF-BB: delivery profile and effects on tendon-derived cells in-vitro. Acta Biomater. **41**, 100–109 (2016)
222. Younesi, M., Knapik, D.M., Cumsky, J., Donmez, B.O., He, P., Islam, A., Learn, G., McClellan, P., Bohl, M., Gillespie, R.J., Akkus, O.: Effects of PDGF-BB delivery from heparinized collagen sutures on the healing of lacerated chicken flexor tendon in vivo. Acta Biomater. **63**, 200–209 (2017)
223. Yue, K., Trujillo-de Santiago, G., Alvarez, M.M., Tamayol, A., Annabi, N., Khademhosseini, A.: Synthesis, properties, and biomedical applications of gelatin methacryloyl (GelMA) hydrogels. Biomaterials **73**, 254–271 (2015)
224. Yunoki, S., Matsuda, T.: Simultaneous processing of fibril formation and cross-linking improves mechanical properties of collagen. Biomacromolecules **9**(3), 879–885 (2008)
225. Zhang, H., Voytik-Harbin, S., Brookes, S., Zhang, L., Wallace, J., Parker, N., Halum, S.: Use of autologous adipose-derived mesenchymal stem cells for creation of laryngeal cartilage. Laryngoscope **128**(4), E123–E129 (2018)
226. Zuhdi, N., Hawley, W., Voehl, V., Hancock, W., Carey, J., Greer, A.: Porcine aortic valves as replacements for human heart valves. Ann. Thorac. Surg. **17**(5), 479–491 (1974)

Roles of Interactions Between Cells and Extracellular Matrices for Cell Migration and Matrix Remodeling

Jing Li, Wonyeong Jung, Sungmin Nam, Ovijit Chaudhuri and Taeyoon Kim

Abstract Cells can sense mechanical properties of surrounding environments and also structurally remodel the environments. Interactions between cells and extracellular matrix (ECM) play a crucial role in diverse cellular behaviors, including migration, growth, and differentiation. Advances in experimental and computational methods enabled us to better understand the molecular bases and underlying mechanisms of the cell-ECM interactions. This chapter provides a comprehensive review regarding how cells sense and remodel ECMs and why such capabilities are of great importance for cell migration. First, the molecular structure, dynamics, and functions of focal adhesions (FAs) formed between cells and ECM are discussed, followed by a brief review about the significance of interactions between FAs and the actin cytoskeleton occurring in the intracellular space. Then, it is discussed how cells remodel surrounding ECMs mechanically and biochemically. Additionally, various experimental and computational methods designed for studying cell migration facilitated by cell-ECM interactions and ECM remodeling are summarized, and findings obtained using these methods are discussed.

Jing Li and Wonyeong Jung are equal contribution.

J. Li · W. Jung · T. Kim (✉)
Weldon School of Biomedical Engineering, Purdue University,
206 S. Martin Jischke Drive, West Lafayette, IN 47907, USA
e-mail: kimty@purdue.edu

J. Li
e-mail: li2221@purdue.edu

W. Jung
e-mail: jung164@purdue.edu

S. Nam · O. Chaudhuri
Department of Mechanical Engineering, Stanford University,
440 Escondido Mall, Stanford, CA 94305, USA
e-mail: sungmin@stanford.edu

O. Chaudhuri
e-mail: chaudhuri@stanford.edu

© Springer Nature Switzerland AG 2020
Y. Zhang (ed.), *Multi-scale Extracellular Matrix Mechanics and Mechanobiology*,
Studies in Mechanobiology, Tissue Engineering and Biomaterials 23,
https://doi.org/10.1007/978-3-030-20182-1_8

1 Introduction

Interactions between cells and extracellular matrix (ECM) play an important role for various cellular behaviors, including migration, growth, and differentiation. During these behaviors, cells are able to sense mechanical forces and properties of their surrounding ECM. Such mechano-sensing ability of cells is of great importance for a wide range of physiological processes, including morphogenesis and angiogenesis [1, 2]. In particular, effects of the stiffness of ECM on diverse cellular processes have been investigated. For example, it was shown that a gradient of ECM stiffness can guide migrating cells toward stiffer regions, which is called durotaxis [3]. In addition, mesenchymal stem cells are differentiated into cells with different phenotypes, depending on the stiffness of surrounding mechanical environments [4].

Cells can also remodel surrounding ECM actively rather than sensing it passively; forces transmitted through focal adhesions (FAs) are capable of significantly remodeling ECM for physiological functions, such as cancer metastasis and wound healing [5–7]. For example, fibroblasts migrate toward a wound site to close it by generating contractile forces [8]. During the initial stage of metastasis called intravasation, cancer-associated fibroblasts make collagen fibers aligned to help tumor cells migrate effectively toward blood vessels via the aligned fibers [9]. In addition, matrix remodeling enables cells to sense distant cells. The long-range sensation plays an important role in many biological processes [10–14]. For example, at the single-cell level, long-range mechanical communication regulates tube formation and the detachment of cells from multicellular aggregates [15].

Therefore, a deep understanding of cell-ECM interactions and cell-induced ECM remodeling is crucial for illuminating mechanisms of a wide range of biological processes. This review article will cover various aspects of the cell-ECM interactions at multiple length-scales. First, force-dependent dynamic behaviors of FAs will be discussed. Then, effects of the actin cytoskeleton on cell-ECM interactions will be explained. Lastly, matrix remodeling and cell migration driven by the cell-ECM interactions will be discussed.

2 Molecular Interactions Between Cells and ECM

Interactions between cells and ECM are based on binding between receptors and ligands at molecular level. There are various types of receptors and ligands depending on types of cells and ECM. We will primarily review the most important adhesion between cells and ECM called focal adhesion (FA) and will discuss how FAs are regulated by the actin cytoskeleton within cells. In addition, computational models and experimental techniques designed for studying cell-ECM interactions will be introduced.

2.1 Focal Adhesions Between Cells and ECM

Cells form physical links between the cytoskeleton and ECM via FAs (Fig. 1). The function, molecular structure, and dynamics of FAs have been studied during recent decades. One of the most important functions of FAs is bi-directional transmission of mechanical force between intracellular and extracellular spaces. This function enables cells to sense mechanical properties of surrounding ECM and also mechanically remodel the ECM. Dozens of proteins are involved with formation of the FA complex, but the main component is a receptor protein called integrin (Fig. 1) [16]. Integrin serves as a pivotal linker as it binds to both the actin cytoskeleton and ligands in ECM, such as fibronectins [17]. A functional unit is a heterodimer consisting of α and β integrin subunits, and binding affinity to ECM ligands depends on types of α and β subunits [18]. For example, integrin with β1 subunit plays an important role for binding to collagen and fibronectin [19]. Binding to ECM ligands causes a conformational change in integrin subunits, resulting in separation of cytoplasmic tails of α and β subunits. The cytoplasmic tail of β integrin interacts with proteins responsible for a signaling cascade of intracellular dynamics [20], such as Rho activity [21]. Rho activation induces formation of FAs [22]. The initial adhesion is associated with talin, which binds to both actin and integrins [23, 24] and can be characterized by accumulation of vinculin which binds talin and actin [25] or by the recruitments of paxillin and phosphoproteins (Fig. 1) [26, 27]. FAs undergo mechanical forces exerted by both intracellular and extracellular forces. The magnitude of forces exerted on individual FAs in cardiac fibroblasts or myocytes was measured to be ~10 nN [28]. The magnitude typically varies depending on types of cells. For example, FAs in myocytes bear higher forces than those in fibroblasts [28]. A bond formed by a single integrin can only withstand very low level of forces (<40 pN) [29, 30]. Thus, to bear high force in FA and maintain the physical links, a large number of integrins are required in each FA site. FAs can be matured by growth in response to mechanical forces [26, 31] or increasing stiffness of ECMs [32, 33]. An increase in force facilitates clustering of integrins and FA-associated proteins, leading to growth of FAs in the direction aligned with the force [28, 34, 35]. Previous studies have shown that FAs are stabilized by force-dependent recruitment of vinculin, leading to further maturation of FAs [26, 31]; it was observed that concentration of vinculin scales linearly with forces [28]. Other FA-associated proteins, including talin and tensin, are known to be stabilized by an increase in forces or ECM rigidity [36–38]. The molecular composition of FAs is one of the criteria to differentiate nascent FAs from matured FAs. Nascent FAs are enriched with tensin but lack the other proteins. By contrast, matured FAs have high levels of paxillin and vinculin with a relatively low level of tensin [39].

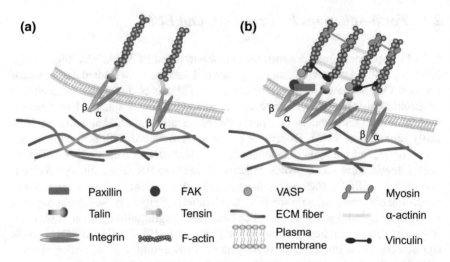

Fig. 1 Formation and maturation of focal adhesions (FAs) facilitated by various proteins. **a** Formation of nascent FAs governed by integrin. Integrin is localized to a plasma membrane, playing a role as pivotal linkers between intracellular cytoskeleton and extracellular matrix (ECM). Integrin consists of α (green) and β (orange) subunits. In the intracellular space, β subunit of integrin actively binds to other actin binding proteins, such as talin and tensin, many of which are mechano-sensitive. **b** Maturation of FAs is characterized by the accumulation of a number of integrin proteins as well as by binding of vinculin to both F-actin and talin. The downstream signaling of integrin proteins requires other FA-associated proteins including paxillin and focal adhesion kinase (FAK). Paxillin helps the maturation process by connecting FAK to F-actin as a substrate for FAK. As crosslinking protein (α-actinin) and myosin motor bind to F-actin with VASP assembling F-actin, the actin cytoskeleton is structurally remodeled to thick bundles

2.2 Roles of the Actin Cytoskeleton for Cell-ECM Interactions

Mechanical forces exerted on FAs originate mostly from molecular interactions between myosin motors and actin filaments (F-actins) in the actin cytoskeleton. Actomyosin contractility has been studied extensively using in vivo experiments, in vitro reconstituted actomyosin networks, and computational studies [40–42]. Since the intracellular forces are transmitted to ECM via FAs for matrix remodeling and cell migration, it is very important to understand how the actin cytoskeleton generates protrusive and contractile forces and structurally remodels itself in response to various mechanical stimuli. Various mechano-sensitive proteins constituting the actin cytoskeleton facilitate the maintenance and remodeling of the cytoskeletal structures and mediate FA dynamics.

A major load-bearing component in cells is the stress fiber (SF) consisting of F-actins cross-linked in parallel [43]. There are three types of SFs in cells: dorsal SF, ventral SF, and transverse arcs (Fig. 2a). Dorsal SFs are relatively short non-contractile SFs formed by rapidly growing barbed ends interacting with FAs near the

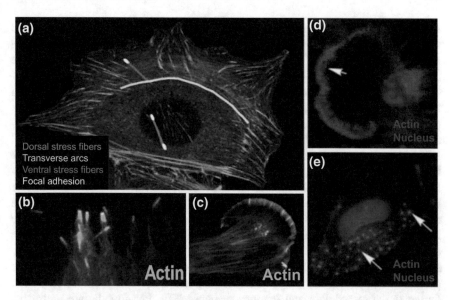

Fig. 2 Examples of actin structures emerging during cell-ECM interactions. **a** Three major types of stress fibers. Stress fibers are thick bundles consisting of F-actins. Dorsal and ventral stress fibers interact directly with focal adhesions, whereas transverse arcs do not interact with FAs. Reprinted with permission from [213]. **b** Filopodia are finger-like protrusive structures composed of cross-linked parallel F-actins. **c** Lamellipodia are protrusions composed of branched actin networks. Reprinted with permission from [214]. **d** Podosomes and **e** invadopodia (white arrows). Podosomes and invadopodia are actin-rich dynamic protrusions formed on a plasma membrane, playing an important role as sites that bind to or degrade ECM. Reprinted with permission from [70]

leading edge [44–46]. Transverse arcs in lamella formed by compaction of branched networks assembled by Arp2/3 in lamellipodia are highly contractile, and they do not directly interact with FAs but are linked to dorsal SFs [45, 47]. Ventral SFs are also contractile due to myosin motors and interact with FAs [45]. Thus, contractile forces generated by SFs can be transmitted to substrates via FAs. Recent experiments showed that laser ablation of SFs disrupts connection between SFs and FAs and thus leads to relaxation of traction stress exerted on a substrate [48]. SFs respond to mechanical cues in various fashions. For example, several experimental studies showed that force polarization would induce anisotropic distribution of FAs, resulting in remodeling of ventral SFs from an initial isotropic structure to a highly aligned one [28, 34, 49–51]. It was found that the aligned structure is capable of generating larger and more stable forces. In addition, it was observed that oscillatory strains can drastically change the direction of aligned SFs [52].

The formation and force-dependent maturation of FAs are significantly affected by actin dynamics. Fast actin retrograde flow in the lamellipodia is known to be a necessary precursor for initial assembly of FAs, and formation of FA sites substantially slows down the otherwise fast flow [53, 54]. Actin polymerization dominates and leads to growth of dorsal SFs in contact with FAs [45]. An experiment using high resolution total internal reflection fluorescence (TIRF) microscopy demonstrated that stability of maturation depends on actin polymerization [55]. The maturation rate scales proportionally with the growth rate of the dorsal SF facilitated by the actin polymerization [55].

Protrusions to surrounding ECMs are driven by various actin structures, such as filopodia, lamellipodia, podosomes, and invadopodia (Fig. 2b–d). Filopodia are protrusive bundle structures elongated by formin and cross-linked by fascin and α-actinin (Fig. 2b) [56–59]. Interestingly, it was suggested that filopodia primarily pull ECM fibers rather than pushing them [60, 61]. Filopodia are of great importance for cell-cell adhesion, migration, and spreading [56, 62–67]. Lamellipodia are sheet-like protrusions composed of dense branched actin networks and found mainly in the leading edge of migrating cells (Fig. 2c) [68, 69]. F-actins in lamellipodia are consistently pulled toward lamella by myosin motors, inducing the actin retrograde flow, and they are eventually compacted into transverse arcs [68]. Podosomes are adhesive structures on the ventral surface of cells that are composed of F-actin, FA-associated proteins, and other proteins, such as Tks5 and Wiskott-Aldrich syndrome protein (WASP) (Fig. 2d) [70]. Podosomes degrade ECMs and facilitate migration during tissue remodeling and immune response [70]. Invadopodia are protrusive F-actin bundles cross-linked by fascin and composed of proteins, such as Arp2/3 and cortactin (Fig. 2e) [70]. Invadopodia degrade a basement membrane by releasing proteases called matrix metalloproteinases (MMPs), thus facilitating cancer cell invasion [71]. All of these structures show mechano-sensitive behaviors that enable them to probe the structural and mechanical properties of surrounding ECMs.

2.3 Experimental Methods for Studying Cell-ECM Interactions

Recent advances in live-cell imaging technique with high resolution microscopy, nano-scale force measurement, biomimetic models, and engineering and synthesis of biomaterials have enabled us to better investigate intrinsic mechanisms of cell-ECM interactions. Traditionally, collagen-coated polyacrylamide gels with variable stiffness have been a robust platform for studying effects of stiffness of a two-dimensional (2D) substrate on cell behaviors [72]. For studying cell behaviors in three-dimensional (3D) matrices, hydrogels made of both synthetic materials (e.g. polyethylene glycol) and biomaterials (e.g. alginate, hyaluronic acid, and polypeptides) have been used [73]. In addition, reconstituted ECMs (e.g. collagen gel, fibrin gel, and reconstituted basement membrane composed of laminin I and collagen IV) have been employed popularly [73, 74]. One advantage of these in vitro materials is that their mechanical properties can be modulated [73, 75, 76]. The easiest way for modulation of stiffness is to vary concentrations of material components. For example, an increase in collagen density results in higher stiffness of collagen gels. However, higher collagen concentration also leads to a decrease in pore size and an increase in ligand density [77]. Since modulating mechanical properties without changes in pore size and ligand density is challenging, several studies have focused on addressing this problem. For example, a study employed interpenetrating networks composed of alginate and reconstituted basement membrane, where ionic crosslinking changes stiffness without significant variations in pore size and ligand density [78]. In collagen-coated polyacrylamide gel, stiffness can be changed by varying density of a cross-linking agent called bis-acrylamide [3, 72]. If reversible crosslinkers are used, viscoelasticity of 3D hydrogels can be tuned [73]. Viscoelastic gels exhibit both viscous and elastic behaviors in response to mechanical stimuli, such as stress and deformation. In addition, one study showed that critical stress beyond which material becomes stiffer can be tuned in polyisocyanopeptide-based hydrogels [79]. Specific applications of diverse bio- and synthetic materials mimicking ECMs were comprehensively reviewed before [73].

In addition, substrates with patterns have been popularly used for studying cell-ECM interactions because the size, density, and ligand of regions with which cells interact can be controlled easily. For instance, micropost arrays and micro-contact printing techniques using polydimethylsiloxane are both widely used due to their capability of allocating contact sites individually and spatially confining cell-ECM interactions. Using these techniques, effects of cell-ECM interactions on cell adhesion [80–82], migration [83], contractility [84–86], division [87], differentiation [88, 89], and cell stiffness [90] were investigated extensively. To study forces at the individual protein level, researchers employed a complex porous substrate fabricated by advanced nanolithography. Ligands are deposited into pores with specific distances at nanometer length-scale. Several studies have revealed that the nano-patterned surface affects cell spreading, adhesion, and other cellular behaviors via regulation of cell-ECM interactions [91–96].

Since the dynamics of FAs and cytoskeletal proteins related to cell-ECM interactions occurs rapidly, it remains one of the challenges to enhance the temporal resolution of live-cell imaging. An early study succeeded to predict a correlation between FA dynamics and ECM deformation with time-lapse differential interference contrast images [97]. Recently, harmonic generation microscopy is extensively applied to analyze the localization of signaling proteins during cancer cell invasion [98–101]. In addition, cellular forces have been measured at various length-scales. One of the most common techniques is traction force microscopy (TFM) that measures traction stress generated by cells using fluorescently labeled particles embedded into an optically-transparent hydrogel [3, 84, 102, 103]. In this technique, displacement of particles induced by cellular forces is measured via imaging, and then the traction stress can be calculated using constitutive relations. TFM-based experiments have shown that traction stress is strongly correlated with cell adhesion strength, cell spreading, and substrate rigidity [104, 105]. Forces involved with protein interactions, such as binding of talin, vinculin, and ECM ligands, have also been probed using atomic force microscopy [23, 106, 107], which measures force by probing a sample with a micron-scale cantilever that acts like a Hookean spring for small deflection. More recently, tethering biosensors which enable visualization of the force distribution on individual adhesion proteins have been developed [108, 109].

The technique was first used to measure tension across vinculin by inserting a tension sensor between head and tail parts of vinculin [108]. The tension sensor has repetitive amino-acid motifs between two fluorophores, which exhibit fluorescence resonance energy transfer (FRET). Measuring FRET efficiency, which depends on a distance between the flurophores, provides estimation of tension across the fluorophores and proteins. Using this technique, a study uncovered complex distribution of tension on individual integrins in a single FA [109].

2.4 Computational Methods for Studying Cell-ECM Interactions

Despite the advances in experimental techniques described above, there are still limitations in spatiotemporal resolutions for measurements. Computational models built based on experimental observations and measurements can help overcome the limitations to uncover working principles of cell-ECM interactions. For example, a finite element (FE) model showed how TFM measurement is correlated to cell stiffness and the energy density of FAs regardless of cell contractility [110]. To better understand how cells sense ECM stiffness during spreading, adhesion, and migration, Chan et al. developed a model that describes bonds between a cell and ECM as a molecular clutch (Fig. 3a) [111]. The molecular clutch model consists of multiple dynamic bonds between F-actin and a substrate which resist actin retrograde flow induced by myosin activities. Results from the molecular clutch model explain distinct retrograde flow rates and traction stress, depending on substrate stiffness (Fig. 3b, c) [112–114]. A stiffer substrate reduces the number of engaged bonds due to fast tension buildup, leading to frictional slippage between the cell and substrate and thus fast actin retrograde flow [115, 116]. By contrast, soft substrate results in transitions between slow and fast actin retrograde flows because tension is developed slowly [111]. The molecular clutch model was extended to account for viscoelastic substrates [117]. Viscoelastic substrates exhibit energy dissipation upon the application and removal of loads. In this study, the authors showed that cells spread more on viscoelastic substrates than on elastic substrates, thus demonstrating that stress relaxation also plays a role for mechano-sensing of cells.

A model simulating the growth of FAs driven by actomyosin contractility reveals a mechanical feedback between actin-based structures and FAs [46]; a branched actin network promotes FA maturation and an increase in traction force in the leading edge of cells, and the resultant higher traction force in the FAs leads to formation of SFs in the same direction. Another study explained the dependence of cellular forces on substrate stiffness by the force-depending walking motion of myosin motors. Myosins walk on F-actin and thus generate force until they become stalled due to high force. On a soft substrate, myosin motors are not stalled well, and there is larger energy dissipation, resulting in lower maximal stress as opposed to stress generated on a stiffer substrate [118]. In addition, effects of non-linear and inelastic properties

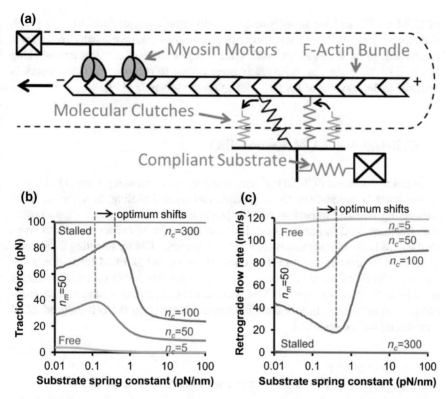

Fig. 3 Molecular clutch model describing mechano-sensitive behaviors of a cell on a substrate. **a** This model accounts for dynamic linkage between the substrate and the cell. Myosin motors generate forces on an F-actin bundle, leading to a retrograde flow. Near the leading edge of the F-actin bundle, physical links called molecular clutches are formed in a reversible manner. The molecular clutches are confined on a plate that is connected to a rigid surface by a spring. It was assumed that the unbinding rate of each molecular clutch varies depending on a force applied to the clutch that is equal to its spring constant multiplied to its displacement with respect to the plate. Reprinted with permission from [215]. **b, c** The molecular clutch model predicts that substrate stiffness would result in three distinct regimes for traction force level and a retrograde flow rate. At low substrate stiffness, the traction force is high, but the flow rate is low. In this regime, multiple molecular clutches can remain bound to the F-actin bundle even when the substrate displacement is large (load), and then they unbind from the F-actin bundle simultaneously due to a highly developed load (fail). By contrast, at high substrate stiffness, the traction force is low, but the flow rate is high. In this regime, each molecular clutch binds to and unbinds from the F-actin bundle rapidly, resulting in frictional slippage. The traction force and flow rate are also affected by the number of molecular clutches (n_c). At $n_c = 300$, the model becomes a stalled system, meaning that clutches never collectively unbind, and F-actin retrograde flow is nearly zero, regardless of substrate stiffness. At $n_c = 5$, the model becomes a free flowing system, meaning that the F-actin bundle flows with velocity close to an unloaded velocity, regardless of substrate stiffness. Reprinted with permission from [111, 216]

of ECM on FA size are investigated by multi-scale computational models [119]. With a non-elastic substrate which exhibits plastic deformation typically under larger mechanical strain at longer time scales due to dissociation of covalent cross-linking bonds [120–123], FA size is inversely proportional to substrate stiffness, which is opposite to dependence of FA size with an elastic substrate [119].

3 Cell-Induced ECM Remodeling

Cells can remodel the ECM to maintain tissue homeostasis and govern cellular processes, such as morphogenesis, migration, and wound healing. In addition, ECM remodeling plays an important role for progression of diseases including fibrosis, tumorigenesis, and tumor invasion. Remodeling of ECM occurs in various ways (Fig. 4). Cells can biochemically remodel surrounding ECM by secreting ECM components or by secreting enzymes that degrade or cross-link ECM fibers. Cells can also remodel ECM mechanically by exerting contractile forces on ECM, resulting in various structural changes in ECM, such as fiber alignment, contraction, translocation, and stiffening. In this section, current literature on the cell-induced ECM remodeling will be reviewed.

3.1 Deposition, Cross-Linking, and Degradation of ECM

Cells maintain tissue homeostasis and regulate ECM compositions by depositing, cross-linking, and degrading ECM (Fig. 4). The ECM components are secreted by cells that reside within the ECM or by cells recruited to the ECM from other tissues [124, 125]. Components of the basement membrane, such as collagen IV, fibronectin (FN), laminin, and heparan sulphate proteoglycan, are synthesized by epithelial cells during morphogenesis [126, 127]. In the interstitial matrix, fibroblasts play a major role for depositing ECM components including collagen, FN, hyaluronic acid (HA), glycosaminoglcans, and proteoglycans [128] within both intact and wounded tissues [124, 129, 130]. In cartilage and bone, chondrocytes and osteoblasts deposit ECM components [131].

ECM components secreted by cells support cell dynamics chemically, by providing cells growth factors and cytokines needed for cell growth and signaling or mechanically, by serving as a physical scaffold or transmitting mechanical signals. For performing the mechanical role, fibrous ECM proteins including collagen and FN form a fiber network which is suitable for serving as a scaffolding structure and transmitting forces. Formation of a fiber network requires fibril formation from soluble molecules in a process known as fibrillogenesis and cross-linking of fibrils. Collagen fibrils are cross-linked by enzymes (e.g. lysyl oxidase (LOX) and LOX family of amin oxidases [100]) or by glycation and transglutamination that are not associated with enzymes [100]. Cross-linking of collagen fibrils enhances connectivity of collagen

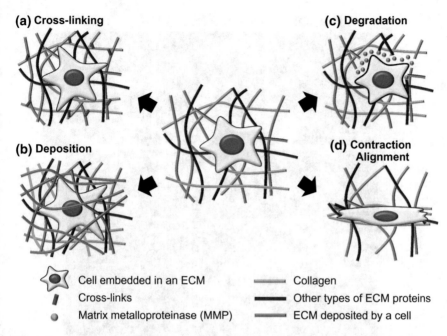

(a) Cross-linking

(c) Degradation

(b) Deposition

(d) Contraction Alignment

Cell embedded in an ECM   Collagen

Cross-links   Other types of ECM proteins

Matrix metalloproteinase (MMP)   ECM deposited by a cell

Fig. 4 Remodeling of ECM by cells. **a** Cells can cross-link ECM fibers, resulting in the enhanced stiffness and elasticity of the ECM. **b** Specific types of cells, such as fibroblast, deposit additional ECM components on a surrounding matrix. This can lead to elevated matrix stiffness and smaller matrix pore size. **c** Cells may biochemically degrade a surrounding ECM by secreting various types of matrix metalloproteinases (MMPs). **d** Mechanical forces generated by cells can structurally remodel the ECM by stretching and aligning ECM fibers

fiber network, enabling it to bear a large fraction of loads exerted on ECM. Cells are also able to degrade ECM for biological processes, such as repair of damaged tissue and invasion of cancer cells. For example, proteases including MMPs, adamalysins, and meprins biochemically break down ECM components [132]. In addition to proteolytic degradation, cells' internalizing and degrading ECM molecules in lysosomes is another way of ECM degradation [128].

Failure in the proper regulation of deposition, cross-linking, and degradation of ECM leads to the abnormal composition and mechanical properties of ECM, which is related to pathological conditions. For example, excessive deposition and cross-linking of collagen increase ECM stiffness, which can result in fibrosis and tumor progression in vivo (Fig. 5a) [101]. In various kinds of cancer including breast, head, and neck cancer, increased level of LOX was reported [128]. In addition, abnormal MMP-induced ECM degradation in tumors forms tube-like structures in ECM, which facilitates tumor cell invasion (Fig. 5b) [133].

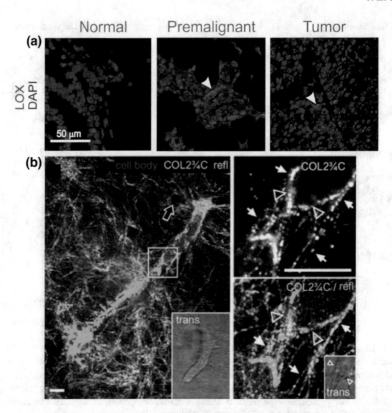

Fig. 5 Experimental observations of cross-linking and degradation of ECM fibers induced by cells. **a** Mammary glands from FVB MMTV-Neu mice stained for lysyl oxidase (LOX, red) and DAPI (nuclei, blue) at different stages of tumor progression. Compared to a normal tissue, premalignant tissue and tumor exhibit a high level of LOX, indicative of excessive ECM cross-linking. Reprinted with permission from [101]. **b** Degraded collagen (COL2³/₄C) induced by HT1080/MT1 cells migrating within a 3D collagen matrix. Proteolytic rear (empty arrowheads) was bordered by partially degraded fibers in parallel alignment (white arrows). Scale bars indicate 10 μm, and a black arrow represents migration direction. Reprinted with permission from [133]

3.2 Contraction, Alignment, and Stiffening of ECM

Forces generated by the actomyosin machinery can be transmitted to the ECM through FAs, which leads to mechanical remodeling of the ECM. For example, highly contractile fibroblasts can significantly remodel the ECM by aligning collagen fibrils and contracting collagen networks. Such mechanical remodeling processes vary the mechanical properties of the ECM, which regulates cellular processes [134]. Therefore, it is of great importance to understand the process of ECM remodeling.

Many studies focus on cell-induced alignment of ECM fibrils due to its importance for many physiological phenomena. For example, long-range communication between cells during angiogenesis and collective cell migration depends on the alignment of the ECM fibrils. It was shown that alignment of collagen fibrils in the direction of forces generated by cells (Fig. 6a) increases force transmission distance between cells to 10 times the cell diameter, enabling long-range cell-cell communication [11, 135]. In addition, ECM fibril alignment was found to enhance the invasion of tumor cells via a positive feedback loop between cell contraction and ECM remodeling [136]; after collagen fibrils are aligned, cells can generate greater force, leading to further fibril alignment [137]. In vitro experiments demonstrated that fibril alignment tends to be irreversible by showing that the alignment remains unchanged even after a cell is removed from the ECM or after cell contractility is inhibited (Fig. 6b, c) [138–140]. Remodeling of ECM consisting of collagen or fibrin cross-linked by non-covalent interactions was shown to be plastic deformation [141]. The plastic deformation of ECM plays a crucial role for cell invasion and migration. Several studies showed that cell migration leads to formation of tube-like structures or expansion of nanometer-scale matrix pores and this plastic deformation enhances migration of other cells later [139, 142].

In addition, cells can stiffen ECM, which helps a tissue maintain its integrity in response to large stress or force. ECM stiffening can occur at different length-scales. At individual fibril level, stiffening of fibrils can arise from their non-linear force-extension behavior [143]. A fibrin protofibril gel is a typical example showing such fibril-level stiffening [12, 144]. Stiffening of the matrix may emerge without the non-linear force-extension behavior of individual fibrils. As mentioned above, fibrils can be aligned in a specific direction due to applied stress or strain, resulting in a transition from bending-dominated deformation to stretching-dominated deformation [12, 145]. Because fibrils are generally able to resist extension better than bending, the transition leads to anisotropic stiffening of the ECM.

Cell-induced ECM stiffening was demonstrated in both in vitro and in vivo studies. It was shown that blood clots composed of fibrin protofibril can stiffen due to alignment of fibrin fibrils induced by contraction of fibroblasts or human mesenchymal stem cells [12, 146]. It was also demonstrated that contractile behaviors of cells stiffen collagen network (Fig. 7a) [147, 148] and matrigel [148]. The ECM stiffening has been reported to regulate stem cell fate [79]. It was shown that when the critical stress for the onset of stress-stiffening was increased, human mesenchymal stem cells preferentially show osteogenesis over adipogenesis [79]. The matrix stiffening can be enhanced by increasing cross-linking density [12] but also affected by viscoelastic and viscoplastic properties of the matrix. For example, a recent in vitro study showed that stiffening at high strain can be less due to stress relaxation originating from force-dependent unbinding of weak cross-links (Fig. 7b) [123].

3.3   Computational Models of ECM Remodeling

Computational models can describe and capture details of cell-ECM interactions, at spatiotemporal scales beyond experimental limits, in order to explain mechanisms of the ECM remodeling. In addition, it is feasible to calculate distribution of mechanical

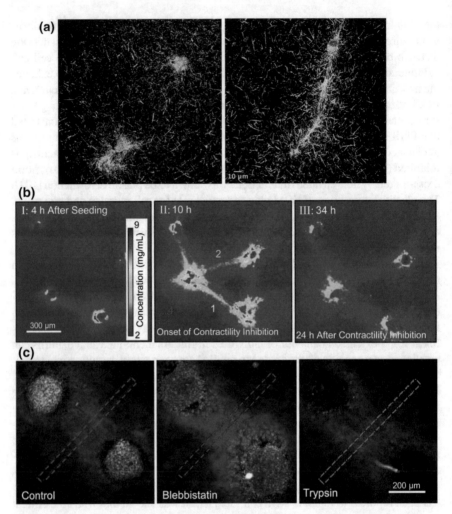

Fig. 6 Experiments demonstrating the cell-induced contraction and alignment of the ECM. **a** A pair of fibroblasts embedded in a collagen gel at 4 h (left) and 24 h (right) after seeding. At 24 h after seeding, collagen fibers between cells are densified and aligned, implying that they sense the existence of other cells located distantly. Reprinted with permission from [11]. **b** Contractile mammary acini in a collagen result in densification of collagen that remains for 24 h after contractility inhibition. **c** Fibroblast spheroids in a collagen gel align collagen fibers, and the alignment persists after contractility is inhibited by blebbistatin or after removal of spheroids via trypsin. **b**, **c** are reprinted with permission from [140]

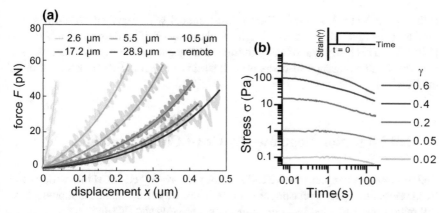

Fig. 7 Complicated mechanical properties of ECMs. **a** A relationship between force and displacement of a collagen gel with MDA-MB-231 cells measured by laser tweezers. These were measured at several locations with different distances from the cells. A curve for "remote" represents locations very far from the cell (>200 μm). The non-linear relationship in all cases indicates the strain-stiffening behaviors of the gel which means gel stiffness increases with larger deformation. Note that the gel tends to be stiffer in locations closer to the cells because cells make their vicinity stiffer by exerting contractile forces. Reprinted with permission from [148]. **b** A collagen gel exhibits enhanced stress relaxation with higher shear strain. The strain-dependent stress relaxation is attributed to force-dependent unbinding of weak cross-links between collagen fibers. Reprinted with permission from [123]

force or stress on the ECM generated by cells using computational models, which cannot be done accurately in experiments due to the heterogeneity of the ECM and the invasive nature of measurement techniques [149].

Several computational models were developed to study effects of forces generated by cells on surrounding ECM modeled as a continuum material. Such a continuum-based ECM model is suitable for tissue-level simulations due to its lower computational cost compared to discrete models. For example, pattern formation during tissue morphogenesis [150] and ECM contraction induced by fibroblasts [151] were simulated using a model where ECM is described as an isotropic viscoelastic material with cells modeled as stress fields. However, the continuum-based ECM model cannot easily account for non-affine deformation of ECM and structural remodeling of ECM such as fibril alignment.

Recently, discrete models have been used to capture the deformation and remodeling of ECMs more accurately. For example, a 2D FE model for a discrete fiber network demonstrated that distance over which force generated by cells propagate increases as a cell's aspect ratio increases (Fig. 8a, b) [152]. To overcome limitations of 2D models, recent studies developed various 3D models. For example, to account for large-scale ECM remodeling, researchers developed fiber network models in the presence of permanent or transient cross-links. A lattice-based model of fiber networks with permanent cross-links was employed to demonstrate stress amplification and the extended range of force transmission in disordered networks with contractile cells (Fig. 8c) [153]. Another fiber network model with transient cross-links was used to investigate the molecular origin of plasticity of ECM deformation (Fig. 8d) [140, 154]. This study found that both force-dependent unbinding of cross-linkers at low strain and elongation of fibers at high strain can lead to plastic deformation of the ECM.

4 Cell Migration Regulated by Cell-ECM Interactions

Cell migration is a representative cell behavior for which cell-ECM interactions and ECM remodeling play a very important role. Thus, we review literature showing how cell migration is affected by interactions between cells and ECM.

4.1 Roles of Adhesion Between Cells and ECM for Cell Migration

Traditional experimental studies on cell motility focused primarily on migration on a 2D substrate because 2D migration experiments were much easier to perform and more suitable for real-time imaging [155, 156]. For 2D migration, formation and maturation of FAs are particularly important (Fig. 9a) [22, 157]. Stabilization and maturation of FAs enable cells to move effectively by transmitting contractile forces generated by the cytoskeleton to the substrate. An experimental study found a high correlation between the size, direction, and orientation of the cell-ECM adhesions and the ECM deformation caused by migrating cells [158]. This demonstrates that intracellular forces transmitted to ECM via the FAs indeed play an important role for cell migration. Once a cell nucleus passes over FA sites after symmetry breaking, the FAs located on the side of the trailing edge are disassembled for net movement (Fig. 9b) [159, 160]. Several factors, including binding affinity, ligand density, integrin expression level, and cytoskeletal dynamics, cooperatively regulate cell migration. Early in vitro studies showed the strong dependence of migration speed on binding affinity between integrin and ligands [161, 162]. High affinity reduced migration speed, regardless of which one was modified between ligand or integrin receptor for reduction of affinity [161]. Density of ligands also affects migration speed; the speed was maximal at optimal ligand density. With higher integrin-ligand affinity, the optimal ligand density was lower. One study using a multi-dimensional model system within microenvironments showed the significance of myosin-dependent FA dynamics for cell migration [163, 164]; cells in the confined geometry need active mechanical coupling between a contractile actomyosin network and FAs to maintain protrusive dynamics at the leading edge for migration.

Recent studies have focused on cell migration within 3D matrices which is more physiologically relevant. Cells migrating within a 3D matrix show very different

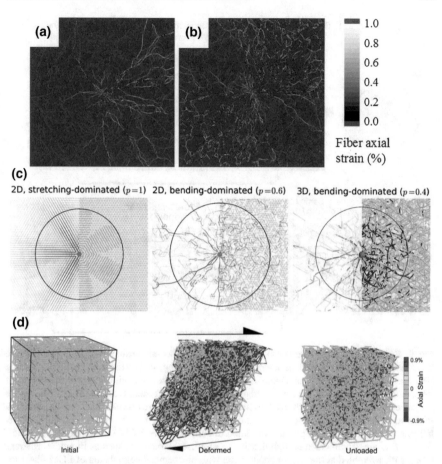

Fig. 8 Computational models designed for investigating mechanical properties of ECMs and interactions between cells and ECMs. **a, b** A two-dimensional (2D) discrete network model composed of randomly oriented fibers and rigid cross-links. A cell with **a** low or **b** high aspect ratio is located at the center of the network, exerting contractile forces to its vicinity. An elongated cell results in anisotropic deformation of the network as well as force propagation over a longer distance. The dimensions of the networks are **a** 100 μm × 100 μm and **b** 170 μm × 170 μm. Reprinted with permission from [152]. **c** 2D and three-dimensional (3D) lattice-based fiber network models. In the 2D cases, network connectivity (p) was varied to let networks to be in a bending-dominated or stretching-dominated regime. The red circles and sphere located at the center of each network represent a force generating unit that mimics actions of contracting cells. Blue and red in the left half of each figure indicate tensile and compressive stress, respectively. Red and green in the right half of each figure show buckled and unbuckled fibers, respectively. Force propagation is better in networks in the bending-dominated regime, and buckling of fibers also helps the force propagation. Reprinted with permission from [153]. **d** A 3D lattice-based fiber network model. This model can account for plastic deformation, so the microscopic axial strain on fibers does not disappear completely even in the unloaded state after shear deformation. Reprinted with permission from [140]

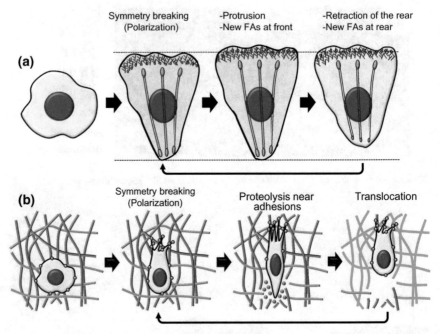

Fig. 9 Roles of focal adhesions (FAs) in cell migration. **a** Two-dimensional cell migration. Before the initiation of migration, a cell undergoes symmetry breaking, leading to the emergence of leading (front) and trailing (rear) edges. Both sides of stress fibers are coupled to matured FAs. Nascent FAs are formed mainly near the leading edge where lamellipodia extend to the surroundings. As the leading edge keeps moving forward, FAs near the trailing edge are disassembled for the retraction of the rear part, followed by formation and maturation of new FAs on the end of stress fibers. **b** Three-dimensional cell migration. Migration of cells within an ECM is also initiated by polarization. FAs are accumulated at high density on protrusive structures, such as filopodia, whereas FAs are disassembled at the rear in conjunction with biochemical degradation of ECM fibers for net translocation of the cell

migratory behaviors compared to those emerging in 2D migration [165]. For example, cells have thinner, more elongated shapes and show more persistent movement in 3D matrices [165]. Also, cells can change substrate compliance more freely in 3D matrices, thus altering integrin distribution that can affect migration [165]. Cukierman et al. designed a 3D matrix derived from mouse embryos or fibroblasts and showed that fibroblasts are attached more strongly to the matrix, compared to conventional 2D matrices [166]. Another difference is that integrin does not exhibit substantial local clustering behaviors that lead to FA sites in a cell adhered on a 2D surface [167]. Adhesions formed on a 3D matrix are more elongated than adhesions formed on 2D substrates [166]. Adhesions between the cell and matrix in a 3D environment regulate cell migration by affecting dynamics of protrusion and retraction at the cell edge as well as by ECM deformation [168, 169]. In contrast to 2D migration, integrin-mediated adhesions are not indispensable for cell migration in 3D environments [170–172]. Even without physical adhesions to the ECM, cell motility can be

driven by fast actomyosin retrograde flow which generates sufficient friction pro-
pelling the cell forward [172, 173]. The friction-driven adhesion-independent cell
migration in 3D environments can also be triggered and sustained via an osmotic
pressure gradient induced by active water transport across a semi-permeable plasma
membrane [174]. However, the efficiency of adhesion-independent cell migration
requires further quantification.

4.2   Effects of ECM Stiffness on Cell Migration

Migration of cells placed on substrates, such as PDMS or collagen gels, is sensitive to ECM stiffness [3]. Dependence of migration speed on ECM stiffness has
been observed in various in vitro experiments designed for studying epithelial-
mesenchymal transition, tumor invasion/metastasis, and capillary morphogenesis
[72, 175–179]. Tumorigenesis is sensitive to the concentration and cross-linking
density of ECM fibers [180]; stiffer ECM results in more stable FAs and more effective invasion of the epithelium [101]. In a study with glioma cells migrating on
polyacrylamide gels with Young's modulus ranging from 0.08 to 119 kPa, it was
found that the cells migrate faster on a stiffer substrate due to elevated cell spreading, maturation of FAs, and stress-fiber formation [179]. Another experiment with
vascular smooth muscle cells on polyacrylamide substrates with Young's modulus
between 1 and 308 kPa showed the biphasic dependence of the migration speed on
substrate stiffness, meaning that an increase in stiffness facilitates migration at lower
stiffness, but a further increase inhibits migration [181]. Optimal substrate stiffness
for the fastest migration with fibronectin density of 0.8 and 8 μg/cm^2 was 51.9 and
21.6 kPa, respectively. The optimal stiffness results in FAs that are stable enough to
allow cells to exert strong traction force and are also transient enough to enable cells
to slide forward fast without significant delay [72]. In addition, a stiffness gradient
regulates cell migration. Cells migrate preferentially toward stiffer substrates, which
is termed durotaxis [3]. Durotaxis is caused by generation of higher traction force on
stiffer regions, and it is enhanced in collective cell migration by force transmission
between cells [182].

Migration of cancer cells embedded within a 3D collagen matrix exhibits biphasic
dependence on collagen concentration [181, 183], meaning that the most efficient
migration occurs at intermediate collagen concentration. A matrix with higher collagen concentration enables cells to exert stronger traction forces on the matrix.
However, if pores become too small due to very high collagen concentration, cells
cannot migrate efficiently through small pores because of their cytoplasmic volume
or nucleus volume [184–186]. Thus, optimal collagen concentration exists for the
most efficient 3D cell migration. Mason et al. varied the stiffness of a collagen gel
between ~175 and ~515 Pa without a significant change in the gel architecture via
glycation before polymerization, showing that endothelial cells spread more with
higher gel stiffness (Fig. 10a) [177].

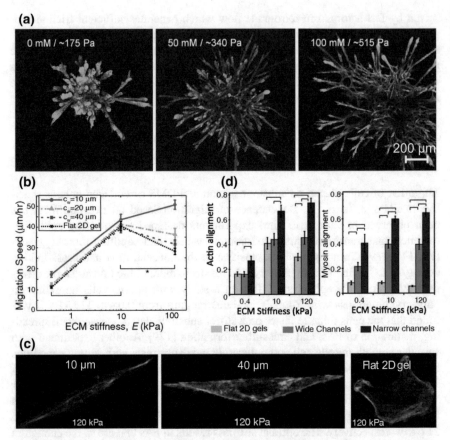

Fig. 10 Effects of matrix stiffness on cell spreading and migration. **a** Spheroids consisting of multiple endothelial cells were grown within three-dimensional collagen gels with tunable stiffness. An increase in stiffness enhanced spreading and branching of cells. Reprinted with permission from [177]. **b** Migration speed of spreading U373-MG human glioma cells placed on channels with various widths and stiffness values. Migration within a narrow channel shows a monotonic increase as ECM stiffness increases. However, a biphasic relationship between migration speed and stiffness emerges for wider channels or an unconfined flat 2D gel. **c** Independent regulation of cancer cell migration via variation of confinement. As the channel width increases, the polarized alignment of stress fibers (green) is reduced. On a flat 2D gel, stress fibers are randomly oriented. **d** Alignments of actin and myosin in stress fibers are affected by ECM stiffness and channel width. **b–d** are reprinted with permission [187]

Interestingly, cells migrating through narrow channels show monotonic dependence of migration speed on ECM stiffness [187]. Those cells migrate faster as ECM stiffness increases in contrast to the biphasic dependence of migration speed on ECM stiffness observed on unconfined ECMs (Fig. 10b). In addition, cells surprisingly migrate faster than those in wide channels or on unconstrained 2D surfaces, which is attributed to an increase in the polarization of cell-ECM traction forces (Fig. 10c, d).

4.3 Influences of Structural Properties of ECM on Cell Migration

ECMs have various structural properties depending on their types and conditions. In addition, since cells exert forces to surrounding ECM during migration, the structural properties of ECMs can change, resulting in alteration of biochemical and mechanical properties of the ECM, such as fiber alignment and porosity. All of these can affect characteristics of cell migration, such as speed and persistency.

It has been shown that dimensionality of ECM critically governs migratory behaviors. Cell migration looks quite different, depending on whether cells migrate on the ECM (2D migration) or within the ECM (3D migration). For example, fibroblasts show a spindle shape with fewer protrusions and fewer stress fibers during the 3D migration, compared to the 2D migration (Fig. 11a) [188]. It was found that there are similarities between the 3D migration and the 1D migration along micro-patterns in terms of significant roles of actomyosin contractility and anterior microtubule [189]. Cells in both cases exhibit rapid uniaxial migration that does not depend on ligand density because aligned fibers in the 3D matrix provide cells with 1D-like environments [189].

Fibers in ECMs are often oriented in an anisotropic fashion. For example, collagen fibers in mammary tumors are aligned perpendicularly to the boundary of tumor cells and stroma [190]. It was shown that aligned collagen fibers can enhance directional cell migration, which was termed contact guidance [191]; alignment of fibers increases the directionality of cell migration by inducing formation of protrusion, adhesion, and polarized actin structures in the direction of alignment (Fig. 11b) [190, 192]. It has been observed that the contact guidance plays an important role for invasion of cancer cells including breast cancer [13] and pancreatic ductal adenocarcinoma [193].

As mentioned earlier, porosity of ECMs also affects cell migration. It was found that cells migrate most effectively in a matrix with pores whose size is similar to or slightly smaller than size of the cells [194]. However, if pore diameter significantly decreases, limited deformability of a cell nucleus prevents cells from migrating through such small pores [195]. It was found that cells cannot squeeze through pores with a size smaller than 3 μm [186]. If pores are larger than the cell size, cells cannot form protrusions and adhesions on ECM well, so cell migration is impeded

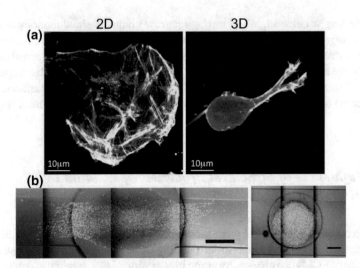

Fig. 11 Influences of dimension and anisotropy of ECMs on cell migration. **a** Cancer cells on a two-dimensional (2D) substrate and within a three-dimensional (3D) matrix. Within the 3D matrix, cells are elongated to a larger extent. Reprinted with permission from [188, 217]. **b** Migration of MDA-MB-231 cells in a matrix with aligned collagen fibers (left) and with randomly oriented collagen fibers (right). It was observed that migration of the cells is much faster in the anisotropic collagen matrix. Scale bars in both cases indicate 500 μm. Reprinted with permission from [190]

[194]. Recently, a study demonstrated that the shape of pores influences the efficiency of cell migration using a micro-fabricated wall with various pore shapes [196]; cells migrate most effectively through pores with long, narrowed shapes oriented along the apicobasal axis of the cells. Cells also modulate pore size during migration. Cells can compact adjacent fibers [197, 198] or degrade ECM components by secreting MMPs [186] to expand pores and thus enhance cell migration. Biochemical degradation of the ECM can further promote migration by changing migration mode. For example, degradation of the ECM leads to a transition from individual to collective migration in case of cancer cells [199]. However, inhibition of biochemical degradation does not prevent cell migration. It was shown that blocking proteolysis of tumor cells results in a transition from mesenchymal to amoeboid migration which does not require ECM degradation [200].

As discussed earlier in Sect. 4.2, concentration of ECM components is also an important factor that can govern cell migration because it affects several ECM properties, such as ECM stiffness, ligand density, and pore size. It was reported that collagen concentration highly affects cell shape and migration mode. In one study, it was shown that cancer cells in a high-density collagen matrix exhibit transdifferentiation and thus migrate collectively [201]. Other studies showed that cancer cells in a densely packed collagen matrix use invadopodia and degrade the ECM via MMP for migration, whereas cancer cells in a less dense collagen matrix form pseudopodial protrusions or blebs for amoeboid migration [99, 201, 202]. In addition, cells can vary collagen density of surrounding ECM, which can enhance invasiveness. In an in vivo study, it was shown that tumors expressing lysyl oxidase-related protein-1 (LOR-1) which increases density of collagen fibers exhibit higher invasiveness [203].

4.4 Computational Models of Cell Migration

Cell migration is an emergent behavior originating from complex interactions between cells and ECM, so various aspects can be further elucidated using computational models [204]. A number of computational models have been developed to study cell migration by accounting for the complex interactions in various ways. Several studies have focused on how ECM geometry affects cell migration. For example, a model based on computational fluid dynamics was used to investigate how ECM fiber alignment affects shear and drag forces that interstitial fluid exerts on cells (Fig. 12a) [205]. It was found that fibers aligned perpendicularly to the interstitial flow decrease shear and drag forces on cells, which can affect cell movement [205]. Other studies focused on how cell-induced ECM remodeling affects migratory behaviors. For example, to simulate 3D cell migration, a continuum model was developed with consideration of adhesions between ligands and receptors and MMP-induced ECM degradation [206]. The model showed that MMP can play two different roles for cell migration by increasing cell speed via removal of steric hindrance and by decreasing cell speed via reduction of traction forces. In addition, a mathematical model with a set of partial differential equations was used to demonstrate that MMP and LOX highly affect migration of cancer cells by degrading and cross-linking collagen fibers, respectively [207]. Both MMP and LOX result in the alignment and bundling of collagen fibers, leading to stiffening of ECM and enhancement of cell migration.

A cell's shape continuously changes during cell migration, affecting the efficiency and strategy of migration significantly. Several models were developed with consideration of the cell shape change. A 2D agent-based model was employed to investigate the effects of matrix geometry on cell migration (Fig. 12b) [208]. In this model, cell membrane and actomyosin cortex were modeled as nodes connected by viscoelastic elements, and an internal cell body was modeled as a viscoelastic material. The membrane nodes interact with highly simplified, non-deformable ECM structures by forming reversible adhesions that break in a force-dependent manner. The model demonstrated that cells can migrate fast by either filopodial protrusion or blebbing, depending on the geometry of a surrounding matrix, which is supported well by in vitro and in vivo experiments. In addition, an FE model was developed to simulate cell-ECM interactions in a microfluidic environment [209]. This model takes into account cell shape change by simplifying a cell into a variable number of voxels; voxels can be added or removed, depending on flow conditions, chemical concentration, and mechanical stress. Mechano-sensing of cells was modeled by incorporating contractility of cortex and cytoplasm that vary depending on ECM

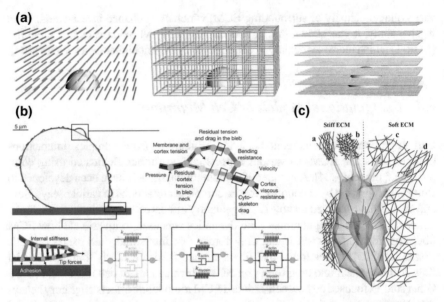

Fig. 12 Computational models for cell migration. **a** Computational fluid dynamics models for cell migration. Cells are modeled as continuum material subject to a fluid flow. Three different configurations of fibers are considered; fibers are aligned perpendicularly to the flow (left), form a cubical lattice (middle), or are aligned in parallel to the flow (right). This model showed that cells can reduce shear and drag forces exerted on them by the flow via alignment of fibers in a direction perpendicular to the interstitial flow. Reprinted with permission from [205]. **b** A two-dimensional cell migration model. A cell is simplified into a structure with actomyosin cortex, membrane, and osmotic pressure. Mechanics of the cortex and membrane is governed by viscoelastic elements. The cell forms filopodial protrusions or blebs, and it can adhere to continuous surfaces or ECM fibers in a reversible fashion. This model showed the relative importance of two types of protrusions and other factors depending on the geometry of extracellular environments. Reprinted with permission from [208]. **c** A three-dimensional migration model. A deformable cell with a nucleus and stress fibers interacts with a three-dimensional matrix composed of ECM fibers. The model showed that mechano-sensing at the tip of filopodia formed on the cell is important for cell migration. Reprinted with permission from [212]

stiffness. This model was able to predict diverse aspects of cell migration, such as velocities, trajectories, mechanical stress, and cell shapes.

A variety of computational models were developed to investigate durotaxis. For example, a 2D agent-based model with cells and fibers simplified into interconnected nodes was developed to demonstrate that ECM fiber alignment can lead to durotactic behaviors of cells [210]. Recently, a highly detailed model with deformable cells embedded in a 3D discrete matrix was developed to show important roles of interactions between filopodia and ECM fibers for durotaxis (Fig. 12c) [211, 212].

5 Summary and Concluding Remarks

In this chapter, we reviewed the mechanism and regulation of interactions between cells and ECM and how cells can migrate and remodel ECM via cell-ECM interactions. First, the molecular bases and cytoskeleton-based regulation of FAs were discussed, and experimental techniques and computational models designed for studying FAs were introduced. In the second section, cell-induced ECM remodeling was discussed. Cells can remodel surrounding ECMs in various ways. Cells deposit, cross-link, degrade, and align ECM components to contract and stiffen ECMs. In the last section, we reviewed literature that showed how cell-ECM interactions facilitate and regulate migratory behaviors of cells.

All of the studies reviewed in this chapter have significantly contributed to advances in understanding of cell-ECM interactions and how the interactions can affect and facilitate biological processes at larger length scales. Nevertheless, there is a long way to go because of many questions that remain unclear. First, roles of mechanical feedbacks between cells and ECMs for matrix remodeling and cell migration just started being considered and studied very recently. Many aspects of the feedback mechanisms are still elusive. For example, while a cell attempts to migrate by mechanically remodeling a surrounding ECM, it is likely that force-sensitive cytoskeletal proteins, such as myosin II and α-actinin, in the cell are spatially redistributed because the cell would feel highly heterogeneous forces transmitted from the ECM. The redistribution enables the cell to exert more anisotropic, focused forces on the ECM, which may enhance the efficiency of migration even more. Despite their potential importance, roles and mechanisms of such mechanical feedbacks have not been studied well yet. In addition, experiments with cells embedded within 3D ECMs have been widely performed for studying 3D cell migration and matrix remodeling relatively recently. Imaging live cells and cytoskeletal dynamics at high spatiotemporal resolution in such experiments remains very challenging. Thus, roles and variations of cytoskeletal structures during ECM remodeling and 3D cell migration have not been fully understood yet. Lastly, although several computational models have been developed in order to understand complex interactions between cells and ECM, most of them have various critical limitations and drastic simplifications, so insights from those models are often limited. It is expected that continuous advances in experimental techniques and computational models will help elucidate mechanisms of the cell-ECM interactions and related biological processes at cell and tissue scales gradually.

References

1. Heisenberg, C.P., Bellaïche, Y.: Forces in tissue morphogenesis and patterning. Cell **153**(5), 948–962 (2013)
2. Ghosh, K., Thodeti, C.K., Dudley, A.C., Mammoto, A., Klagsbrun, M., Ingber, D.E.: Tumor-derived endothelial cells exhibit aberrant Rho-mediated mechanosensing and abnormal angio-

genesis in vitro. Proc. Natl. Acad. Sci. U.S.A. **105**(32), 11305–11310 (2008)

3. Lo, C.M., Wang, H.B., Dembo, M., Wang, Y.L.: Cell movement is guided by the rigidity of the substrate. Biophys. J. **79**(1), 144–152 (2000)

4. Engler, A.J., Sen, S., Sweeney, H.L., Discher, D.E.: Matrix elasticity directs stem cell lineage specification. Cell **126**(4), 677–689 (2006)

5. Kumar, S., Weaver, V.M.: Mechanics, malignancy, and metastasis: the force journey of a tumor cell. Cancer Metastasis Rev. **28**(1–2), 113–127 (2009)

6. Wirtz, D., Konstantopoulos, K., Searson, P.C.: The physics of cancer: the role of physical interactions and mechanical forces in metastasis. Nat. Rev. Cancer **11**(7), 512–522 (2011)

7. Brugues, A., Anon, E., Conte, V., Veldhuis, J.H., Gupta, M., Colombelli, J., Munoz, J.J., Brodland, G.W., Ladoux, B., Trepat, X.: Forces driving epithelial wound healing. Nat. Phys. **10**(9), 683–690 (2014)

8. Li, B., Wang, J.H.: Fibroblasts and myofibroblasts in wound healing: force generation and measurement. J. Tissue Viability **20**(4), 108–120 (2011)

9. Han, W.J., Chen, S.H., Yuan, W., Fan, Q.H., Tian, J.X., Wang, X.C., Chen, L.Q., Zhang, X.X., Wei, W.L., Liu, R.C., Qu, J.L., Jiao, Y., Austin, R.H., Liu, L.Y.: Oriented collagen fibers direct tumor cell intravasation. Proc. Natl. Acad. Sci. U.S.A. **113**(40), 11208–11213 (2016)

10. Harris, A.K., Stopak, D., Wild, P.: Fibroblast Traction as a mechanism for Collagen Morphogenesis. Nature **290**(5803), 249–251 (1981)

11. Ma, X., Schickel, M.E., Stevenson, M.D., Sarang-Sieminski, A.L., Gooch, K.J., Ghadiali, S.N., Hart, R.T.: Fibers in the extracellular matrix enable long-range stress transmission between cells. Biophys. J. **104**(7), 1410–1418 (2013)

12. Winer, J.P., Oake, S., Janmey, P.A.: Non-linear elasticity of extracellular matrices enables contractile cells to communicate local position and orientation. PLoS ONE **4**(7), e6382 (2009)

13. Provenzano, P.P., Eliceiri, K.W., Campbell, J.M., Inman, D.R., White, J.G., Keely, P.J.: Collagen reorganization at the tumor-stromal interface facilitates local invasion. Bmc Med. **4** (2006)

14. Vishwanath, M., Ma, L., Otey, C.A., Jester, J.V., Petroll, W.M.: Modulation of corneal fibroblast contractility within fibrillar collagen matrices. Invest. Ophthalmol. Vis. Sci. **44**(11), 4724–4735 (2003)

15. Guo, C.L., Ouyang, M., Yu, J.Y., Maslov, J., Price, A., Shen, C.Y.: Long-range mechanical force enables self-assembly of epithelial tubular patterns. Proc. Natl. Acad. Sci. U.S.A. **109**(15), 5576–5582 (2012)

16. Huhtala, P., Humphries, M.J., McCarthy, J.B., Tremble, P.M., Werb, Z., Damsky, C.H.: Cooperative signaling by alpha 5 beta 1 and alpha 4 beta 1 integrins regulates metalloproteinase gene expression in fibroblasts adhering to fibronectin. J. Cell Biol. **129**(3), 867–879 (1995)

17. Humphries, J.D., Byron, A., Humphries, M.J.: Integrin ligands at a glance. J. Cell Sci. **119**(19), 3901–3903 (2006)

18. Takada, Y., Ye, X., Simon, S.: The integrins. Genome Biol. **8**(5), 215 (2007)

19. Brockbank, E.C., Bridges, J., Marshall, C.J., Sahai, E.: Integrin β1 is required for the invasive behaviour but not proliferation of squamous cell carcinoma cells in vivo. Br. J. Cancer **92**, 102 (2004)

20. Arnaout, M.A., Mahalingam, B., Xiong, J.P.: Integrin structure, allostery, and bidirectional signaling. Annu. Rev. Cell Dev. Biol. **21**, 381–410 (2005)

21. Tzima, E., del Pozo, M.A., Shattil, S.J., Chien, S., Schwartz, M.A.: Activation of integrins in endothelial cells by fluid shear stress mediates Rho-dependent cytoskeletal alignment. EMBO J. **20**(17), 4639–4647 (2001)

22. Chrzanowska-Wodnicka, M., Burridge, K.: Rho-stimulated contractility drives the formation of stress fibers and focal adhesions. J. Cell Biol. **133**(6), 1403–1415 (1996)

23. del Rio, A., Perez-Jimenez, R., Liu, R., Roca-Cusachs, P., Fernandez, J.M., Sheetz, M.P.: Stretching single talin rod molecules activates vinculin binding. Science **323**(5914), 638–641 (2009)

24. Delon, I., Brown, N.H.: Integrins and the actin cytoskeleton. Curr. Opin. Cell Biol. **19**(1), 43–50 (2007)

25. Burridge, K., Mangeat, P.: An interaction between vinculin and talin. Nature **308**(5961), 744–746 (1984)
26. Galbraith, C.G., Yamada, K.M., Sheetz, M.P.: The relationship between force and focal complex development. J. Cell Biol. **159**(4), 695–705 (2002)
27. DePasquale, J.A., Izzard, C.S.: Accumulation of talin in nodes at the edge of the lamellipodium and separate incorporation into adhesion plaques at focal contacts in fibroblasts. J. Cell Biol. **113**(6), 1351–1359 (1991)
28. Balaban, N.Q., Schwarz, U.S., Riveline, D., Goichberg, P., Tzur, G., Sabanay, I., Mahalu, D., Safran, S., Bershadsky, A., Addadi, L., Geiger, B.: Force and focal adhesion assembly: a close relationship studied using elastic micropatterned substrates. Nat. Cell Biol. **3**(5), 466–472 (2001)
29. Kong, F., García, A.J., Mould, A.P., Humphries, M.J., Zhu, C.: Demonstration of catch bonds between an integrin and its ligand. J. Cell Biol. **185**(7), 1275–1284 (2009)
30. Wang, X., Ha, T.: Defining single molecular forces required to activate integrin and notch signaling. Science **340**(6135), 991–994 (2013)
31. Hirata, H., Sokabe, M., Lim, C.T.: Molecular mechanisms underlying the force-dependent regulation of actin-to-ECM linkage at the focal adhesions. Prog. Mol. Biol. Transl. Sci. **126**, 135–154 (2014)
32. Choquet, D., Felsenfeld, D.P., Sheetz, M.P.: Extracellular matrix rigidity causes strengthening of integrin-cytoskeleton linkages. Cell **88**(1), 39–48 (1997)
33. Katz, B.Z., Zamir, E., Bershadsky, A., Kam, Z., Yamada, K.M., Geiger, B.: Physical state of the extracellular matrix regulates the structure and molecular composition of cell-matrix adhesions. Mol. Biol. Cell **11**(3), 1047–1060 (2000)
34. Riveline, D., Zamir, E., Balaban, N.Q., Schwarz, U.S., Ishizaki, T., Narumiya, S., Kam, Z., Geiger, B., Bershadsky, A.D.: Focal contacts as mechanosensors: externally applied local mechanical force induces growth of focal contacts by an mDia1-dependent and ROCK-independent mechanism. J. Cell Biol. **153**(6), 1175–1185 (2001)
35. Ballestrem, C., Hinz, B., Imhof, B.A., Wehrle-Haller, B.: Marching at the front and dragging behind: differential alpha-V beta 3-integrin turnover regulates focal adhesion behavior. J. Cell Biol. **155**(7), 1319–1332 (2001)
36. Stutchbury, B., Atherton, P., Tsang, R., Wang, D.Y., Ballestrem, C.: Distinct focal adhesion protein modules control different aspects of mechanotransduction. J. Cell Sci. **130**(9), 1612–1624 (2017)
37. Kanchanawong, P., Shtengel, G., Pasapera, A.M., Ramko, E.B., Davidson, M.W., Hess, H.F., Waterman, C.M.: Nanoscale architecture of integrin-based cell adhesions. Nature **468**(7323), 580–584 (2010)
38. Liu, J., Wang, Y., Goh, W.I., Goh, H., Baird, M.A., Ruehland, S., Teo, S., Bate, N., Critchley, D.R., Davidson, M.W., Kanchanawong, P.: Talin determines the nanoscale architecture of focal adhesions. Proc. Natl. Acad. Sci. U.S.A. **112**(35), E4864–E4873 (2015)
39. Zamir, E., Katz, M., Posen, Y., Erez, N., Yamada, K.M., Katz, B.Z., Lin, S., Lin, D.C., Bershadsky, A., Kam, Z., Geiger, B.: Dynamics and segregation of cell-matrix adhesions in cultured fibroblasts. Nat. Cell Biol. **2**(4), 191–196 (2000)
40. Bidone, T.C., Jung, W., Maruri, D., Borau, C., Kamm, R.D., Kim, T.: Morphological transformation and force generation of active cytoskeletal networks. PLoS Comput. Biol. **13**(1), e1005277 (2017)
41. Linsmeier, I., Banerjee, S., Oakes, P.W., Jung, W., Kim, T., Murrell, M.P.: Disordered actomyosin networks are sufficient to produce cooperative and telescopic contractility. Nat. Commun. **7**, 12615 (2016)
42. Li, J., Biel, T., Lomada, P., Yu, Q., Kim, T.: Buckling-induced F-actin fragmentation modulates the contraction of active cytoskeletal networks. Soft Matter **13**(17), 3213–3220 (2017)
43. Pellegrin, S., Mellor, H.: Actin stress fibres. J. Cell Sci. **120**(Pt 20), 3491–3499 (2007)
44. Zigmond, S.H.: Formin-induced nucleation of actin filaments. Curr. Opin. Cell Biol. **16**(1), 99–105 (2004)

45. Hotulainen, P., Lappalainen, P.: Stress fibers are generated by two distinct actin assembly mechanisms in motile cells. J. Cell Biol. **173**(3), 383–394 (2006)
46. Wu, Z., Plotnikov, S.V., Moalim, A.Y., Waterman, C.M., Liu, J.: Two distinct actin networks mediate traction oscillations to confer focal adhesion mechanosensing. Biophys. J. **112**(4), 780–794 (2017)
47. Welch, M.D., Mullins, R.D.: Cellular control of actin nucleation. Annu. Rev. Cell Dev. Biol. **18**, 247–288 (2002)
48. Kumar, S., Maxwell, I.Z., Heisterkamp, A., Polte, T.R., Lele, T.P., Salanga, M., Mazur, E., Ingber, D.E.: Viscoelastic retraction of single living stress fibers and its impact on cell shape, cytoskeletal organization, and extracellular matrix mechanics. Biophys. J. **90**(10), 3762–3773 (2006)
49. Tamada, M., Sheetz, M.P., Sawada, Y.: Activation of a signaling cascade by cytoskeleton stretch. Dev. Cell **7**(5), 709–718 (2004)
50. Kostic, A., Sheetz, M.P.: Fibronectin rigidity response through Fyn and p130Cas recruitment to the leading edge. Mol. Biol. Cell **17**(6), 2684–2695 (2006)
51. Gupta, M., Sarangi, B.R., Deschamps, J., Nematbakhsh, Y., Callan-Jones, A., Margadant, F., Mege, R.M., Lim, C.T., Voituriez, R., Ladoux, B.: Adaptive rheology and ordering of cell cytoskeleton govern matrix rigidity sensing. Nat. Commun. **6**, 7525 (2015)
52. Thoumine, O., Nerem, R.M., Girard, P.R.: Oscillatory shear stress and hydrostatic pressure modulate cell-matrix attachment proteins in cultured endothelial cells. Vitro Cell Dev. Biol. Anim. **31**(1), 45–54 (1995)
53. Alexandrova, A.Y., Arnold, K., Schaub, S., Vasiliev, J.M., Meister, J.J., Bershadsky, A.D., Verkhovsky, A.B.: Comparative dynamics of retrograde actin flow and focal adhesions: formation of nascent adhesions triggers transition from fast to slow flow. PLoS ONE **3**(9), e3234 (2008)
54. Albiges-Rizo, C., Destaing, O., Fourcade, B., Planus, E., Block, M.R.: Actin machinery and mechanosensitivity in invadopodia, podosomes and focal adhesions. J. Cell Sci. **122**(Pt 17), 3037–3049 (2009)
55. Choi, C.K., Vicente-Manzanares, M., Zareno, J., Whitmore, L.A., Mogilner, A., Horwitz, A.R.: Actin and alpha-actinin orchestrate the assembly and maturation of nascent adhesions in a myosin II motor-independent manner. Nat. Cell Biol. **10**(9), 1039–1050 (2008)
56. Steketee, M.B., Tosney, K.W.: Three functionally distinct adhesions in filopodia: shaft adhesions control lamellar extension. J. Neurosci. **22**(18), 8071–8083 (2002)
57. Schafer, C., Borm, B., Born, S., Mohl, C., Eibl, E.M., Hoffmann, B.: One step ahead: role of filopodia in adhesion formation during cell migration of keratinocytes. Exp. Cell Res. **315**(7), 1212–1224 (2009)
58. Romero, S., Quatela, A., Bornschlog, T., Guadagnini, S., Bassereau, P., Guy, T.V.N.: Filopodium retraction is controlled by adhesion to its tip. J. Cell Sci. **125**(21), 4999–5004 (2012)
59. Fierro-Gonzalez, J.C., White, M.D., Silva, J.C., Plachta, N.: Cadherin-dependent filopodia control preimplantation embryo compaction. Nat. Cell Biol. **15**(12), 1424–1433 (2013)
60. Heidemann, S.R., Lamoureux, P., Buxbaum, R.E.: Growth cone behavior and production of traction force. J. Cell Biol. **111**(5 Pt 1), 1949–1957 (1990)
61. Bornschlogl, T.: How filopodia pull: what we know about the mechanics and dynamics of filopodia. Cytoskeleton (Hoboken) **70**(10), 590–603 (2013)
62. Jacquemet, G., Hamidi, H., Ivaska, J.: Filopodia in cell adhesion, 3D migration and cancer cell invasion. Curr. Opin. Cell Biol. **36**, 23–31 (2015)
63. Albuschies, J., Vogel, V.: The role of filopodia in the recognition of nanotopographies. Sci. Rep. **3**, 1658 (2013)
64. Lee, D., Fong, K.P., King, M.R., Brass, L.F., Hammer, D.A.: Differential dynamics of platelet contact and spreading. Biophys. J. **102**(3), 472–482 (2012)
65. Disanza, A., Bisi, S., Winterhoff, M., Milanesi, F., Ushakov, D.S., Kast, D., Marighetti, P., Romet-Lemonne, G., Muller, H.M., Nickel, W., Linkner, J., Waterschoot, D., Ampe, C., Cortellino, S., Palamidessi, A., Dominguez, R., Carlier, M.F., Faix, J., Scita, G.: CDC42

switches IRSp53 from inhibition of actin growth to elongation by clustering of VASP. EMBO J. **32**(20), 2735–2750 (2013)

66. Gardel, M.L., Schneider, I.C., Aratyn-Schaus, Y., Waterman, C.M.: Mechanical integration of actin and adhesion dynamics in cell migration. Annu. Rev. Cell Dev. Biol. **26**, 315–333 (2010)

67. Jahed, Z., Shams, H., Mehrbod, M., Mofrad, M.R.: Mechanotransduction pathways linking the extracellular matrix to the nucleus. Int. Rev. Cell Mol. Biol. **310**, 171–220 (2014)

68. Ponti, A., Machacek, M., Gupton, S.L., Waterman-Storer, C.M., Danuser, G.: Two distinct actin networks drive the protrusion of migrating cells. Science **305**(5691), 1782–1786 (2004)

69. Vallotton, P., Small, J.V.: Shifting views on the leading role of the lamellipodium in cell migration: speckle tracking revisited. J. Cell Sci. **122**(12), 1955 (2009)

70. Murphy, D.A., Courtneidge, S.A.: The'ins' and'outs' of podosomes and invadopodia: characteristics, formation and function. Nat. Rev. Mol. Cell Biol. **12**(7), 413 (2011)

71. Weaver, A.M.: Invadopodia: specialized cell structures for cancer invasion. Clin. Exp. Metas. **23**(2), 97–105 (2006)

72. Pelham Jr., R.J., Wang, Y.: Cell locomotion and focal adhesions are regulated by substrate flexibility. Proc. Natl. Acad. Sci. U.S.A. **94**(25), 13661–13665 (1997)

73. Liu, A.P., Chaudhuri, O., Parekh, S.H.: New advances in probing cell-extracellular matrix interactions. Integr. Biol. (Camb). **9**(5), 383–405 (2017)

74. Terranova, V.P., Hujanen, E.S., Loeb, D.M., Martin, G.R., Thornburg, L., Glushko, V.: Use of a reconstituted basement membrane to measure cell invasiveness and select for highly invasive tumor cells. Proc. Natl. Acad. Sci. U.S.A. **83**(2), 465–469 (1986)

75. McKinnon, D.D., Domaille, D.W., Cha, J.N., Anseth, K.S.: Biophysically defined and cytocompatible covalently adaptable networks as viscoelastic 3D cell culture systems. Adv. Mater. **26**(6), 865–872 (2014)

76. Butcher, D.T., Alliston, T., Weaver, V.M.: A tense situation: forcing tumour progression. Nat. Rev. Cancer **9**(2), 108–122 (2009)

77. Pathak, A., Kumar, S.: Biophysical regulation of tumor cell invasion: moving beyond matrix stiffness. Integr. Biol. (Camb). **3**(4), 267–278 (2011)

78. Chaudhuri, O., Koshy, S.T., Branco da Cunha, C., Shin, J.W., Verbeke, C.S., Allison, K.H., Mooney, D.J.: Extracellular matrix stiffness and composition jointly regulate the induction of malignant phenotypes in mammary epithelium. Nat. Mater. **13**(10), 970–978 (2014)

79. Das, R.K., Gocheva, V., Hammink, R., Zouani, O.F., Rowan, A.E.: Stress-stiffening-mediated stem-cell commitment switch in soft responsive hydrogels. Nat. Mater. **15**(3), 318–325 (2016)

80. Gallant, N.D., Michael, K.E., Garcia, A.J.: Cell adhesion strengthening: contributions of adhesive area, integrin binding, and focal adhesion assembly. Mol. Biol. Cell **16**(9), 4329–4340 (2005)

81. Kita, A., Sakurai, Y., Myers, D.R., Rounsevell, R., Huang, J.N., Seok, T.J., Yu, K., Wu, M.C., Fletcher, D.A., Lam, W.A.: Microenvironmental geometry guides platelet adhesion and spreading: a quantitative analysis at the single cell level. PLoS ONE **6**(10), e26437 (2011)

82. Sim, J.Y., Moeller, J., Hart, K.C., Ramallo, D., Vogel, V., Dunn, A.R., Nelson, W.J., Pruitt, B.L.: Spatial distribution of cell-cell and cell-ECM adhesions regulates force balance while maintaining E-cadherin molecular tension in cell pairs. Mol. Biol. Cell **26**(13), 2456–2465 (2015)

83. Song, J., Shawky, J.H., Kim, Y., Hazar, M., LeDuc, P.R., Sitti, M., Davidson, L.A.: Controlled surface topography regulates collective 3D migration by epithelial-mesenchymal composite embryonic tissues. Biomaterials **58**, 1–9 (2015)

84. Oakes, P.W., Banerjee, S., Marchetti, M.C., Gardel, M.L.: Geometry regulates traction stresses in adherent cells. Biophys. J. **107**(4), 825–833 (2014)

85. Alford, P.W., Nesmith, A.P., Seywerd, J.N., Grosberg, A., Parker, K.K.: Vascular smooth muscle contractility depends on cell shape. Integr. Biol. (Camb). **3**(11), 1063–1070 (2011)

86. Agarwal, A., Farouz, Y., Nesmith, A.P., Deravi, L.F., McCain, M.L., Parker, K.K.: Micropatterning alginate substrates for in vitro cardiovascular muscle on a chip. Adv. Funct. Mater. **23**(30), 3738–3746 (2013)

87. Thery, M., Racine, V., Pepin, A., Piel, M., Chen, Y., Sibarita, J.B., Bornens, M.: The extracellular matrix guides the orientation of the cell division axis. Nat. Cell Biol. **7**(10), 947–953 (2005)

88. Lee, J., Abdeen, A.A., Huang, T.H., Kilian, K.A.: Controlling cell geometry on substrates of variable stiffness can tune the degree of osteogenesis in human mesenchymal stem cells. J. Mech. Behav. Biomed. Mater. **38**, 209–218 (2014)

89. Duffy, R.M., Sun, Y., Feinberg, A.W.: Understanding the role of ECM protein composition and geometric micropatterning for engineering human skeletal muscle. Ann. Biomed. Eng. **44**(6), 2076–2089 (2016)

90. Tee, S.Y., Fu, J., Chen, C.S., Janmey, P.A.: Cell shape and substrate rigidity both regulate cell stiffness. Biophys. J. **100**(5), L25–L27 (2011)

91. Maheshwari, G., Brown, G., Lauffenburger, D.A., Wells, A., Griffith, L.G.: Cell adhesion and motility depend on nanoscale RGD clustering. J. Cell Sci. **113**(Pt 10), 1677–1686 (2000)

92. Cavalcanti-Adam, E.A., Volberg, T., Micoulet, A., Kessler, H., Geiger, B., Spatz, J.P.: Cell spreading and focal adhesion dynamics are regulated by spacing of integrin ligands. Biophys. J. **92**(8), 2964–2974 (2007)

93. Selhuber-Unkel, C., Erdmann, T., Lopez-Garcia, M., Kessler, H., Schwarz, U.S., Spatz, J.P.: Cell adhesion strength is controlled by intermolecular spacing of adhesion receptors. Biophys. J. **98**(4), 543–551 (2010)

94. Heydarkhan-Hagvall, S., Chof, C.H., Dunn, J., Heydarkhan, S., Schenke-Layland, K., Maclellan, W.R., Beygul, R.E.: Influence of systematically varied nano-scale topography on cell morphology and adhesion. Cell Commun. Adhes. **14**(5), 181–194 (2007)

95. Dalby, M.J., Gadegaard, N., Wilkinson, C.D.: The response of fibroblasts to hexagonal nanotopography fabricated by electron beam lithography. J. Biomed. Mater. Res. A. **84**(4), 973–979 (2008)

96. Lamers, E., Walboomers, X.F., Domanski, M., te Riet, J., van Delft, F.C.M.J.M., Luttge, R., Winnubst, L.A.J.A., Gardeniers, H.J.G.E., Jansen, J.A.: The influence of nanoscale grooved substrates on osteoblast behavior and extracellular matrix deposition. Biomaterials. **31**(12), 3307–3316 (2010)

97. Petroll, W.M.: Dynamic assessment of cell-matrix mechanical interactions in three-dimensional culture. Methods Mol. Biol. **370**, 67–82 (2007)

98. Friedl, P., Wolf, K., von Andrian, U.H., Harms, G.: Biological second and third harmonic generation microscopy. Curr. Protoc. Cell Biol. Chapter 4, Unit 4 15 (2007)

99. Giampieri, S., Manning, C., Hooper, S., Jones, L., Hill, C.S., Sahai, E.: Localized and reversible TGFβ signalling switches breast cancer cells from cohesive to single cell motility. Nat. Cell Biol. **11**(11), 1287–1296 (2009)

100. Cox, T.R., Erler, J.T.: Remodeling and homeostasis of the extracellular matrix: implications for fibrotic diseases and cancer. Dis. Model Mech. **4**(2), 165–178 (2011)

101. Levental, K.R., Yu, H., Kass, L., Lakins, J.N., Egeblad, M., Erler, J.T., Fong, S.F., Csiszar, K., Giaccia, A., Weninger, W., Yamauchi, M., Gasser, D.L., Weaver, V.M.: Matrix crosslinking forces tumor progression by enhancing integrin signaling. Cell **139**(5), 891–906 (2009)

102. Kraning-Rush, C.M., Califano, J.P., Reinhart-King, C.A.: Cellular traction stresses increase with increasing metastatic potential. PLoS ONE **7**(2), e32572 (2012)

103. Dembo, M., Wang, Y.L.: Stresses at the cell-to-substrate interface during locomotion of fibroblasts. Biophys. J. **76**(4), 2307–2316 (1999)

104. Weng, S., Fu, J.: Synergistic regulation of cell function by matrix rigidity and adhesive pattern. Biomaterials **32**(36), 9584–9593 (2011)

105. Legant, W.R., Choi, C.K., Miller, J.S., Shao, L., Gao, L., Betzig, E., Chen, C.S.: Multidimensional traction force microscopy reveals out-of-plane rotational moments about focal adhesions. Proc. Natl. Acad. Sci. U.S.A. **110**(3), 881–886 (2013)

106. Sun, Z., Martinez-Lemus, L.A., Hill, M.A., Meininger, G.A.: Extracellular matrix-specific focal adhesions in vascular smooth muscle produce mechanically active adhesion sites. Am. J. Physiol. Cell Physiol. **295**(1), C268–C278 (2008)

107. Stanton, M.M., Parrillo, A., Thomas, G.M., McGimpsey, W.G., Wen, Q., Bellin, R.M., Lambert, C.R.: Fibroblast extracellular matrix and adhesion on microtextured polydimethylsiloxane scaffolds. J. Biomed. Mater. Res. B Appl. Biomater. **103**(4), 861–869 (2015)
108. Grashoff, C., Hoffman, B.D., Brenner, M.D., Zhou, R., Parsons, M., Yang, M.T., McLean, M.A., Sligar, S.G., Chen, C.S., Ha, T., Schwartz, M.A.: Measuring mechanical tension across vinculin reveals regulation of focal adhesion dynamics. Nature **466**(7303), 263–266 (2010)
109. Morimatsu, M., Mekhdjian, A.H., Adhikari, A.S., Dunn, A.R.: Molecular tension sensors report forces generated by single integrin molecules in living cells. Nano Lett. **13**(9), 3985–3989 (2013)
110. Zielinski, R., Mihai, C., Kniss, D., Ghadiali, S.N.: Finite element analysis of traction force microscopy: influence of cell mechanics, adhesion, and morphology. J. Biomech. Eng. **135**(7), 71009 (2013)
111. Chan, C.E., Odde, D.J.: Traction dynamics of filopodia on compliant substrates. Science **322**(5908), 1687–1691 (2008)
112. Owen, L.M., Adhikari, A.S., Patel, M., Grimmer, P., Leijnse, N., Kim, M.C., Notbohm, J., Franck, C., Dunn, A.R.: A cytoskeletal clutch mediates cellular force transmission in a soft, three-dimensional extracellular matrix. Mol. Biol. Cell **28**(14), 1959–1974 (2017)
113. Wang, Y.L.: Flux at focal adhesions: slippage clutch, mechanical gauge, or signal depot. Sci STKE. **2007**(377), pe10 (2007)
114. Aratyn-Schaus, Y., Gardel, M.L.: Transient frictional slip between integrin and the ECM in focal adhesions under myosin II tension. Curr. Biol. **20**(13), 1145–1153 (2010)
115. Hu, K., Ji, L., Applegate, K.T., Danuser, G., Waterman-Storer, C.M.: Differential transmission of actin motion within focal adhesions. Science **315**(5808), 111–115 (2007)
116. Gupton, S.L., Waterman-Storer, C.M.: Spatiotemporal feedback between actomyosin and focal-adhesion systems optimizes rapid cell migration. Cell **125**(7), 1361–1374 (2006)
117. Chaudhuri, O., Gu, L., Darnell, M., Klumpers, D., Bencherif, S.A., Weaver, J.C., Huebsch, N., Mooney, D.J.: Substrate stress relaxation regulates cell spreading. Nat. Commun. **6**, 6364 (2015)
118. Borau, C., Kim, T., Bidone, T., Garcia-Aznar, J.M., Kamm, R.D.: Dynamic mechanisms of cell rigidity sensing: insights from a computational model of actomyosin networks. PLoS ONE **7**(11), e49174 (2012)
119. Cao, X., Ban, E., Baker, B.M., Lin, Y., Burdick, J.A., Chen, C.S., Shenoy, V.B.: Multiscale model predicts increasing focal adhesion size with decreasing stiffness in fibrous matrices. Proc. Natl. Acad. Sci. U.S.A. **114**(23), E4549–E4555 (2017)
120. Muller, K.W., Bruinsma, R.F., Lieleg, O., Bausch, A.R., Wall, W.A., Levine, A.J.: Rheology of semiflexible bundle networks with transient linkers. Phys. Rev. Lett. **112**(23), 238102 (2014)
121. Munster, S., Jawerth, L.M., Leslie, B.A., Weitz, J.I., Fabry, B., Weitz, D.A.: Strain history dependence of the nonlinear stress response of fibrin and collagen networks. Proc. Natl. Acad. Sci. U.S.A. **110**(30), 12197–12202 (2013)
122. Bell, G.I.: Models for the specific adhesion of cells to cells. Science **200**(4342), 618–627 (1978)
123. Nam, S., Hu, K.H., Butte, M.J., Chaudhuri, O.: Strain-enhanced stress relaxation impacts nonlinear elasticity in collagen gels. Proc. Natl. Acad. Sci. **113**(20), 5492 (2016)
124. Frantz, C., Stewart, K.M., Weaver, V.M.: The extracellular matrix at a glance. J. Cell Sci. **123**(24), 4195–4200 (2010)
125. De Wever, O., Demetter, P., Mareel, M., Bracke, M.: Stromal myofibroblasts are drivers of invasive cancer growth. Int. J. Cancer **123**(10), 2229–2238 (2008)
126. Debnath, J., Brugge, J.S.: Modelling glandular epithelial cancers in three-dimensional cultures. Nat. Rev. Cancer **5**(9), 675–688 (2005)
127. Shea, K.S.: Differential deposition of basement membrane components during formation of the caudal neural tube in the mouse embryo. Development **99**(4), 509 (1987)
128. Ford, A.J., Rajagopalan, P.: Extracellular matrix remodeling in 3D: implications in tissue homeostasis and disease progression. **10**(4), e1503 (2018)

129. Bard, J.B., Higginson, K.: Fibroblast-collagen interactions in the formation of the secondary stroma of the chick cornea. J. Cell Biol. **74**(3), 816–827 (1977)
130. McDougall, S., Dallon, J., Sherratt, J., Maini, P.: Fibroblast migration and collagen deposition during dermal wound healing: mathematical modelling and clinical implications. Philos. Trans. A Math. Phys. Eng. Sci. **364**(1843), 1385–1405 (2006)
131. Alberts, B., Johnson, A., Lewis, J., Walter, P., Raff, M., Roberts, K.: Molecular Biology of the Cell, 4th edn. Routledge (2002)
132. Bonnans, C., Chou, J., Werb, Z.: Remodelling the extracellular matrix in development and disease. Nat. Rev. Mol. Cell Biol. **15**(12), 786–801 (2014)
133. Wolf, K., Friedl, P.: Mapping proteolytic cancer cell-extracellular matrix interfaces. Clin. Exp. Metastasis **26**(4), 289–298 (2009)
134. Jansen, K.A., Atherton, P., Ballestrem, C.: Mechanotransduction at the cell-matrix interface. Semin. Cell Dev. Biol. **71**, 75–83 (2017)
135. Rudnicki, M.S., Cirka, H.A., Aghvami, M., Sander, E.A., Wen, Q., Billiar, K.L.: Nonlinear strain stiffening is not sufficient to explain how far cells can feel on fibrous protein gels. Biophys. J. **105**(1), 11–20 (2013)
136. Ahmadzadeh, H., Webster, M.R., Behera, R., Valencia, A.M.J., Wirtz, D., Weeraratna, A.T., Shenoy, V.B.: Modeling the two-way feedback between contractility and matrix realignment reveals a nonlinear mode of cancer cell invasion. Proc. Natl. Acad. Sci. U.S.A. **114**(9), E1617–E1626 (2017)
137. Hall, M.S., Alisafaei, F., Ban, E., Feng, X.Z., Hui, C.Y., Shenoy, V.B., Wu, M.M.: Fibrous nonlinear elasticity enables positive mechanical feedback between cells and ECMs. Proc. Natl. Acad. Sci. U.S.A. **113**(49), 14043–14048 (2016)
138. Petroll, W.M., Cavanagh, H.D., Jester, J.V.: Dynamic three-dimensional visualization of collagen matrix remodeling and cytoskeletal organization in living corneal fibroblasts. Scanning **26**(1), 1–10 (2004)
139. Notbohm, J., Lesman, A., Tirrell, D.A., Ravichandran, G.: Quantifying cell-induced matrix deformation in three dimensions based on imaging matrix fibers. Integr. Biol. **7**(10), 1186–1195 (2015)
140. Ban, E., Franklin, J.M., Nam, S., Smith, L.R., Wang, H.L., Wells, R.G., Chaudhuri, O., Liphardt, J.T., Shenoy, V.B.: Mechanisms of plastic deformation in collagen networks induced by cellular forces. Biophys. J. **114**(2), 450–461 (2018)
141. Nam, S., Lee, J., Brownfield, D.G., Chaudhuri, O.: Viscoplasticity enables mechanical remodeling of matrix by cells. Biophys. J. **111**(10), 2296–2308 (2016)
142. Wisdom, K.M., Adebowale, K., Chang, J., Lee, J.Y., Nam, S., Desai, R., Rossen, N.S., Rafat, M., West, R.B., Hodgson, L., Chaudhuri, O.: Matrix mechanical plasticity regulates cancer cell migration through confining microenvironments. Nat. Commun. **9**(1), 4144 (2018)
143. Storm, C., Pastore, J.J., MacKintosh, F.C., Lubensky, T.C., Janmey, P.A.: Nonlinear elasticity in biological gels. Nature **435**(7039), 191–194 (2005)
144. Brown, A.E., Litvinov, R.I., Discher, D.E., Purohit, P.K., Weisel, J.W.: Multiscale mechanics of fibrin polymer: gel stretching with protein unfolding and loss of water. Science **325**(5941), 741–744 (2009)
145. Onck, P.R., Koeman, T., van Dillen, T., van der Giessen, E.: Alternative explanation of stiffening in cross-linked semiflexible networks. Phys. Rev. Lett. **95**(17), 178102 (2005)
146. Jansen, K.A., Bacabac, R.G., Piechocka, I.K., Koenderink, G.H.: Cells actively stiffen fibrin networks by generating contractile stress. Biophys. J. **105**(10), 2240–2251 (2013)
147. van Helvert, S., Friedl, P.: Strain stiffening of fibrillar collagen during individual and collective cell migration identified by AFM nanoindentation. ACS Appl. Mater. Interfaces. **8**(34), 21946–21955 (2016)
148. Han, Y.L., Ronceray, P., Xu, G.Q., Malandrino, A., Kamm, R.D., Lenz, M., Broedersz, C.P., Guo, M.: Cell contraction induces long-ranged stress stiffening in the extracellular matrix. Proc. Natl. Acad. Sci. U.S.A. **115**(16), 4075–4080 (2018)
149. Polacheck, W.J., Chen, C.S.: Measuring cell-generated forces: a guide to the available tools. Nat. Methods **13**(5), 415–423 (2016)

150. Murray, J.D., Oster, G.F.: Cell traction models for generating pattern and form in morphogenesis. J. Math. Biol. **19**(3), 265–279 (1984)
151. Ramtani, S., Fernandes-Morin, E., Geiger, D.: Remodeled-matrix contraction by fibroblasts: numerical investigations. Comput. Biol. Med. **32**(4), 283–296 (2002)
152. Abhilash, A.S., Baker, B.M., Trappmann, B., Chen, C.S., Shenoy, V.B.: Remodeling of fibrous extracellular matrices by contractile cells: predictions from discrete fiber network simulations. Biophys. J. **107**(8), 1829–1840 (2014)
153. Ronceray, P., Broedersz, C.P., Lenz, M.: Fiber networks amplify active stress. Proc. Natl. Acad. Sci. U.S.A. **113**(11), 2827–2832 (2016)
154. Malandrino, A., Mak, M., Trepat, X., Kamm, R.D.: Non-Elastic Remodeling of the 3D Extracellular Matrix by Cell-Generated Forces. bioRxiv (2017)
155. Abercrombie, M., Heaysman, J.E., Pegrum, S.M.: The locomotion of fibroblasts in culture. I. Movements of the leading edge. Exp Cell Res. **59**(3), 393–398 (1970)
156. Ridley, A.J., Schwartz, M.A., Burridge, K., Firtel, R.A., Ginsberg, M.H., Borisy, G., Parsons, J.T., Horwitz, A.R.: Cell migration: integrating signals from front to back. Science **302**(5651), 1704–1709 (2003)
157. Ridley, A.J., Hall, A.: The small GTP-binding protein rho regulates the assembly of focal adhesions and actin stress fibers in response to growth factors. Cell **70**(3), 389–399 (1992)
158. Petroll, W.M., Ma, L., Jester, J.V.: Direct correlation of collagen matrix deformation with focal adhesion dynamics in living corneal fibroblasts. J. Cell Sci. **116**(8), 1481–1491 (2003)
159. Chen, W.T.: Mechanism of retraction of the trailing edge during fibroblast movement. J. Cell Biol. **90**(1), 187–200 (1981)
160. Vicente-Manzanares, M., Newell-Litwa, K., Bachir, A.I., Whitmore, L.A., Horwitz, A.R.: Myosin IIA/IIB restrict adhesive and protrusive signaling to generate front-back polarity in migrating cells. J. Cell Biol. **193**(2), 381–396 (2011)
161. Huttenlocher, A., Ginsberg, M.H., Horwitz, A.F.: Modulation of cell migration by integrin-mediated cytoskeletal linkages and ligand-binding affinity. J. Cell Biol. **134**(6), 1551–1562 (1996)
162. Palecek, S.P., Loftus, J.C., Ginsberg, M.H., Lauffenburger, D.A., Horwitz, A.F.: Integrin-ligand binding properties govern cell migration speed through cell-substratum adhesiveness. Nature **385**(6616), 537–540 (1997)
163. Doyle, A.D., Kutys, M.L., Conti, M.A., Matsumoto, K., Adelstein, R.S., Yamada, K.M.: Micro-environmental control of cell migration—myosin IIA is required for efficient migration in fibrillar environments through control of cell adhesion dynamics. J. Cell Sci. **125**(9), 2244–2256 (2012)
164. Doyle, A.D., Carvajal, N., Jin, A., Matsumoto, K., Yamada, K.M.: Local 3D matrix microenvironment regulates cell migration through spatiotemporal dynamics of contractility-dependent adhesions. Nat. Commun. **6** (2015)
165. Even-Ram, S., Yamada, K.M.: Cell migration in 3D matrix. Curr. Opin. Cell Biol. **17**(5), 524–532 (2005)
166. Cukierman, E., Pankov, R., Stevens, D.R., Yamada, K.M.: Taking cell-matrix adhesions to the third dimension. Science **294**(5547), 1708–1712 (2001)
167. Fraley, S.I., Feng, Y.F., Krishnamurthy, R., Kim, D.H., Celedon, A., Longmore, G.D., Wirtz, D.: A distinctive role for focal adhesion proteins in three-dimensional cell motility. Nat. Cell Biol. **12**(6), 598–604 (2010)
168. Petrie, R.J., Yamada, K.M.: Multiple mechanisms of 3D migration: the origins of plasticity. Curr. Opin. Cell Biol. **42**, 7–12 (2016)
169. Kang, Y.G., Jang, H., Yang, T.D., Notbohm, J., Choi, Y., Park, Y., Kim, B.M.: Quantification of focal adhesion dynamics of cell movement based on cell-induced collagen matrix deformation using second-harmonic generation microscopy. J. Biomed. Opt. **23**(6) (2018)
170. Lammermann, T., Bader, B.L., Monkley, S.J., Worbs, T., Wedlich-Soldner, R., Hirsch, K., Keller, M., Forster, R., Critchley, D.R., Fassler, R., Sixt, M.: Rapid leukocyte migration by integrin-independent flowing and squeezing. Nature **453**(7191), 51–55 (2008)

171. Bergert, M., Chandradoss, S.D., Desai, R.A., Paluch, E.: Cell mechanics control rapid transitions between blebs and lamellipodia during migration. Proc. Natl. Acad. Sci. U.S.A. **109**(36), 14434–14439 (2012)
172. Liu, Y.J., Le Berre, M., Lautenschlaeger, F., Maiuri, P., Callan-Jones, A., Heuze, M., Takaki, T., Voituriez, R., Piel, M.: Confinement and low adhesion induce fast amoeboid migration of slow mesenchymal cells. Cell **160**(4), 659–672 (2015)
173. Bergert, M., Erzberger, A., Desai, R.A., Aspalter, I.M., Oates, A.C., Charras, G., Salbreux, G., Paluch, E.K.: Force transmission during adhesion-independent migration. Nat. Cell Biol. **17**(4), 524 (2015)
174. Stroka, K.M., Jiang, H., Chen, S.H., Tong, Z., Wirtz, D., Sun, S.X., Konstantopoulos, K.: Water permeation drives tumor cell migration in confined microenvironments. Cell **157**(3), 611–623 (2014)
175. Ghajar, C.M., Chen, X., Harris, J.W., Suresh, V., Hughes, C.C.W., Jeon, N.L., Putnam, A.J., George, S.C.: The effect of matrix density on the regulation of 3-D capillary morphogenesis. Biophys. J. **94**(5), 1930–1941 (2008)
176. Wei, S.C., Fattet, L., Tsai, J.H., Guo, Y.R., Pai, V.H., Majeski, H.E., Chen, A.C., Sah, R.L., Taylor, S.S., Engler, A.J., Yang, J.: Matrix stiffness drives epithelial mesenchymal transition and tumour metastasis through a TWIST1-G3BP2 mechanotransduction pathway. Nat. Cell Biol. **17**(5), 678–688 (2015)
177. Mason, B.N., Starchenko, A., Williams, R.M., Bonassar, L.J., Reinhart-King, C.A.: Tuning three-dimensional collagen matrix stiffness independently of collagen concentration modulates endothelial cell behavior. Acta Biomater. **9**(1), 4635–4644 (2013)
178. Zaman, M.H., Trapani, L.M., Siemeski, A., MacKellar, D., Gong, H.Y., Kamm, R.D., Wells, A., Lauffenburger, D.A., Matsudaira, P.: Migration of tumor cells in 3D matrices is governed by matrix stiffness along with cell-matrix adhesion and proteolysis. Proc. Natl. Acad. Sci. U.S.A. **103**(29), 10889–10894 (2006)
179. Ulrich, T.A., Pardo, E.M.D., Kumar, S.: The mechanical rigidity of the extracellular matrix regulates the structure, motility, and proliferation of glioma cells. Can. Res. **69**(10), 4167–4174 (2009)
180. Provenzano, P.P., Inman, D.R., Eliceiri, K.W., Knittel, J.G., Yan, L., Rueden, C.T., White, J.G., Keely, P.J.: Collagen density promotes mammary tumor initiation and progression. Bmc Med. **6** (2008)
181. Peyton, S.R., Putnam, A.J.: Extracellular matrix rigidity governs smooth muscle cell motility in a biphasic fashion. J. Cell. Physiol. **204**(1), 198–209 (2005)
182. Sunyer, R., Conte, V., Escribano, J., Elosegui-Artola, A., Labernadie, A., Valon, L., Navajas, D., García-Aznar, J.M., Muñoz, J.J., Roca-Cusachs, P., Trepat, X.: Collective cell durotaxis emerges from long-range intercellular force transmission. Science **353**(6304), 1157 (2016)
183. Lang, N.R., Skodzek, K., Hurst, S., Mainka, A., Steinwachs, J., Schneider, J., Aifantis, K.E., Fabry, B.: Biphasic response of cell invasion to matrix stiffness in three-dimensional biopolymer networks. Acta Biomater. **13**, 61–67 (2015)
184. Lautscham, L.A., Kammerer, C., Lange, J.R., Kolb, T., Mark, C., Schilling, A., Strissel, P.L., Strick, R., Gluth, C., Rowat, A.C., Metzner, C., Fabry, B.: Migration in confined 3D environments is determined by a combination of adhesiveness, nuclear volume, contractility, and cell stiffness. Biophys. J. **109**(5), 900–913 (2015)
185. Denais, C.M., Gilbert, R.M., Isermann, P., McGregor, A.L., te Lindert, M., Weigelin, B., Davidson, P.M., Friedl, P., Wolf, K., Lammerding, J.: Nuclear envelope rupture and repair during cancer cell migration. Science **352**(6283), 353–358 (2016)
186. Wolf, K., te Lindert, M., Krause, M., Alexander, S., te Riet, J., Willis, A.L., Hoffman, R.M., Figdor, C.G., Weiss, S.J., Friedl, P.: Physical limits of cell migration: control by ECM space and nuclear deformation and tuning by proteolysis and traction force. J. Cell Biol. **201**(7), 1069–1084 (2013)
187. Pathak, A., Kumar, S.: Independent regulation of tumor cell migration by matrix stiffness and confinement. Proc. Natl. Acad. Sci. U.S.A. **109**(26), 10334–10339 (2012)

188. Hakkinen, K.M., Harunaga, J.S., Doyle, A.D., Yamada, K.M.: Direct comparisons of the morphology, migration, cell adhesions, and actin cytoskeleton of fibroblasts in four different three-dimensional extracellular matrices. Tissue Eng. Part A **17**(5–6), 713–724 (2011)

189. Doyle, A.D., Wang, F.W., Matsumoto, K., Yamada, K.M.: One-dimensional topography underlies three-dimensional fibrillar cell migration. J. Cell Biol. **184**(4), 481–490 (2009)

190. Riching, K.M., Cox, B.L., Salick, M.R., Pehlke, C., Riching, A.S., Ponik, S.M., Bass, B.R., Crone, W.C., Jiang, Y., Weaver, A.M., Eliceiri, K.W., Keely, P.J.: 3D collagen alignment limits protrusions to enhance breast cancer cell persistence. Biophys. J. **107**(11), 2546–2558 (2014)

191. Dickinson, R.B., Guido, S., Tranquillo, R.T.: Biased cell-migration of fibroblasts exhibiting contact guidance in oriented collagen gels. Ann. Biomed. Eng. **22**(4), 342–356 (1994)

192. Ray, A., Lee, O., Win, Z., Edwards, R.M., Alford, P.W., Kim, D.H., Provenzano, P.P.: Anisotropic forces from spatially constrained focal adhesions mediate contact guidance directed cell migration. Nat. Commun. **8**, 14923 (2017)

193. Drifka, C.R., Tod, J., Loeffler, A.G., Liu, Y., Thomas, G.J., Eliceiri, K.W., Kao, W.J.: Periductal stromal collagen topology of pancreatic ductal adenocarcinoma differs from that of normal and chronic pancreatitis. Mod. Pathol. **28**(11), 1470–1480 (2015)

194. Friedl, P., Wolf, K.: Plasticity of cell migration: a multiscale tuning model. J. Cell Biol. **188**(1), 11 (2010)

195. Harada, T., Swift, J., Irianto, J., Shin, J.-W., Spinler, K.R., Athirasala, A., Diegmiller, R., Dingal, P.C.D.P., Ivanovska, I.L., Discher, D.E.: Nuclear lamin stiffness is a barrier to 3D migration, but softness can limit survival. J. Cell Biol. **204**(5), 669 (2014)

196. Green, B.J., Panagiotakopoulou, M., Pramotton, F.M., Stefopoulos, G., Kelley, S.O., Poulikakos, D., Ferrari, A.: Pore shape defines paths of metastatic cell migration. Nano Lett. **18**(3), 2140–2147 (2018)

197. Grinnell, F.: Fibroblasts, myofibroblasts, and wound contraction. J. Cell Biol. **124**(4), 401–404 (1994)

198. Stevenson, M.D., Sieminski, A.L., McLeod, C.M., Byfield, F.J., Barocas, V.H., Gooch, Keith J.: Pericellular conditions regulate extent of cell-mediated compaction of collagen gels. Biophys. J. **99**(1), 19–28 (2010)

199. Wolf, K., Wu, Y.I., Liu, Y., Geiger, J., Tam, E., Overall, C., Stack, M.S., Friedl, P.: Multi-step pericellular proteolysis controls the transition from individual to collective cancer cell invasion. Nat. Cell Biol. **9**(8), 893–904 (2007)

200. Wolf, K., Mazo, I., Leung, H., Engelke, K., von Andrian, U.H., Deryugina, E.I., Strongin, A.Y., Brocker, E.B., Friedl, P.: Compensation mechanism in tumor cell migration: mesenchymal-amoeboid transition after blocking of pericellular proteolysis. J. Cell Biol. **160**(2), 267–277 (2003)

201. Velez, D.O., Tsui, B., Goshia, T., Chute, C.L., Han, A., Carter, H., Fraley, S.I.: 3D collagen architecture induces a conserved migratory and transcriptional response linked to vasculogenic mimicry. Nat. Commun. **8**(1), 1651 (2017)

202. Gligorijevic, B., Bergman, A., Condeelis, J.: Multiparametric classification links tumor microenvironments with tumor cell phenotype. PLoS Biol. **12**(11), e1001995 (2014)

203. Akiri, G., Sabo, E., Dafni, H., Vadasz, Z., Kartvelishvily, Y., Gan, N., Kessler, O., Cohen, T., Resnick, M., Neeman, M.: Lysyl oxidase-related protein-1 promotes tumor fibrosis and tumor progression in vivo. Can. Res. **63**(7), 1657–1666 (2003)

204. He, X., Lee, B., Jiang, Y.: Cell-ECM interactions in tumor invasion. Adv. Exp. Med. Biol. **936**, 73–91 (2016)

205. Pedersen, J.A., Lichter, S., Swartz, M.A.: Cells in 3D matrices under interstitial flow: effects of extracellular matrix alignment on cell shear stress and drag forces. J. Biomech. **43**(5), 900–905 (2010)

206. Chisholm, R.H., Hughes, B.D., Landman, K.A., Zaman, M.H.: Analytic study of three-dimensional single cell migration with and without proteolytic enzymes. Cell Mol. Bioeng. **6**(2) (2013)

207. Edalgo, Y.T.N., Versypt, A.N.F.: Mathematical modeling of metastatic cancer migration through a remodeling extracellular matrix. Processes **6**(5) (2018)

208. Tozluoglu, M., Tournier, A.L., Jenkins, R.P., Hooper, S., Bates, P.A., Sahai, E.: Matrix geometry determines optimal cancer cell migration strategy and modulates response to interventions. Nat. Cell Biol. **15**(7), 751–762 (2013)
209. Borau, C., Polacheck, W.J., Kamm, R.D., Garcia-Aznar, J.M.: Probabilistic Voxel-Fe model for single cell motility in 3D. Silico Cell Tissue Sci. **1**, 2 (2014)
210. Reinhardt, J.W., Krakauer, D.A., Gooch, K.J.: Complex matrix remodeling and durotaxis can emerge from simple rules for cell-matrix interaction in agent-based models. J. Biomech. Eng. **135**(7), 071003-071003-071010 (2013)
211. Kim, M.-C., Whisler, J., Silberberg, Y.R., Kamm, R.D., Asada, H.H.: Cell invasion dynamics into a three dimensional extracellular matrix fibre network. PLoS Comput. Biol. **11**(10), e1004535 (2015)
212. Kim, M.-C., Silberberg, Y.R., Abeyaratne, R., Kamm, R.D., Asada, H.H.: Computational modeling of three-dimensional ECM-rigidity sensing to guide directed cell migration. Proc. Natl. Acad. Sci. U.S.A. **115**(3), E390–E399 (2018)
213. Vallenius, T.: Actin stress fibre subtypes in mesenchymal-migrating cells. Open Biol. **3**(6), 130001 (2013)
214. Yang, C., Czech, L., Gerboth, S., Kojima, S.-I., Scita, G., Svitkina, T.: Novel roles of formin mDia2 in lamellipodia and filopodia formation in motile cells. PLoS Biol. **5**(11), e317 (2007)
215. Bangasser, B.L., Odde, D.J.: Master equation-based analysis of a motor-clutch model for cell traction force. Cell. Mol. Bioeng. **6**(4), 449–459 (2013)
216. Bangasser, B.L., Rosenfeld, S.S., Odde, D.J.: Determinants of maximal force transmission in a motor-clutch model of cell traction in a compliant microenvironment. Biophys. J. **105**(3), 581–592 (2013)
217. Caswell, P.T., Zech, T.: Actin-based cell protrusion in a 3D matrix. Trends Cell Biol. (2018)

Quantification of Cell-Matrix Interaction in 3D Using Optical Tweezers

Satish Kumar Gupta, Jiawei Sun, Yu Long Han, Chenglin Lyu, Tianlei He and Ming Guo

Abstract The behavior of living cells is significantly affected by the mechanical properties of the surrounding soft extracellular matrix (ECM) comprising of various types of biopolymers. More complexity is added as cell-generated forces in turn can mechanically modify their microenvironment. Moreover, these forces can also act as mechanical signals for other cells leading to emergent collective cell dynamics. Bulk measurement techniques are not capable of resolving these local mechanical interactions which are often hidden in 3D, thus, optical tweezers have emerged as a powerful tool to directly characterize microscale mechanics and forces at play. In this chapter, we first introduce a typical experimental setup of optical tweezers and calibration methods that has been widely accepted by the mechanobiology community. Subsequently, we discuss various ways in which optical tweezers can be used to probe mechanics at different length scales such as the cytoplasm at the sub-cellular level, at the level of whole cell and finally explore the cell-cell and cell-matrix interaction. Later, perspectives on the future development of optical tweezers to study cell-matrix interaction is also provided.

Satish Kumar Gupta and Jiawei Sun contribute equally to this work.

S. K. Gupta · J. Sun · Y. L. Han · C. Lyu · T. He · M. Guo (✉)
Department of Mechanical Engineering, Massachusetts Institute of Technology,
Cambridge, MA 02139, USA
e-mail: guom@mit.edu

S. K. Gupta
e-mail: skgupta@mit.edu

J. Sun
e-mail: jiaweis@mit.edu

Y. L. Han
e-mail: ylhan@mit.edu

C. Lyu
e-mail: lvcl16@mails.tsinghua.edu.cn

T. He
e-mail: tianleihe@g.harvard.edu

© Springer Nature Switzerland AG 2020
Y. Zhang (ed.), *Multi-scale Extracellular Matrix Mechanics and Mechanobiology*,
Studies in Mechanobiology, Tissue Engineering and Biomaterials 23,
https://doi.org/10.1007/978-3-030-20182-1_9

1   Introduction

In 1970, Ashkin first reported that one could significantly affect the dynamics of small transparent particles by leveraging the forces of radiation pressure from a focused laser beam [3]. Using the scattering force in the direction of the incident light beam and a gradient force in the direction of the gradient of the beam, it was experimentally demonstrated that one could accelerate, decelerate and even trap micron-sized particles. However, it was only in 1986 when Ashkin and his colleagues demonstrated that a tightly focused light beam was capable of stably holding microscopic beads in three dimensions (3D) which now is typically referred to as optical tweezers [6]. It was originally envisioned for trapping and cooling of atoms [6] however, soon after its invention Ashkin and his colleagues realized its applicability in investigation of biological systems comprising of bacteria [5], viruses [5] and cells [7]. Subsequently, it was adopted and developed by numerous researchers and now optical trapping has emerged as a powerful technique to investigate broad ranging phenomena in living systems from single molecule spectroscopy to cell-matrix interaction [1, 14, 83, 84].

Historically, our understanding of physical properties of cells and their interaction with the surrounding environment has come from light microscopy. The very primary exploration was the observations of the protoplasm structure and its interaction with the surrounding environment. In the late seventeenth century, Van Leeuwenhoek reported the motion of ciliate *Vorticella* as a result of its interaction with the surrounding water [32]. Later, Brownian motion of the particles and organelles were reported in the cytoplasm and suggestions on extraction of viscosity of the protoplasm from the motion were made [52, 53]. Presently, we understand that the Brownian motion cannot be used to extract the mechanical information of the cytoplasm as cells operate out of equilibrium [20, 38, 44, 47, 67, 80, 93] with an exception at high frequencies [47]. These early attempts to understand biological world from a physical perspective were critical in establishing the idea that physical properties and mechanical forces play an important role in biological systems. Therefore, constant efforts have been devoted to develop novel approaches to study the physical properties of the living cells and the mechanical interaction with their microenvironments [11], including atomic force microscopy (AFM) [72], optical tweezers [28, 45], micropipette aspiration [56], magnetic tweezers [12], fluorescence resonance energy transfer (FRET) based force sensors [26] and traction force microscopy (TFM) [76, 81, 94]. Amongst these cutting-edge platforms, optical tweezers exhibit extraordinary advantages in studying the forces, mechanical properties and dynamics in 3D living systems due to its high spatial and force resolution, particularly the capability to perform measurement in 3D environments [83].

Comprehensive reviews on construction, applications and limitations of optical tweezers can be found in previous literature [51, 68, 75, 105]. The main focus of this chapter is kept on applications of optical tweezers in measuring mechanics, dynamics and forces in living cells and their surrounding extracellular matrices (ECM). We will first introduce a typical experimental setup of optical tweezers and calibration methods that is commonly used to study cells and tissues from a physical viewpoint.

Then, we discuss various ways in which optical tweezers can be used to probe mechanics at different length scales starting with the mechanical behavior of the cytoplasm at the sub-cellular level, move to mechanics at the level of whole cell and finally explore the cell-cell and cell-matrix interaction. We also briefly describe a recent method Force Spectrum Microscopy (FSM) [44] that uses optical tweezers to probe the stochastic force fluctuations in the cytoplasm as a result of activity of molecular motors and other active processes as these forces are critical in general operation of the cells. Finally, concluding remarks and perspectives for the future development of optical tweezers to study cell-matrix interaction are addressed.

2   Optical Tweezers Setup

In this section we describe an optical tweezer setup that is typically used for bio-physical studies. A schematic representation of the setup is shown in Fig. 1. The instrument is usually built on an inverted microscope and includes a laser for trapping and a LED source that provides broad-spectrum excitation for fluorescence microscopy. Alternatively, a laser source can also be used for fluorescence. A stage-up kit is usually incorporated to provide sufficient clearance for the rear entry of the laser beam and space for introducing fluorescent light in the same optical path which enables simultaneous imaging and trapping. The laser beam is introduced through the optical path bypassing the fluorescence filter cubes and hence, changing the filters does not affect the intensity, alignment, polarization and general quality of the laser beam. A 1064 nm laser coupled with single mode optical fiber that produces a Gaussian TEM00 mode beam profile with an excellent pointing stability and low power fluctuations is used. Pointing instabilities lead to unwanted displacements of the optical trap position in the specimen plane, whereas power fluctuations lead to temporal variations in the optical trap stiffness. The use of fiber also reduces mechanical noise as the laser source that comprises of electronics with fans and other noise causing elements can be located off the optical table. A Gaussian mode produces an efficient harmonic trap as it focuses the laser to the smallest diameter beam waist. The wavelength of the laser is chosen as 1064 nm to minimize photodamage to biological samples [82]. The beam from the optical fiber is first passed through a $\lambda/2$ wave plate and then a variable polarizing beam splitter (PBS) to achieve a linearly polarized light of desired intensity. To manipulate the trapped particle or more than one particle at once, a number of methods have been developed to create and manipulate planar arrays of optical traps using galvano [96] or piezoelectric scanning mirrors [79], acousto-optic deflectors (AODs) [105], interference of specially designed light beams [19] or the generalized phase contrast method [36]. Here, we describe a setup that uses an AOD to manipulate the trapping laser. The transmitted polarized light from the PBS is shrunk to a beam diameter of ~2 mm using a beam shrinker to reduce the response time of the AOD, elaborated later. The shrunk beam is passed through a two axis AOD driven by an oscillatory drive signal produced by a driver circuit (arbitrary function generator (AFG) in Fig. 1). An AOD is essentially a high optical density crystal in which a traveling sound wave creates a moving diffraction pattern. It comprises of a piezo–electric transducer driven with an oscillatory drive signal that induces a sound wave which propagates through the crystal and creates traveling regions of high and low optical density. A laser beam that is input at roughly the Bragg angle diffracts off this moving grating with most of the power in the first-order diffracted beam. Adjustment of the frequency of the sound wave by changing the drive signal frequency causes changes in the spacing of the diffraction grating and therefore, changing the deflection angle of the first-order diffracted beam. The angular position of the diffracted beam can be changed very quickly by adjustment of the drive frequency with a fast, computer–controlled signal generator which affords fast computer steering of the laser beam. The time response of the AOD is limited only by the time required for the traveling wave to cross the laser beam, and so is equal to the laser beam diameter divided by the acoustic speed which is about 0.21 μs for a 2 mm beam. The purpose of reducing the beam diameter is to achieve a low response time as mentioned earlier, however, here, the limitation is typically set by the analog modulation bandwidth of the driver. After the AOD, the beam is then expanded by deploying a beam expander. This makes the beam size sufficiently large to slightly overfill the back aperture of the objective which can increase the ratio of trapping to scattering force, resulting in improved trapping efficiency [4, 37]. The beam coming out of the beam expander is passed through a relay lens system such that the AOD and back focal plane of the objective are in conjugate planes. The beam through the relay lens system is coupled into the objective using a short pass filter that reflects the laser wavelength, while transmitting the illumination wavelength. An oil immersion high NA objective is used to produce the strong gradient forces necessary to overcome axial scattering force and thus achieve stable trapping. Along with the trapping branch described before, optical tweezers also comprise of a detection branch. Typically, for high bandwidth detection, the back focal plane detection method is adopted wherein, the forward scattered light is collected using a condenser and is detected by a quadrant photo diode (QPD) placed in conjugate with the condenser iris plane. A neutral density (ND) filter is used to achieve a proper intensity of the laser such that the QPD is operated within its linear regime. The objective and condenser lenses are set up in a Keplerian telescope when aligning Koehler illumination to have better illumination of the sample and also it puts back focal plane of the objective and condenser in conjugate planes. In contrast to the imaging detector scheme, laser-based detection requires the incorporation of a dichroic mirror on the condenser side of the microscope to couple out the laser light scattered by the specimen. Therefore, a short pass dichroic mirror that reflects the laser wavelength and transmits visible light is incorporated before the QPD. The back-aperture of the condenser lens is inaccessible, and so a lens is used to image a photodetector into a plane conjugate to the back–focal plane. Signals from the four quadrants of the QPD are passed through a K-cube position sensor detector that generates a left-minus-right x difference signal (x-position), a top-minus-bottom y difference signal (y-position), and a sum signal (average intensity on all four quadrants of the QPD). A data acquisition card (DAQ) is used to acquire these signals using a custom written software.

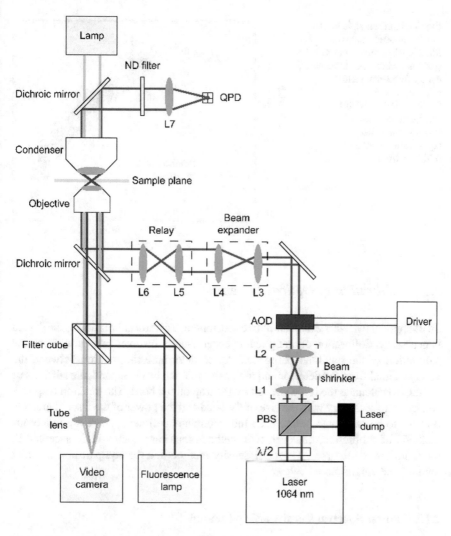

Fig. 1 Schematic of a typical experimental setup. A half-wave plate and variable polarization beam splitter is used to achieve a linearly polarized beam of desired output power. Lens pair (L1, L2) shrinks the beam two-fold and is introduced into the AOD. Lens pair (L3, L4) expands the beam to slightly overfill the back aperture of the objective. Lens pair (L5, L6) is a 1:1 telescope (relay lens system) that images the back focal plane of the objective onto L5. The back apertures of the condenser and objective lenses, the QPD and L5 are in optically conjugate planes. Trapping laser is introduced to the objective through an optical path bypassing the fluorescence filter cube to keep it unaffected while the filter cubes are changed. LED source that provides broad-spectrum excitation is used for fluorescence imaging. Forward scattered light is collected using the condenser for the back focal plane detection using a quadrant photodiode (QPD). The back focal plane of the condenser is imaged on the QPD using the L7 and a ND filter is deployed in between to reduce the light intensity to keep the QPD in the linear regime

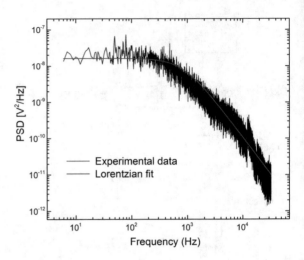

Fig. 2 Representative PSD of a 2 μm bead in the trap fitted to a Lorentzian curve (red line) where the extracted corner frequency yields a trap stiffness of 0.0242 pN/nm. For this measurement, the trap stiffness calculated by equipartition method is 0.0231 pN/nm

2.1 Calibration of Optical Tweezers

To use optical tweezers as a quantitative instrument, a careful calibration of the device is necessary. Calibration of an optical tweezer usually involves two parts. First, the calibration of the quadrant photo detector to determine the relation between the voltage signal from the detector and the position of the bead. Second, the stiffness of the trap to calculate the force exerted by the trap on the bead. The position response and trap stiffness vary with the size of the bead and the power of the laser, therefore, it needs to be calibrated before each independent experiment. There are numerous methods of calibration, however, two methods that have been widely accepted in the literature are the power spectral density method and the equipartition theorem method and are discussed below.

2.1.1   Power Spectral Density (PSD) Method

This method involves calibrating the position response of the QPD and trap stiffness directly from the power spectrum of a trapped bead (Fig. 2). The calibration factor is obtained by observing the thermal fluctuations of beads in a solution taken at the same power settings as for the actual samples of interest. The PSD of a bead in the solution has a Lorentzian form [2, 13, 102]

$$S_{vv}(f) = \rho^2 \frac{k_B T}{\pi^2 \beta (f_0^2 + f^2)} \tag{1}$$

where, $S_{vv}(f)$ is the uncalibrated power spectrum with units of V^2/Hz, f is the frequency, k_B is Boltzmann constant ($\approx 1.38 \times 10^{-23}$ J K^{-1}), β is the drag coefficient equal to $6\pi\eta r$, η is the viscosity of the fluid medium, r is the radius of the bead, T is the absolute temperature, ρ is the linear voltage to displacement calibration factor, and f_0 is the corner frequency. Corner frequency refers to a frequency that divides the power spectrum into two regimes where, for $f < f_0$, $S_{vv}(f)$ is almost constant and for $f > f_0$, $S_{vv}(f)$ falls as $1/f^2$. From, the above equation ρ i.e. the calibration factor is determined which has the units of V/m. Moreover, the trap stiffness k is also determined using the corner frequency using the following equation

$$k = f_0 2\pi\beta. \tag{2}$$

It is worth mentioning that the calibration parameters depend on the beam profile and the polarization of the laser therefore, similar approach is applied to determine the parameters in both directions.

2.1.2 Equipartition Theorem Method

The equipartition theorem method assumes $0.5 \, k_B T$ of thermal energy for each degree of freedom. Energy associated with thermal fluctuations of a particle in an optical trap with stiffness k_x in x-direction equals $0.5k_x\langle x^2\rangle$, where $\langle x^2\rangle$ is the position variance of the trapped particle in the x-direction. By calculating the position variance of the trapped bead experimentally, we can measure the trap stiffness along the x-axis as

$$k_x = k_B T \langle x^2\rangle^{-1}. \tag{3}$$

Likewise, we can measure the trap stiffness in the y-direction k_y using the particle position variance along the y axis. It is worth mentioning that PSD method can be used to calibrate both trap stiffness and responsivity of the QPD, while equipartition method only calibrates the trap stiffness.

3 Quantifying Cell and Matrix Interaction with Optical Tweezers

Accumulating evidence has shown that the cell microenvironment plays a critical role in regulating cellular functions in a wide range of biological processes from embryo development to cancer metastasis [50, 59, 61, 98, 103, 110]. While the biochemical microenvironment has been widely quantified with imaging-based approaches, the mechanical interaction between a cell and its neighbors and the surrounding matrix remains largely unknown, partially due to the challenges in probing these interactions, particularly in 3D. A living cell within tissues generate forces within its cytoplasm and transmits the force to surrounding matrix by forming adhesions, as shown in Fig. 3, and have been widely reviewed [23, 27, 62]. The cell exerts forces and also deforms itself to carry out certain type of biological functions. For instance,

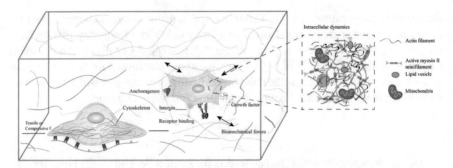

Fig. 3 Living cells actively form adhesion with their surrounding matrix in 3D microenvironment and mechanically interact with their surroundings. Various factors contribute to these physical interactions including mechanical properties of cells and their surrounding matrix, the intracellular dynamics within the cytoplasm

red blood cells need to deform themselves to squeeze through the capillary vessels and deliver oxygen to the body [17] while cancer cells usually generate strong contraction forces and stiffen their surrounding matrix for disease progression [49]. In all these interactions, the mechanical properties of both cell and ECM are important parameters in regulating these physiological processes. Due to the high spatial and force resolution and its ability to carry out measurements in a 3D microenvironment, optical tweezers can be a great tool to quantitatively study these interactions. In this section, we review the recent progresses in measuring cell mechanics, cytoplasmic forces and matrix mechanics using optical tweezers based approaches.

3.1 Cell Mechanics

The mechanical property of cells plays important roles in regulating many key cellular functions, such as mechanotransduction [106], cell migration [66, 91] and even stem cell fate [25, 35, 46]. Thus, characterizing the mechanical properties of living cells and subcellular structures during biological processes is essential for us to understand fundamental cell physiology and subsequently its interaction with the surrounding matrix. To achieve this, optical tweezers can be used to directly measure cell mechanics at different length and time scales. It is considered a powerful platform for its high probing resolution, versatility and non-invasive loading regime.

3.1.1 Cytoplasmic Mechanics

Cytoplasm is a highly heterogeneous complex microenvironment comprising of cytoskeletal network and organelles immersed in cytosol. It is a physical container for countless biochemical reactions that are responsible for almost any cellular

functions. Therefore, physical understanding of the cytoplasm could lend us significant insights into the biological functioning of the cell. Indeed, studies have been
carried out using optical tweezers to probe cytoplasmic mechanics at various length
and time-scales. Since a laser beam can be focused on any specific position in a 3D
sample, it endows optical tweezers the capability to manipulate objects within a cell
and detect resultant forces and displacements. By applying a well-designed loading
method, one can measure mechanical characteristics of the cytoplasm or particular
subcellular structures of interests.

For instance, to study the function of the intermediate filaments in cytoplasm, Guo
et al. [45] performed microrheological measurements using optical tweezers where
bead displacement was recorded under periodic oscillatory input of the laser trap. Two
types of mouse embryonic fibroblasts, wild-type and vimentin$^{-/-}$ were examined and
it was observed that the vimentin intermediate filaments (VIFs) stiffen the cytoplasm
by approximately two times within the frequency range of 1–100 Hz (Fig. 4a–d).
Interestingly, further magnetic microrheology showed that cortical stiffness varied
little with or without VIFs. Together, these quantified results revealed that VIFs
contribute little to cortical stiffness but are a crucial component within the cytoplasm
that significantly contributes to its mechanics [45, 92].

In another article, Gupta and Guo [47] used optical tweezers to perform active
microrheology and study the mechanical response of living cells at a wide range of
frequencies (Fig. 4e). They used a combination of experimentation and theoretical
analysis, to show that intracellular fluctuations are indeed due to non-thermal forces
at relatively long time-scales, however, are dominated solely by thermal forces at
relatively short time-scales. Therefore, they proposed that the mechanical properties
of the cytoplasm can be measured using passive microrheology at high frequencies,
which was previously rendered incorrect as cells operate out of equilibrium. As the
cytoplasm behaves as an equilibrium material at short time-scales, the mean square
displacement of intracellular fluctuations can be used to extract cytoplasmic shear
modulus at high frequencies.

To better understand the mechanical response of the cytoplasm, Hu et al. [58]
combined optical tweezer experiments and scaling analysis to reveal the size- and
speed-dependent resistance in a living mammalian cytoplasm (Fig. 5). Micron-sized
beads of varying diameter a were dragged by laser trap in a unidirectional manner
with a speed V to obtain force-displacement curves. By combining the measurement
with a simple scaling analysis, they showed that the cytoplasm exhibited size-
independent viscoelasticity as long as the effective strain rate V/a was maintained in
a relatively low range (0.1 s^{-1} < V/a < 2 s^{-1}) and exhibited size-dependent poroelasticity at a high effective strain rate regime (5 s^{-1} < V/a < 80 s^{-1}). Specifically, to
identify viscoelasticity or poroelasticity, normalized force-displacement curves with
the same loading parameter combination V/a or Va were plotted together (Fig. 5).
Viscoelasticity describes the mechanical characteristics of materials exhibiting both
elastic and viscous resistance when undergoing deformation. According to the
mechanics of viscoelastic materials, loading curves with same characteristic time
scale V/a should collapse (Fig. 5a). On the other hand, poroelasticity describes
the mechanical features of porous materials filled with liquid where under loading,

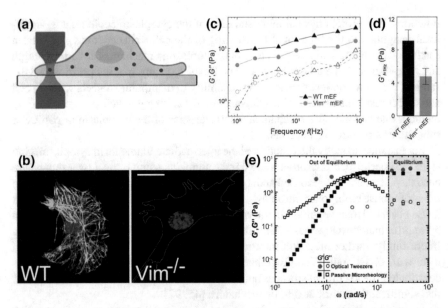

Fig. 4 Microrheology of living cells. **a** Schematic diagram of intracellular optical tweezers measurement. **b** Immunofluorescence using antibodies against vimentin in wild-type (WT, left) and vimentin$^{-/-}$ (right) mouse embryonic fibroblasts. Vimentin in the left panel is shown in green. Cell boundary in the right panel is represented by the yellow line. Scale: 10 µm. **c**, **d** Cytoplasmic elastic moduli G' (solid symbols) and loss moduli G'' (open symbols) as functions of frequency. **e** Comparison of cytoplasmic moduli of A7 cells measured by active microrheology using optical tweezers and passive microrheology. Both methods show an excellent agreement at high frequencies suggesting an equilibrium regime at short time scales. Figures **b–d** are reproduced with permission from [45] and figure **e** is reproduced with permission from [47]

material deformation is determined by solid stresses and fluid flows. According to the mechanics of poroelastic materials, loading curves with same parameter product Va, which has the dimension of diffusivity, should collapse (Fig. 5b). They also showed that the cytoplasmic modulus positively correlated with only V/a in the viscoelastic regime and increased with the bead size at a constant V/a in the poroelastic regime. Finally, they obtained a mechanical state diagram of the living mammalian cytoplasm, which showed that the cytoplasm changes from a viscous fluid to an elastic solid, as well as from compressible material to incompressible material, with increase in the values of two aforementioned dimensionless parameters.

It is worth mentioning that most of the studies report mechanical measurements in terms of a stiffness or a shear modulus assuming that cells are isotropic. However, Gupta et al. [48] used optical tweezers along with particle tracking to show that cell morphology is a critical regulator of the anisotropic behavior of mechanics, dynamics, and forces within the cytoplasm. They found that cells with aspect ratio (AR) ~1 were isotropic; however, when cells broke symmetry, they exhibited significant anisotropy in cytoplasmic mechanics and dynamics. They demonstrated that

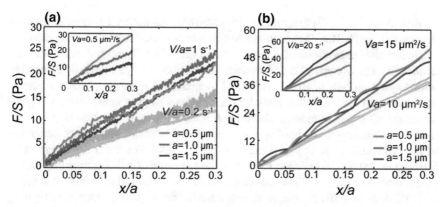

Fig. 5 Normalized force-displacement curves measured in cytoplasm of normal rat kidney epithelial cells. Here the applied force F is divided by cross section area S of a bead. **a** Under low effective strain rate given by bead displacement x divided by bead diameter a, normalized curves with the same speed-diameter ratio V/a are close to each other, while the ones with the same product $V.a$ deviate from each other, suggesting viscoelastic behavior. **b** Under high effective strain rate, normalized curves with the same $V.a$ are close to each other, while the ones with the same V/a deviate from each other, suggesting poroelastic behavior. Figure is reproduced with permission from [58]

the anisotropy in mechanics arises as a consequence of the alignment of cytoskeletal components and the anisotropy in dynamics is mainly due to the generation of anisotropic forces and biased cytoskeletal structures. Therefore, it is important to consider the directional dependence of intracellular mechanics, dynamics and forces under the conditions when the cell shape deviates from isotropicity. These results are particularly important for understanding of biological processes such as migration, differentiation, cancer extravasation and intravasation, etc., where cell polarity is induced.

3.1.2 Deformability of Individual Cells

Deformability of a living cell critically interplays with intra- and extra-cellular environment and regulates the functioning of a cell throughout its cell cycle. This information can lend significant insights in understanding cell functions, cell interaction with extracellular environment and certain types of disease [11, 24, 33, 69, 97, 107]. Indeed, several studies concentrated on deformability of human red blood cells (RBCs) at single-cell level have been carried out using optical tweezers. From the deformation measured under certain loading condition coupled with theoretical or numerical models, one can infer important mechanical parameters like membrane shear modulus, as an index of cell stiffness or deformability, which are known to significantly influence blood flow in microcirculation and serve as important indicators in disease progression [28].

Bronkhorst et al. [17] first proposed the idea of using optical tweezers to measure deformability of RBCs. They manipulated three beads to deform cells into a parachute shape and after release, recorded the process of relaxation. They used relaxation time as an indirect measurement of deformability and reported a reduction in deformability during aging process of RBCs, explained by increased viscosity of cell membrane and cytosol. In this work, flexibility for multiple laser traps to induce preferred types of deformation was leveraged. However, subsequently, uniaxial stretching became the most popular way of loading which is relatively easier and makes better use of the symmetric geometry of RBCs [28, 55, 71, 100]. For instance, Suresh et al. [101] deployed optical tweezers to explore the evolution of cell deformability during progressive maturation of Plasmodium parasite. They observed an increased stiffness of RBC cells at different stages during parasite development. Also investigating malaria, by combining optical tweezers and gene interruption techniques, Mills et al. [78] observed cell stiffening during development of Plasmodium falciparum in host RBCs and investigated the role of a parasite-exported protein, Pf155/Ring-infected erythrocyte surface antigen (RESA), in cell stiffening events.

3.2   Forces in Cytoplasm of Living Cells

Cytoplasm of a living cell is highly dynamic as the intracellular environment is subjected to a wide variety of forces [57, 104]. Majority of these forces can be attributed to the operation of molecular motors such as kinesin and dynein that typically are responsible for driving directional cargo transport along the microtubule tracks and myosin II motors that actively contract actin filaments [104]. The cooperative activity of these motors and other active processes in the cytoplasm can drive critical functions at the level of whole cells such as contraction, division and migration [29, 30, 41, 43, 54]. Nonetheless, the average effect of all the motors and active processes also contribute to an incoherent background of fluctuating forces which is associated with the functional efficiency and the complete metabolic state of the cell [29]. These overall fluctuating forces can give rise to random motion of intracellular components and quantifying it will enable us in characterizing the dynamic state of the cell. In this section, we introduce Force Spectrum Microscopy (FSM) (Fig. 6a, b) that combine independent measurements of the intracellular fluctuating movement of injected tracer particles and mechanics of the cytoplasm with active microrheology using optical tweezers to directly quantify random forces within the cytoplasm of the cell [44, 47]. The method involves calculating the force experienced by the probe particle by adopting the fundamental force displacement Hooke's relation, $f = Kr$ where, f is the driving force, r is the resulting displacement and K is the stiffness of the medium. However, as the effective spring constant for the cytoplasm is frequency dependent and the intracellular fluctuations are random, thus, a quadratic form of the averaged quantities in the frequency domain is considered. Therefore, the frequency spectrum of the intracellular fluctuating forces can be given by:

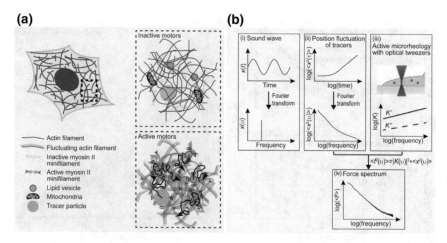

Fig. 6 Graphical representation of FSM. **a** Schematic illustration of intracellular fluctuating forces responsible for enhanced cytoplasmic motion. **b** Basic procedure of FSM. (i) Sound wave in the time can be represented in the frequency domain by taking a Fourier transform to reveal the frequency composition. (ii) First step involves taking the Fourier transform of MSD in the time domain to express it in the frequency domain (iii) The cytoplasmic material property, specifically the spring constant, is measured directly using optical tweezers, also in the frequency domain. (iv) Using Hooke's Law, the randomly intracellular fluctuating force at each frequency is calculated as $f^2(v) = |K(v)|^2 x^2(v)$. Figure reproduced with permission from [44]

$$\langle f^2(v) \rangle = |K(v)|^2 \langle r^2(v) \rangle. \tag{4}$$

where, $\langle r^2(v) \rangle$ is the frequency spectrum of the particle displacement and $K(v)$ is the frequency dependent spring constant.

FSM has been demonstrated to have broad applications for understanding the cytoplasm and its intracellular processes in the context of cell physiology in healthy and diseased states. In particular, FSM has been used to compare the intracellular force spectrum in the benign cells (MCF-10A) to that in the malignant cells (MCF-7) and show that the malignant cells have about three times higher force fluctuations than that to its benign counterpart suggesting that the malignant cells have a more active cytoplasm (Fig. 7c). It is also shown that the movement of tracer particles is enhanced in the malignant cells compared with the benign counterpart (Fig. 7a), while the cytoplasmic stiffness was only about 30% smaller in the malignant MCF-7 cells (Fig. 7b).

3.3 Mechanics of Cell-Cell and Cell-Matrix Adhesions

Cell-cell adhesion plays a crucial role in assembling individual cells into 2D or 3D tissues and regulating tissue functions and dynamics [40, 42, 87, 111]. Optical trap-

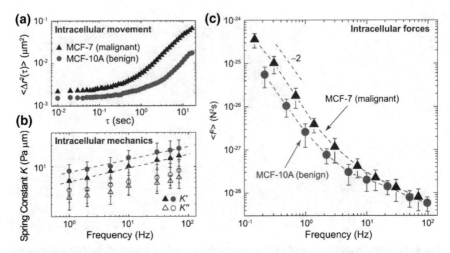

Fig. 7 Intracellular mechanics, dynamics and forces in benign and malignant tumor cells. **a** Two-dimensional MSD $\Delta r^2(\tau)$ of 500 nm tracer particles are plotted against lag time on a log-log scale, in the benign breast cells MCF10A (red circle) and malignant breast tumor cells MCF-7 (black triangle), respectively. The fluctuating movement of tracer particles is stronger in the malignant cells, as compared to the benign cells. **b** Cytoplasmic mechanics measured by active microrheology using optical tweezers. The effective spring constant of the cytoplasm is larger in the MCF-10A (red circle), as compared to the MCF-7 (black triangle). **c** The spectrum of intracellular forces calculated using FSM. The intracellular forces are stronger in the malignant tumor cells MCF-7 (black triangle), than the benign cells MCF-10A (red circle). Figure reproduced with permission from [44]

ping can be a great tool to measure these contacts and to understand the role of these contacts in the physiological processes. Bambardekar et al. [9] used laser to measure the tension of cell junctions in Drosophila embryos during germ-band elongation. A laser beam was focused directly (without probing particles) on cell-cell interface to introduce perpendicular stretching deformation and imaged the deformation with light sheet microscopy, which was modeled with concentrated force acting on an elastic sheet to calculate the surface tension of cell-cell adhesion (on the order of 44 ± 22 pN). The frequency-dependent response of the cell junction was probed with this method and they observed elastic-dominant response at low frequency (≤ 0.5 Hz) and began to observe viscoelastic response at high frequency (>0.5 Hz), which was ascribed to elasticity of the membrane and viscosity of the cytosol. Furthermore, they found the measured tension is highly correlated with the germ-band elongation.

In addition to cell-cell junction, cell-matrix adhesion is also important in many cellular activities like migration [60], differentiation [22] and apoptosis [95]. In a work by Castelain et al. [21], adhesion strength of *Saccharomyces cerevisiae* on a glass substrate was measured using optical tweezers. A laser beam was directly focused near the upper surface of yeast cells. Then starting from a low laser power, they moved the trap back and forth with fixed frequency and amplitude, while slowly increasing the power. The power upon detachment was recorded and transformed to a threshold drag force for an individual cell.

3.3.1 Matrix Mechanics

The mechanical properties of the ECM play an important role in morphogenesis and tumorigenesis [18, 73]. Changes of ECM stiffness are associated with many diseases including atrial and pulmonary fibrosis, infantile cortical hyperostosis, Ehlers–Danlos syndrome and cancer [77, 88, 89]. Therefore, constant efforts have been made to understand the fundamental aspects of these ECM from both structural and mechanical outlooks.

Biopolymers are usually discontinuous and hierarchical in structure at different length scales which further determine the distinct mechanical response when measured with various approaches. For instances, at nanoscale, collagen gels are formed by inter-crosslinked fibers with diameters ranging from 10 to 300 nm [10]; at relatively larger scale, the fibrous network is largely heterogeneous with typical pore sizes within a few to tens of microns [65]. Given this pore size is much larger than cell-ECM adhesions [85], it is expected that the micromechanics of local matrix rather than the macroscopic rheology of the ECM regulate the cellular function [8].

3.3.2 Single Fiber Mechanics

Fibrous networks are abundant in 3D cell microenvironment *in vivo*. On molecular level, cells mechanically interact with the fibers through integrins by forming focal adhesions on 2D surface or clathrin/AP-2 lattices in 3D collagen networks [34]. Thus, a molecular-level understanding of how ECM responds to cell generated stress, or how its mechanical properties are sensitive to spatial geometries, requires a precise quantitative knowledge of the mechanical properties of individual fibers.

Dutov et al. [31] developed a novel method to study the mechanical response of a single collagen fiber of type I collagen (rat-tail) to bending by optical tweezers. Experimentally, a cantilever collagen fiber, that has one end integrated rigidly with the bundle and other end free (Fig.8a), was identified using bright field microscopy and polystyrene microspheres of 1 μm in diameter were injected and attached on the free-end of the fiber through proteoglycans in the solution. Then the microsphere was dragged in the direction perpendicular to the original axis of the fiber, as shown in Fig. 8a, b inset, with a low velocity (0.1 μm/s) to minimize the viscous drag force from the surrounding liquid. The displacements of the microsphere and the forces were recorded, as shown in Fig. 8b. To estimate the stiffness of the fiber, a thin beam Euler-Bernoulli elasticity theory was used:

$$E = \frac{4L^3 F}{3\pi R^4 H} \tag{5}$$

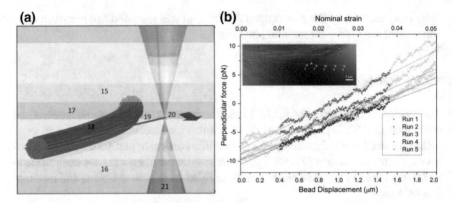

Fig. 8 Measurement of single collagen fiber and fibers bundle mechanics with optical tweezers. **a** Schematic illustrates the experimental settings. A free end-collagen fiber was bended by dragging a 1-μm particle that attached to the fiber using optical tweezers. **b** The resulting force-displacement curves of single collagen fiber from 5 tests. Inset: a phase contrast image shows a free-end single collagen fiber branched out from a collagen bundle. Figure reproduced with permission from [31]

where E is the Young's modulus of the fiber, L is the length of the fiber, F is the resistant force perpendicular to the fiber, R and H are the radius and the deflection of the fiber respectively. The F/H was determined by linear fitting the force-displacement curves measured by optical tweezer while the R was measured by atomic force microscopy (162 ± 20 nm). The average elastic modulus in longitudinal direction varied between 100 and 360 MPa, indicating the heterogeneous nature of collagen mechanics at the level of single fiber.

3.3.3 Spatial Heterogeneity

The local fibrous structure of biopolymers varies in space and time in vivo, leading to a spatial heterogeneity of material properties. This local heterogeneity is usually not detectable by conventional bulk rheology measurement because it gets averaged at large scale. Juliar et al. [64] systematically investigated this effect by comparing the mechanical properties and stabilities of fibrin network at difference scales. The mechanical response at large scale was measured using plate rheology while the one at microscale was done by optical tweezers (Fig. 9a). At large scale, the shear modulus of the network remains nearly constant up to two weeks, with a dramatic decrease in heterogeneity as shown by the error bars in Fig. 9b. To measure the microscale mechanical response, 2 μm silica microbeads were embedded in the fibrin network and oscillated at 50 Hz and an amplitude of 175 nm by optical tweezers. The shear modulus measured at microscale exhibited increased heterogeneity compared to measurements of bulk rheology, as shown in Fig. 9c. These results demonstrate that the mechanical properties of fibrin network are spatially heterogeneous at length

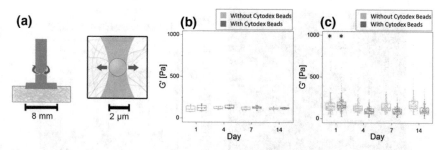

Fig. 9 Comparison of mechanical heterogeneities between bulk rheology and optical tweezers based microrheology. **a** Illustrations of the mechanical measurement at macro-scale using parallel plate rheology (left) and at micro-scale using optical tweezers (right). Shear modulus measured at macro-scale (**b**) and micro-scale (**c**) over 14 days, where micro-rheology reveals significant mechanical heterogeneity both with and without using cytodex beads. Figure reproduced with permission from [64]

scales comparable to cell length, which lays the very fundamental basis to understand how individual cells mechanically interact with their surrounding matrix.

3.3.4 Anisotropy

Anisotropy of native ECM is another property that can have major implications on cell-matrix interaction. Jones et al. [63] developed a pioneering framework to investigate this phenomenon in 3D biological networks using optical tweezers. To do so, optical forces were applied to polystyrene beads (3 μm in diameter) that were dispersed in the collagen network prior to gelation (Fig. 10a). Specifically, the trap was positioned 0.725 μm away from the particle in $+x$ direction and the movement of the bead was recorded by a camera. Its displacements in x- and y-directions were analyzed from a video as shown in Fig. 10b. These two displacements indicate the deformation of the local matrix in the directions along and perpendicular to the applied force. By varying the direction of applied forces, the local network mechanical anisotropy was mapped (Fig. 10c, d). These results demonstrate that the local structures and properties of the collagen network is anisotropic and asymmetric in contrast to its bulk properties. As mentioned previously these local direction dependent physical properties may have implications in the behavior of cells embedded in the 3D network.

3.3.5 Nonlinearity

It is well known that native ECM networks exhibit strong nonlinear mechanical responses with various underlying mechanisms from the entropic elasticity of individual filaments to collective network effects governed by critical phenomena

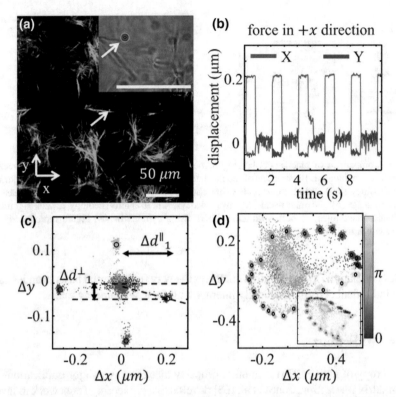

Fig. 10 Local anisotropy measured with optical-tweezers based microrheology. **a** structures of collagen fibers (green) in collagen gels. The white arrows indicate the particles embedded in the matrix. Inset scale: 50 μm. **b** Displacements of the embedded particle in response to a pulsed loading by optical tweezers at +x direction. **c** Displacements of the embedded particle in response to a pulsed loading by optical tweezers at four directions (+x, red; −x, blue, +y, green, −y, pink). **d** Displacements of the embedded particle in response to a pulsed loading by optical tweezers at 24 directions. Figure reproduced with permission from [63]

[15, 16, 39, 70, 86, 99, 109]. Although the nonlinear characteristics have been largely studied with bulk rheology, direction dependent measurement of microscale nonlinear properties within a 3D matrix has recently become accessible by optical tweezers. Han et al. [49] embedded latex particles of 4.5 μm in diameter into collagen network to measure its nonlinear mechanical property deploying optical tweezers. The network had an average pore size of ~2.0 μm therefore the beads were trapped by the collagen fibers maintaining an elastic continuum. The surrounding matrix was deformed by dragging the particle away from its equilibrium position, and resultant forces reflected the local network's restoring force, as shown in Fig. 11a. As the displacement of the particle increased, the force increased nonlinearly. To further quantify the local stiffness of the collagen network under loading, the slope of the force displacement curves (spring constant, pN/μm) was taken and plotted against

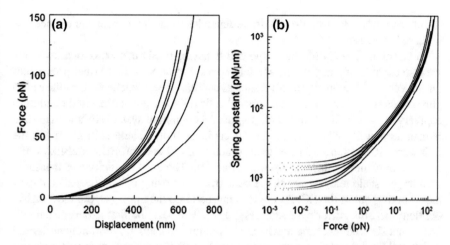

Fig. 11 Nonlinearity of collagen networks probed with optical tweezers. **a** Force-displacement curves of ten measurements; **b** corresponding force-spring constant curves show that the collagen network is stress stiffened. Figure reproduced with permission from [49]

the force, as shown in Fig. 11b. For small forces, the spring constant was a constant; it strongly increased as the force approached 1 pN.

3.4 Cell-Matrix Interaction

Due to the accessibility of local micromechanical properties of ECM, it is now possible to study the cell-matrix interaction quantitatively using optical tweezers . Consequently, some groups have started to directly quantify the cell-matrix interaction in various biological context including cancer invasion and angiogenesis [49, 63, 64]. Cancer progression is associated with a series of alterations in ECM mechanics, however, how cancer cells actively promote these changes remained unclear until recently. By mapping the matrix mechanics near individual cancer cells in 3D collagen networks (Fig. 12a–e), Han el al. [49] found that human breast cancer cells (MDA-MB-231) were able to stiffen their surrounding matrix by two orders of magnitude while the normal mammary epithelial cell line (MCF-10A) were not, as shown in Fig. 12f. This strong stiffening effect were caused by larger contractile forces generated by cancer cells, as inhibition of myosin activity or depolymerization of F-actin decreased it. These results described the mechanical scenario that cancer cells generated a large stress gradient which created a stiffness gradient near the cell. This similar stiffening effects were also observed in other highly contractile cell types such as NIH 3T3 fibroblast [110]. A large strain gradient was evident near a contractile cell in collagen networks (Fig. 12g). By measuring the local compliance of collagen networks, a contractile force dependent increase in local compliance was observed (Fig. 12h).

In addition to studying the single cell behavior in 3D matrix, optical tweezers are also capable of studying the spatiotemporal evolution of ECM micromechanics during various biological processes such as capillary morphogenesis. Capillary morphogenesis is a process where endothelial cells (ECs) invade in the matrix and form capillary vessels in 3D. Juliar et al. [64] used a co-culture system (ECs with normal human dermal fibroblasts) to mimic the capillary morphogenesis in fibrin gel; the ECs were grown on a surface of microbeads on day 1 and formed vessel-like structures in following 14 days, as shown in Fig. 13a. The local elasticity was measured on a length scale relevant for single cells over time using optical tweezers, during the sprouting out of the EC. As the ECs invaded into the fibrin gel, the local matrix elasticity became stiffer over time (Fig. 13b). This was the first experimental evidence that showed real time mechanical dynamics in capillary morphogenesis, and uncovered the role of matrix mechanics and cell-matrix interaction in this process.

3.4.1 Measurement of Cell Generated Stresses in 3D ECM

As discussed above, cell generated forces are critical in regulating various biological processes and constant efforts have been made to measure these forces in physiologically relevant contexts. Comprehensive reviews that summarize and compare available methodologies are available in the literature [90, 108], here, we report an emerging approach to measure the cell-generated stress using optical tweezers. It has been discussed that stiffening of the matrix near living cells was caused by cell contraction in Sect. 3.3. If the increased stiffness is due to the contractile stress, one should be then able to extract the stress distribution from the measured stiffness map. We discuss a recently developed technique Nonlinear Stress Inference Microscopy (NSIM) that leverages this idea [49]. NSIM takes advantage of the nonlinear properties of the biopolymers, since they usually become stiffer when force exceeds a certain threshold. This stiffening behavior could be quantitatively probed using optical tweezers, and used as a dictionary to readout the forces according to the extent of stiffening. The detailed protocol along with the schematic is shown in Fig. 14.

4 Concluding Remarks

Mechanical interactions between cell and its microenvironment play a central role in basic multicellular processes including embryonic development, wound healing and cancer progression. Various approaches including traction force microscopy, magnetic tweezers and bulk rheology have been developed to quantify these interactions separately by measuring cells mechanics, matrix mechanics and contractile forces. However, none of these platforms are able to systematically measure these interactions in a 3D multicellular system at single cell resolution. Spatially and temporally

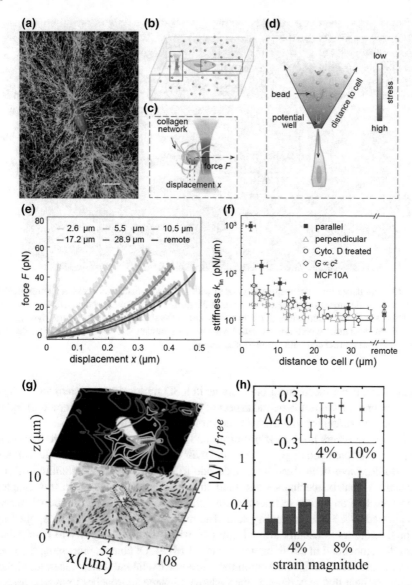

Fig. 12 Cells stiffen their surroundings through active contraction in collagen networks. **a** Image of a MDA-MB-231 cell (blue) in a 3D collagen network. **b–d** Graphics describe the experimental measurement with optical tweezers and the scenario in which cell-generated stress field induces a matrix stiffening gradient. Scale is 10 μm. **e** Local force-displacement curves at locations with different distance to the cell. **f** Local matrix mechanics as a function of distance to the cell. **g** A strain gradient was observed near a contracting cell in collagen networks. **h** A stiffening effect within the matrix was correlated with the observed strain gradient. Figure reproduced with permission from [49, 63]

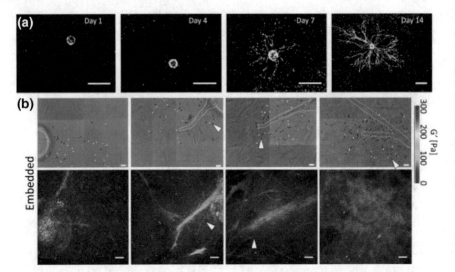

Fig. 13 Capillary morphogenesis is associated with increasing matrix stiffness. **a** Representative images show the formation of capillary vessels in 3D fibrin hydrogel over 14 days. Scale is 500 μm. **b** Stiffness map measured with optical tweezers shows the dynamic spatiotemporal dynamics in local ECM mechanic, where arrows indicate areas of significant stiffness heterogeneity. Scale is 20 μm. Figure reproduced with permission from [64]

quantifying these mechanical interactions in a 3D multicellular system would help us to understand the fundamental aspect of cell-microenvironment interaction and to develop new strategies to mimic and control these interactions in vitro.

Optical tweezers hold great promise to address these problems due to their high resolution in force and displacement, and their capability to measure in 3D. Numerous studies have been done in past decade to enhance our understanding of cell mechanics, matrix mechanics and their interactions, as discussed in this chapter. However, like any other technology, limitations are associated with optical tweezers applied to a 3D biological system. The maximum force that can be applied by optical tweezers is usually around 100 pN, which limits the ability to manipulate particles embedded in stiffer environment (~1 MPa). Of course, increasing the laser power could increase this maximum force to some extent but it also generates higher amount of heat that may damage the system and some approaches to overcome this has been adopted [74]. The throughput of the optical tweezers is another potential problem when large amount of measurements is necessary. Multiple traps using time sharing of the beam and computer-designed diffractive optical element (DOE) alleviates this issue to some extent, however technological advancements in modulation techniques and better design could possibly increase the throughput significantly. Nonetheless, optical trap is a powerful tool to study cell and matrix mechanics and their interaction. Strategies that incorporate other established techniques such as single molecule microscopy, in situ sequencing, gene editing techniques and other

Reference curves - measurements without prestress in the 3D matrix

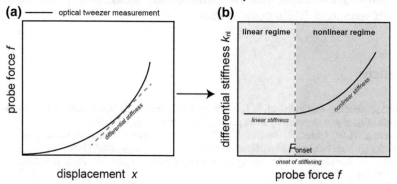

Stiffness measurements in a prestressed 3D matrix and force inference

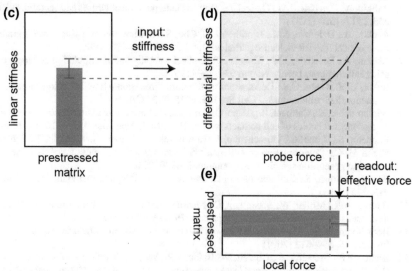

Fig. 14 Schematics illustrate the steps of NSIM. **a** To measure the local prestress in a matrix, a reference curve for the specific matrix is generated using optical tweezers by dragging the particle unidirectionally with a constant speed in a random direction. **b** The slope in the force-displacement relationship is defined as the differential stiffness of the matrix, as indicated by the gray dashed line in the Fig. 14a. Then the differential stiffness of the matrix is calculated, as: $k_{nl} = \frac{df}{dx}$, where f is the applied force and x is the particle displacement. k_{nl} is plotted against f, as shown in Fig. 14 b. This relationship is used later as a reference curve to readout the local forces induced by a contracting cell. **c** Measure the stiffness of the matrix under a pre-stress (such as cell contraction stress) with optical tweezers using method similar to **a**. **d–e** To infer the local stress from the local stiffness, the force value is read from the reference curve obtained in step 2, by finding the force value corresponding to the same stiffness value on the reference curve. Figure reproduced with permission from [49]

biochemical assays with optical tweezers have recently become popular and enable us to better understand 3D biological systems.

References

1. Abbondanzieri, E.A., Greenleaf, W.J., Shaevitz, J.W., Landick, R., Block, S.M.: Direct observation of base-pair stepping by RNA polymerase. Nature **438**, 460 (2005)
2. Allersma, M.W., Gittes, F., Stewart, R.J., Schmidt, C.F.: Two-dimensional tracking of ncd motility by back focal plane interferometry. Biophys. J. **74**, 1074–1085 (1998)
3. Ashkin, A.: Acceleration and trapping of particles by radiation pressure. Phys. Rev. Lett. **24**, 156 (1970)
4. Ashkin, A.: Forces of a single-beam gradient laser trap on a dielectric sphere in the ray optics regime. Biophys. J. **61**, 569–582 (1992)
5. Ashkin, A., Dziedzic, J.M.: Optical trapping and manipulation of viruses and bacteria. Science **235**, 1517–1520 (1987)
6. Ashkin, A., Dziedzic, J.M., Bjorkholm, J., Chu, S.: Observation of a single-beam gradient force optical trap for dielectric particles. Opt. Lett. **11**, 288–290 (1986)
7. Ashkin, A., Dziedzic, J.M., Yamane, T.: Optical trapping and manipulation of single cells using infrared laser beams. Nature **330**, 769 (1987)
8. Baker, B.M., Chen, C.S.: Deconstructing the third dimension—how 3D culture microenvironments alter cellular cues. J Cell Sci **125**, 3015–3024 (2012)
9. Bambardekar, K., Clément, R., Blanc, O., Chardès, C., Lenne, P.-F.: Direct laser manipulation reveals the mechanics of cell contacts in vivo. Proc. Natl. Acad. Sci. **112**, 1416–1421 (2015)
10. Bancelin, S., Aimé, C., Gusachenko, I., Kowalczuk, L., Latour, G., Coradin, T., Schanne-Klein, M.-C.: Determination of collagen fibril size via absolute measurements of second-harmonic generation signals. Nat. Commun. **5**, 4920 (2014)
11. Bao, G., Suresh, S.: Cell and molecular mechanics of biological materials. Nat. Mater. **2**, 715–725 (2003)
12. Bausch, A.R., Möller, W., Sackmann, E.: Measurement of local viscoelasticity and forces in living cells by magnetic tweezers. Biophys. J. **76**, 573–579 (1999)
13. Berg-Sørensen, K., Flyvbjerg, H.: Power spectrum analysis for optical tweezers. Rev. Sci. Instrum. **75**, 594–612 (2004)
14. Bianco, P.R., Brewer, L.R., Corzett, M., Balhorn, R., Yeh, Y., Kowalczykowski, S.C., Baskin, R.J.: Processive translocation and DNA unwinding by individual RecBCD enzyme molecules. Nature **409**, 374 (2001)
15. Broedersz, C., Sheinman, M., MacKintosh, F.: Filament-length-controlled elasticity in 3D fiber networks. Phys Rev Lett **108**, 078102 (2012)
16. Broedersz, C.P., MacKintosh, F.C.: Modeling semiflexible polymer networks. Rev. Mod. Phys. **86**, 995 (2014)
17. Bronkhorst, P., Streekstra, G., Grimbergen, J., Nijhof, E., Sixma, J., Brakenhoff, G.: A new method to study shape recovery of red blood cells using multiple optical trapping. Biophys. J. **69**, 1666 (1995)
18. Brown, R.A.: In the beginning there were soft collagen-cell gels: towards better 3D connective tissue models? Exp. Cell Res. **319**, 2460–2469 (2013)
19. Burns, M.M., Fournier, J.-M., Golovchenko, J.A.: Optical matter: crystallization and binding in intense optical fields. Science **249**, 749–754 (1990)
20. Bursac, P., et al.: Cytoskeletal remodelling and slow dynamics in the living cell. Nat. Mater. **4**, 557 (2005)
21. Castelain, M., Rouxhet, P.G., Pignon, F., Magnin, A., Piau, J.-M.: Single-cell adhesion probed in-situ using optical tweezers: a case study with Saccharomyces cerevisiae. J. Appl. Phys. **111**, 114701 (2012)

22. Chaudhuri, O., et al.: Hydrogels with tunable stress relaxation regulate stem cell fate and activity. Nat. Mater. **15**, 326 (2016)
23. Chen, B., Ji, B., Gao, H.: Modeling active mechanosensing in cell-matrix interactions. Annu. Rev. Biophys. **44**, 1–32 (2015)
24. Chien, S.: Red cell deformability and its relevance to blood flow. Annu. Rev. Physiol. **49**, 177–192 (1987)
25. Chowdhury, F., Na, S., Li, D., Poh, Y.-C., Tanaka, T.S., Wang, F., Wang, N.: Material properties of the cell dictate stress-induced spreading and differentiation in embryonic stem cells. Nat. Mater. **9**, 82–88 (2010)
26. Cost, A.-L., Ringer, P., Chrostek-Grashoff, A., Grashoff, C.: How to measure molecular forces in cells: a guide to evaluating genetically-encoded FRET-based tension sensors. Cell. Mol. Bioeng. **8**, 96–105 (2015)
27. Cukierman, E., Pankov, R., Yamada, K.M.: Cell interactions with three-dimensional matrices. Curr. Opin. Cell Biol. **14**, 633–640 (2002)
28. Dao, M., Lim, C.T., Suresh, S.: Mechanics of the human red blood cell deformed by optical tweezers. J. Mech. Phys. Solids **51**, 2259–2280 (2003)
29. Doyle, A.D., Yamada, K.M.: Cell biology: sensing tension. Nature **466**, 192 (2010)
30. Dufrêne, Y.F., Evans, E., Engel, A., Helenius, J., Gaub, H.E., Müller, D.J.: Five challenges to bringing single-molecule force spectroscopy into living cells. Nat. Methods **8**, 123 (2011)
31. Dutov, P., Antipova, O., Varma, S., Orgel, J.P., Schieber, J.D.: Measurement of elastic modulus of collagen type I single fiber. PloS One **11**, e0145711 (2016)
32. Egerton, F.N.: A history of the ecological sciences, part 19: Leeuwenhoek's microscopic natural history. Bull. Ecol. Soc. Am. **87**, 47–58 (2006)
33. Ehrlicher, A.J., Krishnan, R., Guo, M., Bidan, C.M., Weitz, D.A., Pollak, M.R.: Alpha-actinin binding kinetics modulate cellular dynamics and force generation. Proc. Natl. Acad. Sci. **112**, 6619–6624 (2015)
34. Elkhatib, N., Bresteau, E., Baschieri, F., Rioja, A.L., van Niel, G., Vassilopoulos, S., Montagnac, G.: Tubular clathrin/AP-2 lattices pinch collagen fibers to support 3D cell migration. Science **356**, 4713 (2017)
35. Engler, A.J., Sen, S., Sweeney, H.L., Discher, D.E.: Matrix elasticity directs stem cell lineage specification. Cell **126**, 677–689 (2006)
36. Eriksen, R.L., Daria, V.R., Glückstad, J.: Fully dynamic multiple-beam optical tweezers. Opt. Express **10**, 597–602 (2002)
37. Fällman, E., Axner, O.: Influence of a glass-water interface on the on-axis trapping of micrometer-sized spherical objects by optical tweezers. Appl. Opt. **42**, 3915–3926 (2003)
38. Fodor, É., Guo, M., Gov, N., Visco, P., Weitz, D., van Wijland, F.: Activity-driven fluctuations in living cells. EPL (Europhys. Lett.) **110**, 48005 (2015)
39. Gardel, M., Shin, J., MacKintosh, F., Mahadevan, L., Matsudaira, P., Weitz, D.: Elastic behavior of cross-linked and bundled actin networks. Science **304**, 1301–1305 (2004)
40. Geiger, B., Bershadsky, A., Pankov, R., Yamada, K.M.: Transmembrane crosstalk between the extracellular matrix and the cytoskeleton. Nat. Rev. Mol. Cell Biol. **2**, 793 (2001)
41. Grashoff, C., et al.: Measuring mechanical tension across vinculin reveals regulation of focal adhesion dynamics. Nature **466**, 263 (2010)
42. Gumbiner, B.M.: Cell adhesion: the molecular basis of tissue architecture and morphogenesis. Cell **84**, 345–357 (1996)
43. Gundersen, G.G., Worman, H.J.: Nuclear positioning. Cell **152**, 1376–1389 (2013)
44. Guo, M., et al.: Probing the stochastic, motor-driven properties of the cytoplasm using force spectrum microscopy. Cell **158**, 822–832 (2014)
45. Guo, M., et al.: The role of vimentin intermediate filaments in cortical and cytoplasmic mechanics. Biophys. J. **105**, 1562–1568 (2013)
46. Guo, M., et al.: Cell volume change through water efflux impacts cell stiffness and stem cell fate. Proc. Natl. Acad. Sci. **2017**, 05179 (2017)
47. Gupta, S.K., Guo, M.: Equilibrium and out-of-equilibrium mechanics of living mammalian cytoplasm. J. Mech. Phys. Solids **107**, 284–293 (2017)

48. Gupta, S.K., Li, Y., Guo, M.: Anisotropic mechanics and dynamics of a living mammalian cytoplasm. Soft Matter **15**, 190–199 (2019)
49. Han, Y.L., et al.: Cell contraction induces long-ranged stress stiffening in the extracellular matrix. Proc. Natl. Acad. Sci. **2017**, 22619 (2018)
50. Han, Y.L., et al.: Engineering physical microenvironment for stem cell based regenerative medicine. Drug Discovery Today (2014)
51. Hanes, R.D., Jenkins, M.C., Egelhaaf, S.U.: Combined holographic-mechanical optical tweezers: construction, optimization, and calibration. Rev. Sci. Instrum. **80**, 083703 (2009)
52. Heilbrunn, L.: The physical structure of the protoplasm of sea-urchin eggs. Am. Nat. **60**, 143–156 (1926)
53. Heilbrunn, L.: The viscosity of protoplasm. Q. Rev. Biol. **2**, 230–248 (1927)
54. Heisenberg, C.-P., Bellaïche, Y.: Forces in tissue morphogenesis and patterning. Cell **153**, 948–962 (2013)
55. Henon, S., Lenormand, G., Richert, A., Gallet, F.: A new determination of the shear modulus of the human erythrocyte membrane using optical tweezers. Biophys. J. **76**, 1145–1151 (1999)
56. Hochmuth, R.M.: Micropipette aspiration of living cells. J. Biomech. **33**, 15–22 (2000)
57. Howard, J.: Mechanics of motor proteins and the cytoskeleton. Sunderland, MA: Sinauer Associates (2001)
58. Hu, J., Jafari, S., Han, Y., Grodzinsky, A.J., Cai, S., Guo, M.: Size-and speed-dependent mechanical behavior in living mammalian cytoplasm. Proc. Natl. Acad. Sci. **2017**, 02488 (2017)
59. Huang, S., Ingber, D.E.: Cell tension, matrix mechanics, and cancer development. Cancer Cell **8**, 175–176 (2005)
60. Huttenlocher, A., Sandborg, R.R., Horwitz, A.F.: Adhesion in cell migration. Curr. Opin. Cell Biol. **7**, 697–706 (1995)
61. Ingber, D.E., et al.: Cellular tensegrity: exploring how mechanical changes in the cytoskeleton regulate cell growth, migration, and tissue pattern during morphogenesis. In: International Review of Cytology, vol. 150, pp. 173–224. Elsevier (1994)
62. Jansen, K.A., Atherton, P., Ballestrem, C.: Mechanotransduction at the cell-matrix interface. Semin Cell Dev. Biol. **71**, 75–83 (2017)
63. Jones, C.A.R., Cibula, M., Feng, J., Krnacik, E.A., McIntyre, D.H., Levine, H., Sun, B.: Micromechanics of cellularized biopolymer networks. PNAS **112**, E5117 (2015)
64. Juliar, B.A., Keating, M.T., Kong, Y.P., Botvinick, E.L., Putnam, A.J.: Sprouting angiogenesis induces significant mechanical heterogeneities and ECM stiffening across length scales in fibrin hydrogels. Biomaterials **162**, 99–108 (2018)
65. Lang, N.R., et al.: Estimating the 3D pore size distribution of biopolymer networks from directionally biased data. Biophys J **105**, 1967–1975 (2013)
66. Lange, J.R., Fabry, B.: Cell and tissue mechanics in cell migration. Exp. Cell Res. **319**, 2418–2423 (2013)
67. Lau, A.W., Hoffman, B.D., Davies, A., Crocker, J.C., Lubensky, T.C.: Microrheology, stress fluctuations, and active behavior of living cells. Phys. Rev. Lett. **91**, 198101 (2003)
68. Lee, W.M., Reece, P.J., Marchington, R.F., Metzger, N.K., Dholakia, K.: Construction and calibration of an optical trap on a fluorescence optical microscope. Nat. Protoc. **2**, 3226 (2007)
69. Lekka, M., Laidler, P., Gil, D., Lekki, J., Stachura, Z., Hrynkiewicz, A.: Elasticity of normal and cancerous human bladder cells studied by scanning force microscopy. Eur. Biophys. J. **28**, 312–316 (1999)
70. Lieleg, O., Claessens, M.M., Heussinger, C., Frey, E., Bausch, A.R.: Mechanics of bundled semiflexible polymer networks. Phys. Rev. Lett. **99**, 088102 (2007)
71. Liu, Y.-P., Li, C., Liu, K.-K., Lai, A.C.: The deformation of an erythrocyte under the radiation pressure by optical stretch. J. Biomech. Eng. **128**, 830–836 (2006)
72. Maloney, J.M., Nikova, D., Lautenschläger, F., Clarke, E., Langer, R., Guck, J., Van Vliet, K.J.: Mesenchymal stem cell mechanics from the attached to the suspended state. Biophys. J. **99**, 2479–2487 (2010)

73. Mammoto, T., Ingber, D.E.: Mechanical control of tissue and organ development. Development **137**, 1407–1420 (2010)
74. Mao, H., Arias-Gonzalez, J.R., Smith, S.B., Tinoco Jr., I., Bustamante, C.: Temperature control methods in a laser tweezers system. Biophys. J. **89**, 1308–1316 (2005)
75. Maragò, O.M., Jones, P.H., Gucciardi, P.G., Volpe, G., Ferrari, A.C.: Optical trapping and manipulation of nanostructures. Nat. Nanotechnol. **8**, 807 (2013)
76. Maskarinec, S.A., Franck, C., Tirrell, D.A., Ravichandran, G.: Quantifying cellular traction forces in three dimensions. Proc. Natl. Acad. Sci. **106**, 22108–22113 (2009)
77. Matheson, W., Markham, M.: Infantile cortical hyperostosis. Br. Med. J. **1**, 742 (1952)
78. Mills, J., et al.: Effect of plasmodial RESA protein on deformability of human red blood cells harboring Plasmodium falciparum. Proc. Natl. Acad. Sci. **104**, 9213–9217 (2007)
79. Mio, C., Gong, T., Terray, A., Marr, D.: Design of a scanning laser optical trap for multiparticle manipulation. Rev. Sci. Instrum. **71**, 2196–2200 (2000)
80. Mizuno, D., Tardin, C., Schmidt, C.F., MacKintosh, F.C.: Nonequilibrium mechanics of active cytoskeletal networks. Science **315**, 370–373 (2007)
81. Munevar, S., Y-l, Wang, Dembo, M.: Traction force microscopy of migrating normal and H-ras transformed 3T3 fibroblasts. Biophys. J. **80**, 1744–1757 (2001)
82. Neuman, K.C., Chadd, E.H., Liou, G.F., Bergman, K., Block, S.M.: Characterization of photodamage to *Escherichia coli* in optical traps. Biophys. J. **77**, 2856–2863 (1999)
83. Neuman, K.C., Nagy, A.: Single-molecule force spectroscopy: optical tweezers, magnetic tweezers and atomic force microscopy. Nat. Methods **5**, 491 (2008)
84. Nishizaka, T., Miyata, H., Yoshikawa, H., Si, Ishiwata, Kinosita Jr., K.: Unbinding force of a single motor molecule of muscle measured using optical tweezers. Nature **377**, 251 (1995)
85. O'Brien, F.J., Harley, B., Yannas, I.V., Gibson, L.J.: The effect of pore size on cell adhesion in collagen-GAG scaffolds. Biomaterials **26**, 433–441 (2005)
86. Onck, P., Koeman, T., Van Dillen, T., van der Giessen, E.: Alternative explanation of stiffening in cross-linked semiflexible networks. Phys. Rev. Lett. **95**, 178102 (2005)
87. Parsons, J.T., Horwitz, A.R., Schwartz, M.A.: Cell adhesion: integrating cytoskeletal dynamics and cellular tension. Nat. Rev. Mol. Cell Biol. **11**, 633–643 (2010)
88. Pickup, M.W., Mouw, J.K., Weaver, V.M.: The extracellular matrix modulates the hallmarks of cancer. EMBO Rep. e201439246 (2014)
89. Plodinec, M., et al.: The nanomechanical signature of breast cancer. Nat. Nanotechnol. **7**, 757 (2012)
90. Polacheck, W.J., Chen, C.S.: Measuring cell-generated forces: a guide to the available tools. Nat. Methods **13**, 415 (2016)
91. Recho, P., Putelat, T., Truskinovsky, L.: Mechanics of motility initiation and motility arrest in crawling cells. J. Mech. Phys. Solids **84**, 469–505 (2015)
92. Ridge, K.M., et al.: Methods for determining the cellular functions of Vimentin intermediate filaments. In: Methods in Enzymology, vol. 568, pp. 389–426. Elsevier (2016)
93. Röding, M., Guo, M., Weitz, D.A., Rudemo, M., Särkkä, A.: Identifying directional persistence in intracellular particle motion using Hidden Markov Models. Math. Biosci. **248**, 140–145 (2014)
94. Sabass, B., Gardel, M.L., Waterman, C.M., Schwarz, U.S.: High resolution traction force microscopy based on experimental and computational advances. Biophys. J. **94**, 207–220 (2008)
95. Santini, M.T., Rainaldi, G., Indovina, P.L.: Apoptosis, cell adhesion and the extracellular matrix in the three-dimensional growth of multicellular tumor spheroids. Crit. Rev. Oncol./Hematol. **36**, 75–87 (2000)
96. Sasaki, K., Koshioka, M., Misawa, H., Kitamura, N., Masuhara, H.: Pattern formation and flow control of fine particles by laser-scanning micromanipulation. Opt. Lett. **16**, 1463–1465 (1991)
97. Schmid-Schönbein, H., Volger, E.: Red-cell aggregation and red-cell deformability in diabetes. Diabetes **25**, 897–902 (1976)
98. Serra-Picamal, X., et al.: Mechanical waves during tissue expansion. Nat. Phys. **8**, 628 (2012)

99. Sharma, A., Licup, A.J., Jansen, K.A., Rens, R., Sheinman, M., Koenderink, G.H., MacK-
    intosh, F.C.: Strain-controlled criticality governs the nonlinear mechanics of fibre networks.
    Nat. Phys. **12**, 584–587 (2016)
100. Sleep, J., Wilson, D., Simmons, R., Gratzer, W.: Elasticity of the red cell membrane and its
    relation to hemolytic disorders: an optical tweezers study. Biophys. J. **77**, 3085–3095 (1999)
101. Suresh, S., et al.: Connections between single-cell biomechanics and human disease states:
    gastrointestinal cancer and malaria. Acta Biomater. **1**, 15–30 (2005)
102. Svoboda, K., Block, S.M.: Biological applications of optical forces. Annu. Rev. Biophys.
    Biomol. Struct. **23**, 247–285 (1994)
103. Tambe, D.T., et al.: Collective cell guidance by cooperative intercellular forces. Nature Mater.
    10, 469 (2011)
104. Vale, R.D.: The molecular motor toolbox for intracellular transport. Cell **112**, 467–480 (2003)
105. Visscher, K., Gross, S.P., Block, S.M.: Construction of multiple-beam optical traps with
    nanometer-resolution position sensing. IEEE J. Sel. Top. Quantum Electron. **2**, 1066–1076
    (1996)
106. Wang, N., Butler, J.P., Ingber, D.E.: Mechanotransduction across the cell surface and through
    the cytoskeleton. Science **260**, 1124–1127 (1993)
107. Weed, R.I., LaCelle, P.L., Merrill, E.W.: Metabolic dependence of red cell deformability. J.
    Clin. Invest. **48**, 795–809 (1969)
108. Wu, P.-H., et al.: A comparison of methods to assess cell mechanical properties. Nat. Methods
    15, 491–498 (2018)
109. Wyart, M., Liang, H., Kabla, A., Mahadevan, L.: Elasticity of floppy and stiff random net-
    works. Phys. Rev. Lett. **101**, 215501 (2008)
110. Xue, X. et al.: Mechanics-guided embryonic patterning of neuroectoderm tissue from human
    pluripotent stem cells. Nat. Mater. (2018)
111. Zhu, C., Bao, G., Wang, N.: Cell mechanics: mechanical response, cell adhesion, and molec-
    ular deformation. Annu. Rev. Biomed. Eng. **2**, 189–226 (2000)

Cell-Matrix Interactions in Cardiac Development and Disease

Matthew C. Watson, Erica M. Cherry-Kemmerling and Lauren D. Black III

Abstract Cardiac development is a dynamic process in which cells and tissues experience a multitude of biophysical cues. The extracellular matrix, long thought to have a passive role throughout cardiac development, is now known to actively modulate cells during development via both chemical and mechanical cues. Throughout cardiac development and cardiac disease, the cardiac extracellular matrix remodels, changing its composition, organization, and mechanical properties in response to a variety of cues. Cardiac cells both initiate and respond to these changes in cardiac extracellular matrix properties to maintain homeostasis or promote either normal or pathological tissue development. In order to better understand this interplay, it is necessary to study how cardiac extracellular matrix and cardiac cells interact, and a number of specialized techniques have been utilized to study this complicated interaction. This chapter first provides a review of common techniques used to interrogate the bidirectional interaction between cardiac cells and cardiac extracellular matrix before discussing changes in the phenotype of the major cell populations of the heart in response to changes in cardiac extracellular matrix properties throughout cardiac development and cardiac disease.

M. C. Watson · E. M. Cherry-Kemmerling (✉)
Department of Mechanical Engineering, Tufts University,
200 College Avenue, Robinson Hall, Medford, MA 02155, USA
e-mail: erica.kemmerling@tufts.edu

M. C. Watson · L. D. Black III
Department of Biomedical Engineering, Tufts University,
4 Colby Street, Rm 229, Medford, MA 02155, USA

L. D. Black III (✉)
Cellular, Molecular and Developmental Biology Program, Sackler School for Graduate
Biomedical Sciences, Tufts University School of Medicine, Boston, MA 02111, USA
e-mail: lauren.black@tufts.edu

© Springer Nature Switzerland AG 2020
Y. Zhang (ed.), *Multi-scale Extracellular Matrix Mechanics and Mechanobiology*,
Studies in Mechanobiology, Tissue Engineering and Biomaterials 23,
https://doi.org/10.1007/978-3-030-20182-1_10

1   Introduction

Development is a complicated and continuous process. Interactions between cells and their environment (i.e. mechanics, growth factors, other cells, etc.) orchestrate the proper development of the organism. Cells are dynamic; they proliferate, migrate, and change shape and structure to eventually make up functional tissues and organs. In the context of development, the heart is the first functioning organ in the developing embryo and the primitive linear heart tube begins to contract well before the final formation of the four-chambered heart. In humans, as with most mammals, heart development begins with two separate heart fields present in the cardiac crescent which signal each other through soluble factors and lead to the development of the primitive heart tubes on either side of the flat embryo. As the embryo folds together, these bilateral linear heart tubes fuse to a single tube, which then undergoes a dynamic looping and folding process to become a functioning four chambered heart. During this early process, and throughout cardiac development, cardiac cells and cardiac extracellular matrix (cECM) are exposed to many mechanical forces, causing changes to cell phenotype as well as the remodeling of cECM and cardiac tissue. The cECM, long thought to be a passive scaffold in which cells reside, has within the last few decades been re-classified as a complex signaling milieu, in which, in addition to its structural role, it actively interacts with cells and surrounding tissues via both chemical and mechanical cues [58]. Understanding the dynamic interactions between cardiac cells and cECM in response to a dynamic physical environment is critical to understanding cardiac development and disease. To understand these physical cues that modulate cell-cell interactions and cell-cECM interactions, an understanding of how the biophysical properties of the environment change throughout development and disease is necessary. In the first part of this chapter, we will give an overview of methods used to study and measure the changes in ECM mechanics and ECM composition, before addressing techniques utilized to incorporate these mechanics in both in vitro and in vivo experiments aimed at investigating alterations in cell-ECM interactions in response to changes in the biophysical environment.

Through changes in cECM composition and cECM mechanical properties, cECM is able to communicate with cells through both physical and chemical cues controlling cell migration, differentiation, morphogenesis [39, 117, 127]. For example, compositional differences in fetal cECM and adult cECM significantly impacted the proliferative response of neonatal rat cardiomyocytes to varying degrees; cardiomyocytes cultured on fetal cECM were more proliferative than cardiomyocytes cultured on adult cECM [156]. Dysregulation of cECM composition and mechanical properties can also lead to many pathological conditions throughout development. Such pathologies include congenital endocardial fibroelastosis (EFE), hypertrophy, or cardiac fibrosis [140]. The mammalian ECM is composed of approximately 300 proteins which have been shown to be relatively consistent across species [59, 87, 117]; however, composition and relative abundance of these proteins vary between tissues and organ systems within species. The environment dictates organ development and cell specification, and these compositional differences between organs suggest that ECM plays a direct and dynamic role in development, using mechanical and biochemical cues to communicate with surrounding tissue. Studies related specifically to the heart and cECM show structural and compositional changes to cECM occur during aging and disease, and that these changes can influence cell fate, phenotype, and function [9, 133, 135, 156, 157]. These studies implicate the influence that cECM composition and mechanics have on cardiac development and pathological conditions. cECM is composed of many macromolecules with diverse roles in signaling cardiac cells. Many of these macromolecules have structural and mechanical functions [35] which influence material mechanics, cell adhesion, cell migration, and morphology. These properties affect cytoskeletal organization [117] and a variety of cell functions including proliferation [1, 36], migration [83], and differentiation [31, 64, 65, 92, 127]. Cells that reside in the cECM are responsible for maintaining homeostasis of the heart, and many aspects of this involve functions related to maintenance of the cECM including cECM synthesis, cECM organization, and maintenance and remodeling of cECM in response to developmental aging, disease, or injury [35, 41]. The maintenance or imbalance of cECM in the heart is critical as it can govern both the normal function and as well as the development of pathologies of the heart.

In summary, the interactions between cECM and cardiac cells constitute a complex bidirectional system. This chapter will explore the role that cECM mechanics and composition plays in the dynamic interactions with cardiac cells. We will begin with a brief overview of cardiac development, cardiac cell populations, cECM composition, and how cardiac cells sense and respond to biophysical signals. Then we will provide an overview of some of the techniques used to measure cECM properties, and then provide examples of how researchers have incorporated these properties into in vitro and in vivo models. Finally, we will clarify the biophysical interactions between cECM and major cardiac cell types during fetal and post-natal development, and how these interactions change during disease before concluding with some open questions and potential future directions for the field.

2 Overview of Cardiac Development

2.1 Normal Cardiac Development

Heart development begins soon after gastrulation, and the heart is the first organ to form and begin functioning in the developing vertebrate embryo [3, 154]. The heart is derived from the mesoderm, which lies between the endoderm and ectoderm layers of the early embryo. The cardiac crescent is first formed by cardiac precursor cells originating from the anterior lateral mesoderm [13] (Fig. 1). The cardiac crescent supports the development of the bilateral primitive heart tubes which then fuse into the linear heart tube, comprised of an inner layer of endothelial cells and an outer myocardial layer. The inner endothelial layer and outer myocardial layer are separated

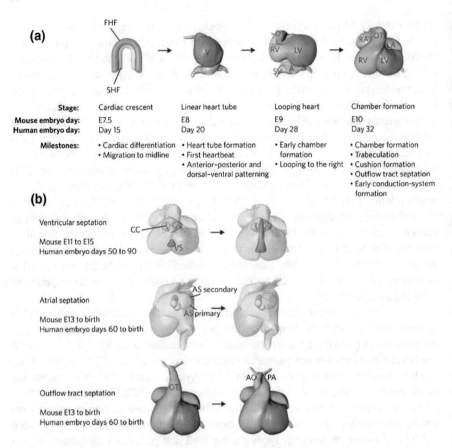

Fig. 1 a Early steps in heart development with corresponding timelines for both the mouse embryo and the human embryo. **b** Maturation of the embryonic heart prior to birth. Adapted from [13]

by acellular cECM called the cardiac jelly. Over the next several weeks, the linear heart tube then undergoes a dynamic asymmetric looping process, beginning with rightward looping for the formation of heart chambers in physiological positions, and a subsequent leftward looping that is necessary for the formation of the great vessels and the outflow tracts. In the human fetus, cardiac crescent formation, heart tube formation, and cardiac looping begin before the end of the first trimester. Heart structure continues to develop and mature until birth.

2.2 Major Cardiac Cell Populations

There are several major cell types that reside within cECM and interact with cECM to form the functioning heart. During embryonic cardiac development, cardiac progenitors differentiate into cardiomyocytes, cardiac fibroblasts, endothelial cells, and many other supportive cells (i.e. smooth muscle cells, resident immune cells, etc.). Early in heart development, cardiac progenitor cells arise from the mesoderm and from the cardiac crescent. Cardiac progenitor cells then migrate and morph to form a beating linear heart tube comprised of an outer layer of early cardiomyocytes and an inner layer of endothelial cells [61, 110]. Cardiomyocytes are responsible for heart function, actuating contractions and modulating heart growth. In the fetal heart, growth of the heart is primarily due to proliferating cardiomyocytes, while in hearts for most of the postnatal time period, growth is primarily due to cardiomyocyte hypertrophy [78]. During and after asymmetric looping, more cell types are present, including cardiac fibroblasts. Although poorly characterized in molecular terms, and in origin, cardiac fibroblasts have a critical role in maintenance of normal cardiac function [164]. Early in heart development, the endocardium and myocardium are responsible for the secretion of cECM, or cardiac jelly; however, in the maturing fetal and postnatal heart, cardiac fibroblasts are primarily responsible for synthesizing, maintaining, and remodeling the cECM to maintain homeostasis [154].

2.3   Major cECM Components

cECM is composed of many types of molecules, including collagen-based or elastic fibers, glycoproteins, and glycosaminoglycans (GAGs) and proteoglycans (PGs) [117, 154]. Given the need for dynamic cell infiltration and remodeling, embryonic cECM is vastly different from adult cECM. Embryonic cECM is less cross-linked and contains components that are softer and more malleable than adult cECM [154]. Among these molecules, collagens are some of the most abundant molecules in cECM. There are 28 types of collagens characterized by different structure and function. Among these collagens, those found in the developing heart include types I, III, IV, V, VI, and XIV [154, 156], with collagen types I and III being the major components of the fibrillar cECM network. They are present early in cardiac morphogenesis and provide mechanical strength and elasticity to the mature heart [68]. Collagen IV is abundant in the basement membrane surrounding cardiomyocytes and is critical for the stability and function of the basement membrane under increasing mechanical load [11, 104]. Though abundant in low amounts, the functions of "minor" collagens (i.e. types V, VI, and XIV) in normal development and cardiac pathologies likely play important roles in structural fiber bundling and crosslinking and warrant further study.

Glycoproteins have diverse structures and functions. Important glycoproteins that contribute to cardiac development include fibronectin, laminins, vitronectin, EMILIN, fibrillins, and fibulins [154]. Fibronectin is present in all tissues of the body and throughout development. Through interactions with other ECM proteins, cell surface receptors, glycosaminoglycans, and other fibronectin molecules [124], fibronectins can mediate cardiac regeneration and a variety of cell processes, including proliferation, migration and cell spreading or morphology [149]. In addition to collagen IV, Laminins are a major component of the basement membrane and are a set of adhesive ligands for mature cardiomyocytes [11]. Vitronectin is found in the developing heart and is a cell-adhesive glycoprotein that can influence cell migration and cell proliferation [108]; however, it does not appear to be necessary for normal embryonic morphogenesis and survival [117]. EMILINs (Elastin Microfibril Interface Located proteINs), which are found throughout heart, develop and contribute to elastic fibers of the cECM [154, 156], although their specific influence on cardiac cells has not yet been studied. Fibrillins also form elastic microfibrils, providing structural templates and support to the developing cECM. Furthermore, fibrillin can interact with cell surface receptors, can interact with other cECM proteins, and can locally regulate growth factors important in cardiac development, such as members of the transforming growth factor beta family (TGF-βs) and bone morphogenetic proteins (BMPs) [113].

Both GAGs and PGs bind water and provide structural integrity to hydrated matrices. Among the important GAGs and PGs present in cECM and cardiac development are hyaluronan, versican, and perlecan. These cECM components are abundant in the cardiac jelly during early heart formation; however, perlecan is also present in neonatal and mature ventricular cECM [156]. Hyaluronan promotes cell migration, cell invasion, and epithelial-to-mesenchymal transition, and is necessary for the development of functioning heart valves. By interacting with hyaluronan, versican plays a role in cECM assembly. Studies show that versican plays a crucial role in cardiac development [95]; however, the mechanism is not yet fully understood. Perlecan is an abundant PG secreted into the pericellular space, where it can modulate the action of signaling molecules controlling cell migration, cell cycle, and differentiation [152]. Fibulins contributes to the basement membrane and elasticity of cECM, and are required during ventricle formation [20].

3   Cell-cECM/Cell-Cell Mechanosensing and Mechanosensitive Pathways

Mechanosensors help cardiac cells to detect mechanical forces and activate signaling cascades, allowing them to interact with both cECM and neighboring cells. The primary mechanism by which cardiac cells attach to the cECM and detect mechanical forces is via integrins. Integrins are a family of large cell-surface receptors, comprised of heterodimeric α and β subunits with large extracellular domains and small intracellular domains [154], which couple the cellular cytoskeleton to the cECM at locations called focal adhesion sites [28]. The large extracellular domains bind to the cECM, while the smaller domains bind to cytoskeletal molecules [74]. There are at least 24 unique integrins, consisting of various combinations of the α and β subunits [143]. Integrins and integrin associated proteins facilitate bidirectional signaling between the cell and cECM, and modulate cell behaviors such as proliferation, spreading, migration, and survival [55, 129]. Integrin expression is not only restricted to particular cell types, and expression is often spatiotemporal, with integrin expression varying with developmental age [17].

The process of cells by which sense the surrounding mechanical environment and respond by converting the mechanical cues to biochemical signals is called mechanotransduction. There are numerous mechanosensitive signaling pathways. We will review a few that are critically important to cardiac development and disease, although others may also play a role. Among these pathways, those involving transforming growth factor β (TGF-β) and bone morphogenic protein (BMP) among the most studied. A detailed review of TGF-β signaling in cardiac development and remodeling has been done by Dobaczewski et al. Briefly, TGF-β signaling or BMP signaling begins when TGF-β ligands or BMP ligands bind to serine/threonine receptor kinases [27]. This phosphorylates cytoplasmic signaling molecules Smad2 and Smad3 for TGF-β signaling and Smad1, Smad5, and Smad9 for BMP signaling. These receptor-activated Smads form complexes with Smad and Smad4 which translocate to the nucleus where they activate or repress gene transcription [27]. This transcriptional regulation can alter a variety of cell processes including cECM deposition, cell proliferation, and cell differentiation [27].

4 Methods for Characterizing cECM Composition, and Investigating Cell-cECM Interactions

4.1 cECM Composition

Over the last decade, many techniques have been developed to measure cECM composition and mechanics ex vivo and in vitro. By using these techniques, accurate matrix mechanics can be incorporated into studies investigating cardiac development or the progression of cardiac disease. Traditionally, the compositional makeup of cECM was assessed via assays for specific components using gene expression or Western blot analysis for protein expression [21, 128]. Advances in proteomics methods now offer improved characterization of ECM composition via compiled libraries of detailed proteins and protein abundance in different tissue types [94]. Liquid chromatography tandem mass spectrometry (LC-MS/MS) offers a way to characterize the abundance of individual cECM components. LC-MS/MS is a combination of liquid chromatography, which physically separates components of a liquid mixture, and mass spectrometry, which is used to measure the mass to charge ratio of charged particles. Recently, our group has used LC-MS/MS to observe cECM composition at different developmental ages [156], demonstrating significant variations in cECM composition throughout development. For example, the most abundant protein in fetal rat cECM is fibronectin, while in the adult rat heart, cECM composition is dominated by collagens, in particular collagen I. These techniques have not only been utilized to study cECM, but also in studies of ECM composition in a wide variety of tissues. For example, an investigation by Ungerleider et al. used mass spectrometry to characterize the abundance of proteins in decellularized porcine skeletal muscle tissue and the abundance of proteins in decellularized human umbilical cord extracellular matrix, and utilized the data to engineer injectable ECM hydrogels as a therapy for patients with peripheral arterial disease [141].

4.2 cECM Mechanics

In tandem with cECM composition, mechanical properties of the cECM govern cell, tissue, and organ behavior, and it is important to measure the mechanical characteristics of these cardiac tissues to advance our knowledge of cardiac development and cardiac diseases. Importantly, cECM composition and mechanics are linked (although not linearly) in that alterations to composition can lead to changes to cECM mechanical properties (e.g. increased Collagen I expression leads to increase bulk tissue stiffness).Temporal changes and pathological cell phenotypes can also change the bulk mechanics of the cardiac tissue and cECM, usually through modulation of ECM expression and crosslinking (for example stiffening of the ventricular walls under increased pressure load) [86]. While cell phenotype dictates cardiac function and cECM mechanics, cardiac cells directly sense mechanics at a cellular scale. A study by Majkut et al. demonstrated that changes to these substrate mechanics affect cardiomyocyte phenotype; the authors observed decreased beating in myocytes cultured on soft and stiff cECM [86]. This study suggests that mechanical forces on the cellular scale are equally important to bulk mechanics in governing cellular phenotype and thus cardiac function. Techniques to study both bulk mechanics and cellular scale mechanics exist, and we will review a few common methods for each.

There have been a variety of methods used to assess the bulk mechanics of cardiac tissue and cECM. Uniaxial and biaxial mechanical testing machines can measure mechanical properties of tissues via controlled stretch [99, 112, 122, 130]. While several studies have been carried out on normal tissue, assessing the mechanics of the cardiac ECM in its native form was difficult until the advent of perfusion decellularization techniques, which allowed the maintenance of the ECM structure and composition after cell removal. In the seminal study by Ott et al., equibiaxial mechanical testing was used to measure stress-strain behaviors in cadaveric and decellularized left ventricles. Decellularized ventricle samples at 40% strain yielded a higher tangential modulus in both tangential and longitudinal directions compared to cadaveric left ventricles. However, the comparison of bulk mechanics between these two sample types does not make much sense. This is because during the decellularization process 90% of the tissue thickness is lost (the cardiomyocytes themselves makeup a significant fraction of adult tissue volume), and the lost volume contributes very little to the overall passive tissue modulus (ECM accounts for most of the tissue mechanical properties in passive conditions). This comparison would be akin to comparing the bulk mechanics of iron rods evenly distributed throughout a large volume of gelatin and comparing it to the bulk mechanics of the iron rods bundled together: the bulk mechanics would most certainly be different, but the mechanics of the individual iron rods would not change. To combat this effect, the authors calculated membrane stiffness of the samples in order to remove the dependence on thickness. Importantly, membrane stiffness at 40% strain did not differ between decellularized and cadaveric tissues, demonstrating importance of scalability between bulk mechanics and cell-scale mechanics [99]. While the bulk mechanical properties of the tissue differed between the decellularized ventricle and the cadaveric ventricle, the mechanics of the individual ECM fibers remained the same for the two cases.

As discussed above it also necessary to measure mechanical properties of cECM and cells at a cellular scale. Atomic force microscopy (AFM) is commonly used to characterize cellular scale mechanics [75, 81]. A study by Lieber et al. used AFM to investigate changes in stiffness of individual young (4 months) and old (30 month) male rat cardiomyocytes. They demonstrated a significant increase in stiffness from 35.1 kPa in young cardiomyocytes to 42.5 kPa in old cardiomyocytes [81]. Additionally, AFM has been used to characterize the stiffness of cardiac tissues [44], and verify the mechanical properties of in vitro models. An additional technique utilized to assess cellular mechanics is traction-force microscopy [64, 101]. Traction force microscopy functions by culturing cardiac cells onto deformable substrates that contain entrapped beads as fiduciary markers. By tracking small displacements in the substrate material via the beads and knowing the stiffness of the substrate, the strains and stresses generated by individual cells can be determined.

4.3 Animal Models to Study Cell-cECM Interactions

Animal models, including zebrafish, chick embryo, mouse, and rat, have greatly contributed to our understanding of the vertebrate heart. In these models, the heart arises from bilateral heart fields of cardiac precursor cell. The type of animal used for cardiac studies remains dependent upon the experiment. Techniques such as high-resolution microscopy (i.e. confocal microscopy), in situ immunostaining and single-cell transcriptome analysis allow for the interrogation of cardiac cell phenotype during development and disease. In recent years, the zebrafish has been a popular model to study cardiac development and disease, given its propensity for cardiac regeneration at any developmental stage [5]. The zebrafish genome has almost been entirely sequenced, and gene function compared to humans is highly conserved, making the zebrafish a decent model to study genetic variations and cardiovascular defects [5]. Furthermore, the zebrafish genome encodes a similar overlap of ECM associated proteins found in other vertebrate organisms [67]. Combined with its high affinity for high-resolution microscopy, the zebrafish offers a useful platform to study the structure and function of ECM proteins [67].

Due to low costs, the lack of requirement for IACUC approval at many institutions and the development of open-egg handling techniques, the embryonic chicken has been used as an animal model to describe early cardiac development and pathologies in ovo and ex ovo [69]. The relatively large size of the in ovo embryo offers a useful platform for studying the differentiation and behavior of cardiac cells [69]. Advances in culturing the chick embryo *ex ovo* allows a more accessible way to view cardiac development, and offers ways for experimental manipulation [69]. A detailed review of the chick embryo as a model of cardiovascular research is provided by Kain et al.

The mouse has become a popular model used to study human heart disease. One reason mouse models are so popular is that the relatively short life span of mice expedites investigations cardiac development and cardiac diseases [114]. Mice are also especially useful in studying genetic defects in humans, as 99% of human genes have direct murine orthologs [114], and the development of transgenic mice lines has accelerated our understanding of the role of genetic abnormalities on human disease. In particular, these models allow for the identification of genes and mechanisms that are necessary for the progression of the disease [49]. Transgenic mice are therefore a useful platform to study the interactions between cECM and disease. For example, in a study using elastin knockout mice, mice are born with stiff, tortuous arteries, and cardiac hypertrophy [147]. Although widely used, the mouse model is a mammalian heart model that is least representative of human heart contractile function [15]. Thus, mouse models are often used as proof-of-principle models which can be later extended into larger animal models [15]. Rat models have also found usefulness in the field mostly due to the aforementioned shorter lifespan, low cost and larger heart size which makes it easier for carrying out surgical procedures as compared to the mouse.

4.4   In Vitro Models to Study Cell-cECM Interactions

In determining the impact of mechanics on cardiac cells, in vitro models offer platforms to study cellular-cECM interactions ex vivo. In order to accurately mimic the physiological environment, it is important for these in vitro models to incorporate cECM components and to mimic the mechanical forces present in vivo, including those caused by substrate stiffness and/or mechanical strain. Two-dimensional polyacrylamide (PAA) hydrogels are frequently used to study the effects of cECM composition and substrate stiffness on monolayers of cardiac cells [4, 8, 52]. Preparation of PAA gels is described elsewhere. Briefly, gels are cast on activated glass coverslips by crosslinking an acrylamide solution with a bisacrylamide (bis) solution. Changing the ratio of acrylamide solution to bis solution controls PAA elasticity, and a wide range of physiological and pathological cardiac stiffnesses (~1–100 kPa) can be achieved [8, 30, 52, 101]. Furthermore, PAA allows cECM and cECM components to be covalently added to the substrate without changing the stiffness of the gel [30]. In a study by Pasqualini et al, neonatal rat cardiomyocytes (NRCMs) were seeded onto fibronectin patterns stamped on PAA with soft (1 kPa), normal (13 kPa), and diseased (90 kPa) stiffnesses. Fluorescent beads were integrated into the PAA gels, and traction force microscopy was used measure stiffness of NRCMs and work of NRCMs [101]. A similar method of incorporating fluorescent microbeads into PAA gels of varying stiffness was used in

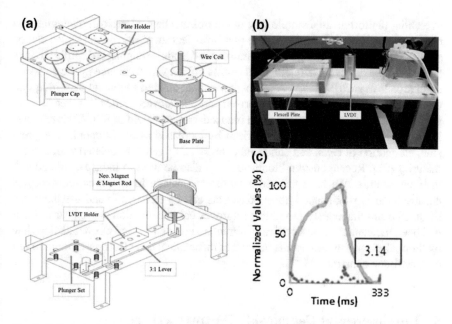

Fig. 2 Schematic (**a**) and images (**b**) of a custom made stretching device capable of applying precise customizable stretch waveforms **c** to cells in culture. Adapted from [77]

a recent study by Wheelwright et al, who interrogated the force output of single cell human induced pluripotent stem cell derived cardiomyocytes (iPSC-CMs) [151].

While hydrogels are an easy way to assess the role of substrate stiffness on cardiac cell phenotype and function, specialized devices are needed to investigate the effects of mechanical strain on cell-cECM interactions (see Fig. 2 for an example taken from [71]). Flexcell® plates, and other similar systems, utilize vacuum systems and motor systems to induce cyclic, bidirectional strain on 2D cultures by applying a vacuum or plunger to flexible-bottomed plates [7, 29, 76]. These strains can be tuned to replicate the strains associated with physiological and pathological hearts [33, 120]. A recent study by Rysä et al. applied bidirectional strain to neonatal rat cardiomyocytes and characterized gene expression [120]. Laser-etched glass plates have been used with a piezoelectric length controller to induce unidirectional stress in cardiomyocytes [16, 109]. A study by Caporizzo et al. utilized a glass plate stretching device to stretch cardiomyocytes while simultaneously imaging cardiomyocyte sarcomere structure. They investigated microtubule buckling in adult rat cardiomyocytes, and it was concluded that microtubules regulate contractility in cardiomyocytes via breakable crosslinks between the cytoskeleton that resist length changes induced by stretch [16].

To better mimic the cardiac mechanical environment in vitro, a combination of stiffness and stretch can be incorporated into three-dimensional engineered cardiac tissue constructs using customizable bioreactor systems [63, 80, 85, 93, 132]. Static stretching platforms are a simple way to control mechanical preload on engineered cardiac tissues [12, 85]. Most of these platforms feature plastic pillars which can be adjusted to induce physiological and pathological loads on engineered cardiac tissues [12, 85]. Uniaxial cyclic stretching platforms offer a way to induce cyclic stress on cardiac cells in vitro [91, 132]. A recent experiment utilized a uniaxial stretching system to stretch embryonic stem cell-derived cardiomyocytes (ESC-CMs) seeded into gelatin-based scaffolds. Cyclic stretch induced a greater yield of ESC-CMs expressing mature cardiac markers characterized by increased cardiac troponin-T expression, elongation of cells, and enhanced expression of genes associated with cardiac maturity [91]. Recent custom bioreactors can also be used to induce cyclic stretch in tissue culture. We have utilized a bioreactor system in which cardiomyocytes encapsulated in ring-shaped fibrin constructs can be electrically and mechanically stimulated simultaneously, promoting improved cardiac function over constructs that had been mechanically or electrically stimulated alone [93, 163]. A detailed review by Stoppel et al. provides details on further mechanical and electrical stimulation in cardiac tissue constructs [132].

5   Developmental Cellular-cECM Interactions

The cECM is a complex, three-dimensional network inhabited primarily by myocytes, fibroblasts, and vascular endothelial cells, as well as significant population of other cells (smooth muscle cells, macrophages, etc.) [116]. The cECM begins to be synthesized and excreted by embryonic cells at very early stages [117], providing a microenvironment rich in biochemical and mechanical cues supportive of proliferation, differentiation, and maturation. Throughout development, changes to the local stiffness, stresses, and strains change the behavior of these residing cardiac cells [53, 117, 156] (Fig. 3). For migration and morphogenesis to occur, the mechanical properties and composition of the early cECM is critical: softer cECM is more adaptive to infiltration and morphogenesis and important for normal heart development. Both fibronectin and collagens are needed for normal embryonic development and dictate cECM bulk mechanics by cross-linking via enzyme activation in the native tissue [96].

Immediately after birth, the heart undergoes dramatic changes in phenotype, including the closing of the foramen ovale as a pressure gradient forms from left to right [136], and a switch from proliferative cells to hypertrophic cells as the primary mechanism of heart growth [148]. cECM must adapt to these changes, particularly to increases in pressure load, and thus, stiffens by changes its composition and organization. A reduction in fibronectin, which is necessary for proliferation and embryonic development, and a replacement increase in collagen is observed in the composition of neonatal rat cECM. By the time the rat heart matures to adulthood, fibronectin is reduced to only 4% of cECM composition [156]. At this stage, the cECM is dominated by collagens, especially collagen I, which increases tissue stiffness in response to increased cardiac load [156] and leads to a more homeostatic

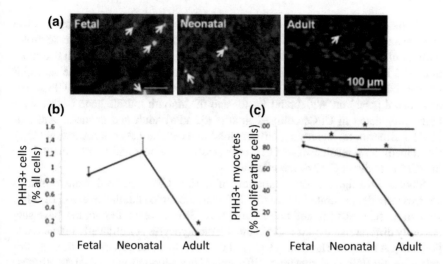

Fig. 3 Characterization of cardiomyocyte proliferation in fetal, neonatal, and adult hearts. **a** Representative images of tissues sections stained for nuclei (blue), cardiac α-actinin (red), and the mitosis marker phosoph-histone H3(PlH3). **b** Quantification of proliferating cells, with a significant decrease in proliferating cells in the adult heart. **c** Quantification of proliferating cardiomyocytes, with the highest proliferation observed in the fetal heart. Adapted from [158]

maintenance of tissue function. The compositional changes of cECM during postnatal development orchestrate changes in the mechanical properties of the cECM that cells sense, and also alter the transmission of mechanical loads from contractile cells to the cECM as well as from hemodynamic pressures in the heart to the cells [86, 126, 127]. In the subsequent sections we detail the impact of cECM changes during normal development on the phenotype and function of cardiac progenitor cells, cardiomyocytes, cardiac fibroblasts and endothelial cells.

5.1 Cardiac Progenitor Cell-cECM Interactions

Cardiac progenitor cells (CPCs) are defined as self-renewing cells present in the embryonic heart with the ability to differentiate into cardiomyocytes, endothelial cells, or cardiac fibroblasts [89]. Among these cardiac progenitor cells, type III tyrosine kinase receptor c-kit+ cardiac progenitor cells are among the most studied. c-kit+ CPC's ability to self-renew has been tested in vitro and it appears to be the most undifferentiated cardiac progenitor population. Additionally, c-kit+ CPCs are present in embryonic and postnatal hearts, although the extent of their differentiation potential depends upon species and developmental stage.

In a study from our lab, c-kit+ CPCs isolated from pediatric patients were seeded into engineered fibrin scaffolds containing adult or neonatal ECM. Scaffolds had elastic moduli of 2, 8, 14, and 32 kPa [155]. Cell viability was significantly lower in neonatal ECM-fibrin scaffolds with 14 and 32 kPa compared to fibrin scaffolds with an elastic modulus of 2 kPa and ECM-fibrin scaffolds with an elastic modulus of 2 kPa [155]. Both ECM developmental age and stiffness of the scaffolds affected cardiovascular gene expression and network formation of c-kit+ CPCs. The endothelial gene von Willebrand Factor and the smooth muscle gene CNN1 were both upregulated in CPCs cultured in stiff (32 kPa) adult and neonatal scaffolds [155]. Furthermore, increasing stiffness inhibited cellular network formation [155]. Combined, these results suggest that both scaffold stiffness and developmental age of ECM may direct CPC differentiation.

Studies investigating differentiation of c-kit+ CPCs derived from postnatal sources have demonstrated successful differentiation into endothelial cells [18, 146], smooth muscle cells [2], and fibroblasts [2, 18, 38]; however, few studies have successfully differentiated c-kit+ CPCs into cardiomyocytes at a clinically relevant rate [18, 146]. A study by Hazeltine et al. split CPCs from tissue culture plastic to hydrogels with physiological substrate stiffnesses. They reported that substrate stiffness did not impact the efficiency of CPC differentiation to cardiomyocytes; however, substrate stiffness promoted cardiac differentiation during early mesoderm specification and promoted mature cardiomyocyte phenotype post differentiation [50]. There is a need for more studies that interrogate the interactions between CPCs and temporal changes to cECM mechanics and composition throughout development, in particular studying ckit+ CPCs from earlier developmental stages.

5.2 Cardiomyocyte-cECM Interactions

Cardiomyocytes (CMs) comprise 30% of the native cells in the adult heart, and the behavior of embryonic CMs are different from the behavior of adult CMs. The human heart is one of the first organs formed in the embryo, and it begins to beat around 22 days post fertilization. As mentioned previously, the embryonic heart tube consists of an inner layer of endothelial cells (ECs) and an outer layer of CMs [56]. In order for the heart tube to form a functional heart by birth, it is necessary for cells to undergo epithelial to mesenchymal transition (EMT) to form cardiac fibroblasts (CFs). CFs are the major cell type responsible for the secretion of cECM; however, in the early embryonic heart, components are secreted by contracting CMs. Components secreted by CMs that promote EMT include laminins and other glycoproteins as well as matrix-associated proteins, including members of the TGF-β family of cytokines and receptors [56].

In addition to matrix deposition, CMs must proliferate and migrate to form a functioning heart. Many studies have investigated paracrine signaling between the epicardium and endocardium as mechanisms promoting CM proliferation. Secreted insulin growth factor (IGF) [79] and canonical Wnt [14] signaling are involved in modulating the numbers of and size of CMs and thus the size of the embryonic heart. Fibroblast growth factors (FGF) and bone morphogenetic proteins (BMPs) also demonstrated a proliferative effect on CMs during embryonic development [62, 159]. Given the abundance of growth factors needed to orchestrate CM proliferation, it is vital to understand the role cECM plays in cell-cell communication. Fan et al. gives a detailed review on interstitial transport in ECM [32]. The diffusion of these growth factors is critical for proper development and morphogenesis, and the cECM plays an important role in regulating not only the diffusion of these molecules, but also their sequestration. Heparan sulfate proteoglycans (HSPGs) are a family of proteins that are a part of the nonstructural ECM. Heparan sulfate (HS) chains are able to bind to a broad range of ligands [98], and thus can hinder the diffusion of growth factors. In contrast, Wnt proteins are hydrophobic molecules and require transportation using lipoprotein particles in the cECM [153]. Recent studies by our group have shown that embryonic cECM promotes CM proliferation [156], suggesting that the composition and ratio of components in cECM are important to understanding CM proliferation and morphogenesis in the embryonic heart.

Embryonic CMs are also exposed to passive and active mechanical forces generated by cECM during heart development. For example, during development, passive cECM elasticity increases from a soft mesoderm tissue (~0.5 kPa) [48] to a semi-stiff tissue (10–15 kPa) [47]. This increased tissue stiffness throughout cardiogenesis is caused by the infiltration and proliferation of cells that deposit and reorganize cECM components. Fibronectin, the most abundant component in fetal cECM [156], is critical for embryonic heart development. Studies in fish and mice suggest that the cell surface receptor integrin $\alpha5\beta1$, which specifically binds to fibronectin, is required for proper symmetry in heart development [111]. Interestingly, studies showed CMs cultured on a matrix with heart-like elasticity could achieve functional contractility [30], and other studies showed an increase in proliferation of CMs cultured on embryonic cECM [156]. Furthermore, embryonic CMs cultured on ECM substrates mimicking the mature heart tissue produced more organized sarcomeres and had greater functionality in their mature state [66]. Temporal effects of cECM stiffness also change the function of CMs and may be important, as in addition to the passive increase in cECM stiffness, contractile forces of beating CMs are resisted by cECM resulting in a change in local stiffness [86, 126, 162]. A recent study by Young et al. utilized dynamic hyaluronic acid hydrogels whose stiffness either changed from ~0.2–8 kPa quickly (<70 h) or slowly (≫100 h) over the course of a week to culture embryonic chick CMs. Myocytes cultured on the more dynamic hydrogels developed myofibrils with a z-disc spacing of 1.8 μm, indicative of a mature myofibrils, while myocytes cultured on the less dynamic hydrogels exhibited lower z-disc spacings and immature sarcomeres [162]. Furthermore, compared to static hydrogels, dynamically stiffening hydrogels exhibited time-dependent up regulation of kinases (e.g. PI3K/AKT, p38 MAPK) that promote CM maturity and survival and a downregulation of GSK3β, an antagonist of cardiomyocyte maturity [162]. These results suggest that CM maturation is at least partially modulated by dynamic mechanosensing at focal adhesions and suggest a combination of static and dynamic changes to cECM stiffness drives CM maturation. Studies have demonstrated that there is an optimal match between the tissue stiffness and local stiffness sensed by CMs which is needed for the optimal assembly of myofibrils and thus CM function [86, 126].

The mechanical load of the heart drastically changes upon birth: substrate stiffness increases to ~10–15 kPa [30] and left ventricular systolic pressure increases to ~14–17 kPa [70, 166]. Furthermore, post-natal cardiomyocyte proliferation drastically decreases compared to fetal cardiomyocytes, and hypertrophy becomes the main mechanism for cardiomyocyte growth as we age [156]. To account for these changes, cardiomyocyte phenotype and interactions with cECM also change to maintain homeostasis. Recent studies investigated the work of ventricular cardiomyocytes on gels with physiological and pathological stiffnesses and cECM components [4, 101]. Neonatal rat ventricular cardiomyocytes cultured on gels of normal stiffness stamped with fibronectin generated higher stresses and maximal cardiac work when compared to cardiomyocytes cultured on either soft or pathological stiffnesses [101]. A similar study investigated a mixed population of neonatal rat ventricular cardiomyocytes and cardiac fibroblasts cultured on 13 kPa gels and 90 kPa gels coated with either fibronectin or laminin, and reported that cross-sectional systolic force remained the same throughout the varying conditions; however, peak systolic work was lower on the 90 kPa gels [4]. These studies suggest that cECM stiffness tightly regulates the contractility of cardiomyocytes. Due to the significant mechanical work cardiomyocytes must output, understanding their energy consumption is critical. Mitochondria contribute to ~35% of cardiomyocyte cell volume, likely due to the significant work output of cardiomyocytes [25, 43]. One study investigated cECM elasticity on metabolic activity of cardiomyocytes and demonstrated that cECM stiffness regulated mitochondrial function [84]. Interestingly, in a similar study, metabolic respiration of ventricular cardiomyocytes was observed to be higher on gels mimicking immature cECM stiffness (1 kPa) compared to cardiomyocytes cultured on gels with a physiological stiffness (13 kPa) [101]. This suggests that there is an inefficient coupling between the contractile work of cardiomyocytes and metabolism of cardiomyocytes. These studies suggest that cECM stiffness dominates over cECM composition with respect to controlling cardiomyocyte contractility and metabolism; however, further studies with more complicated three-dimensional in vitro models could further improve upon the understanding of cECM mechanics and composition and cardiomyocyte metabolism and contractility.

5.3   Cardiac Fibroblast-cECM Interactions

Cardiac fibroblasts (CFs) are a cardiac cell type essential to homeostasis of cECM [140] and CFs constitute a large proportion of the non-cardiomyocyte cell population in the heart [103]. CFs are observed in the developing murine heart by embryonic day 12.5, and throughout development their numbers increase [6]. It is believed that the majority of CFs are formed during development via EMT [24] from the epicardium although some are also thought to be derived from the neural crest. Once epithelial cells undergo EMT, platelet-derived growth factor (PDGF) and transforming growth factor β (TGF-β) promote differentiation into a cardiac fibroblast phenotype [125]; however, these proteins must be tightly regulated, as an overexpression of TGF-β leads to fibrosis from CFs [82]. A recent study demonstrated that decreases in matrix rigidity led to increased TGF-β-mediated apoptosis, while increased matrix rigidity resulted in EMT [77]. In this manner, a negative feedback loop between cECM and CFs regulating matrix stiffness maintains a low level of TGF-β in the microenvironment to maintain homeostasis. This negative feedback loop of matrix deposition and mechanical sensing has also been observed in mesenchymal stem cells (MSCs) during development; MSCs responded to substrate stiffness by changing their extracellular matrix composition [40], which further regulates their phenotype.

While the function of CFs has been described in the adult heart, less is known about CFs' role during embryonic heart development. Compared to adult fibroblasts, embryonic cardiac fibroblasts express more growth factors and secrete a greater abundance of ECM proteins, such as fibronectin, several members of the collagen family, and other glycoproteins and glycans, such as periostin, tenascin C, hyaluronan, and proteoglycan link protein 1 [60]. The cECM components deposited by CFs are largely determined by the CFs' ability to interpret mechanical forces by mechanotransduction [71]. For example, CFs can differentiate into myofibroblasts when exposed to TGF-β and under conditions of high stress [145], increasing the contractility of CFs and the amount of ECM components secreted by CFs. Due to the low stiffness of embryonic cECM compared to adult cECM (~10–15 kPa) [47], an abundance of fibronectin and collagen is necessary to increase the cECM stiffness and prepare for the increased pressure load of blood flow postnatally [129]. Additionally, compared to adult CFs, embryonic CFs do not contribute to observed hypertrophy in cardiomyocytes co-cultured with adult CFs [60]. More work is necessary to understand the role CFs play at different stages of embryonic development.

Postnatal and adult hearts experience dynamic changes in mechanical load at a frequency of around 1 Hz. To maintain homeostasis, CFs sense this mechanical load and respond with regulated matrix turnover. The healthy human heart maintains an elasticity of ~10–15 kPa as measured by atomic force microscopy [42]. Studies have demonstrated that CFs cultured on 5 kPa polydimethylsiloxane (PDMS) or PAA gels displayed radial geometries while CFs cultured on 30 kPa PDMS or PAA gels are polarized [107]. Additionally, both increasing substrate stiffness and increased TGF-β exposure have been shown to induce myofibroblast phenotypes in CFs [145, 158]. Primary CFs cultured on substrates greater than 3 kPa demonstrate a proto-myofibroblast phenotype with the early formation of actin stress fibers [145, 161].

Homeostasis is maintained via proper cell-cell and cell-environment interactions and cyclic strain has been shown to regulate myofibroblast phenotype. A recent study has proposed that cyclic mechanical stretch may upregulate the natriuretic peptide system, potentially reducing CF differentiation to myofibroblast phenotype [145, 150]. Cyclic strain has also been shown to increase CF production of collagen I and fibronectin [45, 145], maintaining homeostasis. While many studies have investigated strain and stiffness independently of CF phenotype, it is necessary to investigate both substrate stiffness and cyclic strain together to better understand conditions in the healthy heart.

Importantly, there are likely significant interactions between passive stiffness and mechanical strain in regulating CF phenotype. A study investigating both substrate elastic moduli and cyclic strain concluded that substrate stiffness had temporary effects on CF remodeling, while cyclic strain and mechanical loading can induce stiffer environments [73, 145]. Similarly, studies on valve interstitial cells (VICs), a CF-like cell in the valve interior, suggest that substrate stiffness and dynamic stretch can promote cell spreading in soft (<3 kPa) environments [138, 145]. A detailed review by van Putten et al. gives further information of CF differentiation and response to mechanical cues [145]; however, more studies are needed to accurately model the beating healthy heart and elucidate mechanisms that maintain homeostasis in the context of mechanical strain.

5.4 Endothelial Cell-cECM Interactions

Like CFs and CMs, ECs are exposed to many passive and active forces during embryonic heart development. Notably, ECs have a direct interface with blood flow and are responsive to flow-induced shear stress [22]. They can also sense mechanics of the surrounding cECM, and increase their spread area and their assembly of actin stress fibers in response to a substrate stiffness of around ~3 kPa [57]. There are also many biophysical cues that affect ECs during endocardial development, such as mechanical forces generated by the stiffness of the cECM [46]. Extracellular proteins of the glycocalyx, sialic acid, and hyaluronic acid contribute to mechanotransducive pathways of ECs [115, 137]. While it is well-established that ECs are mechanosensitive cells, there are many open questions associated with understanding their role in embryonic cardiac development. More studies must focus on interactions between ECs and ECM considering the composition and organization of the ECM and effective transport and motility of signaling molecules. For example, Neuregulin-1 is shown to be secreted by microvasculature ECs and activates both ErbB2 and ErbB4 on CMs inducing kinase cascades which can modulate myocyte cell growth and survival [100, 119, 121]; however, few studies have investigated how mechanics, such as substrate stiffness, wall shear stress, and pressure load may impact EC expression of Neuregulin-1.

6 Pathological Cell-cECM Interactions

As described above, there are many feedback loops that tightly regulate homeostasis during physiological development. However, certain pathologies can induce positive feedback loops leading to drastic remodeling of cECM. Heart failure associated with cardiac fibrosis results from increased loads and increased tissue stiffness [123]. Cardiac fibrosis can be triggered by local tissue injury and increased mechanical load [102, 123]. Reparative cardiac fibrosis occurs after cardiomyocyte necrosis, with increased collagen accumulation at the location of cell death. One example of reparative fibrosis is the fibrosis that occurs after myocardial infarction (MI) [123].

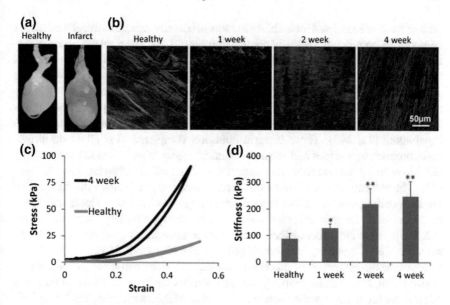

Fig. 4 Characterization decellularized myocardial infarcts. **a** Representative images of decelled healthy and infarcted rat hearts. **b** Representative two photon confocal images of collagen deposition after infarction (red = forward second-harmonic generation, green = backward SHG). **c** Stress strain curves of healthy and unhealthy ventricle tissue. **d** Tangent modulus of healthy and diseased ventricle, demonstrating an increase in stiffness as a function of remodeling time. Taken from [136]

After MI, left ventricular remodeling is induced by inflammation and an immune response, leading to myocardial cell death, and cardiac fibrosis resulting in scarring of the infarct region [102, 105, 106]. First, immune cells release matrix metallo-proteinases (MMPs), which digest and degrade the cECM, allowing for infiltration of more immune cells and the invasion of myofibroblasts [134]. Myofibroblasts are responsible for the deposition of collagens type I and type III that increase the stiffness of the infarct region with remodeling time post-MI [34, 134] (Fig. 4). These alterations in cECM composition and mechanics in response to cardiac fibrosis modulate the phenotype of fibroblasts, cardiomyocytes, and endothelial cells as described in the sections below.

6.1 Myofibroblast-cECM Interactions

While physiological heart tissue typically has a stiffness of ~10–15 kPa [30], fibrotic tissue typically has a stiffness of ~20–100 kPa [135, 144]. In response to this increased matrix stiffness, native fibroblasts and other resident cells, such as smooth muscle cells and perivascular pericytes, transition to become myofibroblasts [54, 72, 144, 158]. Cardiac myofibroblasts are identified by their expression of alpha smooth muscle actin (αSMA) and their ability to more efficiently secrete extracellular matrix components compared to resident cardiac fibroblasts [23]. The transformation of resident cells to myofibroblasts occurs in two stages. During the first stage, fibroblasts transition phenotype to become "proto-myofibroblasts" [139]. Proto-myofibroblasts are characterized by formation of actin stress fibers in the cytoplasm [23, 54]. Additionally, stress fibers terminate in small integrin adhesion complexes, and allow the proto-myofibroblasts to infiltrate the infarct area through enhanced migration capabilities [23, 145]. These integrin adhesion complexes also allow the proto-myofibroblast to transmit and sense mechanical forces to and from the surrounding cECM, inducing the secretion of collagens and fibronectins within the wounded area [23]. The increased deposition of collagen consequently increases the stiffness of the surrounding cECM. In the second stage, large mechanical stiffness and strain cause the small integrin adhesion complexes to develop into large focal adhesions [42, 145]. These large focal adhesion complexes allow incorporation of αSMA into the stress fibers, inducing fully-developed myofibroblasts [42, 145].

Myofibroblasts have many mechanisms which allow them to sense the mechanical environment, although the primary mechanosensors are integrins. Several integrins allow myofibroblasts to sense the mechanical state of the surrounding cECM, including collagen sensing integrins α1β1, α2β1, α11β1, and α1β3 and fibronectin binding integrins α5β1, α8β1, αvβ1, αvβ3, and αvβ5 [88]. The role of these integrins in controlling fibroblast behavior has been reviewed elsewhere [17]. In environments under an increased mechanical load, these integrins can bind specifically to strained cECM proteins. For example, normal cECM is rich in latent TGF-β1. On stiff substrates, myofibroblast contraction mediates the release of TGF-β1 from cECM through the transmission of force via αv integrins [51, 145]. Activated TGF-β1 released from the cECM under mechanical stress and strain further promotes differentiation of fibroblasts to myofibroblasts [158] as well pericardial cells to myofibroblasts via EMT [97]. In addition to myofibroblasts, stiffness and contractility affects cardiomyocyte function. An understanding that myofibroblasts develop in response to compositional changes and mechanical changes to cECM offers potential therapeutic strategies to combat these fibrotic pathologies. For example, therapies targeting the mechanisms by which CFs sense mechanical forces could help to reduce myofibroblast phenotype without impeding the function of cardiomyocytes.

6.2 Cardiomyocyte-cECM Interactions

CMs must respond to the compositional changes in cECM driven by myofibroblasts and mechanical changes induced by cardiac fibrosis to maintain heart function. Recently, substrate stiffness was shown to affect cardiomyocyte action potential duration. The study by Boothe et al. showed that primary isolated rat ventricular myocytes generated the longest action potential duration on hydrogels having a stiffness of 9 kPa, while duration was significantly reduced when the cells were cultured on 25 kPa gels [10]. Furthermore, the voltage at maximum calcium flux decreased in cardiomyocytes seeded on hydrogels with an elasticity greater than 9 kPa [10]. CMs undergo increased cell spreading on stiffer materials (Fig. 5) [30, 118]. Engler et al. investigated single cardiomyocytes seeded on physiological stiffness and pathological stiffnesses. Cardiomyocytes seeded on hydrogels with a physiological stiffnesses (~11 kPa) beat at ~1 Hz, with 20% of the population spontaneously beating. On stiffer substrates, the contraction frequency decayed to less than 1 Hz and spontaneous beating was only observed in 2–8% of the population [30]. It is suggested that single cardiomyocytes can sense matrix stiffness through adhesions that simultaneously sense fast cardiac contractions and slow non-muscle myosin contractions. The combination of these adhesions causes continuous stretching of cardiomyocytes on fibrotic stiffnesses. Furthermore, activation of protein kinase-C (PKC) induces hypertrophy in cardiomyocytes cultured on stiff substrates by activating non-muscle myosin, and both PKC and non-muscle myosin are upregulated after myocardial infarction [30].

In addition to changes to cECM stiffness orchestrated by cardiac fibrosis, compositional changes in cECM may play a role in orchestrating pathological cardiomyocyte phenotype. For example, the pressure overload associated with left ventricular remodeling induces overexpression of collagens and fibronectins, and also increases levels of collagen and fibronectin-binding integrins [17, 90]. Changes in cECM composition and mechanics, as well as changes in integrins, may lead to pathological organ function induced by cardiac fibrosis. Tenascin-C (TNC) is a cECM protein that is present during early cardiac development and is typically repressed in adult tissues; however, in response to pathologies such as reparative cardiac fibrosis and pressure overload, increased levels of TNC have been observed [142]. TNC is capable of detaching CMs from cECM, possibly leading to CM apoptosis and the invasion of myofibroblasts and immune cells [142]. The abundance of many more cECM proteins are likely modulated by pathological stresses that change cECM mechanics, and more studies should be done to elucidate potential therapies targeting these cECM molecules.

6.3 Endothelial Cell-cECM Interactions

Matrix metalloproteinases digest necrotic cardiomyocyte tissue and fibrotic cECM, allowing infiltration of proliferative non-myocyte cells such as CFs and ECs. Latent TGF-β1 is released by the cECM following infarction and induces ECs to undergo endothelial-mesenchymal transition (EndMT) and become myofibroblasts, contributing to the stiffening of the cECM [145, 165]. In addition to the transition of ECs to myofibroblasts, an additional function of ECs is to provide oxygen and nutrients to the native tissue through intricate networks of capillaries. This function is negatively affected by pathological changes in cECM. Endothelial cells cultured on stiff fibronectin or collagen-coated surfaces show increased cell-cell permeability and decreased angiogenic sprouting [160]. A study by Sack et al. showed that the way endothelial cells process vascular endothelial growth factor (VEGF) tethered

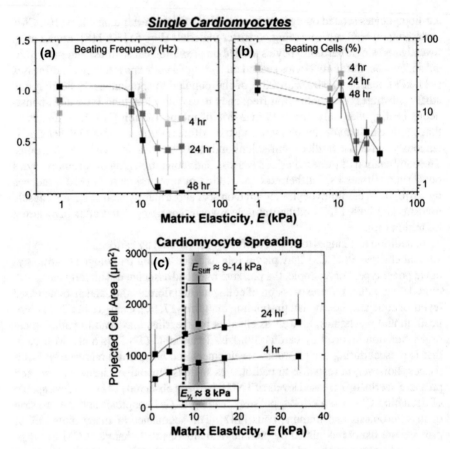

Fig. 5 Beating sensitivity (**a**, **b**) and spreading (**c**) of cardiomyocytes seeded on matrices of varied stiffness. Adapted from [30]

to the cECM is aberrantly modulated by substrate stiffness via downregulation of integrin $\beta 1$ in endothelial cells cultured on stiff substrates.

Many studies have investigated the effects of shear stress on the function of ECs [19, 26, 131], while fewer studies have investigated the effects of both shear stress and substrate stiffness. In one study, it was observed that EC spreading and stiffness increases when cultured on stiff substrates [37]. This study further investigated combinatory flow-induced shear stress and substrate stiffness. Alignment of endothelial cells has been used as a defining feature of vasculature homeostasis [131]. Unlike other studies where ECs aligned with physiological shear stress (1 N/m^2), greater shear stress was needed to align ECs on stiffer substrates [37]. This suggests that substrate stiffness has an important role in modulating EC behavior, and that increased cECM stiffness may have aberrant effects on EC phenotype.

7 Conclusions and Remaining Questions

In summary, throughout development, cardiac cells and cECM are highly interactive. Changes in the mechanics (e.g. substrate stiffness, stretch and strain, pressure load, etc.), composition and organization of cECM cause changes in cardiac cell phenotype which further modulates changes in cECM mechanics and cECM composition and organization in a continuous feedback loop. The complex mechanics of cECM and cardiac tissue modulate cardiac cell responses and orchestrate healthy or pathological behaviors of the cardiac cells and the heart. We have discussed how cECM mechanics change at different developmental ages and how cECM mechanics change during cardiac fibrosis. Furthermore, we have discussed how these changes in cECM mechanics and composition induce changes in the phenotype of several major cardiac cell types (cardiac fibroblasts, cardiomyocytes, and endothelial cells); however, there remain many questions regarding how developmental and pathological changes to cECM mechanics and composition modulate the multicellular interactions between cardiac cells and cECM. A few that we have include:

- Studies of congenital heart disease are mostly reliant on patients and expensive animal models. How do we develop an in vitro system to study pathological mechanics in the human embryo?
- Adult cardiomyocytes dedifferentiate upon isolation, and stem cell derived cardiomyocytes express immature phenotypes. How do we investigate how adult human cardiomyocytes behave in response to different mechanical stresses?
- In addition to collagens and fibronectins, there are hundreds of other proteins that compose cECM. Which proteins are important? How do the abundancies these proteins change cECM mechanics? How do these proteins change cardiac cell phenotype?
- How do cECM mechanics throughout development influence the rest of the cardiac cell population (e.g. immune cells, mesenchymal cells, smooth muscle cells)? How do we determine what populations of cell types to use in in vitro models? How will different populations of cell types alter the cECM mechanics and cECM composition?

Future studies should continue to explore these questions by utilizing new technologies to improve upon existing in vitro platforms, investigate ways to mature and characterize stem cell derived cardiac cells, and explore how changes in cECM properties effect a more proportional heterogeneous cell population which includes the "minor" population of cardiac cells, along with cardiomyocytes and cardiac fibroblasts. Furthermore, the study of the importance of some of the more minor abundance proteins in the cardiac ECM is important for understanding their role in cardiac development and disease and may lead to the development of more ECM-related therapeutic approaches for cardiac maladies. Lastly, while outside the scope of this review, the degradation of the ECM during remodeling in development and disease also generates soluble fragments of the ECM that may signal cells in different ways that the whole proteins. These so-called matrikines or matricryptins may also ace in concert with mechanical cues from the degrading ECM to help drive cell phenotype and should be studied in more detail in subsequent lines of research. Overall, these studies will improve our understanding of the interactions between cardiac cells and cECM, and may elucidate therapies to better treat cardiac diseases.

References

1. Ahmann, K.A., Weinbaum, J.S., Johnson, S.L., et al.: Fibrin degradation enhances vascular smooth muscle cell proliferation and matrix deposition in fibrin-based tissue constructs fabricated in vitro. Tissue Eng. Part A **16**(10), 3261–3270 (2010)
2. Al-Maqtari, T., Hong, K.U., Vajravelu, B.N., et al.: Transcription factor-induced activation of cardiac gene expression in human c-kit+ cardiac progenitor cells. PLoS ONE **12**(3), e0174242–e0174242 (2017). https://doi.org/10.1371/journal.pone.0174242
3. Andrés-Delgado, L., Mercader, N.: Interplay between cardiac function and heart development. Biochim. et biophys. acta **1863**(7 Pt B), 1707–1716 (2016). https://doi.org/10.1016/j.bbamcr. 2016.03.004
4. Ariyasinghe, N.R., Reck, C.H., Viscio, A.A., et al.: Engineering micromyocardium to delineate cellular and extracellular regulation of myocardial tissue contractility. Integr. Biol. **9**(9), 730–741 (2017). https://doi.org/10.1039/C7IB00081B
5. Bakkers, J.: Zebrafish as a model to study cardiac development and human cardiac disease. Cardiovasc. Res. **91**(2), 279–288 (2011). https://doi.org/10.1093/cvr/cvr098
6. Banerjee, I., Fuseler, J.W., Price, R.L., et al.: Determination of cell types and numbers during cardiac development in the neonatal and adult rat and mouse. Am. J. Physiol.-Heart Circulatory Physiol. **293**(3), H1883–H1891 (2007)
7. Banes, A.J., Gilbert, J., Taylor, D., et al.: A new vacuum-operated stress-providing instrument that applies static or variable duration cyclic tension or compression to cells in vitro. J. Cell Sci. **75**(1), 35 (1985)
8. Baugh, L.M., Liu, Z., Quinn, K.P., et al.: Non-destructive two-photon excited fluorescence imaging identifies early nodules in calcific aortic-valve disease. Nat. Biomed. Eng. **1**(11), 914–924 (2017). https://doi.org/10.1038/s41551-017-0152-3
9. Bayomy, A.F., Bauer, M., Qiu, Y., et al.: Regeneration in heart disease—is ECM the key? Life Sci. **91**(17–18), 823–827 (2012)
10. Boothe, S.D., Myers, J.D., Pok, S., et al.: The effect of substrate stiffness on cardiomyocyte action potentials. Cell Biochem. Biophys. **74**(4), 527–535 (2016). https://doi.org/10.1007/s12013-016-0758-1
11. Borg, T.K., Rubin, K., Lundgren, E., et al.: Recognition of extracellular matrix components by neonatal and adult cardiac myocytes. Dev. Biol. **104**(1), 86–96 (1984). https://doi.org/10.1016/0012-1606(84)90038-1
12. Boudou, T., Legant, W.R., Mu, A., et al.: A microfabricated platform to measure and manipulate the mechanics of engineered cardiac microtissues. Tissue Eng. Part A **18**(9–10), 910–919 (2012). https://doi.org/10.1089/ten.tea.2011.0341
13. Bruneau, B.G.: The developmental genetics of congenital heart disease. Nature **451**(943) (2008). https://doi.org/10.1038/nature06801
14. Buikema, J.W., Mady, A.S., Mittal, N.V., et al.: Wnt/β-catenin signaling directs the regional expansion of first and second heart field-derived ventricular cardiomyocytes. Development 099325 (2013)
15. Camacho, P., Fan, H., Liu, Z., et al.: Small mammalian animal models of heart disease. Am. J. Cardiovasc. Dis. **6**(3), 70–80 (2016)
16. Caporizzo, M.A., Chen, C.Y., Salomon, A.K., et al.: Microtubules provide a viscoelastic resistance to myocyte motion. Biophys. J. **115**(9), 1796–1807 (2018). https://doi.org/10.1016/j.bpj.2018.09.019

17. Chen, C., Li, R., Ross, R.S., et al.: Integrins and integrin-related proteins in cardiac fibrosis. J. Mol. Cell. Cardiol. **93**, 162–174 (2016). https://doi.org/10.1016/j.yjmcc.2015.11.010

18. Chen, Z., Zhu, W., Bender, I., et al.: Pathologic stimulus determines lineage commitment of cardiac C-kit(+) Cells. Circulation **136**(24), 2359–2372 (2017). https://doi.org/10.1161/CIRCULATIONAHA.117.030137

19. Chistiakov, D.A., Orekhov, A.N., Bobryshev, Y.V.: Effects of shear stress on endothelial cells: go with the flow. Acta Physiol. **219**(2), 382–408 (2017). https://doi.org/10.1111/apha.12725

20. Cooley, M.A., Fresco, V.M., Dorlon, M.E., et al.: Fibulin-1 is required during cardiac ventricular morphogenesis for versican cleavage, suppression of ErbB2 and Erk1/2 activation, and to attenuate trabecular cardiomyocyte proliferation. Dev. Dyn. (An Off. Publ. Am. Assoc. Anatomists) **241**(2), 303–314 (2012). https://doi.org/10.1002/dvdy.23716

21. Davies, B., d'Udekem, Y., Ukoumunne, O.C., et al.: Differences in extra-cellular matrix and myocyte homeostasis between the neonatal right ventricle in hypoplastic left heart syndrome and truncus arteriosus. Eur. J. Cardiothorac. Surg. **34**(4), 738–744 (2008)

22. Davies, P.F.: Flow-mediated endothelial mechanotransduction. Phsiol. Rev. **75**(3), 519–560 (1995). https://doi.org/10.1152/physrev.1995.75.3.519

23. Davis, J., Molkentin, J.D.: Myofibroblasts: trust your heart and let fate decide. J. Mol. Cell. Cardiol. **70**(9–18 (2014). https://doi.org/10.1016/j.yjmcc.2013.10.019

24. Deb, A., Ubil, E.J.J.o.m., Cardiology, C.: Cardiac fibroblast in development and wound healing. **70**, 47–55 (2014)

25. Dedkova, E.N., Blatter, L.A.: Measuring mitochondrial function in intact cardiac myocytes. J. Mol. Cell. Cardiol. **52**(1), 48–61 (2012). https://doi.org/10.1016/j.yjmcc.2011.08.030

26. Dewey, J.C.F., Bussolari, S.R., Gimbrone, J.M.A., et al.: The dynamic response of vascular endothelial cells to fluid shear stress. J. Biomech. Eng. **103**(3), 177–185 (1981). https://doi.org/10.1115/1.3138276

27. Dobaczewski, M., Chen, W., Frangogiannis, N.G.: Transforming growth factor (TGF)-β signaling in cardiac remodeling. J. Mol. Cell. Cardiol. **51**(4), 600–606 (2011). https://doi.org/10.1016/j.yjmcc.2010.10.033

28. Dostal, D.E., Feng, H., Nizamutdinov, D., et al.: Mechanosensing and regulation of cardiac function. J. Clin. Exp. Cardiol. **5**(6), 314–314 (2014). https://doi.org/10.4172/2155-9880.1000314

29. Ellis, E.F., McKinney, J.S., Willoughby, K.A., et al.: A new model for rapid stretch-induced injury of cells in culture: characterization of the model using astrocytes. J. Neurotrauma **12**(3), 325–339 (1995). https://doi.org/10.1089/neu.1995.12.325

30. Engler, A.J., Carag-Krieger, C., Johnson, C.P., et al.: Embryonic cardiomyocytes beat best on a matrix with heart-like elasticity: scar-like rigidity inhibits beating. **121**(22), 3794–3802 (2008)

31. Engler, A.J., Sen, S., Sweeney, H.L., et al.: Matrix elasticity directs stem cell lineage specification. Cell **126**(4), 677–689 (2006)

32. Fan, D., Creemers, E.E., Kassiri, Z.: Matrix as an interstitial transport system. Circ. Res. **114**(5), 889–902 (2014)

33. Feng, H., Gerilechaogetu, F., Golden, H.B., et al.: p 38α MAPK inhibits stretch-induced JNK activation in cardiac myocytes through MKP-1. Int. J. Cardiol. **203**, 145–155 (2016). https://doi.org/10.1016/j.ijcard.2015.10.109

34. Fomovsky, G.M., Holmes, J.W.: Evolution of scar structure, mechanics, and ventricular function after myocardial infarction in the rat. Am. J. Physiol. Heart Circulatory Physiol. **298**(1), H221–H228 (2010). https://doi.org/10.1152/ajpheart.00495.2009

35. Fomovsky, G.M., Thomopoulos, S., Holmes, J.W.: Contribution of extracellular matrix to the mechanical properties of the heart. J. Mol. Cell. Cardiol. **48**(3), 490–496 (2010)

36. French, K.M., Boopathy, A.V., DeQuach, J.A., et al.: A naturally derived cardiac extracellular matrix enhances cardiac progenitor cell behavior in vitro. Acta Biomater. **8**(12), 4357–4364 (2012)

37. Galie, P.A., van Oosten, A., Chen, C.S., et al.: Application of multiple levels of fluid shear stress to endothelial cells plated on polyacrylamide gels. Lab Chip **15**(4), 1205–1212 (2015). https://doi.org/10.1039/C4LC01236D

38. Gambini, E., Perrucci, G.L., Bassetti, B., et al.: Preferential myofibroblast differentiation of cardiac mesenchymal progenitor cells in the presence of atrial fibrillation. Transl. Res. **192**, 54–67 (2018). https://doi.org/10.1016/j.trsl.2017.11.003

39. Geiger, B., Spatz, J.P., Bershadsky, A.D.: Environmental sensing through focal adhesions. Nat. Rev. Mol. Cell Biol. **10**(1), 21–33 (2009). https://doi.org/10.1038/nrm2593

40. Gershlak, J.R., Resnikoff, J.I., Sullivan, K.E., et al.: Mesenchymal stem cells ability to generate traction stress in response to substrate stiffness is modulated by the changing extracellular matrix composition of the heart during development. Biochem. Biophys. Res. Commun. **439**(2), 161–166 (2013). https://doi.org/10.1016/j.bbrc.2013.08.074

41. Gjorevski, N., Nelson, C.M.: Bidirectional extracellular matrix signaling during tissue morphogenesis. Cytokine Growth Factor Rev. **20**(5–6), 459–465 (2009). https://doi.org/10.1016/j.cytogfr.2009.10.013

42. Goffin, J.M., Pittet, P., Csucs, G., et al.: Focal adhesion size controls tension-dependent recruitment of α-smooth muscle actin to stress fibers. J. Cell Biol. **172**(2), 259–268 (2006). https://doi.org/10.1083/jcb.200506179

43. Gottlieb, R.A., Stotland, A.: MitoTimer: a novel protein for monitoring mitochondrial turnover in the heart. J. Mol. Med. (Berlin, Germany) **93**(3), 271–278 (2015). https://doi.org/10.1007/s00109-014-1230-6

44. Graham, H.K., Hodson, N.W., Hoyland, J.A., et al.: Tissue section AFM: In situ ultrastructural imaging of native biomolecules. Matrix Biol.: J. Int. Soc. Matrix Biol. **29**(4), 254–260 (2010). https://doi.org/10.1016/j.matbio.2010.01.008

45. Gupta, V., Grande-Allen, K.J.: Effects of static and cyclic loading in regulating extracellular matrix synthesis by cardiovascular cells. Cardiovasc. Res. **72**(3), 375–383 (2006). https://doi.org/10.1016/j.cardiores.2006.08.017

46. Haack, T., Abdelilah-Seyfried, S.: The force within: endocardial development, mechanotransduction and signalling during cardiac morphogenesis. Development **143**(3), 373–386 (2016). https://doi.org/10.1242/dev.131425

47. Handorf, A.M., Zhou, Y., Halanski, M.A., et al.: Tissue stiffness dictates development, homeostasis, and disease progression. Organogenesis **11**(1), 1–15 (2015)

48. Happe, C.L., Engler, A.J.: Mechanical forces reshape differentiation cues that guide cardiomyogenesis. Circ. Res. **118**(2), 296–310 (2016). https://doi.org/10.1161/CIRCRESAHA.115.305139

49. Hasenfuss, G.: Animal models of human cardiovascular disease, heart failure and hypertrophy. Cardiovasc. Res. **39**(1), 60–76 (1998). https://doi.org/10.1016/S0008-6363(98)00110-2

50. Hazeltine, L.B., Badur, M.G., Lian, X., et al.: Temporal impact of substrate mechanics on differentiation of human embryonic stem cells to cardiomyocytes. Acta Biomater. **10**(2), 604–612 (2014). https://doi.org/10.1016/j.actbio.2013.10.033

51. Henderson, N.C., Sheppard, D.: Integrin-mediated regulation of TGFβ in fibrosis. Biochimica et Biophysica Acta (BBA). Mol. Basis Dis. **1832**(7), 891–896 (2013). https://doi.org/10.1016/j.bbadis.2012.10.005

52. Heras-Bautista, C.O., Katsen-Globa, A., Schloerer, N.E., et al.: The influence of physiological matrix conditions on permanent culture of induced pluripotent stem cell-derived cardiomyocytes. Biomaterials **35**(26), 7374–7385 (2014). https://doi.org/10.1016/j.biomaterials.2014.05.027

53. Herum, K.M., Choppe, J., Kumar, A., et al.: Mechanical regulation of cardiac fibroblast profibrotic phenotypes. Mol. Biol. Cell **28**(14), 1871–1882 (2017). https://doi.org/10.1091/mbc.E17-01-0014

54. Hinz, B.: The role of myofibroblasts in wound healing. Curr. Res. Transl. Med. **64**(4), 171–177 (2016). https://doi.org/10.1016/j.retram.2016.09.003

55. Howard, C.M., Baudino, T.A.: Dynamic cell–cell and cell–ECM interactions in the heart. J. Mol. Cell. Cardiol. **70**, 19–26 (2014). https://doi.org/10.1016/j.yjmcc.2013.10.006

56. Huang, J.X., Potts, J.D., Vincent, E.B., et al.: Mechanisms of cell transformation in the embryonic heart. Ann. N. Y. Acad. Sci. **752**(1), 317–330 (1995)

57. Humphrey, J.D., Dufresne, E.R., Schwartz, M.A.: Mechanotransduction and extracellular matrix homeostasis. Nat. Rev. Mol. Cell Biol. **15**(12), 802 (2014)
58. Hynes, R.O.: The extracellular matrix: not just pretty fibrils. Science **326**(5957), 1216–1219 (2009). https://doi.org/10.1126/science.1176009
59. Hynes, R.O., Naba, A.: Overview of the matrisome—an inventory of extracellular matrix constituents and functions. Cold Spring Harb. Perspect. Biol. **4**(1), a004903 (2012)
60. Ieda, M., Tsuchihashi, T., Ivey, K.N., et al.: Cardiac fibroblasts regulate myocardial proliferation through β1 integrin signaling. Dev. Cell **16**(2), 233–244 (2009)
61. Ishii, Y., Langberg, J., Rosborough, K., et al.: Endothelial cell lineages of the heart. Cell Tissue Res. **335**(1), 67–73 (2009). https://doi.org/10.1007/s00441-008-0663-z
62. Itoh, N., Ohta, H.: Pathophysiological roles of FGF signaling in the heart. Front. Physiol. **4**, 247–247 (2013). https://doi.org/10.3389/fphys.2013.00247
63. Jackman, C., Li, H., Bursac, N.: Long-term contractile activity and thyroid hormone supplementation produce engineered rat myocardium with adult-like structure and function. Acta Biomater. **78**, 98–110 (2018). https://doi.org/10.1016/j.actbio.2018.08.003
64. Jacot, J.G., Kita-Matsuo, H., Wei, K.A., et al.: Cardiac myocyte force development during differentiation and maturation. Ann. N. Y. Acad. Sci. **1188**(1), 121–127 (2010)
65. Jacot, J.G., Martin, J.C., Hunt, D.L.: Mechanobiology of cardiomyocyte development. J. Biomech. **43**(1), 93–98 (2010)
66. Jacot, J.G., McCulloch, A.D., Omens, J.H.: Substrate stiffness affects the functional maturation of neonatal rat ventricular myocytes. Biophys. J. **95**(7), 3479–3487 (2008)
67. Jessen, J.R.: Recent advances in the study of zebrafish extracellular matrix proteins. Dev. Biol. **401**(1), 110–121 (2015). https://doi.org/10.1016/j.ydbio.2014.12.022
68. Jung, J.P., Squirrell, J.M., Lyons, G.E., et al.: Imaging cardiac extracellular matrices: a blueprint for regeneration. Trends Biotechnol. **30**(4), 233–240 (2012). https://doi.org/10.1016/j.tibtech.2011.12.001
69. Kain, K.H., Miller, J.W.I., Jones-Paris, C.R., et al.: The chick embryo as an expanding experimental model for cancer and cardiovascular research. Dev. Dyn. (An Off. Publ. Am. Assoc. Anatomists) **243**(2), 216–228 (2014). https://doi.org/10.1002/dvdy.24093
70. Kass David, A., Bronzwaer Jean, G.F., Paulus Walter, J.: What mechanisms underlie diastolic dysfunction in heart failure? Circul. Res. **94**(12), 1533–1542 (2004). https://doi.org/10.1161/01.res.0000129254.25507.d6
71. Katsumi, A., Orr, A.W., Tzima, E., et al.: Integrins in mechanotransduction. J. Biol. Chem. **279**(13), 12001–12004 (2004)
72. Kramann, R., Schneider, Rebekka K., DiRocco, Derek P., et al.: Perivascular Gli1+ Progenitors are key contributors to injury-induced organ fibrosis. Cell Stem Cell **16**(1), 51–66 (2015). https://doi.org/10.1016/j.stem.2014.11.004
73. Krishnamurthy, P., Peterson, J.T., Subramanian, V., et al.: Inhibition of matrix metalloproteinases improves left ventricular function in mice lacking osteopontin after myocardial infarction. Mol. Cell. Biochem. **322**(1–2), 53–62 (2009)
74. Lammerding, J.A.N., Kamm, R.D., Lee, R.T.: Mechanotransduction in Cardiac Myocytes. Ann. N. Y. Acad. Sci. **1015**(1), 53–70 (2004). https://doi.org/10.1196/annals.1302.005
75. Lanzicher, T., Martinelli, V., Puzzi, L., et al.: The Cardiomyopathy Lamin A/C D192G mutation disrupts whole-cell biomechanics in cardiomyocytes as measured by atomic force microscopy loading-unloading curve analysis. Sci. Rep. **5**, 13388 (2015). https://doi.org/10.1038/srep13388
76. Lau, J.J., Wang, R.M., Black 3rd, L.D.: Development of an arbitrary waveform membrane stretcher for dynamic cell culture. Ann. Biomed. Eng. **42**(5), 1062–1073 (2014). https://doi.org/10.1007/s10439-014-0976-x
77. Leight, J.L., Wozniak, M.A., Chen, S., et al.: Matrix rigidity regulates a switch between TGF-β1–induced apoptosis and epithelial–mesenchymal transition. Mol. Biol. Cell **23**(5), 781–791 (2012)
78. Leone, M., Magadum, A., Engel, F.B.: Cardiomyocyte proliferation in cardiac development and regeneration: a guide to methodologies and interpretations. Am. J. Physiol. Heart Circ. Physiol. **309**(8), H1237–H1250 (2015). https://doi.org/10.1152/ajpheart.00559.2015

79. Li, P., Cavallero, S., Gu, Y., et al.: IGF signaling directs ventricular cardiomyocyte proliferation during embryonic heart development. Development 054338 (2011)
80. Li, Y., Asfour, H., Bursac, N.: Age-dependent functional crosstalk between cardiac fibroblasts and cardiomyocytes in a 3D engineered cardiac tissue. Acta Biomater. **55**, 120–130 (2017). https://doi.org/10.1016/j.actbio.2017.04.027
81. Lieber, S.C., Aubry, N., Pain, J., et al.: Aging increases stiffness of cardiac myocytes measured by atomic force microscopy nanoindentation. Am. J. Physiol. Heart Circ. Physiol. **287**(2), H645–H651 (2004). https://doi.org/10.1152/ajpheart.00564.2003
82. Lijnen, P., Petrov, V., Fagard, R.: Induction of cardiac fibrosis by transforming growth factor-β 1. Mol. Genet. Metab. **71**(1), 418–435 (2000)
83. Loftis, M.J., Sexton, D., Carver, W.: Effects of collagen density on cardiac fibroblast behavior and gene expression. J. Cell. Physiol. **196**(3), 504–511 (2003)
84. Lyra-Leite, D.M., Andres, A.M., Petersen, A.P., et al.: Mitochondrial function in engineered cardiac tissues is regulated by extracellular matrix elasticity and tissue alignment. Am. J. Physiol. Heart Circ. Physiol. **313**(4), H757–H767 (2017). https://doi.org/10.1152/ajpheart.00290.2017
85. Ma, S.P., Vunjak-Novakovic, G.: Tissue-Engineering for the study of cardiac biomechanics. J. Biomech. Eng. **138**(2), 021010 (2016). https://doi.org/10.1115/1.4032355
86. Majkut, S., Idema, T., Swift, J., et al.: Heart-specific stiffening in early embryos parallels matrix and myosin expression to optimize beating. Curr. Biol.: CB **23**(23), 2434–2439 (2013). https://doi.org/10.1016/j.cub.2013.10.057
87. Mammoto, T., Mammoto, A., Ingber, D.E.: Mechanobiology and developmental control. Ann. Rev. Cell Dev. Biol. **29**, 27–61 (2013)
88. Manso, A.M., Kang, S.-M., Ross, R.S.: Integrins, focal adhesions, and cardiac fibroblasts. J. Investig. Med. **57**(8), 856 (2009)
89. Mauretti, A., Spaans, S., Bax, N.A.M., et al.: Cardiac progenitor cells and the interplay with their microenvironment. Stem cells Int. **2017**, 7471582 (2017). https://doi.org/10.1155/2017/7471582
90. McCain, M.L., Parker, K.K.: Mechanotransduction: the role of mechanical stress, myocyte shape, and cytoskeletal architecture on cardiac function. Pflügers Arch. Eur. J. Physiol. **462**(1), 89 (2011). https://doi.org/10.1007/s00424-011-0951-4
91. Mihic, A., Li, J., Miyagi, Y., et al.: The effect of cyclic stretch on maturation and 3D tissue formation of human embryonic stem cell-derived cardiomyocytes. Biomaterials **35**(9), 2798–2808 (2014). https://doi.org/10.1016/j.biomaterials.2013.12.052
92. Miskon, A., Mahara, A., Uyama, H., et al.: A suspension induction for myocardial differentiation of rat mesenchymal stem cells on various extracellular matrix proteins. Tissue Eng. Part C: Methods **16**(5), 979–987 (2010)
93. Morgan, K.Y., Black 3rd, L.D.: Mimicking isovolumic contraction with combined electromechanical stimulation improves the development of engineered cardiac constructs. Tissue Eng. Part A **20**(11–12), 1654–1667 (2014). https://doi.org/10.1089/ten.TEA.2013.0355
94. Naba, A., Clauser, K.R., Ding, H., et al.: The extracellular matrix: Tools and insights for the "omics" era. Matrix Biol. **49**, 10–24 (2016). https://doi.org/10.1016/j.matbio.2015.06.003
95. Nandadasa, S., Foulcer, S., Apte, S.S.: The multiple, complex roles of versican and its proteolytic turnover by ADAMTS proteases during embryogenesis. Matrix Biol.: J. Int. Soc. Matrix Biol. **35**, 34–41 (2014). https://doi.org/10.1016/j.matbio.2014.01.005
96. Nielsen, S.H., Mouton, A.J., DeLeon-Pennell, K.Y., et al.: Understanding cardiac extracellular matrix remodeling to develop biomarkers of myocardial infarction outcomes. Matrix Biol. (2017). https://doi.org/10.1016/j.matbio.2017.12.001
97. O'Connor, J.W., Riley, P.N., Nalluri, S.M., et al.: Matrix rigidity mediates TGFβ1-induced epithelial-myofibroblast transition by controlling cytoskeletal organization and MRTF-A localization. J. Cell. Physiol. **230**(8), 1829–1839 (2014). https://doi.org/10.1002/jcp.24895
98. Ori, A., Wilkinson, M.C., Fernig, D.G.: A systems biology approach for the investigation of the heparin/heparan sulfate interactome. J. Biol. Chem. jbc. M111. 228114 (2011)

99. Ott, H.C., Matthiesen, T.S., Goh, S.-K., et al.: Perfusion-decellularized matrix: using nature's platform to engineer a bioartificial heart. Nat. Med. **14**, 213 (2008). https://doi.org/10.1038/nm1684

100. Parodi, E.M., Kuhn, B.: Signalling between microvascular endothelium and cardiomyocytes through neuregulin. Cardiovasc. Res. **102**(2), 194–204 (2014). https://doi.org/10.1093/cvr/cvu021

101. Pasqualini, F.S., Agarwal, A., O'Connor, B.B., et al.: Traction force microscopy of engineered cardiac tissues. PLoS ONE **13**(3), e0194706–e0194706 (2018). https://doi.org/10.1371/journal.pone.0194706

102. Perrucci, G.L., Rurali, E., Pompilio, G.: Cardiac fibrosis in regenerative medicine: destroy to rebuild. J. Thorac. Dis. **10**(Suppl 20), S2376–S2389 (2018). https://doi.org/10.21037/jtd.2018.03.82

103. Pinto, A.R., Ilinykh, A., Ivey, M.J., et al.: Revisiting cardiac cellular composition. Circ. Res. **118**(3), 400–409 (2016)

104. Pöschl, E., Schlötzer-Schrehardt, U., Brachvogel, B., et al.: Collagen IV is essential for basement membrane stability but dispensable for initiation of its assembly during early development. Development **131**(7), 1619 (2004). https://doi.org/10.1242/dev.01037

105. Prabhu, S.D.: Post-infarction ventricular remodeling: an array of molecular events. J. Mol. Cell. Cardiol. **38**(4), 547–550 (2005). https://doi.org/10.1016/j.yjmcc.2005.01.014

106. Prabhu, S.D., Frangogiannis, N.G.: The biological basis for cardiac repair after myocardial infarction: from inflammation to fibrosis. Circ. Res. **119**(1), 91–112 (2016). https://doi.org/10.1161/CIRCRESAHA.116.303577

107. Prager-Khoutorsky, M., Lichtenstein, A., Krishnan, R., et al.: Fibroblast polarization is a matrix-rigidity-dependent process controlled by focal adhesion mechanosensing. Nat. Cell Biol. **13**, 1457 (2011). https://doi.org/10.1038/ncb2370

108. Preissner, K.T., Reuning, U.: Vitronectin in vascular context: facets of a multitalented matricellular protein. Semin. Thromb. Hemost. **37**(04), 408–424 (2011). https://doi.org/10.1055/s-0031-1276590

109. Prosser, B.L., Ward, C.W., Lederer, W.J.: X-ROS signaling: rapid mechano-chemo transduction in heart. Science **333**(6048), 1440 (2011)

110. Pucéat, M.: Embryological origin of the endocardium and derived valve progenitor cells: From developmental biology to stem cell-based valve repair. Biochim. et Biophys. Acta (BBA)—Mol. Cell Res. **1833**(4), 917–922 (2013). https://doi.org/10.1016/j.bbamcr.2012.09.013

111. Pulina, M.V., Hou, S.-Y., Mittal, A., et al.: Essential roles of fibronectin in the development of the left–right embryonic body plan. Dev. Biol. **354**(2), 208–220 (2011)

112. Qi, Y., Li, Z., Kong, C.-W., et al.: Uniaxial cyclic stretch stimulates TRPV4 to induce realignment of human embryonic stem cell-derived cardiomyocytes. J. Mol. Cell. Cardiol. **87**, 65–73 (2015). https://doi.org/10.1016/j.yjmcc.2015.08.005

113. Ramirez, F., Rifkin, D.B.: Extracellular microfibrils: contextual platforms for TGFbeta and BMP signaling. Curr. Opin. Cell Biol. **21**(5), 616–622 (2009). https://doi.org/10.1016/j.ceb.2009.05.005

114. Recchia, F.A., Lionetti, V.: Animal models of dilated cardiomyopathy for translational research. Vet. Res. Commun. **31**(1), 35–41 (2007). https://doi.org/10.1007/s11259-007-0005-8

115. Reitsma, S., Slaaf, D.W., Vink, H., et al.: The endothelial glycocalyx: composition, functions, and visualization. Pflugers Archiv. **454**(3), 345–359 (2007). https://doi.org/10.1007/s00424-007-0212-8

116. Rienks, M., Papageorgiou, A.-P., Frangogiannis, N.G., et al.: Myocardial extracellular matrix. An Ever-Changing Diverse Entity **114**(5), 872–888 (2014). https://doi.org/10.1161/circresaha.114.302533

117. Rozario, T., DeSimone, D.W.: The extracellular matrix in development and morphogenesis: a dynamic view. Dev. Biol. **341**(1), 126–140 (2010)

118. Rubin, S.A., Fishbein, M.C., Swan, H.J.C.: Compensatory hypertrophy in the heart after myocardial infarction in the rat. J. Am. Coll. Cardiol. 1(6), 1435–1441 (1983). https://doi.org/10.1016/S0735-1097(83)80046-1
119. Rupert, C.E., Coulombe, K.L.K.: IGF1 and NRG1 enhance proliferation, metabolic maturity, and the force-frequency response in hESC-derived engineered cardiac tissues. Stem Cells Int. 2017, 7648409 (2017). https://doi.org/10.1155/2017/7648409
120. Rysä, J., Tokola, H., Ruskoaho, H.: Mechanical stretch induced transcriptomic profiles in cardiac myocytes. Sci. Rep. 8(1), 4733 (2018). https://doi.org/10.1038/s41598-018-23042-w
121. Sawyer, D.B., Caggiano, A.: Neuregulin-1β for the treatment of systolic heart failure. J. Mol. Cell. Cardiol. 51(4), 501–505 (2011). https://doi.org/10.1016/j.yjmcc.2011.06.016
122. Schürmann, S., Wagner, S., Herlitze, S., et al.: The IsoStretcher: an isotropic cell stretch device to study mechanical biosensor pathways in living cells. Biosens. Bioelectron. 81, 363–372 (2016). https://doi.org/10.1016/j.bios.2016.03.015
123. Segura, A.M., Frazier, O.H., Buja, L.M.: Fibrosis and heart failure. Heart Fail. Rev. 19(2), 173–185 (2014). https://doi.org/10.1007/s10741-012-9365-4
124. Singh, P., Carraher, C., Schwarzbauer, J.E.: Assembly of fibronectin extracellular matrix. Ann. Rev. Cell Dev. Biol. 26, 397–419 (2010). https://doi.org/10.1146/annurev-cellbio-100109-104020
125. Smith, C.L., Baek, S.T., Sung, C.Y., et al.: Epicardial-derived cell epithelial-to-mesenchymal transition and fate specification require PDGF receptor signaling. Circ. Res. 108(12), e15–e26 (2011)
126. Smith, L., Cho, S., Discher, D.E.: Mechanosensing of matrix by stem cells: from matrix heterogeneity, contractility, and the nucleus in pore-migration to cardiogenesis and muscle stem cells in vivo. Semin. Cell Dev. Biol. 71, 84–98 (2017). https://doi.org/10.1016/j.semcdb.2017.05.025
127. Smith, L.R., Cho, S., Discher, D.E.: Stem cell differentiation is regulated by extracellular matrix mechanics. Physiology 33(1), 16–25 (2017)
128. Snider, P., Hinton, R.B., Moreno-Rodriguez, R.A., et al.: Periostin is required for maturation and extracellular matrix stabilization of noncardiomyocyte lineages of the heart. Circ. Res. 102(7), 752–760 (2008)
129. Souders, C.A., Bowers, S.L., Baudino, T.A.: Cardiac fibroblast: the renaissance cell. Circ. Res. 105(12), 1164–1176 (2009)
130. Stelzer, J.E., Moss, R.L.: Contributions of stretch activation to length-dependent contraction in murine myocardium. J. Gen. Physiol. 128(4), 461–471 (2006). https://doi.org/10.1085/jgp.200609634
131. Steward Jr., R., Tambe, D., Hardin, C.C., et al.: Fluid shear, intercellular stress, and endothelial cell alignment. Am. J. Physiol. Cell Physiol. 308(8), C657–C664 (2015). https://doi.org/10.1152/ajpcell.00363.2014
132. Stoppel, W.L., Kaplan, D.L., Black, L.D., 3rd.: Electrical and mechanical stimulation of cardiac cells and tissue constructs. Adv. Drug Deliv. Rev. 96, 135–155 (2016). https://doi.org/10.1016/j.addr.2015.07.009
133. Sullivan, K.E., Black, L.D.: The role of cardiac fibroblasts in extracellular matrix-mediated signaling during normal and pathological cardiac development. J. Biomech. Eng. 135(7), 071001 (2013)
134. Sullivan, K.E., Black, L.D.: The role of cardiac fibroblasts in extracellular matrix-mediated signaling during normal and pathological cardiac development. J. Biomech. Eng. 135(7), 71001 (2013). https://doi.org/10.1115/1.4024349
135. Sullivan, K.E., Quinn, K.P., Tang, K.M., et al.: Extracellular matrix remodeling following myocardial infarction influences the therapeutic potential of mesenchymal stem cells. Stem Cell Res. Therapy 5(1), 14 (2014). https://doi.org/10.1186/scrt403
136. Sylva, M., Hoff, M.J.B.v.d., Moorman, A.F.M.: Development of the human heart. Am. J. Med. Genet. Part A 164(6), 1347–1371 (2014). https://doi.org/10.1002/ajmg.a.35896
137. Tarbell, J.M., Ebong, E.E.: The endothelial glycocalyx: a mechano-sensor and -transducer. Sci. Signal. 1(40), pt8 (2008). https://doi.org/10.1126/scisignal.140pt8

138. Throm Quinlan, A.M., Sierad, L.N., Capulli, A.K., et al.: Combining dynamic stretch and tunable stiffness to probe cell mechanobiology in vitro. PLoS ONE **6**(8), e23272 (2011). https://doi.org/10.1371/journal.pone.0023272

139. Tomasek, J.J., Gabbiani, G., Hinz, B., et al.: Myofibroblasts and mechano-regulation of connective tissue remodelling. Nat. Rev. Mol. Cell Biol. **3**, 349 (2002). https://doi.org/10.1038/nrm809

140. Travers, J.G., Kamal, F.A., Robbins, J., et al.: Cardiac fibrosis: the fibroblast awakens. Circ. Res. **118**(6), 1021–1040 (2016)

141. Ungerleider, J.L., Johnson, T.D., Hernandez, M.J., et al.: Extracellular matrix hydrogel promotes tissue remodeling, arteriogenesis, and perfusion in a rat hindlimb ischemia model. JACC: Basic Transl. Sci. **1**(1–2), 32 (2016). https://doi.org/10.1016/j.jacbts.2016.01.009

142. Valiente-Alandi, I., Schafer, A.E., Blaxall, B.C.: Extracellular matrix-mediated cellular communication in the heart. J. Mol. Cell. Cardiol. **91**, 228–237 (2016). https://doi.org/10.1016/j.yjmcc.2016.01.011

143. van der Flier, A., Sonnenberg, A.: Function and interactions of integrins. Cell Tissue Res. **305**(3), 285–298 (2001). https://doi.org/10.1007/s004410100417

144. van Putten, S., Shafieyan, Y., Hinz, B.: Mechanical control of cardiac myofibroblasts. J. Mol. Cell. Cardiol. **93**, 133–142 (2016). https://doi.org/10.1016/j.yjmcc.2015.11.025

145. van Putten, S., Shafieyan, Y., Hinz, B.: Mechanical control of cardiac myofibroblasts. J. Mol. Cell. Cardiol. **93**, 133–142 (2016)

146. Vicinanza, C., Aquila, I., Scalise, M., et al.: Adult cardiac stem cells are multipotent and robustly myogenic: c-kit expression is necessary but not sufficient for their identification. Cell Death Differ. **24**(12), 2101–2116 (2017). https://doi.org/10.1038/cdd.2017.130

147. Wagenseil Jessica, E., Ciliberto Chris, H., Knutsen Russell, H., et al.: Reduced vessel elasticity alters cardiovascular structure and function in newborn mice. Circ. Res. **104**(10), 1217–1224 (2009). https://doi.org/10.1161/CIRCRESAHA.108.192054

148. Walsh, S., Pontén, A., Fleischmann, B.K., et al.: Cardiomyocyte cell cycle control and growth estimation in vivo—an analysis based on cardiomyocyte nuclei. Cardiovasc. Res. **86**(3), 365–373 (2010). https://doi.org/10.1093/cvr/cvq005

149. Wang, J., Karra, R., Dickson, A.L., et al.: Fibronectin is deposited by injury-activated epicardial cells and is necessary for zebrafish heart regeneration. Dev. Biol. **382**(2), 427–435 (2013). https://doi.org/10.1016/j.ydbio.2013.08.012

150. Watson, C.J., Phelan, D., Xu, M., et al.: Mechanical stretch up-regulates the B-type natriuretic peptide system in human cardiac fibroblasts: a possible defense against transforming growth factor-β mediated fibrosis. Fibrogenesis Tissue Repair **5**(1), 9 (2012). https://doi.org/10.1186/1755-1536-5-9

151. Wheelwright, M., Win, Z., Mikkila, J.L., et al.: Investigation of human iPSC-derived cardiac myocyte functional maturation by single cell traction force microscopy. PLoS ONE **13**(4), e0194909 (2018). https://doi.org/10.1371/journal.pone.0194909

152. Whitelock, J.M., Melrose, J., Iozzo, R.V.: Diverse cell signaling events modulated by perlecan. Biochemistry **47**(43), 11174–11183 (2008). https://doi.org/10.1021/bi8013938

153. Willert, K., Brown, J.D., Danenberg, E., et al.: Wnt proteins are lipid-modified and can act as stem cell growth factors. Nature **423**(6938), 448 (2003)

154. Williams, C., Black, L.D.: The role of extracellular matrix in cardiac development. In: Suuronen, E.J., Ruel, M. (eds.) Biomaterials for Cardiac Regeneration, pp. 1–35. Springer International Publishing, Cham (2015). https://doi.org/10.1007/978-3-319-10972-5_1

155. Williams, C., Budina, E., Stoppel, W.L., et al.: Cardiac extracellular matrix-fibrin hybrid scaffolds with tunable properties for cardiovascular tissue engineering. Acta Biomater. **14**, 84–95 (2015). https://doi.org/10.1016/j.actbio.2014.11.035

156. Williams, C., Quinn, K.P., Georgakoudi, I., et al.: Young developmental age cardiac extracellular matrix promotes the expansion of neonatal cardiomyocytes in vitro. Acta Biomater. **10**(1), 194–204 (2014)

157. Williams, C., Sullivan, K., Black III, L.D.: Partially digested adult cardiac extracellular matrix promotes cardiomyocyte proliferation in vitro. Adv. Healthc. Mater. **4**(10), 1545–1554 (2015)

158. Wipff, P.-J., Rifkin, D.B., Meister, J.-J., et al.: Myofibroblast contraction activates latent TGF-β1 from the extracellular matrix. J. Cell Biol. **179**(6), 1311–1323 (2007)
159. Wu, C.-C., Kruse, F., Vasudevarao, M.D., et al.: Spatially resolved genome-wide transcriptional profiling identifies BMP signaling as essential regulator of zebrafish cardiomyocyte regeneration. Dev. Cell **36**(1), 36–49 (2016). https://doi.org/10.1016/j.devcel.2015.12.010
160. Wu, Y., Al-Ameen, M.A., Ghosh, G.: Integrated effects of matrix mechanics and vascular endothelial growth factor (VEGF) on capillary sprouting. Ann. Biomed. Eng. **42**(5), 1024–1036 (2014). https://doi.org/10.1007/s10439-014-0987-7
161. Yeung, T., Georges, P.C., Flanagan, L.A., et al.: Effects of substrate stiffness on cell morphology, cytoskeletal structure, and adhesion. Cell Motil. Cytoskelet. **60**(1), 24–34 (2005). https://doi.org/10.1002/cm.20041
162. Young, J.L., Kretchmer, K., Ondeck, M.G., et al.: Mechanosensitive kinases regulate stiffness-induced cardiomyocyte maturation. Sci. Rep. **4**, 6425 (2014). https://doi.org/10.1038/srep06425
163. Yuan Ye, K., Sullivan, K.E., Black, L.D.: Encapsulation of cardiomyocytes in a fibrin hydrogel for cardiac tissue engineering. J. Visualized Exp.: JoVE **55**, 3251 (2011). https://doi.org/10.3791/3251
164. Zeisberg Elisabeth, M., Kalluri, R.: Origins of cardiac fibroblasts. Circ. Res. **107**(11), 1304–1312 (2010). https://doi.org/10.1161/CIRCRESAHA.110.231910
165. Zeisberg, E.M., Tarnavski, O., Zeisberg, M., et al.: Endothelial-to-mesenchymal transition contributes to cardiac fibrosis. Nat. Med. **13**, 952 (2007). https://doi.org/10.1038/nm1613
166. Zile, M.R., Gaasch, W.H.: Mechanical loads and the isovolumic and filling indices of left ventricular relaxation. Prog. Cardiovasc. Dis. **32**(5), 333–346 (1990). https://doi.org/10.1016/0033-0620(90)90020-3

Multi-scale Mechanics of Collagen Networks: Biomechanical Basis of Matrix Remodeling in Cancer

J. Ferruzzi, Y. Zhang, D. Roblyer and M. H. Zaman

Abstract The study of fiber network structure and mechanics is key to understanding extracellular matrix (ECM) remodeling in a variety of diseases, including cancer. The tumor microenvironment, which consists of stromal cells and ECM constituents, is altered by tumor cells via biochemical and biomechanical signals in order to support cancer progression. In particular, the tumor ECM displays consistent remodeling phenotypes that have been shown to aid cancer invasion both in vivo and in vitro. In this chapter, we focus on collagen—the most abundant protein in the body that endows the ECM with its structural and mechanical properties. Hydrogels made of reconstituted collagen fibers are commonly used as ECM models to study cell-matrix interactions and cancer cell migration in vitro. Due to their hierarchical organization, collagen networks reveal complex mechanical properties at different length scales. We present a comprehensive review of the experimental and modeling techniques available to investigate the structure and multi-scale mechanics of collagen. We emphasize the nonlinear mechanical properties of collagen from monomers to fiber networks and highlight the different aspects of collagen mechanics investigated using different loading conditions. Improved methods for quantitative imaging and biomechanical modeling are continuously needed to provide a holistic understanding

J. Ferruzzi (✉) · Y. Zhang · D. Roblyer · M. H. Zaman
Department of Biomedical Engineering, Boston University,
44 Cummington Mall, Boston, MA 02215, USA
e-mail: jacopofe@bu.edu

Y. Zhang
e-mail: yanhang@bu.edu

D. Roblyer
e-mail: roblyer@bu.edu

M. H. Zaman
e-mail: zaman@bu.edu

Y. Zhang
Department of Mechanical Engineering, Boston University, Boston, MA, USA

M. H. Zaman
Howard Hughes Medical Institute, Boston University, Boston, MA, USA

© Springer Nature Switzerland AG 2020
Y. Zhang (ed.), *Multi-scale Extracellular Matrix Mechanics and Mechanobiology*,
Studies in Mechanobiology, Tissue Engineering and Biomaterials 23,
https://doi.org/10.1007/978-3-030-20182-1_11

of collagen remodeling in response to cell-generated traction forces and to elucidate the mechanobiological pathways underlying cellular responses to biophysical cues.

1 Introduction

The extracellular matrix (ECM) is a tissue compartment constituted by proteins, glycoproteins, and glycosaminoglycans that assemble to form a complex three-dimensional (3D) network [1]. Such a matrix is endowed with tissue-specific structural features that allow optimal biomechanical function and biochemical signaling through the insoluble matrix. Structural, mechanical, and transport properties of the ECM evolve dynamically due to active remodeling by resident cells and can be impaired by multiple diseases, including cancer [2, 3]. Typically, disease processes initiate as abnormal cell behaviors, progress via imbalanced cell-matrix interactions, and finally manifest themselves as disrupted tissue homeostasis and function. In several cancers, including breast carcinoma, typical steps in the tumorigenic cascade include abnormal proliferation of mammary epithelial cells, rupture of the basement membrane, engagement of the fibrous ECM, and invasion [4]. Invasion from the primary tumor, as individual cells or as multicellular clusters [5], represents the first step towards metastasis, the main cause of cancer-related deaths. Type I collagen represents the most abundant protein in the human body, and the primary load-bearing structural element of the ECM. Cells adhere to, and migrate through, fibrous collagen networks in response to topographic, mechanical, and chemical cues [6] while remodeling of such microenvironment occurs via matrix deposition, proteolysis, and physical engagement [7]. Cancer progression is therefore associated with remodeling of collagen, which causes biomechanical alterations in the matrix and fosters a malignant phenotype [8]. Matrix stiffness, often quantified as linear Young's modulus, has long been regarded as the primary biomechanical factor impacting cell behavior [9–11], despite mechanical properties of tissues are known to be not only non-linear and anisotropic, but also viscoelastic and poroelastic [12]. Reconstituted networks of collagen I have been widely employed both as scaffolds for tissue engineering and as three-dimensional matrices to investigate mechanisms underlying cell-matrix interactions, and are characterized by complex mechanical properties [13]. Both in vivo and in vitro, tumors exert tensile and compressive forces on the matrix due to uncontrolled proliferation and enhanced contractility of cancer cells. Such cell-generated forces engage collagen fibers and—due to the unique properties of fibrillar collagen—lead to matrix remodeling, which in turn facilitates migration, invasion, and metastasis. Hence, collagen's unique structural and mechanical properties underlie remodeling phenotypes observed in several types of cancer as well as the mechanobiological pathways involved in invasion and metastasis. This chapter focuses on collagen structure and mechanics with relevance to tumor-matrix interactions. Specific attention is paid to scale dependency, sources of non-linear behavior, and differences between various modalities of loading. First, we describe biomechanical features of the tumor ECM in vivo and how they can be reproduced in vitro using single cells or multicellular spheroids embedded in 3D collagen networks. We then briefly review techniques to control collagen hydrogel architecture in vitro and present methods for quantitative imaging of fibrous collagen. Second, we describe experimental methods developed to measure mechanical properties of collagen at various length scales, from molecular to tissue level. Third, we review available modeling approaches to simulate the multi-scale mechanics of collagen. We conclude the chapter by discussing how experiments and models are combined to estimate traction forces generated by cancer cells migrating within complex 3D collagen networks.

2 Collagen Remodeling in Breast Cancer

2.1 The Tumor ECM

A distinct feature of many cancers, and particularly of breast carcinomas, is *desmoplasia*, the chronic deposition of collagen in the tumor microenvironment [3]. The increased content of collagen at the tumor margin is usually associated with a stiffer mechanical behavior, which likely arises from a combination of increased density, alignment, and cross-linking of stromal collagen. In breast cancer, collagen density is directly related to mammographic density [14], which is a known risk factor [15], while collagen alignment was found to facilitate cancer invasion and to correlate with patient outcomes [16, 17]. In particular, the group led by Patricia Keely [18, 19] identified three different stages in the so-called tumor-associated collagen signatures (TACS): randomly oriented collagen which retains significant waviness (TACS-1), dense collagen fibers aligned parallel (TACS-2), and perpendicular (TACS-3) with respect to the tumor margin (Fig. 1a). While the TACS-1 phenotype is reflective of the normal breast tissue, TACS-2 is likely caused by abnormal cell proliferation and overall tumor growth, while TACS-3 has been shown to depend on cell contractility [20] and to correlate with increased stiffness upon indentation [21]. Matrix stiffness is primarily controlled by collagen cross-linking [22] and it has been shown that high substrate stiffness can induce a malignant phenotype even in benign cells [11]. ECM densification and stiffening can also increase interstitial fluid pressure in solid tumors, thus generating hypoxia and acting as a barrier to drug delivery [23]. Therefore, remodeling of fibrillar collagen plays an active role in invasion and metastasis by establishing a positive feedback mechanism in which cell contractility and collagen engagement cause tissue stiffening and hypoxia, which in turn elicit further contractility and cell migration by impacting several hallmarks of cancer [8]. Matrix remodeling is defined as a change in matrix structure and mechanical properties [24]. Proteolytic degradation and de novo synthesis are known contributors to matrix remodeling in cancer [25]. However, mechanical remodeling plays a key role: forces are generated at the tumor-stroma boundary as a result of cell proliferation and contractility and alter structure and mechanics of the fibrous matrix, thus leading to tissue remodeling. Tumor-generated forces induce complex mechanical responses in the fibrous ECM, including non-affine fiber deformations, interstitial fluid flow, viscoelastic and plastic deformations. Such complex responses can be investigated by studying cancer cell behavior within reconstituted collagen networks.

2.2 In Vitro Models

The biophysical mechanisms and mechanobiological pathways by which ECM structure, mechanics, and remodeling influence breast cancer invasion can be elucidated using 3D culture models [4, 26–28]. 3D cultures provide a physiologically relevant context to study cell-matrix interactions by fully embedding single cells or cell clusters within natural or synthetic hydrogels. A review of hydrogel materials used for 3D cultures goes beyond the scope of this chapter, but the interested reader is referred to the review by Caliari and Burdick [29]. Here, we will focus only on collagen hydrogels because collagen is the major structural component of the ECM in vivo and collagen monomers self-assemble in vitro into 3D fiber networks which are compatible to culture nearly all cell types. Several groups have adopted the multicellular tumor spheroid model [30] to recreate in vitro key features observed in solid tumors, such as tumor heterogeneity and chemoresistance. Briefly, cancer cells are grown in low attachment conditions, such as hanging drop or non-adherent cultures [31], and self-assemble over the course of 24–48 h into 3D spherical structures containing typically between 1000 and 20,000 cells [32, 33]. Once spheroids reach a critical size, chemical gradients of nutrients, metabolic waste, and oxygen are developed. The resulting hypoxia creates a necrotic core and has been involved in the emergence of cancer stem cells [34], which are thought to contribute to tumor heterogeneity and drug resistance. A realistic chemo-mechanical microenvironment can be reproduced in vitro by embedding spheroids within collagen hydrogels. Spheroid embedding is achieved by allowing liquid collagen to polymerize around individual spheroids, and was used to investigate treatment response [34, 35] as well as spheroid-matrix interactions [36]. Among the various breast cancer cell lines, the epithelial MCF-10A [37] and the mesenchymal MDA-MB-231 [38] are widely used. These cells lines represent models of non-tumorigenic (MCF-10A) and post-metastatic (MDA-MB-231) human mammary epithelial cells, and are often used in comparative studies [39]. 3D cultures of tumor spheroids embedded in collagen hydrogels for 48 h display differential remodeling of the collagen matrix by the different cell lines (Fig. 1b). MCF-10A spheroids proliferate while maintaining a nearly spherical geometry, which results in the generation of compressive forces on the matrix and densification of collagen around the spheroid margin. MDA-MB-231 spheroids, instead, align collagen fibers radially via generation of contractile forces. As a result, single cells separate from the main cell cluster to invade along the aligned collagen. These different types of remodeling are reminiscent, respectively, of the TACS-2 and TACS-3 phenotypes observed in vivo in mouse mammary tumors (Fig. 1a). Generation of compressive and tensile forces by tumor spheroids on the surrounding microenvironment are also

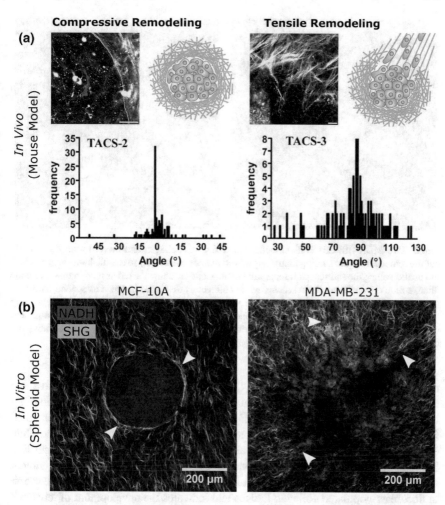

Fig. 1 Examples of compressive and tensile matrix remodeling driven, respectively, by proliferation and contractility. **a** MMT-PyVT mammary tumor explants categorized as TACS-2 (left) and TACS-3 (right) along with the respective histograms of fiber angles with respect to the tumor boundary, where 0° indicates parallel fibers and 90° indicates perpendicular fibers. Modified, with permission from BioMed Central, from Provenzano et al. [16] BMC Medicine. Schematics reprinted by permission from Springer Nature: Provenzano et al. [18] Clinical & Experimental Metastasis, Copyright 2009. **b** Endogenous images generated by multicellular tumor spheroids embedded in 2 mg/mL collagen and acquired using multiphoton microscopy after 48 h in 3D culture. Cells are shown in red as nicotinamide adenine dinucleotide (NADH) autofluorescence, while collagen is displayed in green as second-harmonic generation (SHG). White arrows indicate collagen fibers that are arranged parallel to the edge of MCF-10A spheroids and perpendicular to the edge of MDA-MB-231 spheroids. Note how, in the latter case, cells migrate along the aligned fibers

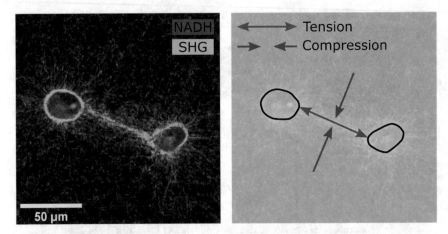

Fig. 2 Two MDA-MB-231 cells embedded in 1 mg/mL collagen and imaged after 48 h in 3D culture remodel the surrounding matrix via adhesion and contractility to establish mechanical communication along the shortest path separating them. Cells are shown in red as nicotinamide adenine dinucleotide (NADH) autofluorescence, while collagen is displayed in green as second-harmonic generation (SHG). Note how the matrix is densified around each cell and along the path separating the two cells (left) thus revealing the presence of tensile and compressive forces that align and compact collagen fibers (right). Note also how collagen is also aligned radially around individual cells

reported elsewhere in the literature [40, 41]. The coexistence of compressive and tensile forces can also be observed at a smaller scale by examining the remodeling of collagen by single MDA-MB-231 cells (Fig. 2). In response to cell-generated forces, collagen fibers simultaneously align along the path of minimum distance while compacting perpendicularly, thus establishing a matrix-mediated mechanical link between neighboring cells in the absence of cell-cell interactions. Matrix alignment between individual cells or multicellular tissue explants is a common observation in fibroblast-populated collagen lattices and indicates the development of tensional homeostasis [42]. Plasticity of the collagen network underlies such remodeling, as significant fiber alignment remains after disrupting the actin cytoskeleton or lysing resident cells [43]. Nevertheless, Vader et al. [44] showed that collagen alignment in presence of local mechanical loads occurs both in control and cross-linked networks, with the latter displaying a fully elastic, rather than visco-plastic, behavior. Therefore, the development of localized mechanical anisotropy and densification is a key feature of collagen network mechanics, which enables long-range transmission of mechanical loads [45].

3 Micro-structure

It is important to understand and be able to control the micro-structure of collagen networks in order to create realistic environments in vitro and expose cells to specific micro-architectural cues. In this section, we present collagen self-assembly from monomers to fibrils and methods to cross-link such fibrils into mechanically competent networks. We will also discuss imaging methods available for quantifying microstructural features of collagen hydrogels used for 3D cultures.

3.1 Collagen Self-assembly

The basic molecular unit of collagen I is a triple helical peptide capable of interacting specifically with other like peptides and self-assemble into fibrils. Collagen self-assembly is an endothermic reaction largely driven by entropic contributions and controlled by electrostatic and hydrophobic interactions. In vivo, collagen is synthesized by cells as a precursor molecule (procollagen), secreted into the extracellular space where it self-assembles into fibrils, and is later arranged as ordered fibers thanks to cellular control and enzymatic activity [46]. In vitro, collagen self-assembly generates network structures that are reminiscent of those observed in vivo but are characterized by random orientations and by a variety of fibrillar sizes, due to lack of cellular oversight and absence of other ECM components (e.g., proteoglycans). Monomeric collagen is usually extracted from rat tails or bovine skin and maintained in acidic solutions, where the net positive charge on the monomers acts against intermolecular aggregation [47]. Polymerization initiates spontaneously at neutral pH and is catalyzed by increasing temperature, with different fibrillar structures obtained following a "neutral start" (increase in pH followed by a rise in temperature) as opposed to a "warm start" (raise in temperature followed by pH neutralization) procedure [48]. Regardless of the procedure, the kinetics of self-assembly and the final fibril structure can both be controlled by tightly regulating concentration, temperature, ionic strength, and pH [49–53]. Measurements of turbidity [49, 54] and confocal microscopy [55] revealed that fiber network formation follows a nucleation and growth process which progresses in three phases: a "lag" phase during which most nucleating structures are formed, a "growth" phase where fibrils aggregate via a combination of linear and lateral fusion, and a "plateau" phase in which fibrils thicken as the monomers in solution are depleted (Fig. 3a). Collagen self-assembles into fibrils which may then associate into fibril bundles. Regardless of their size, herein we will refer to such structures as fibers. Experimentally, the most common approach to control network structure and fiber size is by regulating polymerization temperature and pH. Figure 3b shows that fiber size and pore area can be increased by lowering pH [50, 53] or by decreasing polymerization temperature [52, 55]. Interestingly, polymerization conditions impact fiber size by affecting the

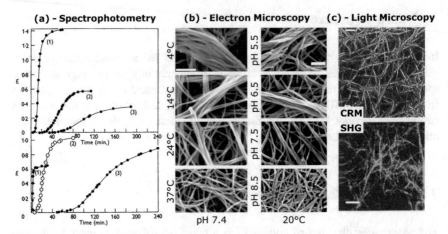

Fig. 3 Methods used to image collagen micro-structure. **a** Spectrophotometry allows one to monitor the temporal evolution of turbidity (indicated as E) during collagen self-assembly. Turbidity measurements show that the kinetics of self-assembly depends on concentration (top, 1—0.084%, 2—0.028%, 3—0.017%) and temperature (bottom, 1—37 °C, 2—25 °C, 3—20 °C), as well as ionic strength and pH. Modified, with permission, from Wood and Keech [49] Biochemical Journal. **b** Scanning electron microscopy images of collagen hydrogels formed under various conditions of temperature (left) and pH (right). A larger fiber size and pore area can be achieved by lowering pH or temperature during self-assembly. Scale bars represent 500 nm. Modified from Raub et al. [52, 53] Biophysical Journal, Copyright (2007, 2008), with permission from The Biophysical Society. **c** Light microscopy images of collagen hydrogels in their hydrated state using confocal reflection microscopy (CRM, top), and multiphoton second-harmonic generation (SHG, bottom). Scale bars represent 5 μm. Modified, with permission, from Brightman et al. [70] Biopolymers, and Zoumi et al. [75] Proceedings of the National Academy of Sciences. Copyright (2002) National Academy of Sciences, U.S.A

linear and lateral aggregation (bundling) of fibrils, while individual fibrils retain a nearly constant size.

3.2   Collagen Cross-Linking

The final steps of network formation involve the generation of cross-links between collagen fibrils that stabilize the network structure. Both weak and covalent intermolecular interactions contribute to cross-linking of the collagen network [47] which underlies its elasticity. The transition from primarily viscous to elastic behavior observed during polymerization has suggested that collagen self-assembly represents a phase transition, and the development of mechanical stiffness can be interpreted using percolation theory [56]. However, further mechanical stiffening of reconstituted collagen scaffolds is required to achieve mechanical properties similar to those measured in excised connective tissues. The stiffness of collagen gels can be tuned using a variety of cross-linking methods, including enzymatic, chemical, and physical techniques. Enzymatic cross-linking is the most natural approach as it recapitulates the final steps of in vivo fibrillogenesis. The main enzymes that can be used to cross-link collagen fibrils in vitro are lysyl oxidase [57, 58] and tissue transglutaminase [59, 60], although they both cause modest changes in tissue mechanical properties. Chemical cross-linking is commonly achieved using glutaraldehyde (GA), a bifunctional molecule that interacts with the ε-amino groups of lysine and hydroxylysine via its aldehyde groups [61–63]. Cytotoxicity represents the main drawback of GA and limits its use for 3D cell cultures. Glycation represents a method for chemical cross-linking of collagen that employs sugars, such as glucose or ribose [64, 65]. It is an appealing alternative to GA because cells or spheroids can be embedded in collagen monomers that were pre-incubated with sugars, and increasing sugar concentrations lead to significant changes in gel mechanics [66, 67]. A physical approach to cross-linking is ultra-violet (UV) light irradiation [68]. For the interested reader, a detailed review of methods available to control collagen architecture and cross-linking was offered by Hapach et al. [69].

3.3 Quantitative Imaging

The generation of physiologically relevant collagen scaffolds via self-assembly can be evaluated by using a variety of quantitative imaging techniques. The oldest method used to image collagen fibrillogenesis is spectrophotometry (Fig. 3a), in which relative changes in the absorption of light—reported as turbidity measurements—were used to track the kinetic details of fibril formation [49]. This method, however, provides no structural details regarding local fibril structure and global network architecture. Electron microscopy (EM) represents the gold standard for visualizing the structure of self-assembled collagen gels. Both transmission EM (TEM) and scanning EM (SEM) are commonly used to image collagen fibril and fiber structure: the former is primarily used to study molecular details such as the periodic banding (D-banding) of collagen fibrils while the latter is used to image supra-molecular arrangement of fibrils into bundles and porous networks (Fig. 3b). The high-resolution power of EM requires sample fixation and dehydration, thus not allowing live imaging. Over the past twenty years, advances in light microscopy have dramatically improved our ability to visualize collagen in vivo and in vitro and quantify biophysical details of cell-matrix interactions. Confocal and multiphoton microscopy represent powerful tools to visualize the time-resolved, three-dimensional distribution of endogenous and exogenous fluorophores. Confocal reflection microscopy (CRM) enables visualization of collagen fibers without the need for fluorescent labeling (contrast is generated by backscattered light [70]) and has been widely employed to visualize collagen-matrix interactions in vitro [43, 71–74]. Multiphoton microscopy has emerged as a tool to image deep within biological tissues while separating second-harmonic generation (SHG) and two-photon fluorescence (TPF). SHG is a signal generated specifically from non-absorptive light-collagen interactions and therefore provides local structural information from collagen without the risk of photobleaching. Instead, TPF depends on non-linear excitation of molecular fluorophores and allows detection of endogenous molecules such as the metabolic co-factor nicotinamide adenine dinucleotide (NADH, cf. Figs. 1 and 2) [75, 76]. Importantly, TPF has been shown to increase after chemical cross-linking of collagen with GA [52], thus suggesting that multiphoton microscopy is not only sensitive to structural properties (e.g., fiber alignment) but also to functional properties (e.g., degree of cross-linking). Due to its capacity for deep-tissue imaging, SHG has become the standard to investigate collagen structure in tumors [16, 77, 78]. The major limitation of light microscopy approaches is that collagen fibrils perpendicular to the imaging plane generate very low CRM and SHG signals with respect to in-plane structures [79, 80], effectively reducing the amount of 3D information that can be extracted using these techniques. On the other hand, these techniques allow imaging of unlabeled collagen scaffolds or native tissues with no need for potential sample-altering preparations (Fig. 3c).

Quantification of network structure implies the characterization of collagen fiber diameter, length, alignment, and density from image data. The most common imaging techniques used to quantify collagen micro-structure—primarily fiber orientation and diameter—include SEM, CRM, and SHG. Fiber orientation has received considerable attention as preferential alignment of collagen fibers underlies the development of mechanical anisotropy in biological tissues. One of the first algorithms, known as OrientationJ, is based on the calculation of a structure tensor for each pixel which defines the predominant orientation as the direction of its largest eigenvector [81]. Since then, other automated algorithms have been proposed which calculate a pixel-wise metric of fiber orientation using various criteria, such as the squared gradient methods [82] or directional statistics [83, 84]. Determination of fiber diameter and length was initially based on manual measurements, that is plotting the image intensity along a line path drawn parallel or perpendicular to the fiber, to quantify fiber size [70]. More sophisticated semi-automated methods have been developed as plugins or graphical user interfaces for image analysis platforms such as ImageJ or MATLAB. In particular, DiameterJ (an ImageJ plugin that includes OrientationJ) was developed based on Euclidean distance transforms and was validated with the goal of analyzing high resolution SEM micrographs [85]. Extraction of the collagen network structure from CRM images with the purpose of predicting mechanical properties of collagen hydrogels was introduced by Stein and colleagues with the FIbeR Extraction (FIRE) algorithm [86]. FIRE extracts network organization by defining nucleation points based on the local maxima of a distance function and by tracing branches that extend from each nucleation point. Bredfeldt et al. [87] found that preprocessing images using a discrete curvelet transform (CT) to denoise raw images improved FIRE-based segmentation of collagen fibers from SHG images of breast cancer tissues. This method, named CT-FIRE, has been released as a software that allows automatic extraction of fiber diameter, length, waviness, and orientation [88]. It should be noted that most of the image analysis methods developed so far work only on 2D images, thus limiting the amount of truly 3D information that can be extracted from image stacks. Accurate quantification of fiber structure is still a challenging task both experimentally and computationally: sample preparation for

TEM/SEM induces shrinkage due to fixation and dehydration whereas CRM/SHG imaging is limited by the ability to detect only fibers with dimensions above the diffraction limit for optical wavelengths ($\sim\lambda/2$ N.A., where λ represents the wavelength of the exciting light and N.A. is the numerical aperture of the objective lens). The mechanical properties of tissues are a function of fibrillar micro-structure, presence of preferential directions of fiber alignment, and degree of cross-linking. Thus, accurate quantification of fiber network structure is a key step towards inferring mechanical properties from structural data.

4 Micro-mechanics

The mechanical behavior of collagen is highly dependent on its underlying micro-structure. In addition, measured properties vary based on the length scale at which they are probed and on the type of imposed loads or deformations (e.g., tensile vs. compressive). Figure 4 summarizes experimentally available methods to test collagen mechanics at different length scales and under various loading conditions. In this section, we will focus on methods developed to investigate collagen micro-mechanics, from individual monomers to fiber networks.

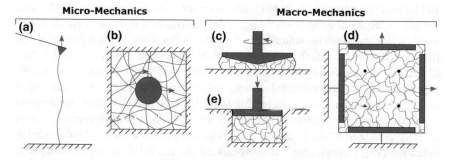

Fig. 4 Schematic representation of the mechanical testing approaches available to investigate collagen mechanics at different length scales. **a** Uniaxial stretching of individual fibrils can be achieved by using an AFM cantilever. **b** Rheological measurements at the microscale can be carried out by embedding beads (1–4 μm in diameter) within the collagenous network and applying linear displacements or rotations externally via lasers (optical tweezers) or magnetic fields (magnetic pulling, magnetic twisting). **c** Bulk rheology, here shown with a cone-plate geometry, characterizes the properties of the sample under shear with the option of simultaneously varying the axial strain. **d** Tensile testing, here shown in a biaxial configuration, applies loads in two perpendicular directions, thereby allowing to characterize the degree of anisotropy due to preferential fiber alignment. Uniaxial tensile testing is a special case of biaxial test, as it requires loading along one axis. **e** Confined compression is a uniaxial test that characterizes both compressive material properties and fluid transport properties of the matrix, provided that a porous indenter is used

4.1 Individual Monomers and Fibrils

A variety of nano- and micro-scale manipulation techniques—including optical tweezers, atomic force microscopy (AFM), and microelectromechanical systems (MEMS)—have been used to probe the mechanics of various molecules, including collagen monomers and fibrils. Procollagen monomers in solution can be idealized as rod-like molecules and direct measurements of force-extension behavior via optical tweezers have shown that the triple helical molecules of collagen I and II are biomechanically flexible, non-linearly elastic, and stiffer than macroscopic tissues [89, 90] (Fig. 5a). Elastic moduli for collagen monomers between hundreds of MPa and tens of GPa have been reported in the literature, with the large variability due to experimental uncertainty [91]. At the fibril level, AFM measurements have shown that collagen monomers arrange helically to form rope-like structures having diameters between 260 and 410 nm and lengths of 4–6 μm, which are not capable of bearing compressive loads [92] and unfold under the action of tensile loads [93]. AFM has been widely used not only for structural measurements but also to probe the mechanical behavior of collagen fibrils by employing mainly two testing modalities. The first involves uniaxial stretching of the fibril by attaching one end to a glass coverslip and the other end to a tipless cantilever beam (Figs. 4a and 5b) [94, 95]. The second one consists of a three-point bending test of fibrils deposited across a microchannel [96, 97]. Interestingly, while the former has revealed nonlinearly elastic, viscoelastic, and inelastic properties of collagen fibrils via force-extension experiments, the latter has been interpreted using beam theory under the assumption of linear elastic behavior, thus leading to the determination of elastic moduli from force-indentation data. Despite the different underlying assumptions, both approaches have revealed that fibril stiffness depends on hydration and cross-linking. Hydration reduces the stresses experienced by the collagen fibrils by increasing their diameter of ~70% with respect to dry conditions, but also by decreasing fibril stiffness: hydrated fibrils are softer (E ~ 300 MPa) with respect to dry fibrils (E ~ 5 GPa) [94] and this phenomenon has been recently attributed to the weak intermolecular forces that seem to be modulated by hydration [98]. Instead, covalent cross-linking via GA increases significantly fibril stiffness, leading to a nearly three-fold increase in the measured Young's modulus [95, 96]. MEMS platforms have allowed investigators to perform controlled uniaxial tensile tests with strain tracking via fluorescent labeling of collagen [99]. In general, uniaxial testing via MEMS leads to a nearly linear stress-strain behavior followed by plastic yielding, and sometimes brittle failure [100–103]. In contrast, tensile testing via AFM leads to marked nonlinear responses [94, 95, 98]. Such discrepancy in material behavior is probably due to methodological differences and should be investigated further, despite the fact that comparable values of liner elastic moduli were obtained using both methods. Viscoelasticity of collagen fibrils was revealed by the presence of hysteresis and strain-rate dependency. Stress-relaxation tests are best described using a two-term Maxwell model [95, 104] in which the time constants have been implicated with sliding of microfibrils (fast time scale) and collagen triple helices (slow time scale).

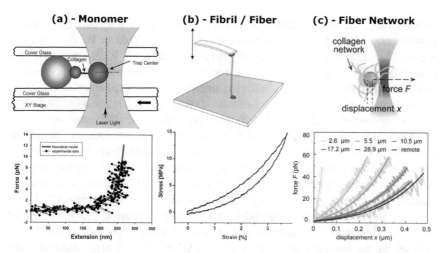

Fig. 5 Micro-mechanical experiments on collagen allow one to characterize its material properties at different length scales. **a** A collagen monomer can be stretched by attaching it to two polystyrene beads, one of which is linked to a larger bead clamped to the microscope stage, while the other is trapped optically using a laser (top). Extension of the molecule is imposed by moving the stage, thus revealing nonlinear force-extension behavior of collagen II (bottom), with similar results for collagen I. Modified from Sun et al. [90] Journal of Biomechanics, Copyright (2004), with permission from Elsevier. **b** Extension of an individual fibril or fiber can be achieved by gluing one of its ends to a glass coverslip and the other end to an AFM tip (top). Uniaxial stretching reveals nonlinear material behavior and hysteresis in hydrated collagen fibrils (bottom). Modified, with permission, from van der Rijt et al. [94] Macromolecular Bioscience. **c** Optical trapping enables controlled linear displacement of beads embedded within a 3D collagen network (top). Local force-displacement curves measured at varying distances from a cancer cell reveal that cell contractility causes stiffening in the collagen network (bottom). Modified, with permission, from Han et al. [115] Proceedings of the National Academy of Sciences

4.2 Fiber Networks

Characterizing the mechanical response of a network of collagen fibers to loads imposed at the microscale is key to understanding the physical mechanisms regulating cell-matrix interactions. For this reason, there is an increasing interest in measuring collagen mechanics at a length scale between 1 and 100 μm, that is at the scale of a cell or a group of cells. In fact, such studies often include cells cultured on top or within collagen gels to investigate the feedback loop between cell adhesion/migration and collagen mechanics. Based on its widespread use, AFM has been used also to indent the surface of collagenous fiber networks. Elastic moduli are extracted from force-indentation responses by using the linear Hertz model from contact mechanics [105], which employs a series of assumptions that are likely not applicable to collagen. In fact, collagen is assumed to behave as a linear, elastic, homogenous, and isotropic material with a Poisson's ratio of 0.5 to enforce incompressibility. Nevertheless, AFM indentation has been used to detect matrix stiffening at the leading edge of cells migrating on top of collagen [106] and of acellular gels polymerized at various temperatures [107]. Elastic moduli range between hundreds of Pa (E ~ 0.2 kPa reported by van Helvert and Friedl [106]) to a few kPa (E ~ 1–9 kPa reported by Xie et al. [107]). The major limitation is that indentation causes complex deformations in the collagen network which make the interpretation of experimental data quite challenging. Moreover, AFM is inherently 2D and does not allow measurement of the 3D response to localized loads like those generated by embedded cells.

The main technique used to probe 3D micromechanical properties of viscoelastic materials is known as microrheology. In analogy with bulk rheology (examined in Sect. 5.1), microrheological measurements are concerned with studying the deformation and flow around microscopic particles that are excited by thermal fluctuations (passive microrheology) or by external force fields (active microrheology) [108]. Tracking the motion of such colloidal particles (often spherical beads) embedded within a soft material enables extraction of its rheological properties, typically summarized by the complex shear modulus $G = G' + iG''$, where G' represents the elastic storage modulus (in-phase component) and G'' the viscous loss modulus (out-of-phase component). For materials that behave like viscoelastic liquids, Brownian motions of the particles elicited by thermal fluctuations are sufficient to extract material properties such as viscosity, under the assumption of linear response to small deformations. However, networks of collagen fibers behave as viscoelastic solids, rigid enough to inhibit the Brownian motion of embedded particles. Hence, active microrheology should be used to impose known forces or deformations and monitor the response of a material that is driven out of its mechanical equilibrium [109]. Focused lasers and magnetic fields are used to apply external forces to embedded beads in a static or sinusoidal fashion, with the latter interrogating material behavior at multiple frequencies. While optical tweezers are designed to apply only linear forces by trapping a bead within a laser beam, magnetic fields can be used to apply both forces (magnetic tweezers) and torques (magnetic twisting), as shown in Fig. 4b. Among these techniques, optical tweezers have been used to probe collagen micromechanics in several studies. Velegol and Lanni [110] used 2 μm polystyrene beads to apply local deformations to collagen gels of concentrations ranging between 0.5 and 2.3 mg/mL. They found that collagen displays a primarily elastic response and that the shear modulus increases with concentration with a mean value of $G = 55$ Pa over the range of concentrations considered in this study. More importantly, its behavior was found to be spatially inhomogeneous and, in a separate study, the authors found that such heterogeneity was associated with the development of mechanical anisotropy [111]. In fact, while solutions of triple helical collagen molecules are spatially homogeneous, self-assembly into collagen networks generates spatial mechanical heterogeneities due to the variety of interactions that are established between the bead displacement and the local fiber orientation [112]. Local stiffness and anisotropy can be assessed by displacing individual beads in two orthogonal directions and it has been shown that, for a given concentration, networks with larger pore areas (polymerized at 21 °C) are locally softer and more anisotropic than denser networks with smaller pores (polymerized at 37 °C) [113]. Opposite findings were reported by Staunton et al. [114] who evaluated the frequency responses of various collagen microarchitectures by tuning collagen concentration and polymerization temperature. Collagen gels with larger pore areas were also associated with longer and thicker fibers and with stiffer properties. Both storage (G') and loss (G'') moduli increase with frequency but present equal magnitudes only at the cross-over frequency (~500 Hz): below collagen displays a primarily elastic behavior ($G' > G''$) while at larger frequencies it displays a liquid-like behavior ($G' < G''$), thus showing features of a glass transition. Recently, Han et al. [115] evaluated the local force-displacement behavior of collagen by displacing optically trapped 4.5 μm beads at a constant velocity of 1 μm/s to measure quasi-static responses. Despite absolute network stiffness was not reported, micromechanical properties of collagen gels at 2 mg/mL were found to be highly nonlinear and to display spatial stiffness gradients generated by contractile cells (Fig. 5c).

Magnetic twisting represents a valid, albeit less frequently used, alternative to optical tweezers. Ferromagnetic beads are first exposed to a strong but brief magnetic field which aligns all the magnetic dipoles in one direction. Subsequent application of a weaker magnetic field in a direction perpendicular to the magnetization direction generates a magnetic torque, which has the same intensity for all the beads within the volume of interest [116]. Magnetic twisting presents additional complications with respect to optical tweezers because of the need to monitor bead rotations instead of in-plane displacements, but it also offers the advantage of probing a large number of beads at once, thus allowing high throughput experiments. With application to collagen gels, magnetic twisting was first employed by Leung et al. [117] who showed a primarily elastic behavior of 2 mg/mL collagen networks tested under static and oscillatory twisting of 4.5 μm iron oxide beads. In this study, the collective rotation of all the beads embedded within a gel was estimated via an in-line magnetometer, thus leading to a global estimate of local material properties. An implementation of magnetic twisting that is well suited to measure spatial heterogeneities relies on tracking lateral displacements of individual, partially embedded, beads [118]. When partially embedded on the surface of a continuum, a bead will undergo lateral displacements upon the application of a twisting torque. This version of magnetic twisting was applied to collagen gels by Li et al. [119] who showed that microscopic twisting is less sensitive than macroscopic shearing to changes in collagen concentration. The major disadvantage here lies in the 2D nature of such measurement and by the need to estimate—via appropriate modeling—the role played by variable degrees of bead embedding on the measured shear moduli. Finally, and to the best of the authors' knowledge, magnetic tweezers have not been used to probe the micromechanics of collagen but only to validate phenomenological constitutive relations determined from macroscopic data [120]. Overall, various microrheological methods have led to estimates of shear moduli for collagen gels in the range of 50–2000 Pa (E ~ 0.15–6 kPa, assuming a Poisson's ratio of 0.5) and are generally consistent with estimates from bulk rheology [112, 114, 117].

4.3 Scale Dependency

The multi-scale nature of collagen mechanics is revealed by micro-mechanical experiments. In fact, the highest elastic moduli are measured in tropocollagen monomers and display a consistent decrease as monomers assemble into fibrils and fibers, and as fibers entangle and cross-link to form networks (Table 1). Such differences are due to molecular and structural effects and are best investigated by combining experimental measurements with mechanistic modeling at different length scales. In fact, the relatively small differences observed between the mechanics of monomers and fibrils has been imputed to the sliding of tropocollagen molecules (cf. Sect. 6.1). More significant differences are observed at the network level, where elastic moduli are six orders of magnitude lower with respect to those of individual fibrils. Such mismatch is based on the fact that network mechanics is dominated by non-affine deformations and nonlinear phenomena such as fiber buckling (cf. Sect. 6.2). A network of cross-linked and entangled fibrils responds to complex loads by aligning in the direction of tension and by collapsing in the direction of compression (Fig. 2), with microscale fiber strains not necessarily matching the applied macroscopic strains. In addition, it is important to note that microrheology techniques used at the network level rely on the continuum assumption [108]. A fiber network can be reliably treated as a continuum only if it exists a clear separation between the length scales of the mechanical probe and of the local network micro-structure. However, beads of diameters of 1–4 μm are often used and have been shown to be effectively immobilized within fibrous networks [121]. However, it remains yet unknown how different bead sizes probe different aspects of collagen network mechanics, thus affecting ultimate results. Finally, ensuring that beads are anchored to the fiber network is of paramount importance to ensure that relevant viscoelastic properties are recovered. Insufficient chemical coupling of particles smaller than the network pore size will show as random walk trajectories, thus estimating fluid viscosity rather than network stiffness [121].

Table 1 Scale dependency of collagen micro-mechanics

Collagen structure	Method	Stiffness (range)	References
Monomer	Optical tweezers	0.35–12.2 GPa	[89–91]
Fibril/fiber	AFM/MEMS	0.30–5 GPa	[94, 96, 97, 99, 102, 103]
Network	AFM/microrheology	0.15–6 kPa	[106, 107, 110, 114, 117]

The range of mechanical properties, measured as elastic moduli under the assumption of linear elastic behavior, are compared at different length scales

5 Macro-mechanics

The primary function of collagen fibers is to provide tensile stiffness to tissues. However, collagenous tissues are often exposed to complex multiaxial loads in vivo, including tension, compression, and shear. This section summarizes the macro-mechanical properties of collagen hydrogels and the different features of collagen biomechanics that can be isolated by using various testing modalities in vitro.

5.1 Shear Rheometry

The macroscopic (bulk) rheological properties of hydrogels are commonly investigated under the application of shear forces or deformations by means of a rheometer (Fig. 4c). The shear modulus (G) is defined as the ratio between shear stress (τ) and shear strain (γ), that is $G = \tau/\gamma$. Under the assumption of linear elasticity, the Young's modulus is connected to the shear modulus by the relation $E = 2G(1 + \nu)$, where ν represents the Poisson's ratio. Also, and still under the assumption of linearity, oscillatory measurements lead to the complex shear modulus $G = G' + i\, G''$ [122], which has the same meaning as in microrheology. Due to its capacity to discern between elastic and viscous contributions, shear rheometry has been used to monitor the evolution of viscoelastic properties during collagen self-assembly [56]. Application of small oscillatory shears reveals that a neutralized collagen solution initially behaves as a Newtonian liquid, with a small but finite G'' and an undetectable G'. As collagen monomers self-assemble into fibrils, and individual fibrils into a network, both components of the shear modulus increase but G' increases faster than G''. Steady-state measurements, usually achieved within 30 min [123], reveal that G' is approximately one order of magnitude larger than G'', which is consistent with a primarily elastic behavior of the collagen network. The group led by Laura Kaufman has extensively characterized the timeline of collagen self-assembly by simultaneously imaging network structure (via confocal microscopy) and probing network mechanics (via shear rheometry), both presenting features of a fluid-solid transition [123–125]. The shear modulus in the linear regime increases with collagen concentration and this relationship has been proposed to follow a power law [123, 124, 126], whereas the relation between structural organization and mechanics is less clear. In fact, while Raub et al. [52, 53] have shown that denser networks made by increasing polymerization temperature or pH are stiffer, Yang et al. [124] have shown the exact opposite, with higher shear moduli for networks with lower densities and thicker fibers. Such mismatch mirrors similar inconsistencies observed from microrheology measurements and emphasizes the need for a structure-based modeling approach to correlate network structure and function.

In addition to exploring the linear elastic regime, rheometry can be used to investigate nonlinear elastic and inelastic properties of collagen gels by means of strain sweep experiments, in which increasing shear strains are applied at a constant frequency. The onset of nonlinear elastic behavior upon shearing is known as strain stiffening and occurs at a concentration-dependent critical strain followed by yield and failure [126]. Strain stiffening is associated with another phenomenon, that is the development of negative normal stresses [127]. This behavior is also known as inverse Poynting effect and consists in the fact that biopolymer networks under torsion contract axially, instead of elongating like observed in linear materials [128]. Both strain stiffening and the generation of negative normal stresses have been explained as a consequence of the nonlinear force-extension behavior of collagen fibers. In particular, the tension-compression nonlinearity causes fibers aligned along the first principal direction to engage, while fibers perpendicular to it buckle under compressive loads and do not contribute to the measured shear moduli [127, 129, 130] (Fig. 6a). This proposed mechanism was further validated by shearing collagen and fibrin gels under axial deformation [131–133]. Contrarily to linear materials such as polyacrylamide, the shear modulus of biopolymer networks is higher under axial tension and lower under axial compression (although it stiffens further under extreme compaction). In addition, the onset of strain stiffening upon shearing occurs earlier under axial tension and later under axial compression [133]. Hence, there is evidence that collagen fibers are not able to bear compressive loads, and the subsequent buckling reduces their contribution to the overall network stiffness upon shearing. Instead, axial tension minimizes buckling and engages more fibers upon shearing, thus resulting in higher shear moduli.

5.2 Uniaxial/Biaxial Tension

Shear rheometry has provided useful data on the multiaxial mechanics of collagen networks, but tensile tests are more informative as they allow one to probe material symmetry and to parametrize constitutive descriptors of material behavior. Among other studies, differences between the two testing modalities were explored by Xu et al. [134], who found that the tangent modulus of collagen measured from tensile testing is three orders of magnitude higher with respect to the storage modulus measured from shear rheometry. Uniaxial properties of collagen gels were investigated in a series of publications from the laboratory of Sherry Voytik-Harbin, and the data set from Roeder et al. [51] has found widespread use in the literature (Fig. 6b). They combined tensile testing with confocal imaging to explore structural rearrangements of collagen fibers under uniaxial loading [51, 135–137]. By varying the gel structure, they observed a nonlinear increase in Cauchy stress with strain and stiffer material properties for increasing collagen concentration and pH. More importantly, they used confocal imaging to track 3D strains both globally (by tracking the width and thickness at the center of the samples [135]) and locally (via digital volume correlation analysis of CRM data [136]). Upon uniaxial loading, the deformation field was found to be rather homogeneous and consistent with a principal state of strain. Microstructurally, uniaxial tension was associated with alignment of the fibers loaded axially and with buckling of the fibers loaded transversely. Both deformation mechanisms

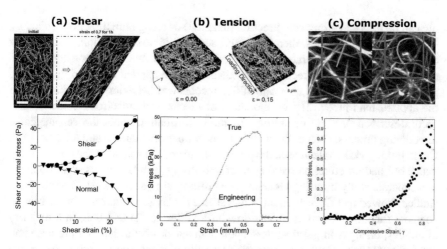

Fig. 6 Macro-mechanical experiments allow one to characterize collagen behavior under different loading conditions. **a** Shear testing causes straightening of fibers aligned along the principal direction of strain, indicated in red, and contemporary buckling of fibers aligned perpendicularly, indicated in blue (top). Scale bar represents 20 µm. Modified, with permission, from Munster et al. [130] Proceedings of the National Academy of Sciences. Alignment and buckling underlie the nonlinear increase in shear stress as well as the development of a negative normal stress (bottom). Modified with permission from Springer Nature: Janmey et al. [127] Nature Materials, Copyright 2007. **b** Uniaxial tension causes alignment of fibers and lateral compaction (top) with subsequent development of nonlinear stress-strain responses and true stresses higher with respect to Engineering stresses due to the reduction of cross-sectional area caused by compaction (bottom). Top figure adapted, with permission, from Voytik-Harbin et al. [135] Microscopy and Microanalysis. Bottom figure republished with permission of ASME, from Roeder et al. [51]; permission conveyed through Copyright Clearance Center, Inc. **c** Unconfined compression testing of fibrin causes bending and subsequent buckling of fibers aligned along the direction of compression (top). The network softens after fibers buckle, and then offers little resistance as it compacts until it finally responds nonlinearly due to extreme densification under large compressive strains (bottom). Modified from Kim et al. [131] Biomaterials, Copyright (2014), with permission from Elsevier. Similar results were reported for collagen by Kim et al. [132]

lead to a dramatic decrease in gel width and thickness, both associated with compaction of fibers and concurrent expulsion of interstitial fluid. These findings are therefore consistent with a highly compressible behavior.

Despite uniaxial stress-strain tests are informative about material behavior (especially if accompanied by information on lateral deformations), a better understanding about the development of material anisotropy can be achieved by exposing collagen gels to biaxial loading (Fig. 4d). In this regard, the development of mechanical anisotropy in fibroblast populated collagen gels was the subject of a series of studies from the laboratory of Jeffrey Holmes [138–141]. They generated squared collagen gels with embedded porous polyethylene bars which allowed transmission of external loads and a relatively homogenous strain field in the center of the sample [138]. Compaction of the gels under the action of fibroblast-generated contractile forces can be avoided by applying external loads and such constraint to collagen remodeling can be imposed along one direction (uniaxial) or along both directions (biaxial). Biaxially constrained gels, therefore, exhibited negligible compaction, random fiber distribution, and isotropic mechanical properties. Instead, uniaxially constrained gels compacted only in the direction perpendicular to the applied load and displayed a preferential alignment of fiber and stiffer mechanical behavior in the direction of the applied load [139]. The development of mechanical anisotropy could be partially described only by including structural information about the cell-mediated remodeling process, such as the distributions of fiber orientations and lengths [140]. Cross-linking via GA decreased the preload strain and the spread of the biaxial strain paths, but had no effect on the degree of anisotropy. Similar results were obtained in a separate study by using Genepin as a cross-linking agent [134]. Overall, these studies showed that fiber orientation and stiffness play different roles in the biomechanical function of tissues. A key drawback is due to the fact that, in a majority of studies, changes in gel thickness upon uniaxial or biaxial loading are not measured directly but calculated under the assumption of incompressibility. As shown by Voytik-Harbin et al. [135], collagen gels are highly compressible due to the fact that the collagen fibers (solid) and the interstitial fluid (liquid) are not physically bound together but undergo relative motions upon mechanical loading. This type of behavior is typical of biphasic materials, that is binary mixtures of solid and liquid phases. The mechanical behavior of biphasic materials can be better investigated under compression.

5.3   Unconfined/Confined Compression

Uniaxial compression is a testing modality widely used to study the biomechanics of articular cartilage since it reproduces the loading conditions experienced by synovial joints and allows one to separate the rheological properties of the synovial fluid from the mechanical properties of collagen [142, 143]. Similarly, compressive testing of collagen hydrogels can be used to estimate the biphasic properties of reconstituted collagen networks. With application to cancer, collagen is compressed by collective proliferation (Fig. 1), and such compressive remodeling has received little attention in the literature. Uniaxial compression testing of cylindrical specimens can be performed via two testing configurations: unconfined compression, in which the axial compression of the specimen is associated with lateral (i.e., radial and circumferential) expansion, and confined compression, in which such lateral expansion of the specimen subjected to compression is impeded by the presence of a confining chamber (Fig. 4e). In confined compression, the loading piston must be made of a porous material so that the fluid can flow in order to preserve its volume.

Unconfined compression testing of collagen, fibrin, and fibrin-collagen composites using a commercial rheometer paired with a confocal microscope has shown that biopolymer networks display non-linear properties over a large range of compressive strains [131, 132, 144] (Fig. 6c). The stress-strain response is linear under low compressive strains, then it softens quite suddenly and reaches a plateau spanning a broad range of strains. Direct visualization of fiber deformations has shown that this plateau is associated with fiber buckling, leading to a compaction of the network with little resistance to the imposed deformation [131]. Such compaction leads to the development of a "compression front", that is a layer of densified matrix near the surface of the piston, which propagates towards the bottom of the gel with further compression [144]. At high deformations, the stress-strain behavior under compression becomes highly nonlinear due to the extreme densification of the network across its thickness. Even at lower strains, material nonlinearities can be observed by increasing the rate of compression, the concentration of collagen, or the degree of cross-linking [145, 146]. Confined compression testing was used to characterize the biphasic properties of collagen hydrogels in a series of studies by the group of Victor Barocas [147–149]. Collagen was polymerized within custom confining chambers and was compressed using a piston made of porous polyethylene. The linear range under compression was found to be below 15% strain. Different mechanical and structural responses were induced in stress-relaxation tests by varying the rate of compression: a slow compression (ramp test) was associated with more uniform deformations and stresses that were maintained for longer times, while a fast compression (step test) was instead associated with network collapse near the piston, higher stresses which decayed rapidly over time [149]. Overall, it was shown that collagen fibers deform heterogeneously upon compression by aligning perpendicular to the direction of applied strain, and store energy primarily via bending [148, 149].

5.4 Choice of Loading Conditions

The complex macro-mechanical behavior of collagen hydrogels is best explored by using more than one testing modality. Ultimately, the loading conditions should be determined on the basis of the behaviors/mechanisms of interest, and not by the ease of implementation. For instance, compression is often used to extract bulk properties from collagen networks, mainly because of its relatively straightforward experimental setup and not because the compressive properties of collagen are of interest [65–67]. On the other hand, tensile testing is rendered quite difficult by the poor structural stability of collagen gels, which display a frustratingly high rate of failure upon uniaxial or biaxial tension. For this reason, most studies employ cell compaction or cross-linking to stiffen the gels before exposing them to tensile loads [139, 141, 147]. Finally, shear rheometry is a standardized technique capable of applying tensile and compressive loads on different populations of fibers while preserving the overall gel volume (Fig. 6c). Stiffness is usually higher when gels are tested in tension, with respect to shear or compression [13], thus highlighting that different loading conditions probe different aspects of network mechanics. Table 2 lists some of the main features of collagen hydrogel mechanics reported in the literature while indicating the loading conditions best suited for probing anisotropy (tension), interstitial fluid flow (compression), and fluid-solid transition during collagen self-assembly (shear). Nonlinearity, instead, is more ubiquitous and presents itself under all testing modalities (Fig. 6). Similarly, viscoelastic and plastic properties have been

Table 2 Various features of collagen macro-mechanics are revealed by using different loading conditions

Mechanical features of collagen hydrogels	Loading conditions		
	Shear	Compression	Tension
Nonlinearity	[126, 127, 130, 133]	[131, 132, 144–146]	[51, 135, 139, 141]
Anisotropy			[134, 139–141]
Interstitial fluid flow		[147, 149, 196]	
Fluid-solid transition	[56, 123–125]		
Viscoelasticity/plasticity	[130, 150–152]	[147, 149]	[153–155]

Nonlinear, but also viscoelastic and plastic behaviors have been probed via shear, compression, and tension testing. On the other hand, other properties have been more selectively assessed using a specific testing modality

reported for collagen hydrogels under shear [130, 150–152], tension [153–155], and compression [147, 149].

6   Mechanical Modeling

Mechanical modeling approaches are needed in order to interpret experimental results and to identify the mechanisms responsible for the observed behaviors. This section summarizes multiple modeling approaches that have been developed to investigate collagen mechanics at different length scales.

6.1   Semiflexible Biopolymers

The key measure of collagen flexibility at the molecular level is its persistence length, that is the length over which the molecule behaves like a straight beam. For length scales larger than the persistence length, the molecule bends under the action of Brownian forces. Based on such definition, the persistence length of a biopolymer chain is related to its bending stiffness (EI) by the following relation:

$$l_p = \frac{EI}{k_B T},$$
(1)

where E is the modulus of elasticity, $I = \pi r^4/4$ is the moment of inertia of a circular cross section of radius r, k_B is the Boltzmann's constant, and T is the temperature. Collagen is part of a class of materials, known as semiflexible biopolymers, which are constituted by undulated filaments presenting contour lengths comparable to their persistence lengths. Their response to axial elongation is controlled by thermal fluctuations for segments longer than l_p (entropic elasticity) and by mechanical strain energy for segments shorter than l_p (energetic elasticity). The entropic elasticity of tropocollagen monomers has been described by using a worm-like chain model, which was originally developed to describe the force-extension behavior of DNA [156]. The force-extension behavior of a worm-like chain is attributed primarily to bending fluctuations: as the chain is stretched, the amplitude of the lateral undulations decreases, and the reaction force increases. Data from optical tweezer experiments [89] and atomistic simulations [157] were fit to worm-like chain models leading to persistence lengths of $l_p \approx 14.5$–16 nm for tropocollagen. Energetic elasticity instead dominates the mechanical behavior of collagen fibrils, which are characterized by increased bending stiffness and higher persistence lengths due to lateral packing of a large number of tropocollagen molecules [157]. The persistence length of collagen I fibrils has also been measured directly from TEM images by estimating the decay of tangent-tangent correlations, which led to values of $l_p = 5$–10 μm [158]. Conversely, the persistence length can be estimated from micromechanical measurements of fibril stiffness (cf. Sect. 4.1) by means of Eq. 1. It should be noted that in Eq. 1 the fibril radius appears at the fourth power, thus leading to large variations in persistence length from experimental uncertainty and to estimates of persistence length that are orders of magnitude higher with respect to direct measurements [159].

6.2 Molecular Dynamics

Molecular dynamics provides a bottom-up approach to investigate the mechanics of collagen monomers and fibrils: the simulated behavior is calculated by incorporating biochemical details rather than fitting empirical parameters from experimental data. Markus Buehler and colleagues described the mechanical behavior of collagen monomers, fibrils, and fibril bundles by imposing interparticle potentials that define atomic and molecular interactions such as covalent bonds, but also ionic, hydrogen, and van der Waals interactions [160–166]. Nonreactive potentials model biochemical bonds as fixed harmonic springs, while reactive potentials allow one to model also the bond breaking and formation processes. Using molecular structures extracted from x-ray diffraction data, Buehler et al. [160] showed that tropocollagen responds to tensile loads by unfolding its three helices, while it buckles under compressive loads (Fig. 7a). The nonlinear behavior in tension arises from the gradual unfolding of the three polypeptide chains and subsequent engagement of atomic bonds. A nonlinear material behavior was demonstrated for tropocollagen with E $\sim$ 7 GPa under small strains and E $\sim$ 40 GPa under large strains, which is thought to represent the material stiffness of the protein backbone. Both reactive and non-reactive potentials captured the tensile properties of the tropocollagen molecule up to 10% strain, although reactive potentials were able to model fracture above 50% strain [161] (Fig. 7a). In addition, in silico creep tests [166] revealed that collagen displays viscoelastic behavior at the molecular level, and that material differences between dry and hydrated tropocollagen could be explained via differences in the density of hydrogen bonds: hydrated molecules form hydrogen bonds mostly with water and thus are mechanically softer with respect to dry molecules in which hydrogen bonds are established exclusively between the polypeptide chains. Collagen fibrils were instead modeled following a hierarchical approach: a tropocollagen molecule was discretized as a collection of periodic cells, which were then organized following the typical staggered arrangement of collagen triple helices (Fig. 7b). The lower stiffness of collagen fibrils with respect to their monomers (cf. Table 1), and its dependence from intramolecular cross-linking are two aspects that were investigated via molecular dynamics simulations. Tang et al. [164] showed that the relatively weak interactions between tropocollagen molecules cause yield of the fibrillar structure and rupture due to sliding between tropocollagen molecules. These findings were independently confirmed by Varma et al. [159], who showed that the lower elastic moduli associated with fibrils and fibril bundles are not due to the specific packing of monomers but likely from slippage of the triple helices. Covalent cross-links were modeled by changing the adhesive forces of intermolecular interactions between the ends of distinct tropocollagen molecules [163, 164]. By increasing cross-link density, the simulated collagen fibril was shown to rupture at larger strains thus revealing a potential mechanism underlying mechanical strengthening due to chemical cross-linking (Fig. 7b).

6.3 Discrete Fiber Networks

Collagen hydrogels represent complex structures endowed with mechanical properties that depend on the nonlinearly elastic and intrinsically anisotropic properties of individual collagen fibers but also on network architecture, fiber density, orientation, and cross-linking. In particular, it has been a topic of debate whether the nonlinear material behavior observed in collagenous tissues originates from the nonlinear properties of individual fibers deforming affinely or arise from non-affine fiber deformations and subsequent gradual engagement of linearly elastic fibers [167–172]. A variety of network models have been developed over the years to describe the mechanics of fiber networks of various dimensionalities (2D vs. 3D), fiber properties (linear vs non-linear), and nodal interactions (pinned or welded joints vs torsional springs). A comprehensive review of network models is offered by Sander et al. [173]. Here, we will present some insights on collagen mechanics that were gained by modeling hydrogels as networks of elastic beams. The general form for the elastic energy per unit volume U stored by the network can be expressed as follows

$$U = \frac{1}{2} \sum_{i=1}^{N} \left[\frac{K_s}{L^i} \left(L^i - L_0^i \right)^2 + \frac{K_b}{L^i} \left(\theta^i - \theta_0^i \right)^2 \right], \tag{2}$$

where N is the total number of fiber segments used to discretize the network. Therefore, L^i and θ^i are the current segment length and the angle formed with an adjacent

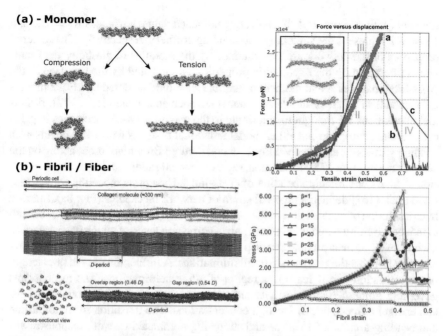

Fig. 7 Mechanics of collagen simulated at the atomistic and molecular level. **a** Mechanical responses to uniaxial loading of an individual tropocollagen molecule. Structurally, a molecule responds to tension by unfolding the three helical polypeptide chains while it buckles under compression (left). Modified, with permission, from Buehler et al. [160] Journal of Materials Research. The tensile response to uniaxial loading can be estimated (right) by using both nonreactive (green line) and reactive (red line) potentials, with the latter being able to predict failure of the molecule due to localized rupture. A mesoscale model (blue line) has the capability of predicting the four phases (I-IV) of the simulated response to tension. Modified, with permission, from Buehler et al. [161] Proceedings of the National Academy of Sciences. Copyright (2006) National Academy of Sciences, U.S.A. **b** Molecular dynamics simulations of a collagen fibril. A coarse-grained model of a collagen molecule is achieved by definition of a periodic cell which can be repeated and arranged in a staggered fashion to reproduce the typical D-banding periodicity and the orthogonal hexagonal pattern of a collagen fibril (left). Reprinted, with permission, from Gautieri et al. [165]. Copyright (2011) by the American Chemical Society. The response of a collagen fibril to uniaxial loading is modulated by its degree of cross-linking (right). Cross-link density and spatial distribution are controlled by a parameter β which modulates the adhesion force between tropocollagen molecules. By increasing β, the simulated collagen fibril ruptures at larger strains thus revealing the nonlinear mechanical behavior of the protein backbone. Modified from Buehler et al. [163] Journal of the Mechanical Behavior of Biomedical Materials, Copyright (2007), with permission from Elsevier

segment (with the subscript "0" indicating unloaded values), whereas $K_s = EA$ and $K_b = EI$ represent the stretching and bending stiffness, respectively. The material properties of the network are thus dictated by the modulus of elasticity E of individual fibers and by their geometric properties, as indicated by the presence of the cross-sectional area A and moment of inertia I. It should be noted that Eq. 2 models the mechanical behavior of athermal networks such as collagen [171, 172]. If thermal fluctuations are thought to contribute to the overall network mechanics (e.g., in cytoskeletal actin), one can either model individual fibers as worm-like chains with bending stiffnesses assigned via Eq. 1, or add random Brownian forces acting on the network nodes [174, 175]. In addition, Eq. 2 can include additional terms, such as torsion of fibers or stretching/bending of cross-links. Using this modeling framework, Head et al. [167] defined a phase diagram of network mechanics (Fig. 8a). Starting from a solution of isolated fibers, one can increase their length, concentration, or both to induce a rigidity percolation transition, in correspondence of which the network as a whole exhibits macroscopic elastic properties. Such mechanical properties are governed by fiber bending and non-affine deformations. A further increase in fiber length or concentration finally leads to a region of affine deformations. The mechanisms governing affine network elasticity are entropic (thermal) for low concentrations, and energetic (mechanical) for high concentrations. The transition from non-affine and bending-dominated to affine and stretching-dominated network responses was confirmed by other studies [168, 169] and associated with the development of nonlinear strain stiffening in networks of linearly elastic beams. In fact, Onck et al. [170] showed that the toe region in the macroscopic stress-strain relationship for a fiber network is caused by non-affine fiber rotations at small strains, thus suggesting that the strain stiffening behavior lies in the network rather than in its individual constituents. In addition, by simulating a fiber network under simple shear, they showed that fibers align along the direction of principal strain (Fig. 8b). This was confirmed by separate modeling studies on collagen [171] and actin [174] and by experiments on fibrin [176], thus suggesting that pre-straining the network (a common step prior to experimental testing) leads to more affine elastic behaviors. Licup et al. [172] found that, when fibers align with the applied load and stretching becomes the dominant modality of energy storage, the network stiffness becomes proportional to the stress (Fig. 8c). Interestingly, this linear dependence between stress and stiffness, which are inherently coupled (the latter is calculated as the derivative of the former with respect to a conjugate metric of strain), is what led Y. C. Fung to postulate an exponential constitutive behavior [177]. Hence, the common assumption of exponential behavior for collagen fibers embedded in soft tissues is also supported by network modeling studies, provided that the tissue operates in the affine regime.

Finite element-based network models have been used primarily to unveil the mechanisms underlying the long-range force transmission properties exhibited by fibrous materials, such as collagen. Cells in collagen can generate stress and strain fields that propagate over distances equal to many times their diameter (cf., Fig. 2). Transmission of stresses far from the cell was found to be caused by the presence of relatively stiff fibers embedded within a nonfibrous material and not by material nonlinearities as previously suggested [178]. Abhilash et al. [179] carried out a

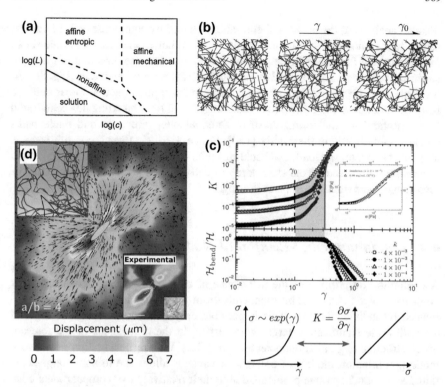

Fig. 8 Simulated networks of elastic beams reveal that nonlinear bulk behavior arises from network architecture as well as individual fiber properties. **a** Phase diagram of network mechanics as a function of fiber concentration (indicated as c) and fiber length (indicated as L). A rigidity percolation transition occurs while increasing c and/or L, and shows emergence of a bulk elastic behavior initially characterized by non-affine, bending-dominated fiber deformations which become progressively more affine, stretching-dominated (and therefore can be calculated locally from the global deformation of the matrix). Entropy characterizes network mechanics at low concentrations, while mechanical effects dominate at high concentrations. Adapted, with permission, from Head et al. [167] Physical Review E. Copyright (2003) by the American Physical Society. **b** A 2D network of randomly aligned fibers under simple shear (γ) shows that alignment of fibers along the direction of maximum principal strain, occurring at a critical strain (γ_0), underlies the transition from non-affine to affine deformations and causes non-linear macroscopic behavior, despite the network is constituted of linearly elastic beams. Adapted, with permission, from Onck et al. [170] Physical Review Letters. Copyright (2005) by the American Physical Society. **c** Bending is the primary modality of energy storage in the network below the critical strain, while stretching dominates at higher strains (top). In the affine regime the network stiffness relates linearly to the stress (insert), which suggests an underlying exponential behavior of the aligned fibers (bottom). Computational and experimental results modified, with permission, from Licup et al. [172] Proceedings of the National Academy of Sciences. **d** Displacement field generated by a cell modeled as an elliptical inclusion (having an aspect ratio a/b = 4, where a and be represent major and minor axes, respectively) contracting within a fiber network. Fibers align along the major axis of the cell and allow long-range force transmission. The heterogeneity of displacements qualitatively matches experimental observations (insert). Modified from Abhilash et al. [179] Biophysical Journal, Copyright (2014), with permission from The Biophysical Society. Experimental data adapted from Gjorevski and Nelson [206] Biophysical Journal, Copyright (2012), with permission from The Biophysical Society

detailed analysis of the network deformations caused by contractile cells of various aspect ratios, modeled as elliptical inclusions. Similar to the case of simple shear, they found that fibers within the network re-orient along the direction of principal strain and that fibers are engaged more as the cell aspect ratio increases (Fig. 8d). At the same time, fibers oriented perpendicularly to the applied traction forces undergo bending and, above the critical load, buckling. Local fiber buckling has been found to be essential for the formation of preferential fiber directions and linear paths connecting cells within the matrix [180]. Hence, intrinsic fiber properties such as alignment under tensile loads and buckling under compressive loads have been found to be essential to recover the slow decay of cell-induced displacements needed for long-range force transmission and sensing.

6.4 Continuum Fiber-Reinforced Models

A continuum description of the mechanics of collagenous tissues considers their volume-averaged behavior, independently from the forces and deformations experienced by individual fibers. This is possible only when considering problems at a length scale significantly larger with respect to the dimensions of the relevant micro-structure (e.g., fiber diameter and pore size). With respect to discrete network models, continuum models of collagen mechanics allow one to solve large-scale initial and boundary value problems due to their relatively low computational cost. A constitutive model is a mathematical description of material behavior under the conditions of interest and, for nonlinearly elastic materials, is generally represented by a scalar strain energy function W indicating the energy stored elastically by the material per unit reference volume [181]. A variety of constitutive models developed in the context of rubber elasticity and composite materials have been used in the literature to describe collagen gels, although most are not able to capture key features of fiber network mechanics. A new strain energy function has been proposed by the group of Vivek Shenoy to describe the fiber alignment and nonlinear stresses generated under the action of cell-generated forces [182]. The functional form is obtained combining an isotropic neo-Hookean model with nonlinear strain stiffening responses along the three principal directions. The neo-Hookean contribution is intended to capture the isotropic fibrous matrix, while the nonlinear contributions represent the subset of fibers that align along the principal directions of strain (cf. Figs. 6a, b and 8b). Such contributions are zero under compressive strains to model fiber buckling, while they increase non-linearly under the action of tensile strains to capture fiber alignment and subsequent strain stiffening. This model was shown to fit well the experimental data set collected by Roeder et al. [51] and, contrarily to simple linear or neo-Hookean materials, displays long-range force transmission properties in response to contraction of spherical or elliptical inclusions simulating contractile cells (Fig. 9a). Albeit motivated by microstructural observations, this model still represents a phenomenological description of collagen mechanics as suggested by the material parameters estimated from fitting experimental data. Wang et al. [182]

estimated a fiber's Young's modulus of 23 kPa for a collagen gel undergoing uniaxial tension. In comparison, Stylianopoulos and Barocas [183] found a fiber's Young's modulus of 79 MPa by fitting the same data set to their multi-scale model based on volume averaging (Fig. 9b). Their approach is based on the definition of representative volume elements (RVEs) centered at the Gauss points of a finite-element mesh that deforms according to the macroscopic strain field. Each RVE consists of a fiber network designed to capture microstructural features of the sample, and its mechanical behavior is governed by the stress-strain behavior of individual fibers as well as by the assumed modality of force transmission between fibers at cross-linked nodes. The force balance within the RVE is solved and volume averaged to calculate the macroscopic stress, which in turn determines the new displacement field [173]. Such approach is inherently multi-scale and leads to the calculation of fiber stiffness values comparable to those measured in isolated fibers experimentally (cf. Table 1) and three orders of magnitudes higher with respect to the estimates from phenomenological models (Fig. 9). The computational cost associated with modeling individual RVEs for the purpose of volume averaging is very high and thus potentially difficult to carry out calculations for complex cell geometries. An intermediate approach between phenomenological and fully microstructural models is offered by the constitutive model originally proposed by Storm et al. [184] based on the molecular theory of semiflexible biopolymers [185]. Such strain energy function can be written as follows,

$$W = \langle \nu\, L\, w(\lambda_f) \rangle_{P(\mathbf{r})}, \tag{3}$$

where ν represents the number of fibers per unit reference volume (and should not to be mistaken for the Poisson's ratio), while L is the contour length of individual unloaded fibers (here assumed to be the same for all fibers). $w(\lambda_f)$ is the strain energy (per unit length) of individual fibers with $\lambda_f = |\mathbf{Fr}|/L$ indicating the fiber stretch, where $\mathbf{F}$ is the macroscopic deformation gradient and $\mathbf{r}$ is the end-to-end fiber vector. In absence of deformations, it is $|\mathbf{r}| = L$. Finally, $P(\mathbf{r})$ represents the probability distribution of fiber vectors in the reference configuration, thus modeling the distribution of fiber orientations (for a random isotropic network $P(\mathbf{r})$ is a sphere of radius L). The operator $\langle \bullet \rangle_{P(\mathbf{r})}$ implements homogeneization of individual fiber contributions via spatial averaging over $P(\mathbf{r})$ to calculate the macroscopic elastic energy storage. Stress and stiffness tensors can be calculated from Eq. 3 via appropriate differentiations [186]. This model was originally developed to show that the strain stiffening behavior of biopolymer networks can be explained by accounting for entropic effects [184], and was later used to describe the uniaxial behavior of fibrin gels [187], the strain-enhanced stress relaxation of fibrin and collagen [150], and to compute cell traction forces in collagen gels [120]. The diversity of behaviors that can be modeled using Eq. 3 lies in the specification of the constitutive behavior at the fiber level via $w(\lambda_f)$. For example, a linear elastic fiber can be modeled simply as $w(\lambda_f) = EA(\lambda_f - 1)^2/2$ with E and A representing the aforementioned fiber's elastic modulus and cross-sectional area. More generally, all the structural parameters

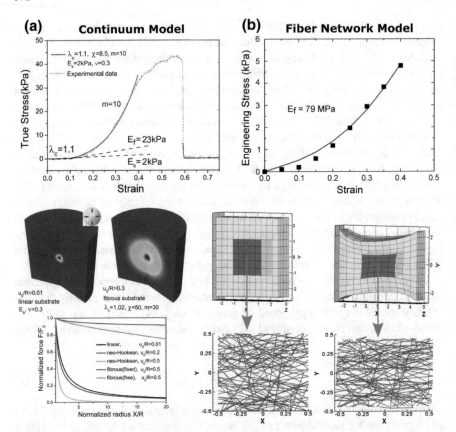

Fig. 9 Estimation of collagen fiber stiffness (E_f) using the macroscopic uniaxial tensile data reported by Roeder et al. [51]. **a** The continuum model proposed by Wang et al. [182] fits the data quite well (top) and leads to $E_f = 23$ kPa. Other parameters include the stiffness of the random matrix of non-aligned fibers (E_b), the critical stretch leading to fiber engagement (λ_c), and the strain hardening exponent (m). Finite element simulations of contracting spherical inclusions show that the constitutive model propagates deformations and forces much further with respect to linear elastic and isotropic neo-Hookean materials (bottom). Modified from Wang et al. [182] Biophysical Journal, Copyright (2014), with permission from The Biophysical Society. **b** The multi-scale model of Stylianopoulos and Barocas [183] leads to $E_f = 79$ MPa (top), which is much closer to values measured experimentally from individual fibers. Instead of relying on constitutive equations, such model simulates fiber networks as representative volume elements (RVEs) within a macroscopic finite element model. Modified from Stylianopoulos and Barocas [183] Computer Methods in Applied Mechanics and Engineering, Copyright (2007), with permission from Elsevier

in Eq. 3 relate to physical properties of the fiber network and thus can be measured directly via quantitative imaging (cf., Sect. 3.3). There is a need for continued validation of the applicability of continuum models because, depending on polymerization conditions, the pores of the collagen matrix can have dimensions comparable to individual cells, thus leading to a break-down of the continuum assumption. Moreover, it should be noted that all the models described in this section were developed using the hyperelastic framework and thus employ the affine assumption, which is known to be a limitation at low strains were the majority of fiber alignment takes place. If deemed necessary, a possibility to account for non-affine fiber motions in a continuum framework is given by the hypoelastic framework originally proposed by Clifford Truesdell [188] and recently used by Morin et al. [189, 190] in the context of arterial mechanics.

6.5 Biphasic Models and Poroelasticity

A common assumption in the biomechanics of soft tissues is that their volume is preserved during physiologically relevant deformations, that is they are incompressible [191, 192]. This assumption is usually justified by the elevated content of water bound to negatively charged proteoglycans, and the interstitial fluid is assumed to move with the solid matrix thus contributing to global tissue mechanics via a hydrostatic pressure term [24]. However, interstitial fluid pressure and flow are known to be increased in tumors due to a leaky vasculature [23] and have been shown to play a role in cancer cell migration [193]. Interstitial flow is therefore expected to impact tissue mechanics at the macroscopic level by causing microscopic frictional interactions between the solid and fluid phases, and by invalidating the incompressibility assumption. With regards to biopolymer networks, it has been shown that fibrous hydrogels are markedly compressible, with large lateral compaction measured upon uniaxial extension of fibrin and collagen gels [135, 187], which has led to the estimation of large Poisson's ratios [44, 120]. This behavior is due to compaction of fibers and contemporary expulsion of interstitial fluid, which can freely move through the hydrogel's porous structure. This type of mechanical behavior is typical of biphasic materials and can be described using Biot's theory of poroelasticity [194] or Truesdell's mixture theory [195]. Both approaches lead to consistent results if the solid and fluid constituents are modeled as individually incompressible and if the mixture is fully saturated. Constitutively, the solid is regarded as elastic while the fluid is inviscid, which means that it is only able to support pressure and not shear stresses. This is because the frictional interactions within the fluid are considered negligible with respect to those between solid and fluid, which represent the primary source of energy loss and therefore account for the time-dependent mechanical response. The biphasic properties of soft tissues should be assessed by inducing volumetric changes in the sample, usually via confined compression testing, and by subsequently tracking the temporal evolution in strain or stress in response to, respectively, creep or stress-relaxation. Biphasic modeling of confined compression experiments was introduced by Van Mow and colleagues to characterize articular cartilage mechanics [143]. In the simplifying case of linear material behavior of the solid, the main biomechanical parameters characterizing the biphasic behavior under confined compression are the aggregate modulus (a measure of uniaxial compressive stiffness of the solid) and the hydraulic permeability (a measure of convective fluid transport properties through the matrix pores). This theory has been applied to collagen hydrogels by Busby et al. [196] who found that increasing collagen concentration from 2 to 4 mg/mL increases the aggregate modulus while decreasing the hydraulic permeability, as a result of a reduced porosity. Despite these differences, the equilibrium stresses are remarkably similar across the various concentrations considered in this study. This effect is likely related to structural rearrangements caused by the visco-plastic deformations experienced by collagen fibers and thus requires further investigation. Extension of the basic theory allows one to account for strain-dependent permeabilities [197], nonlinear material behaviors [198, 199], and intrinsic viscoelasticity of the solid constituent [200]. More studies on the biphasic behavior of collagen gels are needed to gain a better understanding on the structure-function relationship of collagen networks both in terms of material and fluid transport properties, and how these properties are affected by remodeling from single cells and multicellular clusters.

7   Cellular Traction Forces in Collagen

Among the various modeling approaches outlined in the previous sections, continuum models have been employed to estimate cell-generated forces during adhesion and migration. The methodology known as traction force microscopy (TFM) was originally developed to measure the forces generated by single cells migrating on 2D substrates [201], and in recent years has been extended to study single cells and clusters embedded within 3D fibrous hydrogels, including collagen [120, 202, 203]. Direct measurements of matrix deformations due to cellular forces are used in combination with constitutive descriptors of matrix mechanics to calculate traction forces via inverse stress analyses. Both fiber models discussed in Sect. 6.4 have been used to reconstruct the traction forces generated by individual breast carcinoma cells migrating in 3D collagen (Fig. 10). Steinwachs et al. [120] parametrized Eq. 3 using shear rheometry data on acellular gels and validated their model via application of concentrated forces by means of magnetic tweezers. Although the validity of the continuum assumption might seem debatable, they showed that collagen gels behave as a continuum at the cellular length scale (on the order of tens of microns). In a separate study, Hall et al. [203] used the fibrous model developed by Wang et al. [182] to describe the strain fields developed along the cell's major axis due to actomyosin contractility. Overall, these studies found that cells embedded within low collagen concentrations generate higher fiber alignment and higher matrix displacements that propagate for longer distances (Fig. 10). Despite such differences, the cell contractility was found to remain constant and the migratory pattern characterized by cycling between motile and stationary phases [120]. Traction force microscopy in 3D

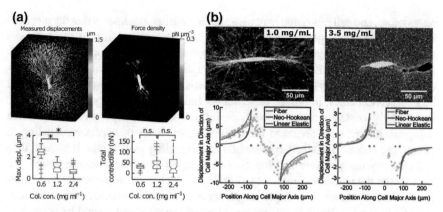

Fig. 10 Traction force microscopy (TFM) on single MDA-MB-231 breast carcinoma cells embedded in collagen hydrogels of various concentrations. **a** Experimentally-measured displacements and distribution of force density calculated using the constitutive model (Eq. 3) proposed by Storm et al. [184] (top). The maximum displacement induced on the matrix by contractile cells decreases for increasing concentrations, while the total contractility remains constant (bottom). Modified with permission from Springer Nature: Steinwachs et al. [120] Nature Methods, Copyright 2015. **b** Consistently, cells within low density matrices generate higher matrix displacements that propagate for longer distances due to fiber alignment (top) The spatial distribution of displacements is described well by the constitutive model developed by Wang et al. [182] (red), but not by linear elastic (blue) and neo-Hookean (black) models. Modified, with permission, from Hall et al. [203] Proceedings of the National Academy of Sciences

environments has the potential to elucidate the biophysical mechanisms underlying cell-matrix interactions and cell migration in cancer, but it requires the development of constitutive models that can capture the evolving matrix structure and mechanics due to remodeling. Such models need not account for fiber alignment and buckling alone, but also for visco-plastic effects [151, 152, 204] and biochemical remodeling, which includes both proteolytic degradation and de novo production of collagen. The lack of structure-based constitutive laws capable of capturing the complex mechanical properties of collagen networks and their evolution due to active remodeling from resident cells currently represents the major limitation of TFM as a tool to investigate cell-matrix mechanosensory interactions in 3D environments [205].

8 Closure

Cancer is a disease initiated by oncogenetic mutations, but its progression is critically influenced by many biomechanical factors, including biophysical cell-matrix interactions and ECM mechanics. 3D cultures of single cells or multicellular spheroids embedded in collagen hydrogels represent appealing models to investigate the complexities of tumor-matrix interactions in vitro. However, there is still much to be learned regarding the mechanisms of cell-matrix adhesion, mechanosensing, matrix remodeling, and cell migration in realistic environments. It is known that the mechanical properties of collagen networks influence cell behavior and, in turn, resident cells remodel the matrix physically and chemically, thus altering its properties. This bidirectional relationship is altered in cancer, and disrupted mechanobiological homeostasis leads ultimately to invasion and metastasis. Given the fundamental role played by matrix mechanics, improved experiments and models of collagen mechanics are increasingly needed to characterize its evolution during tumor progression. Towards this goal, this chapter reviewed the biomechanical basis of matrix remodeling in cancer and summarized the experimental and modeling tools available to modulate and quantify the structure and function of collagen at different length scales. We highlighted the inherent complexity of collagen mechanics, in particular its nonlinear material behavior at multiple length-scales. Additional characteristics, such as viscoelastic and biphasic behaviors, are often disregarded and should be accounted for both experimentally and theoretically. Incorporation of such complexities into a consistent theory for tissue remodeling is needed to develop and test new hypotheses on the mechanobiology of cancer.

Acknowledgements The authors gratefully acknowledge the funding agencies that supported this work: National Institute of Health grants U01 CA202123 (MHZ) and R01 HL098028 (YZ), National Science Foundation grants CMMI 1463390 and CAREER 0954825 (YZ), and Department of Defense grant W81XWH-15-1-0070 (DR).
Conflicts of Interest None.

References

1. Alberts, B., Johnson, A.D., Lewis, J., Morgan, D., Raff, M., Roberts, K., Walter, P.: Molecular Biology of the Cell. W. W. Norton & Company, New York, NY (2014)
2. Cox, T.R., Erler, J.T.: Remodeling and homeostasis of the extracellular matrix: implications for fibrotic diseases and cancer. Dis. Models Mech. **4**, 165–178 (2011). https://doi.org/10.1242/dmm.004077
3. Malik, R., Lelkes, P.I., Cukierman, E.: Biomechanical and biochemical remodeling of stromal extracellular matrix in cancer. Trends Biotechnol. **33**, 230–236 (2015). https://doi.org/10.1016/j.tibtech.2015.01.004
4. Lee, J.Y., Chaudhuri, O.: Regulation of breast cancer progression by extracellular matrix mechanics: insights from 3D culture models. ACS Biomater. Sci. Eng. **4**, 302–313 (2018). https://doi.org/10.1021/acsbiomaterials.7b00071
5. Friedl, P., Wolf, K.: Tumour-cell invasion and migration: diversity and escape mechanisms. Nat. Rev. Cancer **3**, 362–374 (2003). https://doi.org/10.1038/nrc1075
6. Haeger, A., Wolf, K., Zegers, M.M., Friedl, P.: Collective cell migration: guidance principles and hierarchies. Trends Cell Biol. **25**, 556–566 (2015). https://doi.org/10.1016/j.tcb.2015.06.003
7. Wolf, K., te Lindert, M., Krause, M., Alexander, S., te Riet, J., Willis, A.L., Hoffman, R.M., Figdor, C.G., Weiss, S.J., Friedl, P.: Physical limits of cell migration: control by ECM space and nuclear deformation and tuning by proteolysis and traction force. J. Cell Biol. **201**, 1069–1084 (2013). https://doi.org/10.1083/jcb.201210152
8. Pickup, M.W., Mouw, J.K., Weaver, V.M.: The extracellular matrix modulates the hallmarks of cancer. EMBO Rep. **15**, 1243–1253 (2014). https://doi.org/10.15252/embr.201439246

9. Lo, C.-M., Wang, H.-B., Dembo, M., Wang, Y.: Cell movement is guided by the rigidity of the substrate. Biophys. J. **79**, 144–152 (2000). https://doi.org/10.1016/S0006-3495(00)76279-5

10. Engler, A.J., Sen, S., Sweeney, H.L., Discher, D.E.: Matrix elasticity directs stem cell lineage specification. Cell **126**, 677–689 (2006). https://doi.org/10.1016/j.cell.2006.06.044

11. Paszek, M.J., Zahir, N., Johnson, K.R., Lakins, J.N., Rozenberg, G.I., Gefen, A., Reinhart-King, C.A., Margulies, S.S., Dembo, M., Boettiger, D., Hammer, D.A., Weaver, V.M.: Tensional homeostasis and the malignant phenotype. Cancer Cell **8**, 241–254 (2005). https://doi.org/10.1016/j.ccr.2005.08.010

12. Ehret, A.E., Bircher, K., Stracuzzi, A., Marina, V., Zündel, M., Mazza, E.: Inverse poroelasticity as a fundamental mechanism in biomechanics and mechanobiology. Nat. Commun. **8**, 1002 (2017). https://doi.org/10.1038/s41467-017-00801-3

13. Sander, E.A., Barocas, V.H.: Biomimetic collagen tissues: collagenous tissue engineering and other applications. In: Fratzl, P. (ed.) Collagen: Structure and Mechanics, pp. 475–504. Springer US, Boston, MA (2008)

14. Martin, L.J., Boyd, N.F.: Mammographic density Potential mechanisms of breast cancer risk associated with mammographic density: hypotheses based on epidemiological evidence. Breast Cancer Res. **10**, 201 (2008). https://doi.org/10.1186/bcr1831

15. Boyd, N.F., Dite, G.S., Stone, J., Gunasekara, A., English, D.R., McCredie, M.R.E., Giles, G.G., Tritchler, D., Chiarelli, A., Yaffe, M.J., Hopper, J.L.: Heritability of mammographic density, a risk factor for breast cancer. N. Engl. J. Med. **347**, 886–894 (2002). https://doi.org/10.1056/NEJMoa013390

16. Provenzano, P.P., Eliceiri, K.W., Campbell, J.M., Inman, D.R., White, J.G., Keely, P.J.: Collagen reorganization at the tumor-stromal interface facilitates local invasion. BMC Med. **4**, 38 (2006). https://doi.org/10.1186/1741-7015-4-38

17. Conklin, M.W., Eickhoff, J.C., Riching, K.M., Pehlke, C.A., Eliceiri, K.W., Provenzano, P.P., Friedl, A., Keely, P.J.: Aligned collagen is a prognostic signature for survival in human breast carcinoma. Am. J. Pathol. **178**, 1221–1232 (2011). https://doi.org/10.1016/j.ajpath.2010.11.076

18. Provenzano, P.P., Eliceiri, K.W., Keely, P.J.: Multiphoton microscopy and fluorescence lifetime imaging microscopy (FLIM) to monitor metastasis and the tumor microenvironment. Clin. Exp. Metas. **26**, 357–370 (2009). https://doi.org/10.1007/s10585-008-9204-0

19. Conklin, M.W., Keely, P.J.: Why the stroma matters in breast cancer: insights into breast cancer patient outcomes through the examination of stromal biomarkers. Cell Adhes. Migr. **6**, 249–260 (2012). https://doi.org/10.4161/cam.20567

20. Provenzano, P.P., Inman, D.R., Eliceiri, K.W., Trier, S.M., Keely, P.J.: Contact guidance mediated three-dimensional cell migration is regulated by Rho/ROCK-dependent matrix reorganization. Biophys. J. **95**, 5374–5384 (2008). https://doi.org/10.1529/biophysj.108.133116

21. Acerbi, I., Cassereau, L., Dean, I., Shi, Q., Au, A., Park, C., Chen, Y.Y., Liphardt, J., Hwang, E.S., Weaver, V.M.: Human breast cancer invasion and aggression correlates with ECM stiffening and immune cell infiltration. Integr. Biol. **7**, 1120–1134 (2015). https://doi.org/10.1039/C5IB00040H

22. Levental, K.R., Yu, H., Kass, L., Lakins, J.N., Egeblad, M., Erler, J.T., Fong, S.F.T., Csiszar, K., Giaccia, A., Weninger, W., Yamauchi, M., Gasser, D.L., Weaver, V.M.: Matrix crosslinking forces tumor progression by enhancing integrin signaling. Cell **139**, 891–906 (2009). https://doi.org/10.1016/j.cell.2009.10.027

23. Mitchell, M.J., Jain, R.K., Langer, R.: Engineering and physical sciences in oncology: challenges and opportunities. Nat. Rev. Cancer **17**, 659–675 (2017). https://doi.org/10.1038/nrc.2017.83

24. Humphrey, J.D., Rajagopal, K.R.: A constrained mixture model for growth and remodeling of soft tissues. Math. Models Methods Appl. Sci. **12**, 407–430 (2002). https://doi.org/10.1142/S0218202502001714

25. Cox, T.R., Erler, J.T.: Remodeling and homeostasis of the extracellular matrix: implications for fibrotic diseases and cancer. 14

26. Lee, G.Y., Kenny, P.A., Lee, E.H., Bissell, M.J.: Three-dimensional culture models of normal and malignant breast epithelial cells. Nat. Methods **4**, 359–365 (2007). https://doi.org/10.1038/nmeth1015
27. Yamada, K.M., Cukierman, E.: Modeling tissue morphogenesis and cancer in 3D. Cell **130**, 601–610 (2007). https://doi.org/10.1016/j.cell.2007.08.006
28. Fischbach, C., Chen, R., Matsumoto, T., Schmelzle, T., Brugge, J.S., Polverini, P.J., Mooney, D.J.: Engineering tumors with 3D scaffolds. Nat. Methods **4**, 855–860 (2007). https://doi.org/10.1038/nmeth1085
29. Caliari, S.R., Burdick, J.A.: A practical guide to hydrogels for cell culture. Nat. Methods **13**, 405–414 (2016). https://doi.org/10.1038/nmeth.3839
30. Sutherland, R.: Cell and environment interactions in tumor microregions: the multicell spheroid model. Science **240**, 177–184 (1988). https://doi.org/10.1126/science.2451290
31. Soker, S.: Tumor organoids. Springer, Berlin (2017)
32. Ivascu, A., Kubbies, M.: Rapid generation of single-tumor spheroids for high-throughput cell function and toxicity analysis. J. Biomol. Screen. **11**, 922–932 (2006). https://doi.org/10.1177/1087057106292763
33. Charoen, K.M., Fallica, B., Colson, Y.L., Zaman, M.H., Grinstaff, M.W.: Embedded multicellular spheroids as a biomimetic 3D cancer model for evaluating drug and drug-device combinations. Biomaterials **35**, 2264–2271 (2014). https://doi.org/10.1016/j.biomaterials.2013.11.038
34. Reynolds, D.S., Tevis, K.M., Blessing, W.A., Colson, Y.L., Zaman, M.H., Grinstaff, M.W.: Breast cancer spheroids reveal a differential cancer stem cell response to chemotherapeutic treatment. Sci. Rep. **7**, 10382 (2017). https://doi.org/10.1038/s41598-017-10863-4
35. Veelken, C., Bakker, G.-J., Drell, D., Friedl, P.: Single cell-based automated quantification of therapy responses of invasive cancer spheroids in organotypic 3D culture. Methods **128**, 139–149 (2017). https://doi.org/10.1016/j.ymeth.2017.07.015
36. Kopanska, K.S., Alcheikh, Y., Staneva, R., Vignjevic, D., Betz, T.: Tensile forces originating from cancer spheroids facilitate tumor invasion. PLoS ONE **11**, e0156442 (2016)
37. Soule, H.D., Maloney, T.M., Wolman, S.R., Peterson, W.D., Brenz, R., McGrath, C.M., Russo, J., Pauley, R.J., Jones, R.F., Brooks, S.C.: Isolation and characterization of a spontaneously immortalized human breast epithelial cell line, MCF-10. Cancer Res. **50**, 6075–6086 (1990)
38. Cailleau, R., Mackay, B., Young, R.K., Reeves, W.J.: Tissue culture studies on pleural effusions from breast carcinoma patients. Cancer Res. **34**, 10 (1974)
39. The Physical Sciences—Oncology Centers Network: A physical sciences network characterization of non-tumorigenic and metastatic cells. Sci. Rep. **3** (2013). https://doi.org/10.1038/srep01449
40. Cheng, G., Tse, J., Jain, R.K., Munn, L.L.: Micro-environmental mechanical stress controls tumor spheroid size and morphology by suppressing proliferation and inducing apoptosis in cancer cells. PLoS ONE **4**, e4632 (2009). https://doi.org/10.1371/journal.pone.0004632
41. Gjorevski, N., Piotrowski, A.S., Varner, V.D., Nelson, C.M.: Dynamic tensile forces drive collective cell migration through three-dimensional extracellular matrices. Sci. Rep. **5**, 11458 (2015). https://doi.org/10.1038/srep11458
42. Grinnell, F., Petroll, W.M.: Cell motility and mechanics in three-dimensional collagen matrices. Annu. Rev. Cell Dev. Biol. **26**, 335–361 (2010). https://doi.org/10.1146/annurev.cellbio.042308.113318
43. Petroll, W.M., Cavanagh, H.D., Jester, J.V.: Dynamic three-dimensional visualization of collagen matrix remodeling and cytoskeletal organization in living corneal fibroblasts. Scanning **26**, 1–10 (2004)
44. Vader, D., Kabla, A., Weitz, D., Mahadevan, L.: Strain-induced alignment in collagen gels. PLoS ONE **4**, e5902 (2009). https://doi.org/10.1371/journal.pone.0005902
45. Klebe, R.J., Caldwell, H., Milam, S.: Cells transmit spatial information by orienting collagen fibers. Matrix **9**, 451–458 (1990). https://doi.org/10.1016/S0934-8832(11)80014-4
46. Silver, F.H., Freeman, J.W., Seehra, G.P.: Collagen self-assembly and the development of tendon mechanical properties. J. Biomech. **36**, 1529–1553 (2003). https://doi.org/10.1016/S0021-9290(03)00135-0

47. Veis, A., George, A.: Fundamentals of interstitial collagen self-assembly. In: Yurchenco, P.D., Birk, D.E., Mecham, R.P. (eds.) Extracellular Matrix Assembly and Structure, pp. 15–45. Academic Press, San Diego (1994)

48. Holmes, D.F., Capaldi, M.J., Chapman, J.A.: Reconstitution of collagen fibrils in vitro; the assembly process depends on the initiating procedure. Int. J. Biol. Macromol. **8**, 161–166 (1986). https://doi.org/10.1016/0141-8130(86)90020-6

49. Wood, G., Keech, M.K.: The formation of fibrils from collagen solutions 1. The effect of experimental conditions: kinetic and electron-microscope studies. Biochem. J. **75**, 588 (1960)

50. Christiansen, D.L., Huang, E.K., Silver, F.H.: Assembly of type I collagen: fusion of fibril subunits and the influence of fibril diameter on mechanical properties. Matrix Biol. **19**, 409–420 (2000). https://doi.org/10.1016/S0945-053X(00)00089-5

51. Roeder, B.A., Kokini, K., Sturgis, J.E., Robinson, J.P., Voytik-Harbin, S.L.: Tensile mechanical properties of three-dimensional type I collagen extracellular matrices with varied microstructure. J. Biomech. Eng. **124**, 214–222 (2002). https://doi.org/10.1115/1.1449904

52. Raub, C.B., Suresh, V., Krasieva, T., Lyubovitsky, J., Mih, J.D., Putnam, A.J., Tromberg, B.J., George, S.C.: Noninvasive assessment of collagen gel microstructure and mechanics using multiphoton microscopy. Biophys. J. **92**, 2212–2222 (2007). https://doi.org/10.1529/biophysj.106.097998

53. Raub, C.B., Unruh, J., Suresh, V., Krasieva, T., Lindmo, T., Gratton, E., Tromberg, B.J., George, S.C.: Image correlation spectroscopy of multiphoton images correlates with collagen mechanical properties. Biophys. J. **94**, 2361–2373 (2008). https://doi.org/10.1529/biophysj. 107.120006

54. Wood, G.C.: The formation of fibrils from collagen solutions. 2. A mechanism for collagen-fibril formation. Biochem. J. **75**, 598–605 (1960)

55. Zhu, J., Kaufman, L.J.: Collagen I self-assembly: revealing the developing structures that generate turbidity. Biophys. J. **106**, 1822–1831 (2014). https://doi.org/10.1016/j.bpj.2014.03. 011

56. Forgacs, G., Newman, S.A., Hinner, B., Maier, C.W., Sackmann, E.: Assembly of collagen matrices as a phase transition revealed by structural and rheologic studies. Biophys. J. **84**, 1272–1280 (2003)

57. Makris, E.A., Hu, J.C.: Induced collagen cross-links enhance cartilage integration. PLoS ONE **8**, e60719 (2013)

58. Makris, E.A., Responte, D.J., Paschos, N.K., Hu, J.C., Athanasiou, K.A.: Developing functional musculoskeletal tissues through hypoxia and lysyl oxidase-induced collagen cross linking. Proc. Natl. Acad. Sci. **111**, E4832–E4841 (2014). https://doi.org/10.1073/pnas. 1414271111

59. Greenberg, C.S., Birckbichler, P.J., Rice, R.H.: Transglutaminases: multifunctional cross-linking enzymes that stabilize tissues. FASEB J. **5**, 3071–3077 (1991)

60. Orban, J.M., Wilson, L.B., Kofroth, J.A., El-Kurdi, M.S., Maul, T.M., Vorp, D.A.: Crosslinking of collagen gels by transglutaminase. J. Biomed. Mater. Res. Part A **68**, 756–762 (2004)

61. Olde Damink, L.H.H., Dijkstra, P.J., Van Luyn, M.J.A., Van Wachem, P.B., Nieuwenhuis, P., Feijen, J.: Glutaraldehyde as a crosslinking agent for collagen-based biomaterials. J. Mater. Sci. Mater. Med. **6**, 460–472 (1995). https://doi.org/10.1007/bf00123371

62. Sheu, M.-T., Huang, J.-C., Yeh, G.-C., Ho, H.-O.: Characterization of collagen gel solutions and collagen matrices for cell culture. Biomaterials **22**, 1713–1719 (2001). https://doi.org/10. 1016/S0142-9612(00)00315-X

63. Tian, Z., Liu, W., Li, G.: The microstructure and stability of collagen hydrogel cross-linked by glutaraldehyde. Polym. Degrad. Stab. **130**, 264–270 (2016). https://doi.org/10.1016/j. polymdegradstab.2016.06.015

64. Girton, T.S., Oegema, T.R., Tranquillo, R.T.: Exploiting glycation to stiffen and strengthen tissue equivalents for tissue engineering. J. Biomed. Mater. Res. **46**, 87–92 (1999). https:// doi.org/10.1002/(SICI)1097-4636(199907)46:1%3c87::AID-JBM10%3e3.0.CO;2-K

65. Roy, R., Boskey, A., Bonassar, L.J.: Processing of type I collagen gels using nonenzymatic glycation. J. Biomed. Mater. Res. Part A **9999A**, NA–NA (2009). https://doi.org/10.1002/ jbm.a.32231

66. Mason, B.N., Starchenko, A., Williams, R.M., Bonassar, L.J., Reinhart-King, C.A.: Tuning three-dimensional collagen matrix stiffness independently of collagen concentration modulates endothelial cell behavior. Acta Biomater. **9**, 4635–4644 (2013). https://doi.org/10.1016/j.actbio.2012.08.007

67. Bordeleau, F., Mason, B.N., Lollis, E.M., Mazzola, M., Zanotelli, M.R., Somasegar, S., Califano, J.P., Montague, C., LaValley, D.J., Huynh, J., Mencia-Trinchant, N., Negrón Abril, Y.L., Hassane, D.C., Bonassar, L.J., Butcher, J.T., Weiss, R.S., Reinhart-King, C.A.: Matrix stiffening promotes a tumor vasculature phenotype. Proc. Natl. Acad. Sci. **114**, 492–497 (2017). https://doi.org/10.1073/pnas.1613855114

68. Fujimori, E.: Ultraviolet light-induced change in collagen macromolecules. Biopolymers **3**, 115–119 (1965)

69. Hapach, L.A., VanderBurgh, J.A., Miller, J.P., Reinhart-King, C.A.: Manipulation of in vitro collagen matrix architecture for scaffolds of improved physiological relevance. Phys. Biol. **12**, 061002 (2015). https://doi.org/10.1088/1478-3975/12/6/061002

70. Brightman, A.O., Rajwa, B.P., Sturgis, J.E., McCallister, M.E., Robinson, J.P., Voytik-Harbin, S.L.: Time-lapse confocal reflection microscopy of collagen fibrillogenesis and extracellular matrix assembly in vitro. Biopolymers **54**, 222–234 (2000). https://doi.org/10.1002/1097-0282(200009)54:3%3c222:AID-BIP80%3e3.0.CO;2-K

71. Friedl, P., Maaser, K., Klein, C.E., Niggemann, B., Krohne, G., Zänker, K.S.: Migration of highly aggressive MV3 melanoma cells in 3-dimensional collagen lattices results in local matrix reorganization and shedding of α2 and β1 integrins and CD44. Cancer Res. **57**, 2061–2070 (1997)

72. Kaufman, L.J., Brangwynne, C.P., Kasza, K.E., Filippidi, E., Gordon, V.D., Deisboeck, T.S., Weitz, D.A.: Glioma expansion in collagen I matrices: analyzing collagen concentration-dependent growth and motility patterns. Biophys. J. **89**, 635–650 (2005). https://doi.org/10.1529/biophysj.105.061994

73. Harjanto, D., Maffei, J.S., Zaman, M.H.: Quantitative analysis of the effect of cancer invasiveness and collagen concentration on 3D matrix remodeling. PLoS ONE **6**, e24891 (2011). https://doi.org/10.1371/journal.pone.0024891

74. Carey, S.P., Kraning-Rush, C.M., Williams, R.M., Reinhart-King, C.A.: Biophysical control of invasive tumor cell behavior by extracellular matrix microarchitecture. Biomaterials **33**, 4157–4165 (2012). https://doi.org/10.1016/j.biomaterials.2012.02.029

75. Zoumi, A., Yeh, A., Tromberg, B.J.: Imaging cells and extracellular matrix in vivo by using second-harmonic generation and two-photon excited fluorescence. Proc. Natl. Acad. Sci. **99**, 11014–11019 (2002). https://doi.org/10.1073/pnas.172368799

76. Georgakoudi, I., Quinn, K.P.: Optical imaging using endogenous contrast to assess metabolic state. Annu. Rev. Biomed. Eng. **14**, 351–367 (2012). https://doi.org/10.1146/annurev-bioeng-071811-150108

77. Brown, E., McKee, T., diTomaso, E., Pluen, A., Seed, B., Boucher, Y., Jain, R.K.: Dynamic imaging of collagen and its modulation in tumors in vivo using second-harmonic generation. Nat. Med. **9**, 796 (2003)

78. Nadiarnykh, O., LaComb, R.B., Brewer, M.A., Campagnola, P.J.: Alterations of the extracellular matrix in ovarian cancer studied by Second Harmonic Generation imaging microscopy. BMC Cancer **10**, 94 (2010)

79. Jawerth, L.M., Münster, S., Vader, D.A., Fabry, B., Weitz, D.A.: A blind spot in confocal reflection microscopy: the dependence of fiber brightness on fiber orientation in imaging biopolymer networks. Biophys. J. **98**, L1–L3 (2010). https://doi.org/10.1016/j.bpj.2009.09.065

80. Raub, C.B., Tromberg, B.J., George, S.C.: Second-Harmonic Generation imaging of self-assembled collagen gels. In: Pavone, F.S., Campagnola, P.J. (eds.) Second Harmonic Generation imaging. pp. 11, 1–27. CRC Press, Boca Raton, FL (2013)

81. Rezakhaniha, R., Agianniotis, A., Schrauwen, J.T.C., Griffa, A., Sage, D., Bouten, C.V.C., van de Vosse, F.N., Unser, M., Stergiopulos, N.: Experimental investigation of collagen waviness and orientation in the arterial adventitia using confocal laser scanning microscopy. Biomech. Model. Mechanobiol. **11**, 461–473 (2012). https://doi.org/10.1007/s10237-011-0325-z

82. Sun, M., Bloom, A.B., Zaman, M.H.: Rapid quantification of 3D collagen fiber alignment and fiber intersection correlations with high sensitivity. PLoS ONE **10**, e0131814 (2015). https://doi.org/10.1371/journal.pone.0131814

83. Liu, Z., Quinn, K.P., Speroni, L., Arendt, L., Kuperwasser, C., Sonnenschein, C., Soto, A.M., Georgakoudi, I.: Rapid three-dimensional quantification of voxel-wise collagen fiber orientation. Biomed. Opt. Express. **6**, 2294 (2015). https://doi.org/10.1364/BOE.6.002294

84. Liu, Z., Pouli, D., Sood, D., Sundarakrishnan, A., Hui Mingalone, C.K., Arendt, L.M., Alonzo, C., Quinn, K.P., Kuperwasser, C., Zeng, L., Schnelldorfer, T., Kaplan, D.L., Georgakoudi, I.: Automated quantification of three-dimensional organization of fiber-like structures in biological tissues. Biomaterials **116**, 34–47 (2017). https://doi.org/10.1016/j.biomaterials.2016.11.041

85. Hotaling, N.A., Bharti, K., Kriel, H., Simon, C.G.: DiameterJ: a validated open source nanofiber diameter measurement tool. Biomaterials **61**, 327–338 (2015). https://doi.org/10.1016/j.biomaterials.2015.05.015

86. Stein, A.M., Vader, D.A., Jawerth, L.M., Weitz, D.A., Sander, L.M.: An algorithm for extracting the network geometry of three-dimensional collagen gels. J. Microsc. **232**, 463–475 (2008). https://doi.org/10.1111/j.1365-2818.2008.02141.x

87. Bredfeldt, J.S., Liu, Y., Pehlke, C.A., Conklin, M.W., Szulczewski, J.M., Inman, D.R., Keely, P.J., Nowak, R.D., Mackie, T.R., Eliceiri, K.W.: Computational segmentation of collagen fibers from second-harmonic generation images of breast cancer. J. Biomed. Opt. **19**, 016007 (2014). https://doi.org/10.1117/1.JBO.19.1.016007

88. Liu, Y., Keikhosravi, A., Mehta, G.S., Drifka, C.R., Eliceiri, K.W.: Methods for quantifying fibrillar collagen alignment. In: Rittié, L. (ed.) Fibrosis, pp. 429–451. Springer, New York (2017)

89. Sun, Y.-L., Luo, Z.-P., Fertala, A., An, K.-N.: Direct quantification of the flexibility of type I collagen monomer. Biochem. Biophys. Res. Commun. **295**, 382–386 (2002). https://doi.org/10.1016/S0006-291X(02)00685-X

90. Sun, Y.-L., Luo, Z.-P., Fertala, A., An, K.-N.: Stretching type II collagen with optical tweezers. J. Biomech. **37**, 1665–1669 (2004). https://doi.org/10.1016/j.jbiomech.2004.02.028

91. An, K.-N., Sun, Y.-L., Luo, Z.-P.: Flexibility of type I collagen and mechanical property of connective tissue. Biorheology **41**, 239–246 (2004)

92. Bozec, L., van der Heijden, G., Horton, M.: Collagen fibrils: nanoscale ropes. Biophys. J. **92**, 70–75 (2007). https://doi.org/10.1529/biophysj.106.085704

93. Graham, J.S., Vomund, A.N., Phillips, C.L., Grandbois, M.: Structural changes in human type I collagen fibrils investigated by force spectroscopy. Exp. Cell Res. **299**, 335–342 (2004). https://doi.org/10.1016/j.yexcr.2004.05.022

94. van der Rijt, J.A.J., van der Werf, K.O., Bennink, M.L., Dijkstra, P.J., Feijen, J.: Micromechanical testing of individual collagen fibrils. Macromol. Biosci. **6**, 697–702 (2006). https://doi.org/10.1002/mabi.200600063

95. Yang, L., van der Werf, K.O., Dijkstra, P.J., Feijen, J., Bennink, M.L.: Micromechanical analysis of native and cross-linked collagen type I fibrils supports the existence of microfibrils. J. Mech. Behav. Biomed. Mater. **6**, 148–158 (2012). https://doi.org/10.1016/j.jmbbm.2011.11.008

96. Yang, L., van der Werf, K.O., Koopman, B.F.J.M., Subramaniam, V., Bennink, M.L., Dijkstra, P.J., Feijen, J.: Micromechanical bending of single collagen fibrils using atomic force microscopy. J. Biomed. Mater. Res. Part A **82A**, 160–168 (2007). https://doi.org/10.1002/jbm.a.31127

97. Yang, L., van der Werf, K.O., Fitié, C.F.C., Bennink, M.L., Dijkstra, P.J., Feijen, J.: Mechanical properties of native and cross-linked type I collagen fibrils. Biophys. J. **94**, 2204–2211 (2008). https://doi.org/10.1529/biophysj.107.111013

98. Andriotis, O.G., Desissaire, S., Thurner, P.J.: Collagen fibrils: nature's highly tunable nonlinear springs. ACS Nano **12**, 3671–3680 (2018). https://doi.org/10.1021/acsnano.8b00837

99. Eppell, S., Smith, B., Kahn, H., Ballarini, R.: Nano measurements with micro-devices: mechanical properties of hydrated collagen fibrils. J. R. Soc. Interface **3**, 117–121 (2006). https://doi.org/10.1098/rsif.2005.0100

100. Shen, Z.L., Dodge, M.R., Kahn, H., Ballarini, R., Eppell, S.J.: Stress-strain experiments on individual collagen fibrils. Biophys. J. **95**, 3956–3963 (2008). https://doi.org/10.1529/biophysj.107.124602

101. Shen, Z.L., Dodge, M.R., Kahn, H., Ballarini, R., Eppell, S.J.: In vitro fracture testing of submicron diameter collagen fibril specimens. Biophys. J. **99**, 1986–1995 (2010). https://doi.org/10.1016/j.bpj.2010.07.021

102. Liu, Y., Ballarini, R., Eppell, S.J.: Tension tests on mammalian collagen fibrils. Interface Focus. **6**, 20150080 (2016). https://doi.org/10.1098/rsfs.2015.0080

103. Liu, J., Das, D., Yang, F., Schwartz, A.G., Genin, G.M., Thomopoulos, S., Chasiotis, I.: Energy dissipation in mammalian collagen fibrils: cyclic strain-induced damping, toughening, and strengthening. Acta Biomater. **80**, 217–227 (2018). https://doi.org/10.1016/j.actbio.2018.09.027

104. Shen, Z.L., Kahn, H., Ballarini, R., Eppell, S.J.: Viscoelastic properties of isolated collagen fibrils. Biophys. J. **100**, 3008–3015 (2011). https://doi.org/10.1016/j.bpj.2011.04.052

105. Radmacher, M.: Studying the mechanics of cellular processes by atomic force microscopy. In: Methods in Cell Biology, pp. 347–372. Academic Press (2007)

106. van Helvert, S., Friedl, P.: Strain stiffening of fibrillar collagen during individual and collective cell migration identified by AFM nanoindentation. ACS Appl. Mater. Interfaces **8**, 21946–21955 (2016). https://doi.org/10.1021/acsami.6b01755

107. Xie, J., Bao, M., Bruekers, S.M.C., Huck, W.T.S.: Collagen gels with different fibrillar microarchitectures elicit different cellular responses. ACS Appl. Mater. Interfaces **9**, 19630–19637 (2017). https://doi.org/10.1021/acsami.7b03883

108. Squires, T.M., Mason, T.G.: Fluid mechanics of microrheology. Annu. Rev. Fluid Mech. **42**, 413–438 (2010). https://doi.org/10.1146/annurev-fluid-121108-145608

109. Wilson, L.G., Poon, W.C.K.: Small-world rheology: an introduction to probe-based active microrheology. Phys. Chem. Chem. Phys. **13**, 10617 (2011). https://doi.org/10.1039/c0cp01564d

110. Velegol, D., Lanni, F.: Cell traction forces on soft biomaterials. I. Microrheology of type I collagen gels. Biophys. J. **81**, 1786–1792 (2001). https://doi.org/10.1016/s0006-3495(01)75829-8

111. Parekh, A., Velegol, D.: Collagen gel anisotropy measured by 2-D laser trap microrheometry. Ann. Biomed. Eng. **35**, 1231–1246 (2007). https://doi.org/10.1007/s10439-007-9273-2

112. Shayegan, M., Forde, N.R.: Microrheological characterization of collagen systems: from molecular solutions to fibrillar gels. PLoS ONE **8**, e70590 (2013). https://doi.org/10.1371/journal.pone.0070590

113. Jones, C.A.R., Cibula, M., Feng, J., Krnacik, E.A., McIntyre, D.H., Levine, H., Sun, B.: Micromechanics of cellularized biopolymer networks. PNAS **112**, E5117–E5122 (2015). https://doi.org/10.1073/pnas.1509663112

114. Staunton, J.R., Vieira, W., Fung, K.L., Lake, R., Devine, A., Tanner, K.: Mechanical properties of the tumor stromal microenvironment probed in vitro and ex vivo by in situ-calibrated optical trap-based active microrheology. Cell. Mol. Bioeng. **9**, 398–417 (2016). https://doi.org/10.1007/s12195-016-0460-9

115. Han, Y.L., Ronceray, P., Xu, G., Malandrino, A., Kamm, R.D., Lenz, M., Broedersz, C.P., Guo, M.: Cell contraction induces long-ranged stress stiffening in the extracellular matrix. Proc. Natl. Acad. Sci. **115**, 4075–4080 (2018). https://doi.org/10.1073/pnas.1722619115

116. Lele, T.P., Sero, J.E., Matthews, B.D., Kumar, S., Xia, S., Montoya-Zavala, M., Polte, T., Overby, D., Wang, N., Ingber, D.E.: Tools to study cell mechanics and mechanotransduction. In: Methods in Cell Biology, pp. 441–472. Academic Press (2007)

117. Leung, L.Y., Tian, D., Brangwynne, C.P., Weitz, D.A., Tschumperlin, D.J.: A new microrheometric approach reveals individual and cooperative roles for TGF-β1 and IL-1β in fibroblast-mediated stiffening of collagen gels. FASEB J. **21**, 2064–2073 (2007). https://doi.org/10.1096/fj.06-7510com

118. Fabry, B., Maksym, G.N., Shore, S.A., Moore, P.E., Panettieri Jr., R.A., Butler, J.P., Fredberg, J.J.: Time course and heterogeneity of contractile responses in cultured human airway smooth

muscle cells. J. Appl. Physiol. **91**, 986–994 (2001). https://doi.org/10.1152/jappl.2001.91.2.986

119. Li, H., Xu, B., Zhou, E.H., Sunyer, R., Zhang, Y.: Multiscale measurements of the mechanical properties of collagen matrix. ACS Biomater. Sci. Eng. **3**, 2815–2824 (2017). https://doi.org/10.1021/acsbiomaterials.6b00634

120. Steinwachs, J., Metzner, C., Skodzek, K., Lang, N., Thievessen, I., Mark, C., Münster, S., Aifantis, K.E., Fabry, B.: Three-dimensional force microscopy of cells in biopolymer networks. Nat. Meth. Advance online publication (2015). https://doi.org/10.1038/nmeth.3685

121. Valentine, M.T., Perlman, Z.E., Gardel, M.L., Shin, J.H., Matsudaira, P., Mitchison, T.J., Weitz, D.A.: Colloid surface chemistry critically affects multiple particle tracking measurements of biomaterials. Biophys. J. **86**, 4004–4014 (2004). https://doi.org/10.1529/biophysj.103.037812

122. Janmey, P.A., Georges, P.C., Hvidt, S.: Basic rheology for biologists. In: Methods in Cell Biology, pp. 1–27. Academic Press (2007)

123. Yang, Y., Kaufman, L.J.: Rheology and confocal reflectance microscopy as probes of mechanical properties and structure during collagen and collagen/hyaluronan self-assembly. Biophys. J. **96**, 1566–1585 (2009). https://doi.org/10.1016/j.bpj.2008.10.063

124. Yang, Y., Leone, L.M., Kaufman, L.J.: Elastic moduli of collagen gels can be predicted from two-dimensional confocal microscopy. Biophys. J. **97**, 2051–2060 (2009). https://doi.org/10.1016/j.bpj.2009.07.035

125. Tran-Ba, K.-H., Lee, D.J., Zhu, J., Paeng, K., Kaufman, L.J.: Confocal rheology probes the structure and mechanics of collagen through the sol-gel transition. Biophys. J. **113**, 1882–1892 (2017). https://doi.org/10.1016/j.bpj.2017.08.025

126. Motte, S., Kaufman, L.J.: Strain stiffening in collagen I networks. Biopolymers **99**, 35–46 (2013). https://doi.org/10.1002/bip.22133

127. Janmey, P.A., McCormick, M.E., Rammensee, S., Leight, J.L., Georges, P.C., MacKintosh, F.C.: Negative normal stress in semiflexible biopolymer gels. Nat. Mater. **6**, 48–51 (2007). https://doi.org/10.1038/nmat1810

128. Unterberger, M.J., Holzapfel, G.A.: Advances in the mechanical modeling of filamentous actin and its cross-linked networks on multiple scales. Biomech. Model. Mechanobiol. **13**, 1155–1174 (2014). https://doi.org/10.1007/s10237-014-0578-4

129. Wen, Q., Basu, A., Janmey, P.A., Yodh, A.G.: Non-affine deformations in polymer hydrogels. Soft Matter **8**, 8039 (2012). https://doi.org/10.1039/c2sm25364j

130. Munster, S., Jawerth, L.M., Leslie, B.A., Weitz, J.I., Fabry, B., Weitz, D.A.: Strain history dependence of the nonlinear stress response of fibrin and collagen networks. Proc. Natl. Acad. Sci. **110**, 12197–12202 (2013). https://doi.org/10.1073/pnas.1222787110

131. Kim, O.V., Litvinov, R.I., Weisel, J.W., Alber, M.S.: Structural basis for the nonlinear mechanics of fibrin networks under compression. Biomaterials **35**, 6739–6749 (2014). https://doi.org/10.1016/j.biomaterials.2014.04.056

132. Kim, O.V., Litvinov, R.I., Chen, J., Chen, D.Z., Weisel, J.W., Alber, M.S.: Compression-induced structural and mechanical changes of fibrin-collagen composites. Matrix Biol. **60–61**, 141–156 (2017). https://doi.org/10.1016/j.matbio.2016.10.007

133. van Oosten, A.S.G., Vahabi, M., Licup, A.J., Sharma, A., Galie, P.A., MacKintosh, F.C., Janmey, P.A.: Uncoupling shear and uniaxial elastic moduli of semiflexible biopolymer networks: compression-softening and stretch-stiffening. Sci. Rep. **6**, 19270 (2016). https://doi.org/10.1038/srep19270

134. Xu, B., Chow, M.-J., Zhang, Y.: Experimental and modeling study of collagen scaffolds with the effects of crosslinking and fiber alignment. Int. J. Biomater. **2011**, 1–12 (2011). https://doi.org/10.1155/2011/172389

135. Voytik-Harbin, S.L., Roeder, B.A., Sturgis, J.E., Kokini, K., Robinson, J.P.: Simultaneous mechanical loading and confocal reflection microscopy for three-dimensional microbiomechanical analysis of biomaterials and tissue constructs. Microsc. Microanal. **9**, 74–85 (2003). https://doi.org/10.1017/S1431927603030046

136. Roeder, B.A.: Local, three-dimensional strain measurements within largely deformed extra-cellular matrix constructs. J. Biomech. Eng. **126**, 699 (2005). https://doi.org/10.1115/1.1824127

137. Roeder, B.A., Kokini, K., Voytik-Harbin, S.L.: Fibril microstructure affects strain transmission within collagen extracellular matrices. J. Biomech. Eng. **131**, 031004 (2009). https://doi.org/10.1115/1.3005331

138. Knezevic, V., Sim, A.J., Borg, T.K., Holmes, J.W.: Isotonic biaxial loading of fibroblast-populated collagen gels: a versatile, low-cost system for the study of mechanobiology. Biomech. Model. Mechanobiol. **1**, 59–67 (2002). https://doi.org/10.1007/s10237-002-0005-0

139. Thomopoulos, S., Fomovsky, G.M., Holmes, J.W.: The development of structural and mechanical anisotropy in fibroblast populated collagen gels. J. Biomech. Eng. **127**, 742 (2005). https://doi.org/10.1115/1.1992525

140. Thomopoulos, S., Fomovsky, G.M., Chandran, P.L., Holmes, J.W.: Collagen fiber alignment does not explain mechanical anisotropy in fibroblast populated collagen gels. J. Biomech. Eng. **129**, 642 (2007). https://doi.org/10.1115/1.2768104

141. Chandran, P.L., Paik, D.C., Holmes, J.W.: Structural mechanism for alteration of collagen gel mechanics by glutaraldehyde crosslinking. Connect. Tissue Res. **53**, 285–297 (2012). https://doi.org/10.3109/03008207.2011.640760

142. Mow, V., Lai, W.: Recent developments in synovial joint biomechanics. SIAM Rev. **22**, 275–317 (1980). https://doi.org/10.1137/1022056

143. Mow, V.C., Kuei, S.C., Lai, W.M., Armstrong, C.G.: Biphasic creep and stress relaxation of articular cartilage in compression: theory and experiments. J. Biomech. Eng. **102**, 73–84 (1980). https://doi.org/10.1115/1.3138202

144. Kim, O.V., Liang, X., Litvinov, R.I., Weisel, J.W., Alber, M.S., Purohit, P.K.: Foam-like compression behavior of fibrin networks. Biomech. Model. Mechanobiol. **15**, 213–228 (2016). https://doi.org/10.1007/s10237-015-0683-z

145. Ramtani, S., Takahashi-Iñiguez, Y., Helary, C., Geiger, D., Guille, M.M.G.: Mechanical behavior under unconfined compression loadings of dense fibrillar collagen matrices mimetic of living tissues. J. Mech. Med. Biol. **10**, 35–55 (2010). https://doi.org/10.1142/S0219519410003290

146. Lane, B.A., Harmon, K.A., Goodwin, R.L., Yost, M.J., Shazly, T., Eberth, J.F.: Constitutive modeling of compressible type-I collagen hydrogels. Med. Eng. Phys. **53**, 39–48 (2018). https://doi.org/10.1016/j.medengphy.2018.01.003

147. Knapp, D.M., Barocas, V.H., Moon, A.G., Yoo, K., Petzold, L.R., Tranquillo, R.T.: Rheology of reconstituted type I collagen gel in confined compression. J. Rheol. **41**, 971–993 (1997). https://doi.org/10.1122/1.550817

148. Girton, T.S., Barocas, V.H., Tranquillo, R.T.: Confined compression of a tissue-equivalent: collagen fibril and cell alignment in response to anisotropic strain. J. Biomech. Eng. **124**, 568 (2002). https://doi.org/10.1115/1.1504099

149. Chandran, P.L., Barocas, V.H.: Microstructural mechanics of collagen gels in confined compression: poroelasticity, viscoelasticity, and collapse. J. Biomech. Eng. **126**, 152 (2004). https://doi.org/10.1115/1.1688774

150. Nam, S., Hu, K.H., Butte, M.J., Chaudhuri, O.: Strain-enhanced stress relaxation impacts nonlinear elasticity in collagen gels. PNAS. 201523906 (2016). https://doi.org/10.1073/pnas.1523906113

151. Nam, S., Lee, J., Brownfield, D.G., Chaudhuri, O.: Viscoplasticity enables mechanical remodeling of matrix by cells. Biophys. J. **111**, 2296–2308 (2016). https://doi.org/10.1016/j.bpj.2016.10.002

152. Ban, E., Franklin, J.M., Nam, S., Smith, L.R., Wang, H., Wells, R.G., Chaudhuri, O., Liphardt, J.T., Shenoy, V.B.: Mechanisms of plastic deformation in collagen networks induced by cellular forces. Biophys. J. **114**, 450–461 (2018). https://doi.org/10.1016/j.bpj.2017.11.3739

153. Xu, B., Li, H., Zhang, Y.: Understanding the viscoelastic behavior of collagen matrices through relaxation time distribution spectrum. Biomatter **3**, e24651 (2013). https://doi.org/10.4161/biom.24651

154. Xu, B., Li, H., Zhang, Y.: An experimental and modeling study of the viscoelastic behavior of collagen gel. J. Biomech. Eng. **135**, 054501 (2013). https://doi.org/10.1115/1.4024131

155. Pryse, K.M., Nekouzadeh, A., Genin, G.M., Elson, E.L., Zahalak, G.I.: Incremental mechanics of collagen gels: new experiments and a new viscoelastic model. Ann. Biomed. Eng. **31**, 1287–1296 (2003). https://doi.org/10.1114/1.1615571

156. Bustamante, C., Marko, J.F., Siggia, E.D., Smith, S.: Entropic elasticity of lambda-phage DNA. Science **265**, 1599–1600 (1994)

157. Buehler, M.J., Wong, S.Y.: Entropic elasticity controls nanomechanics of single tropocollagen molecules. Biophys. J. **93**, 37–43 (2007). https://doi.org/10.1529/biophysj.106.102616

158. Sivakumar, L., Agarwal, G.: The influence of discoidin domain receptor 2 on the persistence length of collagen type I fibers. Biomaterials **31**, 4802–4808 (2010). https://doi.org/10.1016/j.biomaterials.2010.02.070

159. Varma, S., Orgel, J.P.R.O., Schieber, J.D.: Nanomechanics of type I collagen. Biophys. J. **111**, 50–56 (2016). https://doi.org/10.1016/j.bpj.2016.05.038

160. Buehler, M.J.: Atomistic and continuum modeling of mechanical properties of collagen: elasticity, fracture, and self-assembly. J. Mater. Res. **21**, 1947–1961 (2006). https://doi.org/10.1557/jmr.2006.0236

161. Buehler, M.J.: Nature designs tough collagen: Explaining the nanostructure of collagen fibrils. Proc. Natl. Acad. Sci. **103**, 12285–12290 (2006). https://doi.org/10.1073/pnas.0603216103

162. Buehler, M.J.: Molecular architecture of collagen fibrils: a critical length scale for tough fibrils. Curr. Appl. Phys. **8**, 440–442 (2008). https://doi.org/10.1016/j.cap.2007.10.058

163. Buehler, M.J.: Nanomechanics of collagen fibrils under varying cross-link densities: atomistic and continuum studies. J. Mech. Behav. Biomed. Mater. **1**, 59–67 (2008). https://doi.org/10.1016/j.jmbbm.2007.04.001

164. Tang, Y., Ballarini, R., Buehler, M.J., Eppell, S.J.: Deformation micromechanisms of collagen fibrils under uniaxial tension. J. R. Soc. Interface **7**, 839–850 (2010). https://doi.org/10.1098/rsif.2009.0390

165. Gautieri, A., Vesentini, S., Redaelli, A., Buehler, M.J.: Hierarchical structure and nanomechanics of collagen microfibrils from the atomistic scale up. Nano Lett. **11**, 757–766 (2011). https://doi.org/10.1021/nl103943u

166. Gautieri, A., Vesentini, S., Redaelli, A., Buehler, M.J.: Viscoelastic properties of model segments of collagen molecules. Matrix Biol. **31**, 141–149 (2012). https://doi.org/10.1016/j.matbio.2011.11.005

167. Head, D.A., Levine, A.J., MacKintosh, F.C.: Distinct regimes of elastic response and deformation modes of cross-linked cytoskeletal and semiflexible polymer networks. Phys. Rev. E **68**, 061907 (2003). https://doi.org/10.1103/physreve.68.061907

168. Head, D.A., Levine, A.J., MacKintosh, F.C.: Deformation of cross-linked semiflexible polymer networks. Phys. Rev. Lett. **91**, 108102 (2003). https://doi.org/10.1103/physrevlett.91.108102

169. Wilhelm, J., Frey, E.: Elasticity of stiff polymer networks. Phys. Rev. Lett. **91**, 108103 (2003). https://doi.org/10.1103/physrevlett.91.108103

170. Onck, P.R., Koeman, T., van Dillen, T., van der Giessen, E.: Alternative explanation of stiffening in cross-linked semiflexible networks. Phys. Rev. Lett. **95**, 178102 (2005). https://doi.org/10.1103/physrevlett.95.178102

171. Stein, A.M., Vader, D.A., Weitz, D.A., Sander, L.M.: The micromechanics of three-dimensional collagen-I gels. Complexity **16**, 22–28 (2011). https://doi.org/10.1002/cplx.20332

172. Licup, A.J., Münster, S., Sharma, A., Sheinman, M., Jawerth, L.M., Fabry, B., Weitz, D.A., MacKintosh, F.C.: Stress controls the mechanics of collagen networks. Proc. Natl. Acad. Sci. **112**, 9573–9578 (2015). https://doi.org/10.1073/pnas.1504258112

173. Sander, E.A., Stein, A.M., Swickrath, M.J., Barocas, V.H.: Out of many, one: modeling schemes for biopolymer and biofibril networks. In: Dumitrica, T. (ed.) Trends in Computational Nanomechanics: Transcending Length and Time Scales, pp. 557–602. Springer Netherlands, Dordrecht (2010)

174. Kim, T., Hwang, W., Lee, H., Kamm, R.D.: Computational analysis of viscoelastic properties of crosslinked actin networks. PLoS Comput. Biol. **5**, e1000439 (2009). https://doi.org/10.1371/journal.pcbi.1000439

175. Kim, T., Hwang, W., Kamm, R.D.: Computational analysis of a cross-linked actin-like network. Exp. Mech. **49**, 91–104 (2009). https://doi.org/10.1007/s11340-007-9091-3

176. Wen, Q., Basu, A., Winer, J.P., Yodh, A., Janmey, P.A.: Local and global deformations in a strain-stiffening fibrin gel. New J. Phys. **9**, 428 (2007). https://doi.org/10.1088/1367-2630/9/11/428

177. Fung, Y.C.: Elasticity of soft tissues in simple elongation. Am. J. Physiol. Legacy Content. **213**, 1532–1544 (1967)

178. Ma, X., Schickel, M.E., Stevenson, M.D., Sarang-Sieminski, A.L., Gooch, K.J., Ghadiali, S.N., Hart, R.T.: Fibers in the extracellular matrix enable long-range stress transmission between cells. Biophys. J. **104**, 1410–1418 (2013). https://doi.org/10.1016/j.bpj.2013.02.017

179. Abhilash, A.S., Baker, B.M., Trappmann, B., Chen, C.S., Shenoy, V.B.: Remodeling of fibrous extracellular matrices by contractile cells: predictions from discrete fiber network simulations. Biophys. J. **107**, 1829–1840 (2014). https://doi.org/10.1016/j.bpj.2014.08.029

180. Notbohm, J., Lesman, A., Rosakis, P., Tirrell, D.A., Ravichandran, G.: Microbuckling of fibrin provides a mechanism for cell mechanosensing. J. R. Soc. Interface **12**, 20150320 (2015). https://doi.org/10.1098/rsif.2015.0320

181. Holzapfel, G.A.: Nonlinear Solid Mechanics: A Continuum Approach for Engineering. Wiley, Chichester; New York (2000)

182. Wang, H., Abhilash, A.S., Chen, C.S., Wells, R.G., Shenoy, V.B.: Long-range force transmission in fibrous matrices enabled by tension-driven alignment of fibers. Biophys. J. **107**, 2592–2603 (2014). https://doi.org/10.1016/j.bpj.2014.09.044

183. Stylianopoulos, T., Barocas, V.H.: Volume-averaging theory for the study of the mechanics of collagen networks. Comput. Methods Appl. Mech. Eng. **196**, 2981–2990 (2007). https://doi.org/10.1016/j.cma.2006.06.019

184. Storm, C., Pastore, J.J., MacKintosh, F.C., Lubensky, T.C., Janmey, P.A.: Nonlinear elasticity in biological gels. Nature **435**, 191–194 (2005). https://doi.org/10.1038/nature03521

185. MacKintosh, F.C., Käs, J., Janmey, P.A.: Elasticity of semiflexible biopolymer networks. Phys. Rev. Lett. **75**, 4425–4428 (1995). https://doi.org/10.1103/PhysRevLett.75.4425

186. Liu, H., Sun, W.: Computational efficiency of numerical approximations of tangent moduli for finite element implementation of a fiber-reinforced hyperelastic material model. Comput. Methods Biomech. Biomed. Eng. **19**, 1171–1180 (2016). https://doi.org/10.1080/10255842.2015.1118467

187. Brown, A.E.X., Litvinov, R.I., Discher, D.E., Purohit, P.K., Weisel, J.W.: Multiscale mechanics of fibrin polymer: gel stretching with protein unfolding and loss of water. Science **325**(5941), 741–744 (2009)

188. Truesdell, C.: Remarks on hypo-elasticity. J. Res. Natl. Bur. Stand. Sect. B Math. Math. Phys. **67**, 141 (1963). https://doi.org/10.6028/jres.067b.011

189. Morin, C., Avril, S., Hellmich, C.: The fiber reorientation problem revisited in the context of Eshelbian micromechanics: theory and computations: the fiber reorientation problem revisited in the context of Eshelbian micromechanics: theory and computations. PAMM **15**, 39–42 (2015). https://doi.org/10.1002/pamm.201510011

190. Morin, C., Avril, S., Hellmich, C.: Non-affine fiber kinematics in arterial mechanics: a continuum micromechanical investigation. ZAMM J. Appl. Math. Mech./Z. Angew. Math. Mech. **98**, 2101–2121 (2018). https://doi.org/10.1002/zamm.201700360

191. Fung, Y.C.: Biomechanics: Mechanical Properties of Living Tissues. Springer, New York (1993)

192. Humphrey, J.D., O'Rourke, S.L.: An Introduction to Biomechanics: Solids and Fluids, Analysis and Design. Springer, New York (2015)

193. Polacheck, W.J., Charest, J.L., Kamm, R.D.: Interstitial flow influences direction of tumor cell migration through competing mechanisms. PNAS **108**, 11115–11120 (2011). https://doi.org/10.1073/pnas.1103581108

194. Malandrino, A., Moeendarbary, E.: Poroelasticity of living tissues. In: Reference Module in Biomedical Sciences. Elsevier (2017)

195. Ateshian, G.A.: Mixture theory for modeling biological tissues: illustrations from articular cartilage. In: Holzapfel, G.A., Ogden, R.W. (eds.) Biomechanics: Trends in Modeling and Simulation, pp. 1–51. Springer International Publishing, Cham (2017)

196. Busby, G.A., Grant, M.H., MacKay, S.P., Riches, P.E.: Confined compression of collagen hydrogels. J. Biomech. **46**, 837–840 (2013). https://doi.org/10.1016/j.jbiomech.2012.11.048

197. Lai, W.M., Mow, V.C.: Drag-induced compression of articular cartilage during a permeation experiment. Biorheology **17**, 111–123 (1980)

198. Holmes, M.H.: Finite deformation of soft tissue: analysis of a mixture model in uni-axial compression. J. Biomech. Eng. **108**, 372–381 (1986). https://doi.org/10.1115/1.3138633

199. Kwan, M.K., Lai, W.M., Mow, V.C.: A finite deformation theory for cartilage and other soft hydrated connective tissues—I. Equilibrium results. J. Biomech. **23**, 145–155 (1990). https://doi.org/10.1016/0021-9290(90)90348-7

200. Setton, L.A., Zhu, W., Mow, V.C.: The biphasic poroviscoelastic behavior of articular cartilage: role of the surface zone in governing the compressive behavior. J. Biomech. **26**, 581–592 (1993). https://doi.org/10.1016/0021-9290(93)90019-B

201. Dembo, M., Oliver, T., Ishihara, A., Jacobson, K.: Imaging the traction stresses exerted by locomoting cells with the elastic substratum method. Biophys. J. **70**, 2008–2022 (1996). https://doi.org/10.1016/S0006-3495(96)79767-9

202. Hall, M.S., Long, R., Feng, X., Huang, Y., Hui, C.-Y., Wu, M.: Toward single cell traction microscopy within 3D collagen matrices. Exp. Cell Res. **319**, 2396–2408 (2013). https://doi.org/10.1016/j.yexcr.2013.06.009

203. Hall, M.S., Alisafaei, F., Ban, E., Feng, X., Hui, C.-Y., Shenoy, V.B., Wu, M.: Fibrous nonlinear elasticity enables positive mechanical feedback between cells and ECMs. PNAS **113**, 14043–14048 (2016). https://doi.org/10.1073/pnas.1613058113

204. Wisdom, K.M., Adebowale, K., Chang, J., Lee, J.Y., Nam, S., Desai, R., Rossen, N.S., Rafat, M., West, R.B., Hodgson, L., Chaudhuri, O.: Matrix mechanical plasticity regulates cancer cell migration through confining microenvironments. Nat. Commun. **9**, 4144 (2018). https://doi.org/10.1038/s41467-018-06641-z

205. Stout, D.A., Bar-Kochba, E., Estrada, J.B., Toyjanova, J., Kesari, H., Reichner, J.S., Franck, C.: Mean deformation metrics for quantifying 3D cell–matrix interactions without requiring information about matrix material properties. Proc. Natl. Acad. Sci. **113**, 2898–2903 (2016). https://doi.org/10.1073/pnas.1510935113

206. Gjorevski, N., Nelson, C.M.: Mapping of mechanical strains and stresses around quiescent engineered three-dimensional epithelial tissues. Biophys. J. **103**, 152–162 (2012). https://doi.org/10.1016/j.bpj.2012.05.048

Author Index

© Springer Nature Switzerland AG 2020
Y. Zhang (ed.), *Multi-scale Extracellular Matrix Mechanics and Mechanobiology*,
Studies in Mechanobiology, Tissue Engineering and Biomaterials 23,
https://doi.org/10.1007/978-3-030-20182-1

Subject Index

Printed in the United States
By Bookmasters